Conic Sections

All conic sections have an equation of the form $Ax^2 + Bx + Cy^2 + Dy + E = 0$ where either A or C is nonzero. The particular conic sections are listed below. See graphs in back endpapers.

Conic Section	Characteristic
Parabola	Either A or C is 0, but not both.
Circle	$A = C \neq 0$
Ellipse	$A \neq C$, $AC > 0$
Hyperbola	$AC < 0$

Function

A function is a correspondence or rule that assigns to each element of one set (the domain) exactly one element from another set (the range).

Composition of Functions

$(g \circ f)(x) = g[f(x)]$

One-to-One Function

f is a one-to-one function if $a \neq b$ implies $f(a) \neq f(b)$.

Inverse Function

Let f be a one-to-one function. The inverse function, f^{-1}, can be found by exchanging x and y in the equation for f, then solving for y if possible.

Value of e

$e \approx 2.718281828$

Definition of Logarithm

$y = \log_a x$ if and only if $x = a^y$.

Properties of Logarithms

(a) $\log_a xy = \log_a x + \log_a y$ (b) $\log_a \dfrac{x}{y} = \log_a x - \log_a y$

(c) $\log_a x^r = r \log_a x$ (d) $\log_a a = 1$

(e) $\log_a 1 = 0$

Arithmetic Sequences

nth term: $a_n = a_1 + (n - 1)d$

Sum of n terms: $S_n = \dfrac{n}{2}[a_1 + a_n]$ or $S_n = \dfrac{n}{2}[2a_1 + (n - 1)d]$

Geometric Sequences

nth term: $a_n = a_1 r^{n-1}$

Sum of n terms: $S_n = \dfrac{a_1(1 - r^n)}{1 - r}, \quad r \neq 1$

Sum of an infinite geometric sequence: $S_\infty = \dfrac{a_1}{1 - r}, \quad -1 < r < 1$

Binomial Theorem

For any positive integer n, $(x + y)^n = x^n + \dbinom{n}{n-1} x^{n-1}y + \dbinom{n}{n-2} x^{n-2}y^2 + \ldots + \dbinom{n}{1} xy^{n-1} + y^n$.

Permutations

The number of arrangements of n things taken r at a time is $P(n, r) = \dfrac{n!}{(n - r)!}$.

Combinations

The number of ways to choose r things from a group of n things is $\dbinom{n}{r} = \dfrac{n!}{(n - r)!r!}$.

Properties of Probability

For any events E and F:
1. $0 \leq P(E) \leq 1$
2. P (a certain event) $= 1$
3. P (an impossible event) $= 0$
4. $P(E') = 1 - P(E)$
5. $P(E$ or $F) = P(E \cup F) = P(E) + P(F) - P(E \cap F)$.

College Algebra

5th Edition

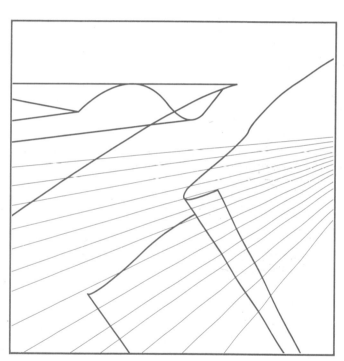

College Algebra

5th Edition

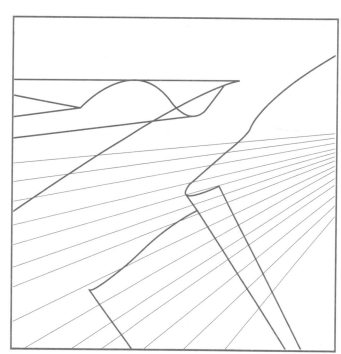

Margaret L. Lial
American River College

Charles D. Miller

HarperCollins*Publishers*

To the Student

If you need further help with algebra, you may want to get copies of both the *Study Guide* and the *Student Solutions Manual* that accompany this textbook. The additional worked-out examples and practice problems that these books provide can help you study and understand the course material. Your college bookstore either has these books or can order them for you.

Cover Acknowledgment

Photograph of polarized light shining through folded pieces of clear acetate. Photo by Scott, Foresman staff.

Library of Congress Cataloging-in-Publication Data

Lial, Margaret L.
  College algebra / Margaret L. Lial, Charles D. Miller.—5th ed.
    p. cm.
  ISBN 0–673–38245–1
  1. Algebra. I. Miller, Charles David. II. Title.
QA154.2.L5 1988
512.9—dc19

88–21074
CIP

56-RRC-939291

Preface

The Fifth Edition of *College Algebra* is designed for a one-semester or one-quarter course that prepares students for work in trigonometry, statistics, finite mathematics, calculus, discrete mathematics, or courses in computer science.

This text has been written with the assumption that students have had an earlier course in algebra. For students whose knowledge of algebra is not as solid as it might be, a chapter of review topics has been included. In fact, Chapter 2 on equations will be considered a review for some students. A diagnostic pretest covering the first two chapters that can be used to determine the necessary topics for review is provided in the *Instructor's Guide*.

Key Features

Examples   Over 350 detailed examples clearly illustrate concepts and techniques. Second color identifies pertinent steps within examples and highlights explanatory side comments.

Applications   The text features a rich selection of applications from many different fields of study. This edition includes several new applications to business and finance.

Intuitive Treatment   Throughout the book, properties are stated simply and intuitively. A formal theorem-proof format has been downplayed.

Exercises

Graded Exercises   The wide range of difficulty in the exercise sets affords students ample practice with drill problems. Students are then eased gradually through problems of increasing difficulty to problems that will challenge outstanding students. A list of challenging exercises is given in the *Instructor's Guide* so that instructors can use this feature as they prefer. More than 4800 exercises, including approximately 4450 drill problems and 350 word problems, are provided in the text.

Calculator Exercises   A large number of exercise sets feature calculator problems. These optional exercises are identified by colored exercise numbers. Brief discussions of the difficulties that can arise when students use calculators also are presented in the text as appropriate. In addition, a discussion of calculator error and significant digits addressed to the student follows the Table of Contents.

Chapter Review Exercises   A lengthy set of review exercises is given at the end of each chapter and reviews each section of the text thoroughly. A total of about 900 review exercises provides opportunity for mastery of the material before students take an examination. In addition, a list of key words and formulas keyed to the appropriate sections is given at the end of each chapter.

Sample Chapter Tests   Each chapter ends with a sample chapter test that students can use as a practice test. Six additional versions of each chapter test are given in the *Instructor's Guide*.

Answers   Answers to odd-numbered exercises and review exercises and to all chapter test questions are given at the back of the text.

Second Color

Second color is used pedogogically in the following ways.

- Screens set off key definitions, formulas, and procedures, helping students review easily.
- Color side comments within examples explain the structure of the problem.
- For clarity, the end of each example is indicated with a color symbol ✚

Content Features

The **algebra review** constitutes a new Chapter 1. Operations on the real numbers, absolute value, exponents, polynomials, factoring, rational expressions, and radicals are covered.

Chapter 2 covers solution of **equations and inequalities in one variable.** A new section on applications of quadratic equations has been added.

Graphing is a main feature of the text. Starting with number lines in Chapter 1 and lines in the plane in Chapter 3, graphing is emphasized throughout. In Chapter 3, general techniques of graphing are discussed, and graphing of particular functions is treated in the next three chapters.

Chapters 3 and 4 have been rearranged to focus on **relations** and **functions.** The concept of functions is crucial to this course. After studying relations in Chapter 3, where many of the basic concepts are presented, Chapter 4 deals entirely with functions so students can concentrate on learning this material without having to cope with the introduction of other new concepts at the same time.

Polynomial and rational functions have been combined with the material on zeros of polynomials to form a new Chapter 5. This places all the work on polynomials before exponential and logarithmic functions.

The emphasis in Chapter 6, on **exponential and logarithmic functions,** is now on their functional properties and applications. The use of calculators has been integrated into this material. Discussion of logarithmic tables to find function values of logarithms has been placed in an appendix.

The material on **matrices** and **systems** has been rearranged to provide a smoother flow of coverage. Systems of equations and inequalities are discussed in Chapter 7. Then, in Chapter 8, matrices and their application to solving systems of equations are presented.

The remaining algebraic topics, on sequences, the Binomial Theorem, mathematical induction, counting theory, and probability are collected in Chapter 9.

The ***Instructor's Guide*** provides a diagnostic pretest with answers, suggested assignments, six alternative forms of chapter tests in a format that is ready for duplication, a test bank for each chapter, two forms of a final examination, and answers for all these test instruments. The solutions to all even-numbered exercises are given in a separate ***Instructor's Solution Manual,*** and an ***Answer Manual*** gives answers to all exercises in the text.

A ***Study Guide,*** in a semiprogrammed format, includes a pretest and posttest for each chapter, plus exercises that give additional practice and reinforcement for students.

A ***Student's Solution Manual*** has solutions to all the odd-numbered exercises in the text. Some students may want to use this book as an additional source of examples.

The ***Scott, Foresman Test Generator for Mathematics*** is a computerized testing system, available in both Apple and IBM, that covers all of the important topics in *College Algebra*. The system enables instructors to select questions by chapter, section, or objective, or to choose a ready-made test for each chapter. Instructors may generate tests in multiple-choice or open-response formats, produce up to 25 versions of a test, and scramble question order while printing. The Test Generator also features a preview option that allows instructors to view questions before printing and to regenerate variables or skip a question if desired.

A set of ***computer-assisted tutorials,*** based on a mastery learning approach, has been developed for this text. Available for Apple and IBM computers, each tutorial module covers one of the important topics in *College Algebra*. Each module presents a tutorial lesson, followed by drill and practice problems that are randomly generated. Each module also contains one or more computer "tools" that assist students with problem solving or demonstrate essential concepts in *College Algebra*.

Audiotapes that cover all topics in the text are available at no charge to users of the book. Students who need help with a particular topic or who have missed class find that these tapes help them master the material.

College Algebra, Fifth Edition, is part of a series of texts, including *Beginning Algebra,* Fifth Edition; *Intermediate Algebra,* Fifth Edition; *Fundamentals of College Algebra,* Second Edition; *Trigonometry,* Fourth Edition; and *Algebra and Trigonometry,* Fourth Edition. The use of a series of texts increases student understanding by offering continuity of notation, definitions, and format.

Thanks are due to the many users of the previous editions of this book who were kind enough to share their experiences. This revision has benefited from their comments and suggestions.

Thanks are also due to those who reviewed all or part of the revised manuscript and gave us many helpful suggestions: Jasper Adams, Stephen F. Austin State University; Richard Armstrong, St. Louis Community College at Florissant

Valley; Susan L. Friedman, Bernard M. Baruch College, The City University of New York; Samuel Graff, John Jay College of Criminal Justice, The City University of New York; Shirley S. Hagewood, Austin Peay State University; Ruth Hunt, University of Missouri; Bill Jordan, Seminole Community College; Julie Monte, Daytona Beach Community College, South Campus; Rebecca Stamper, Western Kentucky University; and Peggy Vandiver, San Jacinto College North.

Appreciation also goes to the staff at Scott, Foresman and Company, who did an excellent job in working toward publication.

Contents

Calculation and Approximation A significant digit is a digit obtained by actual measurement. A number that represents the result of counting, or a number that results from theoretical work and is not the result of a measurement, is an **exact number.** There are fifty states in the United States, so 50 is an exact number. The number of states is not 49 3/4 or 50 1/4; nor is the number 50 used here to represent "some number between 45 and 55." In the formula for the perimeter of a rectangle, $P = 2L + 2W$, the 2's are obtained from the definition of perimeter, and are exact numbers.

The following chart shows some numbers, the number of significant digits in each number, and the range represented by each number.

Number	Number of Significant Digits	Range Represented by Number
29.6	3	29.55 to 29.65
1.39	3	1.385 to 1.395
.000096	2	.0000955 to .0000965
.03	1	.025 to .035
100.2	4	100.15 to 100.25

Not all digits in a measurement are significant; some of the digits serve only as placeholders. For example, the measurement 19.00 m is a measurement to the nearest hundredth meter. What about the measurement 93,000 m? Does it represent a measurement to the nearest meter, to the nearest ten meters, hundred meters, or thousand meters? We cannot tell by the way the number is written. To get around this problem, write the number in **scientific notation,** the product of a number between 1 and 10 and a power of 10. Depending on what is known about the accuracy of the measurement, write 93,000 using scientific notation as follows.

Measurement to Nearest	Scientific Notation	Number of Significant Digits
Meter	9.3000×10^4	5
Ten meters	9.300×10^4	4
Hundred meters	9.30×10^4	3
Thousand meters	9.3×10^4	2

Virtually all values of trigonometric functions are approximations, and virtually all measurements are approximations. To perform calculations on such approximate numbers, follow the rules given below.

Calculation with Significant Digits

For *adding* and *subtracting,* round the answer so that the last digit you keep is in the rightmost column in which all the numbers have significant digits.

For *multiplying* or *dividing,* round the answer to the least number of significant digits found in any of the given numbers.

For *powers* and *roots,* round the answer so that it has the same number of significant digits as the number whose power or root you are finding.

Calculator Errors   A calculator can store only so many digits in its memory. Because of this, numbers that have more digits than can be stored must be rounded. For example, 1/3 is not stored as the exact fraction 1/3, but rather as a decimal, perhaps .3333333333333. Since this rounded form of 1/3 is used, errors can occur in calculations. To see how this happens, use a calculator to divide 1 by 3, and then multiply the result by 3. You should get 1 (exactly), but many machines produce

$$(1 \div 3) \times 3 = \left(\frac{1}{3}\right) \times 3 = .9999999999.$$

Some machines round this result to 1; however, the machine does not treat the number internally as 1. To see this, subtract 1 from the result above; you should get 0 but probably will not.

Another calculator error results when numbers of greatly different size are used in addition. For example,

$$10^9 + 10^{-5} - 10^9 = 10^{-5}.$$

However, most calculators would give

$$10^9 + 10^{-5} - 10^9 = 0.$$

These calculator errors seldom occur in realistic problems, but you should know what is happening if they do occur.

College Algebra

5th Edition

Algebraic Expressions

Today, people in fields ranging from accounting to zoology use algebra to solve a variety of problems. Equations, graphs, and functions are important tools in business, science, and other areas. As preparation for discussing these topics of algebra, in this chapter we review the properties of real numbers and the basic rules of algebra.

1.1 The Real Numbers

Sets of Numbers The idea of counting goes back into the mists of antiquity. When people first counted they used only the **natural numbers,** written in set notation as

$$\{1, 2, 3, 4, 5, \ldots\}.$$

Much more recent is the idea of counting *no* object—that is, the idea of the number 0. Including 0 with the set of natural numbers gives the set of **whole numbers.**

$$\{0, 1, 2, 3, 4, 5, \ldots\}$$

(These and other sets of numbers are summarized later in this section.)

About 500 years ago, people came up with the idea of counting backwards, from 4 to 3 to 2 to 1 to 0. There seemed no reason not to continue this process, calling the new numbers -1, -2, -3, and so on. Including these numbers with the set of whole numbers gives the very useful set of **integers,**

$$\{. \ . \ . \ , \ -4, \ -3, \ -2, \ -1, \ 0, \ 1, \ 2, \ 3, \ . \ . \ .\}.$$

The integer -7 might represent a temperature 7 degrees below zero, a debt of 7 dollars, or a loss of 7 yards in a football game. When thinking of time, the integer -7 might represent a time 7 seconds ago.

Integers can be shown pictorially with a **number line.** (A number line is similar to a thermometer on its side.) To construct a number line, draw a horizontal line and choose any point on the line to represent 0. Then choose any point to the right of 0 and label it 1. The distance from 0 to 1 sets up a unit measure that can be used to locate other points to the right of 1, which are labeled 2, 3, 4, 5, and so on, and corresponding points to the left of 0, labeled -1, -2, -3, -4, and so on. As an example, the elements of the set $\{-3, \ -1, \ 0, \ 1, \ 3, \ 5\}$ are located on the number line in Figure 1.1.

Figure 1.1

The integers 1, 2, 3, 4, and so on (those to the right of 0 on the number line) are called **positive integers,** while -1, -2, -3, and so on (to the left of 0) are **negative integers.** The integer 0 is neither positive nor negative. Positive and negative numbers are sometimes called **signed numbers.**

As you have seen in a previous algebra course, if two integers are added, subtracted, or multiplied, the answer is an integer. However, the result of dividing two integers need not be an integer. For example,

$$\frac{15}{4}, \qquad -\frac{11}{7}, \qquad \text{and} \qquad \frac{28}{9}$$

are not integers. The result of dividing two integers, with the second integer not zero, is called a *rational number.* By definition, the **rational numbers** are the elements of the set

$$\left\{ \frac{p}{q} \ \middle| \ p, \ q \text{ are integers and } q \neq 0 \right\}.$$

This definition, which is given in *set-builder notation,* is read "the set of all elements p/q such that p and q are integers and $q \neq 0$." Examples of rational numbers include 3/4, $-5/8$, 7/2, and $-14/9$. All integers are rational numbers, since any integer can be written as the quotient of itself and 1. For example, $6 = 6/1$, while $-8 = -8/1$, and $0 = 0/1$. The set of integers is a subset of the set of rational numbers.

Rational numbers can be located on a number line by a process of subdivision. For example, 5/8 can be located by dividing the interval from 0 to 1 into 8 equal parts, then labeling the fifth part 5/8. Several rational numbers are located on the number line in Figure 1.2.

Locate $\frac{5}{8}$ by dividing the interval from 0 to 1 into 8 equal parts.

The set of all numbers that correspond to points on a number line is called the set of **real numbers.** The set of real numbers is shown pictorially in Figure 1.3.

Figure 1.3

While all rational numbers are real numbers, not all numbers that correspond to points on a number line are rational. Figure 1.4 shows the diagonal of a square, 1 foot on a side. By the Pythagorean formula from geometry, the length of this diagonal is

$$\sqrt{1^2 + 1^2} = \sqrt{2} \text{ ft.}$$

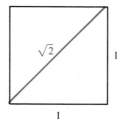

Figure 1.4

As shown in Exercise 87, $\sqrt{2}$ cannot be written as the quotient of two integers, so $\sqrt{2}$ is a real number that is not rational. A real number that is not rational is called an **irrational number.** The set of irrational numbers includes $\sqrt{3}$ and $\sqrt{5}$ but not $\sqrt{1}$, $\sqrt{4}$, $\sqrt{9}$, . . . , which equal 1, 2, 3, . . . , and hence are rational numbers. Another irrational number is π, which is approximately equal to 3.14159. Using a calculator or a square root table shows that $\sqrt{2} \approx 1.414$ (where $\approx$ is read "is approximately equal to") and $\sqrt{5} \approx 2.236$. Using these approximations, the numbers in the set $\{-2/3, 0, \sqrt{2}, \sqrt{5}, \pi, 4\}$ can be located on a number line as shown in Figure 1.5.

Figure 1.5

Real numbers can also be defined in another way, in terms of decimals. Using repeated subdivisions, any number can be located (at least in theory) as a point on a number line. By this process, the set of real numbers can be defined as the set of all decimals. Further work should show that the set of rational numbers is the set of all decimals that repeat or terminate. For example,

$$.25 \qquad\qquad = 1/4,$$
$$.833333\ .\ .\ . \qquad\qquad = 5/6,$$
$$.076923076923\ .\ .\ . \quad = 1/13,$$

and so on. The set of irrational numbers is then the set of decimals that do *not* repeat or terminate. For example,

$$\sqrt{2} = 1.414213562373\ .\ .\ . \qquad \text{and} \qquad \pi = 3.141592654359.\ .\ .\ .$$

The sets of numbers discussed so far are summarized below.

Real numbers	$\{x \mid x$ corresponds to a point on a number line$\}$
Integers	$\{.\ .\ .\ ,\ -3,\ -2,\ -1,\ 0,\ 1,\ 2,\ 3,\ .\ .\ .\}$
Rational numbers	$\left\{\dfrac{p}{q} \,\middle|\, p \text{ and } q \text{ are integers and } q \neq 0\right\}$
Irrational numbers	$\{x \mid x$ is real but not rational$\}$
Whole numbers	$\{0, 1, 2, 3, 4, .\ .\ .\}$
Natural numbers	$\{1, 2, 3, 4, .\ .\ .\}$

Example 1   Let set $A = \{-8, -6, -3/4, 0, 3/8, 1/2, 1, \sqrt{2}, \sqrt{5}, 6, 9/0\}$. List the elements from set A that belong to each of the sets of numbers just discussed.

(a) All elements of A are real numbers except 9/0. Division by 0 is meaningless, so 9/0 is not a number.

(b) The irrational numbers are $\sqrt{2}$ and $\sqrt{5}$.

(c) The rational numbers are -8, -6, $-3/4$, 0, 3/8, 1/2, 1, and 6.

(d) The integers are -8, -6, 0, 1, and 6.

(e) The whole numbers are 0, 1, and 6.

(f) The natural numbers in set A are 1 and 6.   ✥

The relationships among the various sets of real numbers are shown in Figure 1.6.

Figure 1.6

Operations on Real Numbers   There are four fundamental operations on real numbers: addition, subtraction, multiplication, and division. Addition is indicated with the symbol $+$. Subtraction is written with the symbol $-$, as in $8 - 2 = 6$. Multiplication is written in a variety of ways. All the symbols 2×8, $2 \cdot 8$, $2(8)$, and $(2)(8)$ represent the product of 2 and 8, or 16. When writing products involving variables, no operation symbols may be necessary: $2x$ represents the product of 2 and x, while xy indicates the product of x and y. Division of real numbers a and b ($b \neq 0$) is written $a \div b$, or more commonly, a/b.

Exponential notation is used to write the products of repeated factors. For example, the product $2 \cdot 2 \cdot 2$ can be written as 2^3, where the 3 shows that three factors of 2 appear in the product. The notation a^n is defined as follows.

Definition of
a^n

> If n is any positive integer and a is any real number, then
>
> $$a^n = a \cdot a \cdot a \cdots a,$$
>
> where a appears n times.

The integer n is the **exponent,** and a is the **base.** (Read a^n as "a to the nth power," or just "a to the nth.")

Example 2 Evaluate each exponential, and identify the base and the exponent.

(a) $4^3 = 4 \cdot 4 \cdot 4 = 64$; the base is 4 and the exponent is 3
(b) $(-6)^2 = (-6)(-6) = 36$; the base is -6 and the exponent is 2
(c) $-6^2 = -(6 \cdot 6) = -36$; the base is 6 and the exponent is 2
(d) $4 \cdot 3^2 = 4 \cdot 3 \cdot 3 = 36$; the base is 3 and the exponent is 2
   Notice that it is *not correct* to write $4 \cdot 3^2$ as $4 \cdot 3 \cdot 4 \cdot 3$. ✤

When a problem involves more than one operation symbol, use the following *order of operations.*

Order of
Operations

> *If parentheses, square brackets, or fraction bars are present:*
>
> **1.** Work separately above and below each fraction bar.
> **2.** Use the rules below within each set of parentheses or square brackets. Start with the innermost and work outward.
>
> *If no parentheses are present:*
>
> **1.** Simplify all powers and roots, working from left to right.
> **2.** Do any multiplications or divisions in the order in which they occur, working from left to right.
> **3.** Do any additions or subtractions in the order in which they occur, working from left to right.

Example 3 Use the order of operations given above to evaluate each of the following.

(a) $6 \div 3 + 2^3 \cdot 5 = 6 \div 3 + 8 \cdot 5 = 2 + 8 \cdot 5 = 2 + 40 = 42$
(b) $(8 + 6) \div 7 \cdot 3 + (-6) = 14 \div 7 \cdot 3 + (-6)$
$$= 2 \cdot 3 + (-6)$$
$$= 6 + (-6)$$
$$= 0$$

(c) $\dfrac{-(-3)^3 + (-5)}{2(-8) - 5(3)} = \dfrac{-(-27) + (-5)}{2(-8) - 5(3)} = \dfrac{27 + (-5)}{-16 - 15} = \dfrac{22}{-31} = -\dfrac{22}{31}$ ✤

Example 4   Use the order of operations to evaluate each expression if $x = -2$, $y = 5$, and $z = -3$.

(a) $-4x^2 - 7y + 4z$

Replace x with -2, y with 5, and z with -3.

$$-4x^2 - 7y + 4z = -4(-2)^2 - 7(5) + 4(-3)$$
$$= -4(4) - 7(5) + 4(-3)$$
$$= -16 - 35 - 12$$
$$= -63$$

(b) $\dfrac{2(x-5)^2 + 4y}{z+4} = \dfrac{2(-2-5)^2 + 4(5)}{-3+4}$

$$= \dfrac{2(-7)^2 + 20}{1}$$
$$= 2(49) + 20$$
$$= 118 \quad \clubsuit$$

Properties of Real Numbers   There are several properties that describe how real numbers behave. We will discuss each property and then summarize them at the end of the section. The first of these properties, the *commutative properties,* say that two numbers may be added or multiplied in any order:

$$4 + (-12) = -12 + 4, \qquad 8(-5) = -5(8),$$
$$-9 + (-1) = -1 + (-9), \qquad (-6)(-3) = (-3)(-6),$$

and so on. Generalizing, the **commutative properties** say that for all real numbers a and b,

$$a + b = b + a \quad \text{and} \quad ab = ba.$$

Example 5   The following statements illustrate the commutative properties. Notice that the order of the numbers changes from one side of the equals sign to the other.

(a) $6 + x = x + 6$

(b) $(6 + x) + 9 = (x + 6) + 9$

(c) $(6 + x) + 9 = 9 + (6 + x)$

(d) $4 \cdot x = x \cdot 4$

(e) $5 \cdot (9 \cdot 8) = (9 \cdot 8) \cdot 5$

(f) $5 \cdot (9 \cdot 8) = 5 \cdot (8 \cdot 9) \quad \clubsuit$

By the *associative properties,* if three numbers are to be added or multiplied, either the first two numbers or the last two may be "associated." For example, the sum of the three numbers -9, 8, and 7 may be found in either of two ways:

$$-9 + (8 + 7) = -9 + 15 = 6,$$

or
$$(-9 + 8) + 7 = -1 + 7 = 6.$$

Also,

$$5(-3 \cdot 2) = 5 \cdot -6 = -30$$

or
$$(5 \cdot -3)2 = (-15)2 = -30.$$

In summary, the **associative properties** say that for all real numbers a, b, and c,

$$(a + b) + c = a + (b + c) \qquad \text{and} \qquad (ab)c = a(bc).$$

Example 6   The following statements illustrate the associative properties. Here the order of the numbers does not change, but the placement of parentheses (the association of the numbers) does change.

(a) $4 + (9 + 8) = (4 + 9) + 8$

(b) $3(9x) = (3 \cdot 9)x$

(c) $(\sqrt{3} + \sqrt{7}) + 2\sqrt{6} = \sqrt{3} + (\sqrt{7} + 2\sqrt{6})$ ✤

It is a common error to confuse the associative and commutative properties. To avoid this error, check the order of the terms: with the commutative properties the order changes from one side of the equals sign to the other; with the associative properties the order does not change, but the grouping does.

Example 7   This example shows a list of statements using the same symbols. Notice the difference between the commutative and associative properties.

Commutative properties	*Associative properties*
$(x + 4) + 9 = (4 + x) + 9$	$(x + 4) + 9 = x + (4 + 9)$
$7 \cdot (5 \cdot 2) = (5 \cdot 2) \cdot 7$	$7 \cdot (5 \cdot 2) = (7 \cdot 5) \cdot 2$ ✤

Example 8   Simplify each expression, using the commutative and associative properties as needed.

(a) $6 + (9 + x) = (6 + 9) + x = 15 + x$ Associative property

(b) $\dfrac{5}{8}(16y) = \left(\dfrac{5}{8} \cdot 16\right)y = 10y$ Associative property

(c) $(-10p)\left(\dfrac{6}{5}\right) = \dfrac{6}{5}(-10p)$ Commutative property

$\qquad\qquad = \left[\dfrac{6}{5}(-10)\right]p$ Associative property

$\qquad\qquad = -12p$ ✤

The *identity properties* show special properties of the numbers 0 and 1. The sum of 0 and any real number a is a itself:

$$0 + 4 = 4, \qquad -5 + 0 = -5,$$

and so on. The number 0 preserves the identity of a number under addition, making 0 the **identity element for addition.**

The number 1, the **identity element for multiplication,** preserves the identity of a number under multiplication, since the product of 1 and any number a is a:

$$5 \cdot 1 = 5, \qquad 1\left(-\frac{2}{3}\right) = -\frac{2}{3},$$

and so on. In summary, the **identity properties** say that for every real number a, there exists a unique real number 0 such that

$$a + 0 = a \qquad \text{and} \qquad 0 + a = a,$$

and there exists a unique real number 1 such that

$$a \cdot 1 = a \qquad \text{and} \qquad 1 \cdot a = a.$$

Example 9   The following statements illustrate the identity properties.

(a) $-12 \cdot 1 = -12$

(b) $\dfrac{8}{5} + 0 = \dfrac{8}{5}$

(c) $\pi \cdot 1 = \pi \qquad \text{and} \qquad \pi + 0 = \pi$ ✚

The sum of the numbers 5 and -5 is the identity element for addition, 0, just as the sum of $-2/3$ and $2/3$ is 0. In fact, for any real number a there is a real number, written $-a$, such that the sum of a and $-a$ is the identity element 0, or

$$a + (-a) = 0.$$

The number $-a$ is the **additive inverse** or **negative** of a.

Do not confuse the *negative of a number* with a *negative number*. Since a is a variable, it can represent either a positive or a negative number. The negative of a, written $-a$, can also be either a negative or a positive number (or zero). Do not make the common mistake of thinking that $-a$ *must* be a negative number. For example, if a is -3, then $-a$ is $-(-3) = 3$.

For each real number a (except 0) there is a real number $1/a$ such that the product of a and $1/a$ is the identity element for multiplication, 1, or

$$a \cdot \frac{1}{a} = 1 \qquad \text{and} \qquad \frac{1}{a} \cdot a = 1, \qquad a \neq 0.$$

The number $1/a$ is called the **multiplicative inverse** or **reciprocal** of the number a. Every real number except 0 has a reciprocal.

The existence of $-a$ and of $1/a$ comes from the **inverse properties,** which say that for every real number a, there exists a unique real number $-a$ such that

$$a + (-a) = 0 \quad \text{and} \quad -a + a = 0,$$

and for every nonzero real number a, there exists a unique real number $1/a$ such that

$$a \cdot \frac{1}{a} = 1 \quad \text{and} \quad \frac{1}{a} \cdot a = 1.$$

Example 10   The following statements illustrate the inverse properties.

(a) $9 + (-9) = 0$

(b) $-15 + 15 = 0$

(c) $\dfrac{4}{3} \cdot \dfrac{3}{4} = 1$

(d) $-8 \cdot \left(\dfrac{1}{-8}\right) = 1$

(e) $\dfrac{1}{\sqrt{5}} \cdot \sqrt{5} = 1$

(f) There is no real number x such that $0 \cdot x = 1$. In fact, 0 is the only real number without an inverse for multiplication. ❖

Distributive Property   The area of the entire region shown in Figure 1.7 can be found in two ways. One way is to multiply the length of the base of the entire region, or $5 + 3 = 8$, by the width of the region:

$$4(5 + 3) = 4(8) = 32.$$

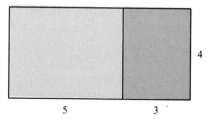

5   3

4

Figure 1.7

Another way to find the area of the region is to add the areas of the smaller rectangles on the left and right:

$$4(5) = 20 \qquad \text{and} \qquad 4(3) = 12,$$

with total area

$$4(5) + 4(3) = 20 + 12 = 32,$$

the same result. The equal results from finding the area in two ways,

$$4(5 + 3) = 4(5) + 4(3),$$

illustrate the **distributive property,** which says that for all real numbers a, b, and c,

$$a(b + c) = ab + ac.$$

Using a commutative property, the distributive property can be rewritten as $(b + c)a = ba + ca$. Another form of the distributive property, $a(b - c) = ab - ac$, comes from the definition of subtraction as "adding the negative." Also, the distributive property can be extended to include more than two numbers in the sum. For example,

$$
\begin{aligned}
9(5x + y + 4z) &= 9(5x) + 9y + 9(4z) \\
&= 45x + 9y + 36z.
\end{aligned}
$$

The distributive property is one of the key properties of the real numbers, because it is used to change products to sums and sums to products.

Example 11   The following statements illustrate the distributive property.

(a) $9(6 + 4) = 9 \cdot 6 + 9 \cdot 4$

(b) $3(x + y) = 3x + 3y$

(c) $-(m - 4n) = -1 \cdot (m - 4n) = -m + 4n$

(d) $7p + 21 = 7p + 7 \cdot 3 = 7(p + 3)$

(e) $\dfrac{1}{3}\left(\dfrac{4}{5}m - \dfrac{3}{2}n - 27\right) = \dfrac{1}{3}\left(\dfrac{4}{5}m\right) - \dfrac{1}{3}\left(\dfrac{3}{2}n\right) - \dfrac{1}{3}(27)$

$$= \dfrac{4}{15}m - \dfrac{1}{2}n - 9 \quad \clubsuit$$

A summary of the properties of the real numbers follows.

Properties of the Real Numbers	For all real numbers a, b, and c:

Commutative properties
$$a + b = b + a$$
$$ab = ba$$

Associative properties
$$(a + b) + c = a + (b + c)$$
$$(ab)c = a(bc)$$

Identity properties

There exists a unique real number 0 such that
$$a + 0 = a \quad \text{and} \quad 0 + a = a.$$

There exists a unique real number 1 such that
$$a \cdot 1 = a \quad \text{and} \quad 1 \cdot a = a.$$

Inverse properties

There exists a unique real number $-a$ such that
$$a + (-a) = 0 \quad \text{and} \quad (-a) + a = 0.$$

If $a \neq 0$, there exists a unique real number $1/a$ such that
$$a \cdot \frac{1}{a} = 1 \quad \text{and} \quad \frac{1}{a} \cdot a = 1.$$

Distributive property $a(b + c) = ab + ac$

1.1 Exercises

Let set $B = \{-6, -12/4, -5/8, -\sqrt{3}, 0, 1/4, 1, 2\pi, 3, \sqrt{12}\}$. List all the elements of B that are elements of the following sets. See Example 1.

1. Natural numbers

2. Whole numbers

3. Integers

4. Rational numbers

5. Irrational numbers

6. Real numbers

For Exercises 7–12, choose all words from the following list that apply: **(a)** *natural number,* **(b)** *whole number,* **(c)** *integer,* **(d)** *rational number,* **(e)** *irrational number,* **(f)** *real number,* **(g)** *meaningless. See Example 1.*

7. 12

8. 0

9. $-\dfrac{3}{4}$

10. $\dfrac{5}{9}$

11. $\sqrt{8}$

12. $-\sqrt{2}$

Simplify each of the following. Assume that all variables represent nonzero real numbers. In Exercises 17–24, write answers with exponents. See Example 2.

13. 3^4 **14.** 4^3 **15.** -3^5 **16.** -2^6

17. $(-2)^5 \cdot (-2)^2$ **18.** $3^8 \cdot 3^4$ **19.** $(2m^2)(4m^5)$ **20.** $-6p^3 \cdot 5p^2$

21. $\dfrac{a^2}{a^6}$ **22.** $\dfrac{b^9}{b^5}$ **23.** $\dfrac{27k^7}{3k^4}$ **24.** $\dfrac{15r^6}{25r^8}$

Use the order of operations to evaluate each of the following. See Example 3.

25. $9 - 4^2 - (-12)$ **26.** $8^2 - (-4) + 11$

27. $16(-9) - 4$ **28.** $-15 - 3(-8)$

29. $-2 \cdot 5 + 12 \div 3$ **30.** $9 \cdot 3 - 16 \div 4$

31. $-4(9 - 8) + (-7)(2)^3$ **32.** $6(-5) - (-3)(2)^4$

33. $(4 - 2^3)(-2 + \sqrt{25})$ **34.** $[-3^2 - (-2)] [\sqrt{16} - 2^3]$

35. $\left(-\dfrac{2}{9} - \dfrac{1}{4}\right) - \left[-\dfrac{5}{18} - \left(-\dfrac{1}{2}\right)\right]$ **36.** $\left[-\dfrac{5}{8} - \left(-\dfrac{2}{5}\right)\right] - \left(\dfrac{3}{2} - \dfrac{11}{10}\right)$

37. $\dfrac{-8 + (-4)(-6) \div 12}{4 - (-3)}$ **38.** $\dfrac{15 \div 5 \cdot 4 \div 6 - 8}{-6 - (-5) - 8 \div 2}$

*39. $\dfrac{-9.23(5.87) + 6.993}{1.225(-8.601) - 148(.0723)}$ **40.** $\dfrac{189.4(3.221) - 9.447(-8.772)}{4.889[3.117 - 8.291(3.427)]}$

Evaluate each of the following if $p = -2$, $q = 4$, and $r = -5$. See Example 4.

41. $-3(p + 5q)$ **42.** $2(q - r)$ **43.** $\dfrac{q + r}{q + p}$

44. $\dfrac{3q}{3p - 2r}$ **45.** $\dfrac{\dfrac{q}{4} - \dfrac{r}{5}}{\dfrac{p}{2} + \dfrac{q}{2}}$ **46.** $\dfrac{\dfrac{3r}{10} - \dfrac{5p}{2}}{q + \dfrac{2r}{5}}$

Evaluate each of the following if $a = -4.706$, $b = 1.983$, and $c = -.942$. Round all answers to the nearest thousandth. See Example 4.

47. $.498b + \dfrac{c}{.273}$ **48.** $1.426(3a - 8b)$ **49.** $\dfrac{3.091c + .769a}{.842b + 5}$ **50.** $\dfrac{.309b + .574c}{.9a - 2}$

Identify the properties that are illustrated in each of the following statements. Some will require more than one property. Assume that all variables represent real numbers. See Examples 5–7 and 9–11.

51. $8 \cdot 9 = 9 \cdot 8$ **52.** $3 + (-3) = 0$

53. $8 + (12 + 6) = (8 + 12) + 6$ **54.** $[9(-3)] \cdot 2 = 9[(-3) \cdot 2]$

55. $6 \cdot 12 + 6 \cdot 15 = 6(12 + 15)$ **56.** $8(m + 4) = 8m + 8 \cdot 4$

---

*Colored numbers identify problems that can be solved more quickly with the help of a calculator.

57. $(x + 6) \cdot \left(\dfrac{1}{x + 6}\right) = 1$, if $x + 6 \neq 0$

58. $\dfrac{2 + m}{2 - m} \cdot \dfrac{2 - m}{2 + m} = 1$, if $m \neq 2$ or -2

59. $(7 - y) + 0 = 7 - y$

60. $[9 + (-9)] \cdot 5 = 5 \cdot 0$

61. $(-5 + 7)(3 + 4) = (-5 + 7) \cdot 3 + (-5 + 7) \cdot 4$

62. $x \cdot \dfrac{1}{x} + x \cdot \dfrac{1}{x} = x\left(\dfrac{1}{x} + \dfrac{1}{x}\right)$, if $x \neq 0$

Use the distributive property to rewrite sums as products and products as sums. See Example 11.

63. $6(x + y)$

64. $9(r - s)$

65. $-3(z - y)$

66. $-2(m + n)$

67. $-(8r - k)$

68. $-(-2 + 5m)$

69. $a(r + s - t)$

70. $p(q - w + x)$

Use the various properties of real numbers to simplify each of the following expressions.

71. $9(4m)$

72. $-6(-3r)$

73. $\dfrac{5}{3}(12y)$

74. $\dfrac{10}{11}(22z)$

75. $\left(\dfrac{3}{4}r\right)(-12)$

76. $\left(-\dfrac{5}{8}p\right)(-24)$

77. $\dfrac{2}{3}(12y - 6z + 18q)$

78. $-\dfrac{1}{4}(20m + 8y - 32z)$

79. $\dfrac{3}{8}\left(\dfrac{16}{9}y + \dfrac{32}{27}z - \dfrac{40}{9}\right)$

80. $\dfrac{2}{3}\left(\dfrac{9}{4}a + \dfrac{15}{14}b - \dfrac{27}{20}\right)$

Give numerical examples to show that each of the following is false.

81. "Commutative property of subtraction," $a - b = b - a$

82. "Commutative property of division," $a/b = b/a$

83. "Associative property of subtraction," $(a - b) - c = a - (b - c)$

84. "Associative property of division," $(a/b)/c = a/(b/c)$

85. By the distributive property of multiplication over addition, $a(b + c) = ab + ac$. Is there a distributive property of addition over multiplication? That is, does

$$a + (b \cdot c) = (a + b)(a + c)$$

for all real numbers a, b, and c? To find out, try various sample values of a, b, and c.

86. Is there an identity property for subtraction? That is, does a real number k exist such that

$$a - k = a \quad \text{and} \quad k - a = a$$

for every real number a? Is there an identity property for division?

87. This exercise is designed to show that $\sqrt{2}$ is irrational. Give a reason for each of steps (a) through (i).

There are two possibilities:
(1) A rational number a/b exists such that $(a/b)^2 = 2$.
(2) There is no such rational number.

We work with assumption (1). If it leads to a contradiction, then we will know that (2) must be correct. Start by assuming that a rational number a/b exists with $(a/b)^2 = 2$. Assume also that a/b is written in lowest terms.

(a) Since $(a/b)^2 = 2$, we must have $a^2/b^2 = 2$, or $a^2 = 2b^2$.

(b) $2b^2$ is an even number.

(c) Therefore, a^2, and a itself, must be even numbers.

(d) Since a is an even number, it must be a multiple of 2. That is, we can find a natural number c such that $a = 2c$. This changes $a^2 = 2b^2$ into $(2c)^2 = 2b^2$.

(e) Therefore, $4c^2 = 2b^2$ or $2c^2 = b^2$.

(f) $2c^2$ is an even number.

(g) This makes b^2 an even number, so that b must be even.

(h) We have reached a contradiction. Show where the contradiction occurs.

(i) Since assumption (1) leads to a contradiction, we are forced to accept assumption (2), which says that $\sqrt{2}$ is irrational.

88. Go through steps similar to those in Exercise 87 to show that $\sqrt{3}$ is irrational.

1.2   Order and Absolute Value

Knowing the order relationship between real numbers helps in determining which of two given numbers is smaller. Figure 1.8 shows a number line with the points corresponding to several different numbers marked on the line. A number that corresponds to a particular point on a line is called the **coordinate** of the point. For example, the leftmost marked point in Figure 1.8 has coordinate -4. The correspondence between points on a line and the real numbers is called a **coordinate system** for the line. (From now on, the phrase "the point on a number line with coordinate a" will be abbreviated as "the point with coordinate a," or simply "the point a.")

Figure 1.8

If the real number a is to the left of the real number b on a number line, then **a is less than b,** written $a < b$. If a is to the right of b, then **a is greater than b,** written $a > b$. (Remember that the "point" of the symbol goes toward the smaller number.)

As an alternative to this geometric definition of "is less than" or "is greater than," there is an algebraic definition: if a and b are two real numbers and if the difference $a - b$ is positive, then $a > b$. If $a - b$ is negative, then $a < b$. The geometric and algebraic statements of order are summarized as follows.

Statement	Geometric form	Algebraic form
$a > b$	a is to the right of b	$a - b$ is positive
$a < b$	a is to the left of b	$a - b$ is negative

Example 1   Part (a) of this example shows how to identify the smaller of two numbers with the geometric approach, and part (b) uses the algebraic approach.

(a) In Figure 1.8, $-\sqrt{5}$ is to the left of 2/3, so

$$-\sqrt{5} < \frac{2}{3}.$$

Since 2/3 is to the right of $-\sqrt{5}$,

$$\frac{2}{3} > -\sqrt{5}.$$

Figure 1.8 also suggests that $2 < \pi$, while $\pi < \sqrt{20}$.

(b) The difference $2/3 - (-11/7)$ is positive, showing that

$$\frac{2}{3} > -\frac{11}{7}.$$

The difference $-11/7 - 2/3$ is negative, showing that

$$-\frac{11}{7} < \frac{2}{3}. \quad \clubsuit$$

The following variations on $<$ and $>$ are often used.

Symbol	Meaning
$\leq$	is less than or equal to
$\geq$	is greater than or equal to
$\not<$	is not less than
$\not>$	is not greater than

Statements involving these symbols, as well as $<$ and $>$, are called **inequalities.**

Example 2   The list below shows several statements and the reason why each is true.

Statement	Reason
$8 \leq 10$	$8 < 10$
$8 \leq 8$	$8 = 8$
$-9 \geq -14$	$-9 > -14$
$-8 \not> -2$	$-8 < -2$
$4 \not< 2$	$4 > 2$ $\clubsuit$

The expression $a < b < c$ says that b is *between* a and c, since

$$a < b < c$$

means $\qquad\qquad a < b \qquad \text{and} \qquad b < c.$

In the same way, $\qquad\qquad a \leq b \leq c$

means $\qquad\qquad a \leq b \qquad \text{and} \qquad b \leq c.$

When writing these "between" statements, make sure that both inequality symbols point in the same direction. For example,

$$\text{both } 2 < 7 < 11 \text{ and } 5 > 4 > -1$$

are true statements, but $3 < 5 > 2$ is meaningless. Generally, it is best to rewrite statements such as $5 > 4 > -1$ as $-1 < 4 < 5$, which is the order of these numbers on the number line.

The following *properties of order* give the basic properties of $<$ and $>$.

Properties of Order

For all real numbers a, b, and c:

Transitive property $\qquad$ If $a < b$ and $b < c$, then $a < c$.

Addition property $\qquad$ If $a < b$, then $a + c < b + c$.

Multiplication property $\qquad$ If $a < b$, and if $c > 0$, then $ac < bc$.

$\qquad\qquad\qquad\qquad\qquad$ If $a < b$, and if $c < 0$, then $ac > bc$.

In these properties, replacing $<$ with $>$ and $>$ with $<$ results in equivalent properties.

Example 3   **(a)** By the transitive property, if $3z < k$ and $k < p$, then $3z < p$.

(b) By the addition property, any real number can be added to both sides of an inequality. For example, adding -2 to both sides of

$$x + 2 < 5$$

gives

$$x + 2 + (-2) < 5 + (-2)$$
$$x < 3.$$

This process is explained in more detail in Chapter 2.

(c) While any number may be *added* to both sides of an inequality, more care must be used when *multiplying* both sides by a number. For example, multiplying both sides of

$$\frac{1}{2}x < 5$$

by 2 gives

$$2\left[\frac{1}{2}x\right] < 2 \cdot 5$$
$$x < 10,$$

by the first part of the multiplication property. On the other hand, multiplying both sides of

$$-\frac{3}{5}r \geq 9$$

by $-5/3$ gives

$$-\frac{5}{3}\left(-\frac{3}{5}r\right) \leq -\frac{5}{3} \cdot 9$$
$$r \leq -15.$$

Here the $\geq$ symbol has changed to $\leq$, using the second part of the multiplication property. ✤

Absolute Value   The distance on the number line from a number to 0 is called the **absolute value** of that number. The absolute value of the number a is written $|a|$. For example, the distance on the number line from 9 to 0 is 9, as is the distance from -9 to 0. (See Figure 1.9.) Therefore,

$$|9| = 9 \qquad \text{and} \qquad |-9| = 9.$$

Distance is 9.   Distance is 9.

−9   0   9

Figure 1.9

Example 4   **(a)** $|-4| = 4$

(b) $|2\pi| = 2\pi$

(c) $\left|-\frac{5}{8}\right| = \frac{5}{8}$

(d) $-|8| = -(8) = -8$

(e) $-|-2| = -(2) = -2$ ✤

The definition of the absolute value of the real number a can be stated as follows.

Definition of Absolute Value

$$|a| = \begin{cases} a \text{ if } a \geq 0 \\ -a \text{ if } a < 0 \end{cases}$$

The second part of this definition requires some thought. If a is a negative number, that is, if $a < 0$, then $-a$ is positive. Thus, for a *negative* number a,

$$|a| = -a.$$

For example, if $a = -5$, then $|a| = |-5| = -(-5) = 5$.

Example 5   Write each of the following without absolute value bars.

(a) $|-8 + 2|$
  Since $-8 + 2 = -6$, $|-8 + 2| = |-6| = 6$.

(b) $|\sqrt{5} - 2|$
  Since $\sqrt{5} > 2$ (see Table 1 in Appendix B or use a calculator),
  $\sqrt{5} - 2 > 0$, and $|\sqrt{5} - 2| = \sqrt{5} - 2$.

(c) $|\pi - 4|$
  Here, $\pi < 4$, so that $\pi - 4 < 0$, and

$$|\pi - 4| = -(\pi - 4)$$
$$= -\pi + 4$$
$$= 4 - \pi.$$

(d) $|m - 2|$ if $m < 2$
  If $m < 2$, then $m - 2 < 0$, so

$$|m - 2| = -(m - 2)$$
$$= -m + 2 \quad \text{or} \quad 2 - m. \quad \clubsuit$$

The definition of absolute value can be used to prove the following properties of absolute value.

Properties of Absolute Value

For all real numbers a and b:

$$|a| \geq 0 \qquad\qquad |-a| = |a| \qquad\qquad |a| \cdot |b| = |ab|$$

$$\left|\frac{a}{b}\right| = \frac{|a|}{|b|} \quad (b \neq 0) \qquad |a + b| \leq |a| + |b| \quad \text{(the triangle inequality)}$$

Example 6   **(a)** $|-15| = 15 \geq 0$

(b) $|-10| = 10$ and $|10| = 10$, so $|-10| = |10|$.

(c) If $x \geq 0$, $|5| \cdot |x| = |5x| = 5x$.

(d) $\left|\dfrac{2}{y}\right| = \dfrac{|2|}{|y|} = \dfrac{2}{|y|}$

(e) For $a = 3$ and $b = -7$:

$|a + b| = |3 + (-7)| = |-4| = 4$

$|a| + |b| = |3| + |-7| = 3 + 7 = 10$

$|a + b| < |a| + |b|$

(f) For $a = 2$ and $b = 12$:

$|a + b| = |2 + 12| = |14| = 14$

$|a| + |b| = |2| + |12| = 2 + 12 = 14$

$|a + b| = |a| + |b|$ ✚

Sometimes it is necessary to evaluate expressions involving absolute value. The next example shows how this is done.

Example 7   Let $x = -6$ and $y = 10$. Evaluate each expression.

(a) $|2x - 3y| = |2(-6) - 3(10)|$

$= |-12 - 30|$

$= |-42|$

$= 42$

(b) $\dfrac{|x| - |3y|}{|xy|} = \dfrac{|-6| - |3(10)|}{|-6(10)|}$

$= \dfrac{|-6| - |30|}{|-60|}$

$= \dfrac{6 - 30}{60}$

$= -\dfrac{2}{5}.$ ✚

1.2   Exercises

Write the following numbers in numerical order, from smallest to largest. Use Table 1 in Appendix B or a calculator, as necessary. See Example 1.

1. $-9, -2, 3, -4, 8$

2. $7, -6, 0, -2, -3$

3. $|-8|, -|9|, -|-6|$

4. $-|-9|, -|7|, -|-2|$

5. $\sqrt{8}, -4, -\sqrt{3}, -2, -5, \sqrt{6}, 3$

6. $\sqrt{2}, -1, 4, 3, \sqrt{8}, -\sqrt{6}, \sqrt{7}$

7. $3/4, \sqrt{2}, 7/5, 8/5, 22/15$

8. $-9/8, -3, -\sqrt{3}, -\sqrt{5}, -9/5, -8/5$

9. $|-8 + 2|, -|3|, -|-2|, -|-2| + (-3), -|-8| - |-6|$

10. $-2 - |-4|, -3 + |2|, -4, -5 + |-3|$

*For each of the given inequalities, multiply both sides by the indicated number to get an
equivalent inequality. See Example 3(c).*

11. $-2 < 5$,   3
12. $-7 < 1$,   2
13. $-6 < -4$, -5

14. $-9 < -8$, -4
15. $-11 \geq -22$, $-1/11$
16. $-2 \geq -14$, $-1/2$

17. $3x < 9$, $1/3$
18. $5x > 45$, $1/5$
19. $-9k > 63$, $-1/9$

20. $-3p \leq 24$, $-1/3$
21. $(-2/5)k < -10$, $-5/2$
22. $(-3/5)a \geq -9$, $-5/3$

Let $x = -4$ and $y = 2$. Evaluate each of the following. See Examples 4, 5, and 7.

23. $|2x|$
24. $|-3y|$
25. $|x - y|$

26. $|2x + 5y|$
27. $|3x + 4y|$
28. $|-5y + x|$

29. $|-4x + y| - |y|$
30. $|-8y + x| - |x|$
31. $\dfrac{|x| + 2|y|}{5 + x}$

32. $\dfrac{2|x| - |y + 2|}{|x| \cdot |y|}$
33. $\dfrac{x| -2 + y + |x| |}{|x + 3|}$
34. $\dfrac{(y - 3)| 5 - |x| + 2|y + 4| |}{2 - |y - 1|}$

Write each of the following without absolute value bars. See Examples 4 and 5.

35. $|-6|$
36. $-|-2|$
37. $-|-8| + |-2|$

38. $3 - |-4|$
39. $|8 - \sqrt{50}|$
40. $|2 - \sqrt{3}|$

41. $|\sqrt{7} - 5|$
42. $|\sqrt{2} - 3|$
43. $|\pi - 3|$

44. $|\pi - 5|$
45. $|x - 4|$,   if $x > 4$
46. $|y - 3|$,   if $y < 3$

47. $|2k - 8|$,   if $k < 4$
48. $|3r - 15|$,   if $r > 5$
49. $|7m - 56|$,   if $m < 4$

50. $|2k - 7|$,   if $k > 4$
51. $|-8 - 4m|$,   if $m > -2$
52. $|6 - 5r|$,   if $r < -2$

53. $|x - y|$,   if $x < y$
54. $|x - y|$,   if $x > y$
55. $|3 + x^2|$

56. $|x^2 + 4|$
57. $|-1 - p^2|$
58. $|-r^4 - 16|$

*Justify each of the following statements by giving the correct property from this section.
Assume that all variables represent real numbers. See Example 3.*

59. If $2k < 8$, then $k < 4$.
60. If $x + 8 < 15$, then $x < 7$.

61. If $-4x < 24$, then $x > -6$.
62. If $x < 5$ and $5 < m$, then $x < m$.

63. If $m > 0$, then $9m > 0$.
64. If $k > 0$, then $8 + k > 8$.

65. $|8 + m| \leq |8| + |m|$
66. $|k - m| \leq |k| + |-m|$
67. $|8| \cdot |-4| = |-32|$

68. $|12 + 11r| \geq 0$
69. $\left|\dfrac{-12}{5}\right| = \dfrac{|-12|}{|5|}$
70. $\left|\dfrac{6}{5}\right| = \dfrac{|6|}{|5|}$

Under what conditions are the following statements true?

71. $|x| = |y|$
72. $|x + y| = |x| + |y|$
73. $|x + y| = |x| - |y|$

74. $|x| \leq 0$
75. $|x - y| = |x| - |y|$
76. $||x + y|| = |x + y|$

Evaluate each of the following if x is a nonzero real number.

77. $\dfrac{|x|}{x}$
78. $\left|\dfrac{|x|}{x}\right|$

1.3 Integer Exponents

Positive integer exponents were introduced in Section 1.1. Work with exponents can be simplified by using various properties of exponents. By definition, the notation a^m (where m is a positive integer and a is a real number) means that a appears as a factor m times. In the same way, a^n (where n is a positive integer) means that a appears as a factor n times. In the product $a^m \cdot a^n$, the number a would appear $m + n$ times.

Product Rule

> For all positive integers m and n and every real number a:
>
> $$a^m \cdot a^n = a^{m+n}$$

Example 1 **(a)** $y^4 \cdot y^7 = y^{4+7} = y^{11}$

(b) $(6z^5)(9z^3)(2z^2) = 6 \cdot 9 \cdot 2 \cdot z^{5+3+2} = 108z^{10}$

(c) $(2k^m)(k^{1+m}) = 2k^{m+(1+m)} = 2k^{1+2m}$ (if m is a positive integer)   ✛

To get a similar rule for division, we must first define a negative exponent. By the definition of exponent, if $a \neq 0$,

$$\frac{a^3}{a^7} = \frac{a \cdot a \cdot a}{a \cdot a \cdot a \cdot a \cdot a \cdot a \cdot a} = \frac{1}{a \cdot a \cdot a \cdot a} = \frac{1}{a^4}.$$

In the product rule, $a^m \cdot a^n = a^{m+n}$, the exponents are *added*. This suggests that we should *subtract* exponents when dividing. Subtracting exponents gives

$$\frac{a^3}{a^7} = a^{3-7} = a^{-4}.$$

The only way to keep these results consistent is to define a^{-4} as $1/a^4$. This definition can be generalized as follows.

Definition of a^{-n}

> If a is a nonzero real number and n is any integer, then
>
> $$a^{-n} = \frac{1}{a^n}.$$

Example 2 **(a)** $4^{-2} = \frac{1}{4^2} = \frac{1}{16}$

(b) $2^{-3} = \frac{1}{2^3} = \frac{1}{8}$

(c) $\left(\dfrac{2}{5}\right)^{-3} = \dfrac{1}{\left(\dfrac{2}{5}\right)^3} = \dfrac{1}{\dfrac{8}{125}} = \dfrac{125}{8}$

(d) $-4^{-2} = -\dfrac{1}{4^2} = -\dfrac{1}{16}$ ✚

Part (c) of Example 2 showed that

$$\left(\frac{2}{5}\right)^{-3} = \frac{125}{8} = \left(\frac{5}{2}\right)^3.$$

This result can be generalized. If $a \neq 0$ and $b \neq 0$, then

$$\left(\frac{a}{b}\right)^{-n} = \left(\frac{b}{a}\right)^n,$$

for any integer n.

The quotient rule for exponents follows from the definition of exponents, as shown above.

Quotient Rule

> For all integers m and n and all nonzero real numbers a:
>
> $$\frac{a^m}{a^n} = a^{m-n}$$

In the following examples, assume that all variables represent nonzero real numbers.

Example 3   **(a)** $\dfrac{12^5}{12^2} = 12^{5-2} = 12^3$

(b) $\dfrac{a^5}{a^8} = a^{5-8} = a^{-3} = \dfrac{1}{a^3}$

(c) $\dfrac{16m^9}{12m^{11}} = \dfrac{4}{3m^2}$

(d) $\dfrac{25r^7z^5}{10r^9z} = \dfrac{5z^4}{2r^2}$

(e) $\dfrac{x^{5y}}{x^{3y}} = x^{5y-3y} = x^{2y}$, if y is an integer   ✚

What happens to the property $a^m/a^n = a^{m-n}$ if m and n are the same? For example,

$$\frac{k^5}{k^5} = k^{5-5} = k^0.$$

If k is a nonzero real number, then $k^5/k^5 = 1$. To be consistent, k^0 must be defined to equal 1.

Definition of a^0	For any nonzero real number a: $$a^0 = 1$$

The symbol 0^0 is not defined.

Example 4   (a) $3^0 = 1$

(b) $(-4)^0 = 1$

(c) $-4^0 = -1$

(d) $-(-4)^0 = -1$

(e) $(7r)^0 = 1,$ if $r \neq 0$ ✚

The expression $(2^5)^3$ can be written as

$$(2^5)^3 = 2^5 \cdot 2^5 \cdot 2^5.$$

By a generalization of the product rule for exponents, this product is

$$(2^5)^3 = 2^{5+5+5} = 2^{15}.$$

The same exponent could have been obtained by multiplying 3 and 5. This example suggests the first of the **power rules** given below. The others are found in a similar way.

Power Rules	For all integers m and n and all real numbers a and b: $$(a^m)^n = a^{mn} \qquad (ab)^m = a^m b^m \qquad \left(\frac{a}{b}\right)^m = \frac{a^m}{b^m}, \quad (b \neq 0)$$

Example 5   (a) $(5^3)^{-2} = 5^{3(-2)} = 5^{-6} = \dfrac{1}{5^6}$

(b) $(3^4 x^2)^3 = (3^4)^3 (x^2)^3 = 3^{4(3)} x^{2(3)} = 3^{12} x^6$

(c) $\left(\dfrac{2^5}{b^4}\right)^3 = \dfrac{(2^5)^3}{(b^4)^3} = \dfrac{2^{15}}{b^{12}},$ if $b \neq 0$

(d) $\left(\dfrac{5^{-2}}{q^3}\right)^{-4} = \dfrac{5^8}{q^{-12}} = \dfrac{5^8}{\dfrac{1}{q^{12}}} = 5^8 \dfrac{q^{12}}{1} = 5^8 q^{12},$ if $q \neq 0$ ✚

The properties of exponents are summarized here.

Properties of Exponents

For all real numbers a and b and all integers m and n:

Product rule $a^m \cdot a^n = a^{m+n}$

Quotient rule $\dfrac{a^m}{a^n} = a^{m-n}$ $(a \neq 0)$

Power rules $(a^m)^n = a^{mn}$

$(ab)^m = a^m b^m$

$\left(\dfrac{a}{b}\right)^m = \dfrac{a^m}{b^m}$ $(b \neq 0)$

Example 6   Use the properties of exponents to simplify each expression. Write answers without negative exponents. Assume that all variables represent nonzero real numbers.

(a) $3x^{-2}(4^{-1}x^{-5})^2 = 3x^{-2}(4^{-2}x^{-10})$

$= 3 \cdot 4^{-2} \cdot x^{-2+(-10)}$

$= 3 \cdot 4^{-2} \cdot x^{-12}$

$= \dfrac{3}{16x^{12}}$

(b) $\dfrac{5m^{-3}}{10m^{-5}} = \dfrac{5}{10}\, m^{-3-(-5)}$

$= \dfrac{1}{2}\, m^2$ or $\dfrac{m^2}{2}$

(c) $\dfrac{12p^3q^{-1}}{8p^{-2}q} = \dfrac{12}{8} \cdot \dfrac{p^3}{p^{-2}} \cdot \dfrac{q^{-1}}{q}$

$= \dfrac{3}{2} \cdot p^{3-(-2)}q^{-1-1}$ $(q = q^1)$

$= \dfrac{3}{2}\, p^5 q^{-2}$

$= \dfrac{3p^5}{2q^2}$

(d) $\dfrac{(3x^2)^{-1}(3x^5)^{-2}}{(3^{-1}x^{-2})^2} = \dfrac{3^{-1}x^{-2}3^{-2}x^{-10}}{3^{-2}x^{-4}}$

$= \dfrac{3^{-1+(-2)}x^{-2+(-10)}}{3^{-2}x^{-4}} = \dfrac{3^{-3}x^{-12}}{3^{-2}x^{-4}}$

$= 3^{-3-(-2)}x^{-12-(-4)} = 3^{-1}x^{-8}$

$= \dfrac{1}{3x^8}$ ✚

Scientific Notation   One application of exponents comes from science, where calculations often involve very large or very small numbers. To avoid working with many zeros, scientists write such numbers in *scientific notation*.

Scientific Notation	In **scientific notation,** a number is written as the product of a number a whose absolute value is at least 1, but less than 10, and a power of 10, or $$a \times 10^n,$$ for an appropriate integer power n.

Example 7   Write each number in scientific notation.

(a) 125,000

The number 1.25 is between 1 and 10. In scientific notation, 125,000 will be written as the product of 1.25 and some power of 10. To go from 1.25 to 125,000 requires moving the decimal point 5 places to the right, so

$$125,000 = 1.25 \times 10^5.$$

(b) $-7900 = -7.9 \times 10^3$

(c) $.003 = 3 \times 10^{-3}$

(d) $.0000056 = 5.6 \times 10^{-6}$

(e) $8.739 = 8.739 \times 10^0$ ✤

One advantage of scientific notation is that some calculations can be simplified using the properties of exponents. This process is shown in the next example.

Example 8   Find each of the following.

(a) (28,000,000)(.000012)

Write each number in scientific notation, then multiply.

$$
\begin{aligned}
(28,000,000)(.000012) &= (2.8 \times 10^7)(1.2 \times 10^{-5}) \\
&= (2.8 \times 1.2) \times 10^{7+(-5)} \\
&= 3.36 \times 10^2 \\
&= 336
\end{aligned}
$$

(b) $\dfrac{.082}{.0000097} = \dfrac{8.2 \times 10^{-2}}{9.7 \times 10^{-6}}$

$$
\begin{aligned}
&= \frac{8.2}{9.7} \times 10^{-2-(-6)} \\
&= .85 \times 10^4 \qquad \text{The quotient is rounded to the nearest} \\
&= 8500 \quad ✤ \qquad \text{hundredth}
\end{aligned}
$$

Another use of scientific notation comes when writing **significant digits.** While the fine points of significant digits are not covered in this book, some examples are shown here.

Number	Significant digits
9	1
.00271	3
2.00094	6

Significant digits are unclear in a number like 8000. Is it measured to the nearest one, ten, hundred, or thousand? If the number is written in scientific notation, the number of significant digits can be identified as follows. (Notice that the zeros after the decimal point are counted as significant digits when a number is written in scientific notation.)

Number	Significant digits	Measured to nearest
8.000×10^3	4	One
8.00×10^3	3	Ten
8.0×10^3	2	Hundred
8×10^3	1	Thousand

1.3 Exercises

Simplify each of the following. Write results with no negative exponents. Assume that all variables represent nonzero real numbers. See Examples 1–6.

1. $(-4)^{-3}$

2. $(-5)^{-2}$

3. $\left(\dfrac{1}{2}\right)^{-3}$

4. $\left(\dfrac{2}{3}\right)^{-2}$

5. $\left(\dfrac{5}{x}\right)^{-4}$

6. $\left(\dfrac{n}{3}\right)^{-2}$

7. 7^0

8. $\left(\dfrac{2}{3}\right)^0$

9. $-(-7^0)$

10. $-\left(\dfrac{2}{3}\right)^0$

11. $(2^2)^5$

12. $(6^4)^3$

13. $(x^5y^4)^3$

14. $(m^3n^9)^2$

15. $\left(\dfrac{p^4}{q}\right)^2$

16. $\left(\dfrac{r^8}{s^2}\right)^3$

17. $(z^{-3}y^3)^{-3}$

18. $(y^{-3}x^{-5})^{-1}$

19. $(2a^3b^{-3})^3$

20. $(5p^4q^{-1})^{-2}$

21. $\left(\dfrac{r^{-2}}{s^{-5}}\right)^{-3}$

22. $\left(\dfrac{p^{-1}}{q^{-5}}\right)^{-2}$

23. $\left(\dfrac{a}{b^{-3}}\right)^{-1}$

24. $\left(\dfrac{m^{-1}}{n}\right)^{-2}$

25. $(-2x^{-3}y^2)^{-2}$

26. $(-3p^2q^{-2}s^{-1})^7$

27. $(6^{-2})(6^{-5})(6^3)$

28. $(4^8)(4^{-10})(4^2)$

29. $(m^{10})(m^{-4})(m^{-5})$

30. $(r^{-3})(r^{-2})(r^4)$

31. $\dfrac{6m^2n^4}{2mn^5}$

32. $\dfrac{3x^5y^{10}}{9xy^5z^4}$

33. $(3m)^2(-2m)^3(-3m^4)$

34. $(-p^3)(2p^2)^3(-2p^4)$

35. $\dfrac{2^{-1}x^3y^{-3}}{xy^{-2}}$

36. $\dfrac{5^{-2}m^2y^{-5}}{5^2m^{-1}y^{-2}}$

37. $\dfrac{(4 + s)^3(4 + s)^{-2}}{(4 + s)^{-5}}$

38. $\dfrac{(m + n)^4(m + n)^{-2}}{(m + n)^{-5}}$

39. $\dfrac{(2x^3)^{-2}(2^2x^5)^{-1}}{(2\,x^4)^{-3}}$

40. $\dfrac{(5p^2q)^{-1}(5p^3q^{-2})^2}{5(pq)^{-3}(p^4q^{-2})}$

41. $\dfrac{(r^2s^3t^4)^{-2}(r^3s^2t)^{-1}}{(rst)^{-3}(r^2st^2)^3}$

42. $\dfrac{(ab^2c)^{-3}(a^2bc)^{-2}}{(abc^2)^4}$

43. $\dfrac{[k^2(p + q)^4]^{-1}}{k^{-4}(p + q)^3}$

44. $\dfrac{t^5(x + y)^4t^{-8}}{(x + y)^{-2}t^{-5}}$

Evaluate each expression, letting $a = 2$, $b = -3$, and $c = 0$.

45. $a^7b^9c^5$

46. $12a^{15}b^6c^8$

47. $a^b + b^a$

48. $b^c + a^c$

49. $(3a^3 + 5b^2)^c$

50. $-(4b^5 - 2a^3)^c$

51. $a^{-1} + b^{-1}$

52. $b^{-2} - a$

53. $\dfrac{2b^{-1} - 3a^{-1}}{a + b^2}$

54. $\dfrac{3a^2 - b^2}{b^{-3} + 2a^{-1}}$

55. $\left(\dfrac{a}{3}\right)^{-1} + \left(\dfrac{b}{2}\right)^{-2}$

56. $\left(\dfrac{2b}{5}\right)^2 - 3\left(\dfrac{a^{-1}}{4}\right)$

Express each of the following numbers in scientific notation. See Example 7.

57. 69,300

58. 5000

59. 12,700,000

60. 6,000,000,000

61. .0001

62. .00792

63. .054

64. .965

Express each of the following numbers without exponents. See Example 7.

65. 8.2×10^5

66. 3.7×10^3

67. 2.98×10^2

68. 1.7×10^8

69. 3.61×10^{-2}

70. 4.02×10^{-5}

71. 6.15×10^{-3}

72. 8.76×10^{-1}

Use scientific notation to evaluate each of the following. See Example 8.

73. $(4600)(.00092)$

74. $(.87)(.0004)$

75. $(.00002)(.00009)$

76. $\dfrac{.0000042}{.0006}$

77. $\dfrac{28}{.0004} + \dfrac{(.003)(50)}{.0006}$

78. $\dfrac{50}{.0000025} + \dfrac{(.024)(900)}{.00006}$

Calculators with scientific notation capability display the number 1.46×10^{12} as follows.

$$1.46 \qquad 12$$

Use a calculator to evaluate each of the following. Write all answers in scientific notation rounded to three significant digits.

79. $(1.66 \times 10^4)(2.93 \times 10^3)$

80. $(6.92 \times 10^7)(8.14 \times 10^5)$

81. $(4.08 \times 10^{-9})(3.172 \times 10^4)$

82. $(9.113 \times 10^{12})(8.72 \times 10^{-11})$

83. $\dfrac{-(4.389 \times 10^4)(2.421 \times 10^{-2})}{1.76 \times 10^{-9}}$

84. $\dfrac{(1.392 \times 10^{10})(5.746 \times 10^{-8})}{4.93 \times 10^{-15}}$

Use scientific notation to work the following exercises.

85. The distance to the sun is 9.3×10^7 mi. How long would it take a rocket, traveling at 2.9×10^3 mph, to reach the sun?

86. A *light-year* is the distance that light travels in one year. Find the number of miles in a light-year if light travels 1.86×10^5 mi per sec.

87. Use the information given in the previous two exercises to find the number of minutes necessary for light from the sun to reach the earth.

88. A computer can do one addition in 1.4×10^{-7} sec. How long would it take the computer to do a trillion (10^{12}) calculations? Give the answer in seconds and then in hours.

89. Assume that the volume of the earth is 5×10^{14} cu m and that the volume of a bacterium is 4×10^{-16} cu m. If the earth could be packed full of bacteria, how many would it contain?

90. Our galaxy is approximately 1.2×10^{17} km across. Suppose a spaceship could travel at 1.5×10^5 km per sec (half the speed of light). Find the approximate number of years needed for the spaceship to cross the galaxy.

Find each of the following products or quotients. Assume that all variables appearing in exponents represent integers and all variables are nonzero.

91. $(2k^m)(k^{1-m})$

92. $(y^m)(y^{1+m})$

93. $5p^r(6p^{3-2r})$

94. $(-z^{2r})(4z^{r+3})$

95. $\dfrac{k^{y-5}}{k^{y+2}}$

96. $\dfrac{4a^{m+2}}{2a^{m-1}}$

97. $\dfrac{(b^2)^y}{(2b^y)^3}$

98. $\left(\dfrac{3p^a}{2m^b}\right)^c$

99. $\dfrac{(2m^n)^2(-4m^{2+n})}{8m^{4n}}$

1.4 Polynomials

An **algebraic expression** is any combination of variables or constants joined by the basic operations of addition, subtraction, multiplication, division (except by zero), or extraction of roots. The simplest algebraic expressions, *polynomials,* are discussed in this section.

The product of a real number and one or more variables raised to powers is called a **term.** The real number is called the **numerical coefficient,** or just the **coefficient.** The coefficient in $-3m^4$ is -3, while the coefficient in $-p^2$ is -1. **Like terms** are terms with the same variable raised to the same power. For example, $-13x^3$, $4x^3$, and $-x^3$ are like terms, while $6y$ and $6y^2$ are not.

A **polynomial** is defined as a finite sum of terms, with only nonnegative integer exponents permitted on the variables. If the terms of a polynomial contain only the variable x, then the polynomial is called a **polynomial in x.** (Polynomials in other variables are defined similarly.) Examples of polynomials include

$$5x^3 - 8x^2 + 7x - 4, \qquad 9p^5 - 3, \qquad 8r^2, \qquad \text{and} \qquad 6.$$

The expression $9x^2 - 4x - 6/x$ is not a polynomial because of $-6/x$. Since $-6/x = -6x^{-1}$, the polynomial has a term with an exponent that is a negative integer.

The highest exponent in a polynomial in one variable is the **degree** of the polynomial. A nonzero constant is said to have degree 0. (The polynomial 0 has no degree.) For example, $3x^6 - 5x^2 + 2x + 3$ is a polynomial of degree 6.

A polynomial can have more than one variable. A term containing more than one variable has degree equal to the sum of all the exponents appearing on the variables in the term. For example, $-3x^4y^3z^5$ is of degree $4 + 3 + 5 = 12$. The degree of a polynomial in more than one variable is equal to the highest degree of any term appearing in the polynomial. By this definition, the polynomial

$$2x^4y^3 - 3x^5y + x^6y^2$$

is of degree 8 because of the x^6y^2 term.

A polynomial containing exactly three terms is called a **trinomial;** one containing exactly two terms is a **binomial;** and a single-term polynomial is called a **monomial.** For example, $7x^9 - 8x^4 + 1$ is a trinomial of degree 9.

Example 1 The list below shows several polynomials, gives the degree of each, and identifies each as a monomial, binomial, trinomial, or none of these.

Polynomial	Degree	Type
$9p^7 - 4p^3 + 8p^2$	7	Trinomial
$29x^{11} + 8x^{15}$	15	Binomial
$-10r^6 s^8$	14	Monomial
$5a^3 b^7 - 3a^5 b^5 + 4a^2 b^9 - a^{10}$	11	None of these

Since the variables used in polynomials represent real numbers, a polynomial represents a real number. This means that all the properties of the real numbers mentioned in this chapter hold for polynomials. In particular, the distributive property holds, so

$$3m^5 - 7m^5 = (3 - 7)m^5 = -4m^5.$$

Thus, polynomials are added by adding coefficients of like terms; polynomials are subtracted by subtracting coefficients of like terms. (Notice how the distributive property was used to suggest these definitions.)

Example 2 Add or subtract, as indicated.
(a) $(2y^4 - 3y^2 + y) + (4y^4 + 7y^2 + 6y)$
$$= (2 + 4)y^4 + (-3 + 7)y^2 + (1 + 6)y$$
$$= 6y^4 + 4y^2 + 7y$$
(b) $(6r^4 + 2r^2) + (3r^3 + 9r) = 6r^4 + 3r^3 + 2r^2 + 9r$
(c) $(-3m^3 - 8m^2 + 4) - (m^3 + 7m^2 - 3)$
$$= (-3 - 1)m^3 + (-8 - 7)m^2 + [4 - (-3)]$$
$$= -4m^3 - 15m^2 + 7$$
(d) $8m^4 p^5 - 9m^3 p^5 + (11m^4 p^5 + 15m^3 p^5) = 19m^4 p^5 + 6m^3 p^5$

As shown in parts (a)–(c) of Example 2, polynomials in one variable are often written with their terms in *descending order,* so the term of highest degree is first, the one with the next highest degree is next, and so on.

The distributive property, together with the properties of exponents, can also be used to find the product of two polynomials. For example, to find the product of $3x - 4$ and $2x^2 - 3x + 5$, treat $3x - 4$ as a single expression and use the distributive property as follows.

$$(3x - 4)(2x^2 - 3x + 5) = (3x - 4)(2x^2) - (3x - 4)(3x) + (3x - 4)(5)$$

Now use the distributive property three separate times on the right of the equals sign to get

$$(3x - 4)(2x^2 - 3x + 5)$$
$$= (3x)(2x^2) - 4(2x^2) - (3x)(3x) - (-4)(3x) + (3x)5 - 4(5)$$
$$= 6x^3 - 8x^2 - 9x^2 + 12x + 15x - 20$$
$$= 6x^3 - 17x^2 + 27x - 20.$$

It is sometimes more convenient to write such a product vertically, as shown below.

$$
\begin{array}{r}
2x^2 - 3x + 5 \\
3x - 4 \\
\hline
- 8x^2 + 12x - 20 \\
6x^3 - 9x^2 + 15x \\
\hline
6x^3 - 17x^2 + 27x - 20
\end{array}
\quad \text{Add in columns}
$$

The FOIL method is a convenient way to find the product of two binomials. FOIL (for First, Outside, Inside, Last) gives the pairs of terms to be multiplied to get the product, as shown in the next examples.

Example 3   Find each product.

$$
\begin{array}{cccc}
\mathbf{F} & \mathbf{O} & \mathbf{I} & \mathbf{L}
\end{array}
$$

(a) $(6m + 1)(4m - 3) = (6m)(4m) + (6m)(-3) + 1(4m) + 1(-3)$
$$= 24m^2 - 14m - 3$$

(b) $(2x + 7)(2x - 7) = 4x^2 - 14x + 14x - 49$
$$= 4x^2 - 49$$

(c) $(2k^n - 5)(k^n + 3) = 2k^{2n} + 6k^n - 5k^n - 15$
$$= 2k^{2n} + k^n - 15 \quad \text{(if } n \text{ is an integer)} \quad \clubsuit$$

In parts (a) and (c) of Example 3, the product of two binomials was a trinomial, while in part (b) the product of two binomials was a binomial. The product of two binomials of the form $x + y$ and $x - y$ is always a binomial. Check by multiplying that the following is true.

Difference of Two Squares	$(x + y)(x - y) = x^2 - y^2$

This product is called the **difference of two squares.**

Example 4   **(a)** $(3p + 11)(3p - 11)$
Using the pattern discussed above, replace x with $3p$ and y with 11.

$$(3p + 11)(3p - 11) = (3p)^2 - 11^2 = 9p^2 - 121$$

(b) $(5m^3 - 3)(5m^3 + 3) = (5m^3)^2 - 3^2 = 25m^6 - 9$

(c) $(9k - 11r^3)(9k + 11r^3) = (9k)^2 - (11r^3)^2 = 81k^2 - 121r^6 \quad \clubsuit$

The **squares of binomials** are also special products. The products $(x + y)^2$ and $(x - y)^2$ are shown below.

Squares of Binomials	$$(x + y)^2 = x^2 + 2xy + y^2$$ $$(x - y)^2 = x^2 - 2xy + y^2$$

Example 5 (a) $(2m + 5)^2 = (2m)^2 + 2(2m)(5) + (5)^2$
$$= 4m^2 + 20m + 25$$

(b) $(3x - 7y^4)^2 = (3x)^2 - 2(3x)(7y^4) + (7y^4)^2$
$$= 9x^2 - 42xy^4 + 49y^8 \quad \clubsuit$$

Another pair of special products gives the cubes of binomials. Verify by multiplication that the following patterns are correct.

Cubes of Binomials	$$(x + y)^3 = x^3 + 3x^2y + 3xy^2 + y^3$$ $$(x - y)^3 = x^3 - 3x^2y + 3xy^2 - y^3$$

Example 6 (a) $(m + 4n)^3$

Replace x with m and y with $4n$ in the pattern for $(x + y)^3$ to get

$$(m + 4n)^3 = m^3 + 3m^2(4n) + 3m(4n)^2 + (4n)^3$$
$$= m^3 + 12m^2n + 48mn^2 + 64n^3.$$

(b) $(5k - 2z^5)^3 = (5k)^3 - 3(5k)^2(2z^5) + 3(5k)(2z^5)^2 - (2z^5)^3$
$$= 125k^3 - 150k^2z^5 + 60kz^{10} - 8z^{15} \quad \clubsuit$$

It is important to memorize and be able to apply these special products. (Both the squares of binomials and the cubes of binomials are special cases of the binomial theorem, discussed in the last chapter of this book.)

Quotients of Polynomials   To find the quotients of a polynomial P and a monomial Q, multiply P by the reciprocal of Q.

$$\frac{P}{Q} = P \cdot \frac{1}{Q}, \quad \text{if } Q \neq 0$$

Example 7 (a) $\dfrac{2m^5 - 6m^3}{2m^3} = (2m^5 - 6m^3) \cdot \dfrac{1}{2m^3} = \dfrac{2m^5}{2m^3} - \dfrac{6m^3}{2m^3} = m^2 - 3$

Since the quotient $m^2 - 3$ is a polynomial, $2m^5 - 6m^3$ is divisible by $2m^3$.

(b) $\dfrac{3y^6x^3 - 6y^3x^6 + 8y^5x}{3y^3x^3} = \dfrac{3y^6x^3}{3y^3x^3} - \dfrac{6y^3x^6}{3y^3x^3} + \dfrac{8y^5x}{3y^3x^3} = y^3 - 2x^3 + \dfrac{8y^2}{3x^2}$

Here the quotient is not a polynomial, so $3y^6x^3 - 6y^3x^6 + 8y^5x$ is *not* divisible by $3y^3x^3$. $\clubsuit$

To divide one polynomial by another, use a **division algorithm.** (An **algorithm** is an orderly procedure for performing an operation.) The division algorithm for polynomials is very similar to that used for dividing whole numbers, as the following examples show.

Example 8   Divide $6q^3 - 17q^2 + 22q - 23$ by $2q - 3$.

Begin by writing both polynomials with exponents in descending order.

$$2q - 3 \overline{)6q^3 - 17q^2 + 22q - 23}$$

Step 1   The quotient of $6q^3$ and $2q$ is $3q^2$. Multiply $3q^2$ and $2q - 3$, getting $6q^3 - 9q^2$.

$$
\begin{array}{r}
3q^2 \hphantom{xxxxxxxxxxxxx} \\
2q - 3 \overline{)\, 6q^3 - 17q^2 + 22q - 23\,} \\
\underline{6q^3 - 9q^2} \hphantom{xxxxxxxxxx}
\end{array}
$$

$\dfrac{6q^3}{2q} = 3q^2$

$3q^2(2q - 3) = 6q^3 - 9q^2$

Step 2   Subtract $6q^3 - 9q^2$ from $6q^3 - 17q^2$ by changing the signs on $6q^3 - 9q^2$ and adding. Then bring down $22q$.

$$
\begin{array}{r}
3q^2 \hphantom{xxxxxxxxxxxx} \\
2q - 3 \overline{)\, 6q^3 - 17q^2 + 22q - 23\,} \\
\underline{-6q^3 + 9q^2} \hphantom{xxxxxxx} \\
-8q^2 + 22q \hphantom{xxx}
\end{array}
$$

Change signs and add

Step 3   The quotient of $-8q^2$ and $2q$ is $-4q$. Multiply $-4q$ and $2q - 3$.

$$
\begin{array}{r}
3q^2 - 4q \hphantom{xxxxxxxxx} \\
2q - 3 \overline{)\, 6q^3 - 17q^2 + 22q - 23\,} \\
\underline{6q^3 - 9q^2} \hphantom{xxxxxxxxxx} \\
-8q^2 + 22q \hphantom{xxx} \\
\underline{-8q^2 + 12q} \hphantom{xxx}
\end{array}
$$

$\dfrac{-8q^2}{2q} = -4q$

$-4q(2q - 3) = -8q^2 + 12q$

Step 4   Subtract $-8q^2 + 12q$ from $-8q^2 + 22q$. Bring down -23.

$$
\begin{array}{r}
3q^2 - 4q \hphantom{xxxxxxxxx} \\
2q - 3 \overline{)\, 6q^3 - 17q^2 + 22q - 23\,} \\
\underline{6q^3 - 9q^2} \hphantom{xxxxxxxxxx} \\
-8q^2 + 22q \hphantom{xxx} \\
\underline{8q^2 - 12q} \hphantom{xxx} \\
10q - 23 \hphantom{x}
\end{array}
$$

Change signs and add

Step 5 The quotient of $10q$ and $2q$ is 5. Multiply 5 and $2q - 3$.

$$
\begin{array}{r}
3q^2 - 4q + 5 \leftarrow \qquad \dfrac{10q}{2q} = 5 \\
2q - 3 \overline{\smash{)}\, 6q^3 - 17q^2 + 22q - 23} \\
\underline{6q^3 - 9q^2} \qquad\qquad\qquad \\
-8q^2 + 22q \qquad\qquad \\
\underline{-8q^2 + 12q} \qquad\qquad \\
10q - 23 \qquad \\
\underline{10q - 15} \leftarrow 5(2q - 3) = 10q - 15
\end{array}
$$

Step 6 Subtract $10q - 15$ from $10q - 23$, getting -8.

$$
\begin{array}{r}
3q^2 - 4q + 5 \\
2q - 3 \overline{\smash{)}\, 6q^3 - 17q^2 + 22q - 23} \\
\underline{6q^3 - 9q^2} \qquad\qquad\qquad \\
-8q^2 + 22q \qquad\qquad \\
\underline{-8q^2 + 12q} \qquad\qquad \\
10q - 23 \qquad \\
\underline{-10q + 15} \qquad \text{Change signs and add} \\
-8 \qquad
\end{array}
$$

The quotient is $3q^2 - 4q + 5$, with remainder -8. Write this result as

$$
\frac{6q^3 - 17q^2 + 22q - 23}{2q - 3} = 3q^2 - 4q + 5 + \frac{-8}{2q - 3}. \quad \clubsuit
$$

Example 9 Divide $3x^3 - 2x^2 - 150$ by $x - 4$.

The polynomial $3x^3 - 2x^2 - 150$ has a missing term, the term where the variable is x. When dividing a polynomial with a missing term, it is convenient to allow for that term by leaving a space, as shown below.

$$
\begin{array}{r}
3x^2 + 10x + 40 \\
x - 4 \overline{\smash{)}\, 3x^3 - 2x^2 \qquad\quad - 150} \\
\underline{3x^3 - 12x^2} \qquad\qquad\qquad \\
10x^2 \qquad\qquad \\
\underline{10x^2 - 40x} \qquad\qquad \\
40x - 150 \\
\underline{40x - 160} \\
10
\end{array}
$$

This result is written as follows:

$$
\frac{3x^3 - 2x^2 - 150}{x - 4} = 3x^2 + 10x + 40 + \frac{10}{x - 4}. \quad \clubsuit
$$

By the process of division shown above, if a polynomial P is divided by a polynomial D of lower degree, the result is a **quotient** polynomial Q and a **remainder** polynomial R such that

$$P = D \cdot Q + R,$$

where the degree of R is less than the degree of D. To see this, note that dividing P by D gives

$$D \overline{)P} \quad \dfrac{Q + R/D}{}.$$

As a check, the product of D and $Q + R/D$ should give P:

$$P = D\left(Q + \dfrac{R}{D}\right) = D \cdot Q + R.$$

1.4   Exercises

Identify each of the following as a polynomial *or* not a polynomial. *For each polynomial, give the degree and identify as* monomial, binomial, trinomial, *or* none of these. *See Example 1.*

1. $5x^{11}$

2. $-9y^{12} + y^2$

3. $8p^5q + 6pq$

4. $2a^6 - 5a^2 + 4a$

5. $\sqrt{2}x + \sqrt{3}x^6$

6. $-\sqrt{7}m^5n^2 + 2\sqrt{3}m^4n^2$

7. $\dfrac{1}{3}r^2s - \dfrac{3}{5}r^4s^2 + rs^3$

8. $\dfrac{3}{10}p^7 - \dfrac{2}{7}p^5$

9. $\dfrac{5}{p} + \dfrac{2}{p^2} - \dfrac{5}{p^3}$

10. $\dfrac{q^2}{6} - \dfrac{q^5}{9} + \dfrac{q^8}{10}$

11. $5\sqrt{z} + 2\sqrt{z^3} - 5\sqrt{z^5}$

12. $\dfrac{8}{\sqrt{y}} - \dfrac{6}{\sqrt{y^3}} + \dfrac{3}{\sqrt{y^5}}$

Find each of the following sums and differences. See Example 2.

13. $(3x^2 - 4x + 5) + (-2x^2 + 3x - 2)$

14. $(4m^3 - 3m^2 + 5) + (-3m^3 - m^2 + 5)$

15. $(12y^2 - 8y + 6) - (3y^2 - 4y + 2)$

16. $(8p^2 - 5p) - (3p^2 - 2p + 4)$

17. $(3a^2 - 2a) + (4a^2 + 3a + 1) - (a^2 + 2)$

18. $(6m^4 - 3m^2 + m) - (2m^3 + 5m^2 + 4m) + (m^2 - m)$

19. $(5b^2 - 4b + 3) - (2b^2 + b) - (3b + 4)$

20. $-(8x^3 + x - 3) + (2x^3 + x^2) - (4x^2 + 3x - 1)$

21. $-3(4q^2 - 3q + 2) + 2(-q^2 + q - 4)$

22. $2(3r^2 + 4r + 2) - 3(-r^2 + 4r - 5)$

Find each of the following products. See Examples 3–6.

23. $p(4p - 6) + 2(3p - 8)$

24. $m(5m - 2) + 9(5 - m)$

25. $-y(y^2 - 4) + 6y^2(2y - 3)$

26. $-z^3(9 - z) + 4z(2 + 3z)$

27. $(4r - 1)(7r + 2)$

28. $(5m - 6)(3m + 4)$

29. $\left(3x - \dfrac{2}{3}\right)\left(5x + \dfrac{1}{3}\right)$

30. $\left(2m - \dfrac{1}{4}\right)\left(3m + \dfrac{1}{2}\right)$

31. $\left(\dfrac{2}{5}y + \dfrac{1}{8}z\right)\left(\dfrac{3}{5}y + \dfrac{1}{2}z\right)$

32. $\left(\dfrac{3}{4}r - \dfrac{2}{3}s\right)\left(\dfrac{5}{4}r + \dfrac{1}{3}s\right)$

33. $(5r + 2)(5r - 2)$

34. $(6z + 5)(6z - 5)$

35. $(4m + 2n)^2$

36. $(a - 6b)^2$

37. $[(2p - 3) + q]^2$

38. $[(4y - 1) + z]^2$

39. $[(3q + 5) - p][(3q + 5) + p]$

40. $[(9r - s) + 2][(9r - s) - 2]$

41. $(2z - 1)^3$

42. $(3m + 2)^3$

43. $(5r + 3t^2)^3$

44. $(2z^4 - 3y)^3$

45. $4x^2(3x^3 + 2x^2 - 5x + 1)$

46. $2b^3(b^2 - 4b + 3)$

47. $5m(3m^3 - 2m^2 + m - 1)$

48. $4y^3(y^3 + 2y^2 - 6y + 3)$

49. $(2z - 1)(-z^2 + 3z - 4)$

50. $(x - 1)(x^2 - 1)$

51. $(2m + 1)(4m^2 - 2m + 1)$

52. $(k + 2)(12k^3 - 3k^2 + k + 1)$

53. $(m - n + k)(m + 2n - 3k)$

54. $(r - 3s + t)(2r - s + t)$

55. $(a - b + 2c)^2$

56. $(k - y + 3m)^2$

Perform each of the following divisions. See Examples 7–9.

57. $\dfrac{4m^3 - 8m^2 + 16m}{2m}$

58. $\dfrac{30k^5 - 12k^3 + 18k^2}{6k^2}$

59. $\dfrac{25x^2y^4 - 15x^3y^3 + 40x^4y^2}{5x^2y^2}$

60. $\dfrac{-8r^3s - 12r^2s^2 + 20rs^3}{4rs}$

61. $\dfrac{6y^2 + y - 2}{2y - 1}$

62. $\dfrac{16r^2 + 2r - 3}{2r + 1}$

63. $\dfrac{20p^2 + 11p - 3}{4p + 3}$

64. $\dfrac{32m^2 + 20m - 3}{8m - 1}$

65. $\dfrac{8z^2 + 14z - 20}{2z + 5}$

66. $\dfrac{6a^2 - 13a - 18}{2a + 1}$

67. $\dfrac{2x^3 - 11x^2 + 19x - 10}{2x - 5}$

68. $\dfrac{3p^3 - 11p^2 + 5p + 3}{3p + 1}$

69. $\dfrac{15x^3 + 11x^2 + 20}{3x + 4}$

70. $\dfrac{4r^3 + 2r^2 - 14r + 15}{2r + 5}$

71. $\dfrac{3m^4 - 14m^3 + 22m^2 - 22m + 6}{m - 3}$

72. $\dfrac{5a^4 + 4a^3 - a^2 + 19a - 10}{a + 2}$

73. $\dfrac{x^4 + 2x^3 + 2x^2 - 2x - 3}{x^2 - 1}$

74. $\dfrac{2y^5 + y^3 - 2y^2 - 1}{2y^2 + 1}$

75. $\dfrac{4z^5 - 4z^2 - 5z + 3}{2z^2 + z + 1}$

76. $\dfrac{12z^4 - 25z^3 + 35z^2 - 26z + 10}{4z^2 - 3z + 5}$

77. $\dfrac{6x^3 - 17x^2y + 14xy^2 - 3y^3}{2x - 3y}$

78. $\dfrac{20p^3 - 43p^2q + 45pq^2 - 24q^3}{4p - 3q}$

79. $\dfrac{2y^4 - 8y^3z + 13y^2z^2 - 10yz^3 + 3z^4}{2y^2 - 4yz + 3z^2}$

80. $\dfrac{10a^4 + 3a^3b + 8a^2b^2 + ab^3 + 2b^4}{5a^2 - ab + 2b^2}$

Perform the indicated operations.

81. $(6p + 5q)(3p - 7q)$

82. $(2p - 1)(3p^2 - 4p + 5)$

83. $(3x - 4y)^3$

84. $(r^5 - r^3 + r) + (3r^5 - 4r^4 + r^3 + 2r)$

85. $(6k - 3)^2$

86. $\dfrac{6r^6 - 31r^5 + 22r^4 + 40r^3 + 14r^2 + 27r - 6}{2r^3 - 5r^2 - 3}$

87. $\dfrac{12x^2 - 7x - 12}{4x - 5}$

88. $(2m + n)^3$

89. $(4x + 3y)(4x - 3y)$

90. $(4y^2 - 2y + 7) - (6y - 9)$

91. $(3p - 1)(9p^2 + 3p + 1)$

92. $\dfrac{16a^6 + 24a^5 - 48a^4}{8a^2}$

93. $(p^3 - 4p^2 + p) - (3p^2 + 2p + 7)$

94. $\dfrac{12p^2 - 13p - 16}{3p - 2}$

95. $\dfrac{15x^4 + 30x^3 + 12x^2}{3x}$

96. $(2z + y)(3z - 4y)$

97. $\dfrac{6p^6 - 13p^5 - 8p^4 - 2p^3 + 3p^2 - 6p + 2}{3p^3 + p^2 - 1}$

98. $(7m + 2n)(7m - 2n)$

99. $(x^2 + 4x) + (3x^3 - 4x^2 + 2x + 2)$

100. $(3p + 5)^2$

In each of the following, find the coefficient of x^3 without finding the entire product.

101. $(x^2 + 4x)(-3x^2 + 4x - 1)$

102. $(4x^3 - 2x + 5)(x^3 + 2)$

103. $(1 + x^2)(1 + x)$

104. $(3 - x)(2 - x^2)$

105. $x^2(4 - 3x)^2$

106. $-4x^2(2 - x)(2 + x)$

Find each of the following products. Assume that all variables used in exponents represent integers.

107. $(k^m + 2)(k^m - 2)$

108. $(y^x - 4)(y^x + 4)$

109. $(b^r + 3)(b^r - 2)$

110. $(q^y + 4)(q^y + 3)$

111. $(3p^x + 1)(p^x - 2)$

112. $(2^a + 5)(2^a + 3)$

113. $(m^x - 2)^2$

114. $(z^r + 5)^2$

115. $(q^p - 5p^q)^2$

116. $(3y^x - 2x^y)^2$

117. $(3k^a - 2)^3$

118. $(r^x - 4)^3$

Suppose one polynomial has degree m and another has degree n, where m and n are natural numbers with $n < m$. Find the degree of the following for the polynomials.

119. Sum

120. Difference

121. Product

122. What would be the degree of the square of the polynomial of degree *m*?

123. Derive a formula for $(x + y + z)^2$.

124. Derive a formula for $(x + y + z)(x + y - z)$.

1.5 Factoring Polynomials

The process of finding polynomials whose product equals a given polynomial is called **factoring.** For example, since $4x + 12 = 4(x + 3)$, both 4 and $x + 3$ are called **factors** of $4x + 12$. Also, $4(x + 3)$ is the **factored form** of $4x + 12$. A polynomial that cannot be written as a product of two polynomials with integer coefficients is a **prime** or **irreducible polynomial.** A polynomial is **factored completely** when it is written as a product of prime polynomials with integer coefficients.

One way to factor a polynomial is to use the distributive property. For example, to factor $6x^2y^3 + 9xy^4 + 18y^5$, look for a monomial that is the greatest common factor of each term. This polynomial has $3y^3$ as greatest common factor. By the distributive property,

$$6x^2y^3 + 9xy^4 + 18y^5 = (3y^3)(2x^2) + (3y^3)(3xy) + (3y^3)(6y^2)$$
$$= 3y^3(2x^2 + 3xy + 6y^2.)$$

Example 1 Factor the greatest common factor from each polynomial.

(a) $9y^5 + y^2$

The greatest common factor is y^2.

$$9y^5 + y^2 = y^2 \cdot 9y^3 + y^2 \cdot 1$$
$$= y^2 (9y^3 + 1)$$

Do not make the common error of forgetting to include the 1. Since $y^2(9y^3) \neq 9y^5 + y^2$, the 1 is essential in the answer.

(b) $6x^2t + 8xt + 12t = 2t(3x^2 + 4x + 6)$

(c) $14m^4(m + 1) - 28m^3(m + 1) - 7m^2(m + 1)$

The greatest common factor is $7m^2(m + 1)$. Use the distributive property.

$$14m^4(m + 1) - 28m^3(m + 1) - 7m^2(m + 1)$$
$$= [7m^2(m + 1)](2m^2 - 4m - 1)$$
$$= 7m^2(m + 1)(2m^2 - 4m - 1) \quad \clubsuit$$

Factoring by Grouping When a polynomial has more than three terms, it can sometimes be factored by a method called **factoring by grouping.** For example, to factor

$$ax + ay + 6x + 6y,$$

collect the terms into two groups so that each group has a common factor.

$$ax + ay + 6x + 6y = (ax + ay) + (6x + 6y)$$

Factor each group, getting

$$ax + ay + 6x + 6y = a(x + y) + 6(x + y).$$

The quantity $(x + y)$ is now a common factor, which can be factored out, producing

$$ax + ay + 6x + 6y = (x + y)(a + 6).$$

It is not always obvious which terms should be grouped. Experience and repeated trials are the most reliable tools for factoring by grouping.

Example 2 Factor by grouping.

(a) $mp^2 + 7m + 3p^2 + 21$

Group the terms as follows.

$$mp^2 + 7m + 3p^2 + 21 = (mp^2 + 7m) + (3p^2 + 21)$$

Use the greatest common factor for each part.

$$(mp^2 + 7m) + (3p^2 + 21) = m(p^2 + 7) + 3(p^2 + 7)$$
$$= (p^2 + 7)(m + 3)$$

(b) $2y^2 - 2z - ay^2 + az$

Grouping terms as above gives

$$2y^2 - 2z - ay^2 + az = (2y^2 - 2z) + (-ay^2 + az)$$
$$= 2(y^2 - z) + a(-y^2 + z).$$

The expression $-y^2 + z$ is the negative of $y^2 - z$, so the terms should be grouped as follows.

$$2y^2 - 2z - ay^2 + az = (2y^2 - 2z) - (ay^2 - az)$$
$$= 2(y^2 - z) - a(y^2 - z)$$
$$= (y^2 - z)(2 - a). \quad \clubsuit$$

Factoring Trinomials   For a trinomial of degree 2, such as $kx^2 + mx + n$, where k, m, and n are integers, any factorization will be of the form

$$(ax + b)(cx + d),$$

with a, b, c, and d integers. Multiplying out the product $(ax + b)(cx + d)$ gives

$$(ax + b)(cx + d) = acx^2 + (ad + bc)x + bd,$$

which equals $kx^2 + mx + n$ only if

$$ac = k, \qquad ad + bc = m, \qquad \text{and} \qquad bd = n. \qquad (*)$$

By this result, a trinomial $kx^2 + mx + n$ can be factored by finding four integers a, b, c, and d satisfying the conditions given in equations (*). If no such numbers exist, the trinomial is prime.

Example 3   Factor each polynomial.

(a) $y^2 - 5y + 6$

To factor this polynomial, integers a, b, c, and d are needed such that

$$y^2 - 5y + 6 = (ay + b)(cy + d).$$

By equations (*) above, these integers satisfy the conditions $ac = 1$, $ad + bc = -5$, and $bd = 6$. The condition $ac = 1$ requires that $a = 1$ and $c = 1$. If both a and c equal 1, then $ad + bc = -5$ becomes $d + b = -5$. Now integers b and d must be found such that $b + d = -5$ and $bd = 6$. Inspection leads to the integers -2 and -3. Either b or d can equal either -2 or -3. Finally,

$$y^2 - 5y + 6 = (y - 2)(y - 3).$$

(b) $6p^2 - 7p - 5$

To find integers a, b, c, and d so that

$$6p^2 - 7p - 5 = (ap + b)(cp + d),$$

look for integers a, b, c, and d such that $ac = 6$, $ad + bc = -7$, and $bd = -5$. To find these integers, try various possibilities. Since $ac = 6$, let $a = 2$ and $c = 3$. Since $bd = -5$, let $b = -5$ and $d = 1$, giving

$$(2p - 5)(3p + 1) = 6p^2 - 13p - 5. \qquad \text{Incorrect}$$

Try

$$(3p - 5)(2p + 1) = 6p^2 - 7p - 5. \qquad \text{Correct}$$

Finally, $6p^2 - 7p - 5$ factors as $(3p - 5)(2p + 1)$. ✦

The trinomials of Example 3 can also be factored using grouping, instead of the trial and error approach used above. For example, to factor $6p^2 - 7p - 5$ by this method, find two numbers whose sum is -7 and whose product is $6(-5) = -30$. The numbers -10 and 3 satisfy these conditions. Writing

$$-7p \text{ as } -10p + 3p,$$

the trinomial $6p^2 - 7p - 5$ can be factored by grouping.

$$6p^2 - 10p + 3p - 5 = 2p(3p - 5) + 1(3p - 5)$$
$$= (3p - 5)(2p + 1)$$

Example 4   Factor each polynomial by grouping.

(a) $8x^2 + 10x - 3$

Two numbers with a sum of 10 and a product of $8(-3) = -24$ are -2 and 12.

$$8x^2 + 10x - 3 = 8x^2 - 2x + 12x - 3$$
$$= 2x(4x - 1) + 3(4x - 1)$$
$$= (4x - 1)(2x + 3)$$

(b) $4x^3 + 6x^2r - 10xr^2$

The greatest common factor of this polynomial is $2x$.

$$4x^3 + 6x^2r - 10xr^2 = 2x(2x^2 + 3xr - 5r^2)$$

After factoring out $2x$, factor the trinomial $2x^2 + 3xr - 5r^2$. Find factors of $2(-5) = -10$ that have a sum of 3: -2 and 5 are the required factors. Then

$$2x^2 + 3xr - 5r^2 = 2x^2 - 2xr + 5xr - 5r^2$$
$$= 2x(x - r) + 5r(x - r)$$
$$= (x - r)(2x + 5r),$$

and

$$4x^3 + 6x^2r - 10xr^2 = 2x(2x^2 + 3xr - 5r^2)$$
$$= 2x(x - r)(2x + 5r).$$

(c) $r^2 + 6r + 7$

It is not possible to find integers, a, b, c, and d so that

$$r^2 + 6r + 7 = (ar + b)(cr + d);$$

therefore, $r^2 + 6r + 7$ is a prime polymomial.   ✇

Each of the special patterns of multiplication given earlier can be used in reverse to get a pattern for factoring. One of the most common of these is shown below.

Difference of Two Squares	$$x^2 - y^2 = (x + y)(x - y)$$

Example 5   Factor each of the following polynomials.

(a) $4m^2 - 9$

First, recognize that $4m^2 - 9$ is the difference of two squares, since $4m^2 = (2m)^2$ and $9 = 3^2$. Use the pattern for the difference of two squares with $2m$ replacing x and 3 replacing y. Doing this gives

$$4m^2 - 9 = (2m)^2 - 3^2$$
$$= (2m + 3)(2m - 3).$$

(b) $144r^2 - 25s^2 = (12r + 5s)(12r - 5s)$

(c) $256k^4 - 625m^4$

Use the difference of two squares pattern twice, as follows:

$$256k^4 - 625m^4 = (16k^2)^2 - (25m^2)^2$$
$$= (16k^2 + 25m^2)(16k^2 - 25m^2)$$
$$= (16k^2 + 25m^2)(4k + 5m)(4k - 5m).$$

(d) $(a + 2b)^2 - 4c^2 = (a + 2b)^2 - (2c)^2$
$$= [(a + 2b) + 2c][(a + 2b) - 2c]$$
$$= (a + 2b + 2c)(a + 2b - 2c)$$

(e) $y^{4q} - z^{2q} = (y^{2q} + z^q)(y^{2q} - z^q)$ (if q is a positive integer)   ✇

Perfect square trinomials can be factored as follows.

Perfect Square Trinomials	$$x^2 + 2xy + y^2 = (x + y)^2$$ $$x^2 - 2xy + y^2 = (x - y)^2$$

Example 6   Factor each polynomial.

(a) $16p^2 - 40pq + 25q^2$

Since $16p^2 = (4p)^2$ and $25q^2 = (5q)^2$, use the second pattern shown above with $4p$ replacing x and $5q$ replacing y to get

$$16p^2 - 40pq + 25q^2 = (4p)^2 - 2(4p)(5q) + (5q)^2$$
$$= (4p - 5q)^2.$$

Make sure that the middle term of the trinomial being factored, $-40pq$ here, is twice the product of the two terms in the binomial $(4p - 5q)$.

$$-40pq = 2(4p)(-5q)$$

(b) $169x^2 + 104xy^2 + 16y^4 = (13x + 4y^2)^2$

(c) $x^2 - 6x + 9 - y^4$

Group the first three terms to get a perfect square trinomial. Then use the difference of squares pattern.

$$x^2 - 6x + 9 - y^4 = (x^2 - 6x + 9) - y^4$$
$$= (x - 3)^2 - (y^2)^2$$
$$= [(x - 3) + y^2][(x - 3) - y^2]$$
$$= (x - 3 + y^2)(x - 3 - y^2) \quad \clubsuit$$

Two other special results of factoring are listed below.

Difference and Sum of Cubes	**Difference of two cubes** $x^3 - y^3 = (x - y)(x^2 + xy + y^2)$ **Sum of two cubes** $x^3 + y^3 = (x + y)(x^2 - xy + y^2)$

Example 7   Factor each polynomial.

(a) $x^3 + 27$

Notice that $27 = 3^3$, so the expression is a sum of two cubes and

$$x^3 + 27 = x^3 + 3^3 = (x + 3)(x^2 - 3x + 9)$$

(b) $m^3 - 64n^3$

Since $64n^3 = (4n)^3$, the given polynomial is a difference of two cubes. To factor, use the first pattern in the box above, replacing x with m and y with $4n$.

$$m^3 - 64n^3 = m^3 - (4n)^3$$
$$= (m - 4n)[m^2 + m(4n) + (4n)^2]$$
$$= (m - 4n)(m^2 + 4mn + 16n^2)$$

(c) $8q^6 + 125p^9$

Write $8q^6$ as $(2q^2)^3$ and $125p^9$ as $(5p^3)^3$, so that the given polynomial is a sum of two cubes.

$$8q^6 + 125p^9 = (2q^2)^3 + (5p^3)^3$$
$$= (2q^2 + 5p^3)[(2q^2)^2 - (2q^2)(5p^3) + (5p^3)^2]$$
$$= (2q^2 + 5p^3)(4q^4 - 10q^2p^3 + 25p^6) \quad \clubsuit$$

Method of Substitution   Sometimes a polynomial can be factored by substituting one expression for another. The next example shows this **method of substitution.**

Example 8   Factor each polynomial.

(a) $6z^4 - 13z^2 - 5$

Replace z^2 with y, so that $y^2 = (z^2)^2 = z^4$. This replacement gives

$$6z^4 - 13z^2 - 5 = 6y^2 - 13y - 5.$$

Factor $6y^2 - 13y - 5$ as

$$6y^2 - 13y - 5 = (2y - 5)(3y + 1).$$

Replacing y with z^2 gives

$$6z^4 - 13z^2 - 5 = (2z^2 - 5)(3z^2 + 1).$$

(b) $10(2a - 1)^2 - 19(2a - 1) - 15$

Replacing $2a - 1$ with m gives

$$10m^2 - 19m - 15 = (5m + 3)(2m - 5).$$

Now replace m with $2a - 1$ and simplify.

$$10(2a - 1)^2 - 19(2a - 1) - 15 = [5(2a - 1) + 3][2(2a - 1) - 5]$$
$$= (10a - 5 + 3)(4a - 2 - 5)$$
$$= (10a - 2)(4a - 7)$$
$$= 2(5a - 1)(4a - 7)$$

(c) $(2a - 1)^3 + 8$

Let $2a - 1 = K$ to get

$$(2a - 1)^3 + 8 = K^3 + 8$$
$$= K^3 + 2^3$$
$$= (K + 2)(K^2 - 2K + 2^2).$$

Replacing K with $2a - 1$ gives

$$(2a - 1)^3 + 8 = (2a - 1 + 2)[(2a - 1)^2 - 2(2a - 1) + 4]$$
$$= (2a + 1)(4a^2 - 4a + 1 - 4a + 2 + 4)$$
$$= (2a + 1)(4a^2 - 8a + 7). \quad \clubsuit$$

Sometimes it is useful to factor out a common factor that has a negative exponent on the variable.

Example 9   Factor $9x^{-2} - 6x^{-3}$ by factoring out the variable with the smallest exponent in addition to any common numerical factor.

The smallest exponent here is -3. Since 3 is a common numerical factor, factor out $3x^{-3}$.

$$9x^{-2} - 6x^{-3} = 3x^{-3}(3x - 2)$$

Check by multiplying on the left.   ✛

1.5 Exercises

Factor the greatest common factor from each polynomial. See Example 1.

1. $12mn - 8m$

2. $3pq - 18pqr$

3. $12r^3 + 6r^2 - 3r$

4. $5k^2g - 25kg - 30kg^2$

5. $4k^2m^3 + 8k^4m^3 - 12k^2m^4$

6. $28r^4s^2 + 7r^3s - 35r^4s^3$

7. $2(a + b) + 4m(a + b)$

8. $4(y - 2)^2 + 3(y - 2)$

9. $(2y - 3)(y + 2) + (y + 5)(y + 2)$

10. $(6a - 1)(a + 2) + (6a - 1)(3a - 1)$

11. $(5r - 6)(r + 3) - (2r - 1)(r + 3)$

12. $(3z + 2)(z + 4) - (z + 6)(z + 4)$

13. $2(m - 1) - 3(m - 1)^2 + 2(m - 1)^3$

14. $5(a + 3)^3 - 2(a + 3) + (a + 3)^2$

Factor each of the following by grouping. See Example 2.

15. $rq + 2r + 2qs + 4s$

16. $zx + 2yx + 3z + 6y$

17. $6st + 9t - 10s - 15$

18. $10ab - 6b + 35a - 21$

19. $rt^3 + rs^2 - pt^3 - ps^2$

20. $2m^4 + 6 - am^4 - 3a$

21. $6p^2 - 14p + 15p - 35$

22. $8r^2 - 10r + 12r - 15$

23. $20z^2 - 8zx - 45zx + 18x^2$

24. $16a^2 + 10ab - 24ab - 15b^2$

25. $15 - 5m^2 - 3r^2 + m^2r^2$

26. $6y + 3yz^2 - 4x^2 - 2x^2z^2$

Factor each trinomial. See Examples 3 and 4.

27. $y^2 + y - 72$

28. $z^2 + 9z + 20$

29. $s^2 + 2st - 35t^2$

30. $n^2 - 12np + 35p^2$

31. $6a^2 - 48a - 120$

32. $8h^2 - 24h - 320$

33. $3m^3 + 12m^2 + 9m$

34. $9y^4 - 54y^3 + 45y^2$

35. $12r^2 + 24r - 15$

36. $12p^2 + p - 20$

37. $6k^2 + 5kp - 6p^2$

38. $14m^2 + 11mr - 15r^2$

39. $5a^2 - 7ab - 6b^2$

40. $12s^2 + 11st - 5t^2$

41. $9x^2 - 6x^3 + x^4$

42. $30a^2 + am - m^2$

43. $24a^4 + 10a^3b - 4a^2b^2$

44. $18x^5 + 15x^4z - 75x^3z^2$

Factor each difference of two squares. See Example 5.

45. $p^2 - 100$

46. $m^2 - 81$

47. $9a^2 - 16$

48. $16q^2 - 25$

49. $25s^4 - 9t^2$

50. $36z^2 - 81y^4$

51. $x^2y^2 - 25a^2$

52. $9p^2 - q^2r^2$

53. $(a + b)^2 - 16$

54. $(p - 2q)^2 - 100$

55. $p^4 - 625$

56. $m^4 - 81$

Factor each perfect square trinomial. See Example 6.

57. $9m^2 - 12m + 4$

58. $16p^2 - 40p + 25$

59. $32a^2 - 48ab + 18b^2$

60. $20p^2 - 100pq + 125q^2$

61. $4x^2y^2 + 28xy + 49$

62. $9m^2n^2 - 12mn + 4$

63. $(a - 3b)^2 - 6(a - 3b) + 9$

64. $(2p + q)^2 - 10(2p + q) + 25$

Factor each sum or difference of cubes. See Example 7.

65. $8 - a^3$

66. $r^3 + 27$

67. $125x^3 - 27$

68. $8m^3 - 27n^3$

69. $27y^9 + 125z^6$

70. $27z^3 + 729y^3$

71. $(r + 6)^3 - 216$

72. $(b + 3)^3 - 27$

73. $27 - (m + 2n)^3$

74. $125 - (4a - b)^3$

Factor each of the following, using the method of substitution. See Example 8.

75. $m^4 - 3m^2 - 10$

76. $a^4 - 2a^2 - 48$

77. $7(3k - 1)^2 + 26(3k - 1) - 8$

78. $6(4z - 3)^2 + 7(4z - 3) - 3$

79. $9(a - 4)^2 + 30(a - 4) + 25$

80. $20(4 - p)^2 - 3(4 - p) - 2$

Factor by any method.

81. $21m^2 + 13mn + 2n^2$

82. $m^2 + 4mp + 4p^2 - q^2$

83. $a^3(r + s) + b^2(r + s)$

84. $4b^2 + 4bc + c^2 - 16$

85. $(2y - 1)^2 - 4(2y - 1) + 4$

86. $x^2 + xy - 5x - 5y$

87. $8r^2 - 3rs + 10s^2$

88. $p^4(m - 2n) + q(m - 2n)$

89. $36a^2 + 60a + 25$

90. $6x^2 + 7x + 3$

91. $64 - x^6$

92. $4z^2 + 28z + 49$

93. $6p^4 + 7p^2 - 3$

94. $1000x^3 + 343y^3$

95. $b^2 + 8b + 16 - a^2$

96. $125m^6 - 216$

97. $q^2 + 6q + 9 - p^2$

98. $12m^2 + 16mn - 35n^2$

99. $216p^3 + 125q^3$

100. $20y^2 + 39yx - 11x^2$

101. $4p^2 + 3p - 1$

102. $100r^2 - 169s^2$

103. $144z^2 + 121$

104. $(3a + 5)^2 - 18(3a + 5) + 81$

105. $(x + y)^2 - (x - y)^2$

106. $4z^4 - 7z^2 - 15$

Factor the variable with the smallest exponent, together with any common numerical factor, from each of the following expressions. See Example 9.

107. $p^{-4} - p^{-2}$

108. $m^{-1} + 3m^{-5}$

109. $12k^{-3} + 4k^{-2} - 8k^{-1}$

110. $6a^{-5} - 10a^{-4} + 18a^{-2}$

111. $100p^{-6} - 50p^{-2} + 75p^2$

112. $32y^{-3} + 48y - 64y^2$

Factor each of the following polynomials. Assume that all variables used in exponents represent positive integers.

113. $m^2 - mn^p - 2n^{2p}$

114. $r^2 + rs^q - 6s^{2q}$

115. $6z^{2a} - z^ax^b - x^{2b}$

116. $5p^{2c} + 13p^cq^c - 6q^{2c}$

117. $9a^{4k} - b^{8k}$

118. $16y^{2c} - 25x^{4c.}$

119. $4y^{2a} - 12y^a + 9$

120. $25x^{4c} - 20x^{2c} + 4$

121. $6(m + p)^{2k} + (m + p)^k - 15$

122. $8(2k + q)^{4z} - 2(2k + q)^{2z} - 3$

Find a value of b or c that will make the following expressions perfect square trinomials.

123. $4z^2 + bz + 81$

124. $9p^2 + bp + 25$

125. $100r^2 - 60r + c$

126. $49x^2 + 70x + c$

1.6 Rational Expressions

An expression that is the quotient of two algebraic expressions (with denominator not 0) is called a **fractional expression.** The most common fractional expressions are those that are the quotients of two polynomials; these are called **rational expressions.** Since fractional expressions involve quotients, it is important to keep track of the values of the variable that satisfy the requirement that no denominator be 0. For example, $x \neq -2$ in the rational expression

$$\frac{x + 6}{x + 2}$$

because replacing x with -2 makes the denominator equal 0. Similarly, in

$$\frac{(x + 6)(x + 4)}{(x + 2)(x + 4)}$$

$x \neq -2$ and $x \neq -4$.

The restrictions on the variable are found by determining the values that make the denominator equal to zero. In the second example above, finding the values of x that make $(x + 2)(x + 4) = 0$ requires using the property that $ab = 0$ if and only if $a = 0$ or $b = 0$, as follows.

$$(x + 2)(x + 4) = 0$$
$$x + 2 = 0 \quad \text{or} \quad x + 4 = 0$$
$$x = -2 \quad \text{or} \quad x = -4$$

Just as the fraction 6/8 is written in lowest terms as 3/4, rational expressions must also be written in lowest terms. This is done with the fundamental principle.

Fundamental Principle

$$\frac{ac}{bc} = \frac{a}{b} \quad (b \neq 0, c \neq 0)$$

Example 1 Write each expression in lowest terms.

(a) $\dfrac{2p^2 + 7p - 4}{5p^2 + 20p}$

Factor the numerator and denominator to get

$$\frac{2p^2 + 7p - 4}{5p^2 + 20p} = \frac{(2p - 1)(p + 4)}{5p(p + 4)}$$

By the fundamental principle,

$$\frac{2p^2 + 7p - 4}{5p^2 + 20p} = \frac{2p - 1}{5p}.$$

In the original expression p cannot be 0 or -4, because $5p^2 + 20p \neq 0$, so this result is valid only for values of p other than 0 and -4. From now on, we shall always assume such restrictions when reducing rational expressions.

(b) $\dfrac{6 - 3k}{k^2 - 4}$

Factor to get

$$\frac{6 - 3k}{k^2 - 4} = \frac{3(2 - k)}{(k + 2)(k - 2)}.$$

The factors $2 - k$ and $k - 2$ have opposite signs. Because of this, multiply numerator and denominator by -1, as follows.

$$\frac{6 - 3k}{k^2 - 4} = \frac{3(2 - k)(-1)}{(k + 2)(k - 2)(-1)}$$

Since $(k - 2)(-1) = -k + 2$, or $2 - k$,

$$\frac{6 - 3k}{k^2 - 4} = \frac{3(2 - k)(-1)}{(k + 2)(2 - k)},$$

giving

$$\frac{6 - 3k}{k^2 - 4} = \frac{-3}{k + 2}.$$

Working in an alternative way would lead to the equivalent result

$$\frac{3}{-k - 2}. \quad \clubsuit$$

Rational expressions are multiplied and divided using definitions from earlier work with fractions.

Definitions of Multiplication and Division	$\dfrac{a}{b} \cdot \dfrac{c}{d} = \dfrac{ac}{bd}$ $\dfrac{a}{b} \div \dfrac{c}{d} = \dfrac{a}{b} \cdot \dfrac{d}{c} \quad \left(\text{if } \dfrac{c}{d} \neq 0\right)$

Example 2   Multiply or divide, as indicated.

(a) $\dfrac{2y^2}{9} \cdot \dfrac{27}{8y^5} = \dfrac{2y^2 \cdot 27}{9 \cdot 8y^5}$

$$= \frac{2 \cdot 9 \cdot 3y^2}{2 \cdot 9 \cdot 4y^2 \cdot y^3}$$

$$= \frac{3}{4y^3}$$

The product was written in lowest terms in the last step.

(b) $\dfrac{3m^2 - 2m - 8}{3m^2 + 14m + 8} \cdot \dfrac{3m + 2}{3m + 4} = \dfrac{(m - 2)(3m + 4)}{(m + 4)(3m + 2)} \cdot \dfrac{3m + 2}{3m + 4}$

$$= \dfrac{(m - 2)(3m + 4)(3m + 2)}{(m + 4)(3m + 2)(3m + 4)}$$

$$= \dfrac{m - 2}{m + 4}$$

(c) $\dfrac{5}{8m + 16} \div \dfrac{7}{12m + 24} = \dfrac{5}{8(m + 2)} \div \dfrac{7}{12(m + 2)}$

$$= \dfrac{5}{8(m + 2)} \cdot \dfrac{12(m + 2)}{7}$$

$$= \dfrac{5 \cdot 12(m + 2)}{8 \cdot 7(m + 2)}$$

$$= \dfrac{15}{14}$$

(d) $\dfrac{3p^2 + 11p - 4}{24p^3 - 8p^2} \div \dfrac{9p + 36}{24p^4 - 36p^3} = \dfrac{(p + 4)(3p - 1)}{8p^2(3p - 1)} \div \dfrac{9(p + 4)}{12p^3(2p - 3)}$

$$= \dfrac{(p + 4)(3p - 1)(12p^3)(2p - 3)}{8p^2(3p - 1)(9)(p + 4)}$$

$$= \dfrac{12p^3(2p - 3)}{9 \cdot 8p^2}$$

$$= \dfrac{p(2p - 3)}{6} \quad \clubsuit$$

Adding and subtracting rational expressions also depends on definitions from earlier work with fractions.

Definitions of Addition and Subtraction	$\dfrac{a}{b} + \dfrac{c}{d} = \dfrac{ad + bc}{bd}$ $\dfrac{a}{b} - \dfrac{c}{d} = \dfrac{ad - bc}{bd}$

In practice, rational expressions are normally added or subtracted after rewriting all the rational expressions with a common denominator found with the steps given below.

Finding a Common Denominator	**1.** Write each denominator as a product of prime factors. **2.** Form a product of all the different prime factors. Each factor should have as exponent the *highest* exponent that appears on that factor.

Example 3   Add or subtract, as indicated.

(a) $\dfrac{5}{9x^2} + \dfrac{1}{6x}$

Write each denominator as a product of prime factors, as follows.

$$9x^2 = 3^2 \cdot x^2$$
$$6x = 2^1 \cdot 3^1 x^1$$

For the common denominator, form the product of all the prime factors, with each factor having the highest exponent that appears on it. Here the highest exponent on 2 is 1, while both 3 and x have a highest exponent of 2. The common denominator is

$$2^1 \cdot 3^2 \cdot x^2 = 18x^2.$$

Now use the fundamental principle to write both of the given expressions with this denominator, then add.

$$\dfrac{5}{9x^2} + \dfrac{1}{6x} = \dfrac{5 \cdot 2}{9x^2 \cdot 2} + \dfrac{1 \cdot 3x}{6x \cdot 3x}$$

$$= \dfrac{10}{18x^2} + \dfrac{3x}{18x^2}$$

$$= \dfrac{10 + 3x}{18x^2}$$

(b) $\dfrac{y + 2}{y^2 - y} - \dfrac{3y}{2y^2 - 4y + 2}$

Factor each denominator, giving

$$\dfrac{y + 2}{y^2 - y} - \dfrac{3y}{2y^2 - 4y + 2} = \dfrac{y + 2}{y(y - 1)} - \dfrac{3y}{2(y - 1)^2}.$$

The common denominator, by the method above, is $2y(y - 1)^2$. Write each rational expression with this denominator and subtract, as follows.

$$\dfrac{y + 2}{y(y - 1)} - \dfrac{3y}{2(y - 1)^2} = \dfrac{(y + 2) \cdot 2(y - 1)}{y(y - 1) \cdot 2(y - 1)} - \dfrac{3y \cdot y}{2(y - 1)^2 \cdot y}$$

$$= \dfrac{2(y^2 + y - 2)}{2y(y - 1)^2} - \dfrac{3y^2}{2y(y - 1)^2}$$

$$= \dfrac{2y^2 + 2y - 4 - 3y^2}{2y(y - 1)^2}$$

$$= \dfrac{-y^2 + 2y - 4}{2y(y - 1)^2}$$

(c) $\dfrac{3}{(x - 1)(x + 2)} - \dfrac{4}{(x + 3)(x - 4)}$

The common denominator here is $(x - 1)(x + 2)(x + 3)(x - 4)$. Write each fraction with this common denominator, then perform the subtraction.

$$\frac{3}{(x - 1)(x + 2)} - \frac{4}{(x + 3)(x - 4)}$$

$$= \frac{3(x + 3)(x - 4)}{(x - 1)(x + 2)(x + 3)(x - 4)} - \frac{4(x - 1)(x + 2)}{(x + 3)(x - 4)(x - 1)(x + 2)}$$

$$= \frac{3(x^2 - x - 12) - 4(x^2 + x - 2)}{(x - 1)(x + 2)(x + 3)(x - 4)}$$

$$= \frac{3x^2 - 3x - 36 - 4x^2 - 4x + 8}{(x - 1)(x + 2)(x + 3)(x - 4)}$$

$$= \frac{-x^2 - 7x - 28}{(x - 1)(x + 2)(x + 3)(x - 4)} \quad \clubsuit$$

Example 4 Simplify $\dfrac{(x + y)^{-1}}{x^{-1} + y^{-1}}$. Write the result with only positive exponents.

Begin by using the definition of a negative integer exponent. Then perform the indicated operations.

$$\frac{(x + y)^{-1}}{x^{-1} + y^{-1}} = \frac{\dfrac{1}{x + y}}{\dfrac{1}{x} + \dfrac{1}{y}}$$

$$= \frac{\dfrac{1}{x + y}}{\dfrac{y + x}{xy}}$$

$$= \frac{1}{x + y} \cdot \frac{xy}{x + y} = \frac{xy}{(x + y)^2} \quad \clubsuit$$

Complex Fractions Any quotient of two rational expressions is called a **complex fraction**. Complex fractions often can be simplified by the methods shown in the following example.

Example 5 Simplify each complex fraction.

(a) $\dfrac{6 - \dfrac{5}{k}}{1 + \dfrac{5}{k}}$

Multiply both numerator and denominator by the common denominator k.

$$\frac{k\left(6 - \dfrac{5}{k}\right)}{k\left(1 + \dfrac{5}{k}\right)} = \frac{6k - k\left(\dfrac{5}{k}\right)}{k + k\left(\dfrac{5}{k}\right)} = \frac{6k - 5}{k + 5}$$

(b) $\dfrac{\dfrac{a}{a + 1} + \dfrac{1}{a}}{\dfrac{1}{a} + \dfrac{1}{a + 1}}$

Multiply both numerator and denominator by the common denominator of all the fractions, in this case $a(a + 1)$.

$$\frac{\dfrac{a}{a + 1} + \dfrac{1}{a}}{\dfrac{1}{a} + \dfrac{1}{a + 1}} = \frac{\left(\dfrac{a}{a + 1} + \dfrac{1}{a}\right) a(a + 1)}{\left(\dfrac{1}{a} + \dfrac{1}{a + 1}\right) a(a + 1)}$$

$$= \frac{a^2 + (a + 1)}{(a + 1) + a} = \frac{a^2 + a + 1}{2a + 1}$$

As an alternate method of solution, first perform the indicated additions in the numerator and denominator, and then divide.

$$\frac{\dfrac{a}{a + 1} + \dfrac{1}{a}}{\dfrac{1}{a} + \dfrac{1}{a + 1}} = \frac{\dfrac{a^2 + 1(a + 1)}{a(a + 1)}}{\dfrac{1(a + 1) + 1(a)}{a(a + 1)}}$$

$$= \frac{\dfrac{a^2 + a + 1}{a(a + 1)}}{\dfrac{2a + 1}{a(a + 1)}}$$

$$= \frac{a^2 + a + 1}{a(a + 1)} \cdot \frac{a(a + 1)}{2a + 1}$$

$$= \frac{a^2 + a + 1}{2a + 1} \quad \clubsuit$$

1.6 Exercises

For each of the following, give the restrictions on the variable.

1. $\dfrac{x - 2}{x + 6}$

2. $\dfrac{x + 5}{x - 3}$

3. $\dfrac{2x}{5x - 3}$

4. $\dfrac{6x}{2x - 1}$

5. $\dfrac{-8}{x^2 + 1}$

6. $\dfrac{3x}{3x^2 + 7}$

Write each of the following in lowest terms. See Example 1.

7. $\dfrac{25p^3}{10p^2}$

8. $\dfrac{14z^3}{6z^2}$

9. $\dfrac{8k + 16}{9k + 18}$

10. $\dfrac{20r + 10}{30r + 15}$

11. $\dfrac{3(t + 5)}{(t + 5)(t - 3)}$

12. $\dfrac{-8(y - 4)}{(y + 2)(y - 4)}$

13. $\dfrac{8x^2 + 16x}{4x^2}$

14. $\dfrac{36y^2 + 72y}{9y}$

15. $\dfrac{m^2 - 4m + 4}{m^2 + m - 6}$

16. $\dfrac{r^2 - r - 6}{r^2 + r - 12}$

17. $\dfrac{8m^2 + 6m - 9}{16m^2 - 9}$

18. $\dfrac{6y^2 + 11y + 4}{3y^2 + 7y + 4}$

Find each of the following products or quotients. See Example 2.

19. $\dfrac{x(y + 2)}{4} \div \dfrac{x}{2}$

20. $\dfrac{mn}{m + n} \div \dfrac{p}{m + n}$

21. $\dfrac{15p^3}{9p^2} \div \dfrac{6p}{10p^2}$

22. $\dfrac{3r^2}{9r^3} \div \dfrac{8r^3}{6r}$

23. $\dfrac{2k + 8}{6} \div \dfrac{3k + 12}{2}$

24. $\dfrac{5m + 25}{10} \cdot \dfrac{12}{6m + 30}$

25. $\dfrac{9y - 18}{6y + 12} \cdot \dfrac{3y + 6}{15y - 30}$

26. $\dfrac{12r + 24}{36r - 36} \div \dfrac{6r + 12}{8r - 8}$

27. $\dfrac{x^2 + x}{5} \cdot \dfrac{25}{xy + y}$

28. $\dfrac{3m - 15}{4m - 20} \cdot \dfrac{m^2 - 10m + 25}{12m - 60}$

29. $\dfrac{4a + 12}{2a - 10} \div \dfrac{a^2 - 9}{a^2 - a - 20}$

30. $\dfrac{6r - 18}{9r^2 + 6r - 24} \cdot \dfrac{12r - 16}{4r - 12}$

31. $\dfrac{p^2 - p - 12}{p^2 - 2p - 15} \cdot \dfrac{p^2 - 9p + 20}{p^2 - 8p + 16}$

32. $\dfrac{x^2 + 2x - 15}{x^2 + 11x + 30} \cdot \dfrac{x^2 + 2x - 24}{x^2 - 8x + 15}$

33. $\dfrac{k^2 - k - 6}{k^2 + k - 12} \cdot \dfrac{k^2 + 3k - 4}{k^2 + 2k - 3}$

34. $\dfrac{n^2 - n - 6}{n^2 - 2n - 8} \div \dfrac{n^2 - 9}{n^2 + 7n + 12}$

35. $\dfrac{m^2 + 3m + 2}{m^2 + 5m + 4} \div \dfrac{m^2 + 5m + 6}{m^2 + 10m + 24}$

36. $\dfrac{y^2 + y - 2}{y^2 + 3y - 4} \div \dfrac{y^2 + 3y + 2}{y^2 + 4y + 3}$

37. $\dfrac{2m^2 - 5m - 12}{m^2 - 10m + 24} \div \dfrac{4m^2 - 9}{m^2 - 9m + 18}$

38. $\dfrac{6n^2 - 5n - 6}{6n^2 + 5n - 6} \cdot \dfrac{12n^2 - 17n + 6}{12n^2 - n - 6}$

39. $\left(1 + \dfrac{1}{x}\right)\left(1 - \dfrac{1}{x}\right)$

40. $\left(3 + \dfrac{2}{y}\right)\left(3 - \dfrac{2}{y}\right)$

41. $\dfrac{x^3 + y^3}{x^2 - y^2} \cdot \dfrac{x + y}{x^2 - xy + y^2}$

42. $\dfrac{8y^3 - 125}{4y^2 - 20y + 25} \cdot \dfrac{2y - 5}{y}$

43. $\dfrac{x^3 + y^3}{x^3 - y^3} \cdot \dfrac{x^2 - y^2}{x^2 + 2xy + y^2}$

44. $\dfrac{x^2 - y^2}{(x - y)^2} \cdot \dfrac{x^2 - xy + y^2}{x^2 - 2xy + y^2} \div \dfrac{x^3 + y^3}{(x - y)^4}$

Perform each of the following additions or subtractions. See Example 3.

45. $\dfrac{8}{r} + \dfrac{6}{r}$

46. $\dfrac{3}{y} + \dfrac{4}{y}$

47. $\dfrac{3}{2k} + \dfrac{5}{3k}$

48. $\dfrac{8}{5p} + \dfrac{3}{4p}$

49. $\dfrac{2}{3y} - \dfrac{1}{4y}$

50. $\dfrac{6}{11z} - \dfrac{5}{2z}$

51. $\dfrac{a + 1}{2} - \dfrac{a - 1}{2}$

52. $\dfrac{y + 6}{5} - \dfrac{y - 6}{5}$

53. $\dfrac{3}{p} + \dfrac{1}{2}$

54. $\dfrac{9}{r} - \dfrac{2}{3}$

55. $\dfrac{2}{y} - \dfrac{1}{4}$

56. $\dfrac{6}{11} + \dfrac{3}{a}$

57. $\dfrac{1}{6m} + \dfrac{2}{5m} + \dfrac{4}{m}$

58. $\dfrac{8}{3p} + \dfrac{5}{4p} + \dfrac{9}{2p}$

59. $\dfrac{1}{y} + \dfrac{1}{y + 1}$

60. $\dfrac{2}{3(x - 1)} + \dfrac{1}{4(x - 1)}$

61. $\dfrac{2}{a + b} - \dfrac{1}{2(a + b)}$

62. $\dfrac{3}{m} - \dfrac{1}{m - 1}$

63. $\dfrac{1}{a + 1} - \dfrac{1}{a - 1}$

64. $\dfrac{1}{x + z} + \dfrac{1}{x - z}$

65. $\dfrac{m + 1}{m - 1} + \dfrac{m - 1}{m + 1}$

66. $\dfrac{2}{x - 1} + \dfrac{1}{1 - x}$

67. $\dfrac{3}{a - 2} - \dfrac{1}{2 - a}$

68. $\dfrac{q}{p - q} - \dfrac{q}{q - p}$

69. $\dfrac{x + y}{2x - y} - \dfrac{2x}{y - 2x}$

70. $\dfrac{m - 4}{3m - 4} + \dfrac{3m + 2}{4 - 3m}$

71. $\dfrac{1}{a^2 - 5a + 6} - \dfrac{1}{a^2 - 4}$

72. $\dfrac{-3}{m^2 - m - 2} - \dfrac{1}{m^2 + 3m + 2}$

73. $\dfrac{1}{x^2 + x - 12} - \dfrac{1}{x^2 - 7x + 12} + \dfrac{1}{x^2 - 16}$

74. $\dfrac{2}{2p^2 - 9p - 5} + \dfrac{p}{3p^2 - 17p + 10} - \dfrac{2p}{6p^2 - p - 2}$

75. $\dfrac{3a}{a^2 + 5a - 6} - \dfrac{2a}{a^2 + 7a + 6}$

76. $\dfrac{2k}{k^2 + 4k + 3} + \dfrac{3k}{k^2 + 5k + 6}$

77. $\dfrac{4x - 1}{x^2 + 3x - 10} + \dfrac{2x + 3}{x^2 + 4x - 5}$

78. $\dfrac{3y + 5}{y^2 - 9y + 20} + \dfrac{2y - 7}{y^2 - 2y - 8}$

Perform all indicated operations and write all answers with positive integer exponents.
See Example 4.

79. $\dfrac{a^{-1} + b^{-1}}{(ab)^{-1}}$

80. $\dfrac{p^{-1} - q^{-1}}{(pq)^{-1}}$

81. $\dfrac{r^{-1} + q^{-1}}{r^{-1} - q^{-1}} \cdot \dfrac{r - q}{r + q}$

82. $\dfrac{xy^{-1} + yx^{-1}}{x^2 + y^2}$

83. $(a + b)^{-1}(a^{-1} + b^{-1})$

84. $(m^{-1} + n^{-1})^{-1}$

85. $\dfrac{x - 9y^{-1}}{(x - 3y^{-1})(x + 3y^{-1})}$

86. $\dfrac{(m + n)^{-1}}{m^{-2} - n^{-2}}$

Perform the indicated operations. See Example 5.

87. $\dfrac{1 + \dfrac{1}{x}}{1 - \dfrac{1}{x}}$

88. $\dfrac{2 - \dfrac{2}{y}}{2 + \dfrac{2}{y}}$

89. $\dfrac{\dfrac{1}{x + 1} - \dfrac{1}{x}}{\dfrac{1}{x}}$

90. $\dfrac{\dfrac{1}{y + 3} - \dfrac{1}{y}}{\dfrac{1}{y}}$

91. $\dfrac{1 + \dfrac{1}{1-b}}{1 - \dfrac{1}{1+b}}$

92. $m - \dfrac{m}{m + \dfrac{1}{2}}$

93. $\dfrac{m - \dfrac{1}{m^2 - 4}}{\dfrac{1}{m + 2}}$

94. $\dfrac{\dfrac{3}{p^2 - 16} + p}{\dfrac{1}{p - 4}}$

95. $\dfrac{\dfrac{8}{z^2 - 25z} + 3}{\dfrac{5}{z(z + 1)}}$

96. $\dfrac{9 - \dfrac{1}{r^2 - r}}{\dfrac{3}{r(r + 2)}}$

97. $\left(\dfrac{3}{p - 1} - \dfrac{2}{p + 1}\right)\left(\dfrac{p - 1}{p}\right)$

98. $\left(\dfrac{y}{y^2 - 1} - \dfrac{y}{y^2 - 2y + 1}\right)\left(\dfrac{y - 1}{y + 1}\right)$

99. $\dfrac{\dfrac{1}{x + h} - \dfrac{1}{x}}{h}$

100. $\dfrac{1}{h}\left(\dfrac{1}{(x + h)^2 + 9} - \dfrac{1}{x^2 + 9}\right)$

1.7 Rational Exponents

The definition of a^n can be extended to rational values of n in such a way that the properties of exponents are still valid. For example, by one of the power rules, if m and n are integers,

$$(a^m)^n = a^{mn}$$

Assume $n \neq 0$ and replace m with $1/n$ to get

$$(a^{1/n})^n = a^{(1/n)n} = a^1 = a,$$

so that a is the nth power of $a^{1/n}$. For consistency with past work, $a^{1/n}$ must be defined as an **nth root** of a. That is, a definition of $a^{1/n}$ must satisfy the condition

$$(a^{1/n})^n = a.$$

The definition of $a^{1/n}$ depends on whether n is odd or even and whether a is positive or negative. For example, $16^{1/2}$ is a 2nd root, or square root, of 16. Since $4^2 = 16$ and $(-4)^2 = 16$, there are two square roots of 16, namely 4 and -4. The notation $16^{1/2}$ is reserved for the positive root, so

$$16^{1/2} = 4.$$

Also, $-16^{1/2} = -4$, because $-16^{1/2}$ means $-(16^{1/2})$. There is no real number square root of -16, so $(-16)^{1/2}$ is not a real number.

The 3rd root, or cube root, of a can be written as $a^{1/3}$. For example,

$$216^{1/3} = 6 \qquad \text{because} \qquad 6^3 = 216,$$

and $\qquad (-8)^{1/3} = -2 \qquad \text{because} \qquad (-2)^3 = -8.$

Note that the cube root of a positive number is positive, and the cube root of a negative number is negative.

In summary, $a^{1/n}$ is defined as follows.

Definition of $a^{1/n}$	If a is a real number, and n is a positive integer, then the **nth root** of a, written $a^{1/n}$, is defined as follows.	

n	$a \geq 0$	$a < 0$
Even	$a^{1/n}$ is the positive real number such that $$(a^{1/n})^n = a.$$	$a^{1/n}$ is not a real number.
Odd	$a^{1/n}$ is the real number such that $$(a^{1/n})^n = a.$$	

Example 1 (a) $36^{1/2} = 6$ because $6^2 = 36$

(b) $-100^{1/2} = -10$

(c) $-(225)^{1/2} = -15$

(d) $625^{1/4} = 5$

(e) $(-1296)^{1/4}$ is not a real number, but $-1296^{1/4} = -6$

(f) $(-27)^{1/3} = -3$

(g) $-32^{1/5} = -2$ ✛

What about more general rational exponents? The notation $a^{m/n}$ should be defined so that all the past rules for exponents still hold. For the power rule to hold, $(a^{1/n})^m$ must equal $a^{m/n}$. Therefore, $a^{m/n}$ is defined as follows.

Definition of $a^{m/n}$	For all integers m, all positive integers n, and all real numbers a for which $a^{1/n}$ exists: $$a^{m/n} = (a^{1/n})^m$$

Example 2 (a) $125^{2/3} = (125^{1/3})^2 = 5^2 = 25$

(b) $32^{7/5} = (32^{1/5})^7 = 2^7 = 128$

(c) $81^{3/2} = (81^{1/2})^3 = 9^3 = 729$

(d) $(-4)^{5/2}$ does not exist because $(-4)^{1/2}$ does not exist. ✛

By starting with $(a^{1/n})^m$ and $(a^m)^{1/n}$, and raising each expression to the nth power, it can be shown that $(a^{1/n})^m$ is equal to $(a^m)^{1/n}$. This means that $a^{m/n}$ could be defined in either of the following ways.

For all real numbers a and positive integers n for which $a^{1/n}$ exists:

$$a^{m/n} = (a^{1/n})^m \quad \text{or} \quad a^{m/n} = (a^m)^{1/n}$$

Now $a^{m/n}$ can be evaluated in either of two ways: as $(a^{1/n})^m$ or as $(a^m)^{1/n}$. It is usually easier to find $(a^{1/n})^m$. For example, $27^{4/3}$ can be evaluated in either of two ways:

$$27^{4/3} = (27^{1/3})^4 = 3^4 = 81$$
$$27^{4/3} = (27^4)^{1/3} = 531{,}441^{1/3} = 81.$$

The form $(27^{1/3})^4$ is easier to evaluate.

It can be shown that all the earlier results concerning integer exponents also apply to rational exponents.

Rules for Exponents	Let r and s be rational numbers. The results below are valid for all real numbers a and b for which all indicated powers exist, and as long as no denominators are 0.

$$a^r \cdot a^s = a^{r+s} \qquad (ab)^r = a^r \cdot b^r \qquad (a^r)^s = a^{rs}$$

$$\frac{a^r}{a^s} = a^{r-s} \qquad \left(\frac{a}{b}\right)^r = \frac{a^r}{b^r} \qquad a^{-r} = \frac{1}{a^r}$$

Example 3 **(a)** $\dfrac{27^{1/3} \cdot 27^{5/3}}{27^3} = \dfrac{27^{1/3 + 5/3}}{27^3}$

$$= \frac{27^2}{27^3} = 27^{2-3} = 27^{-1} = \frac{1}{27}$$

(b) $81^{5/4} \cdot 4^{-3/2} = (3^4)^{5/4}(2^2)^{-3/2} = 3^5 \cdot 2^{-3} = \dfrac{3^5}{2^3}$ or $\dfrac{243}{8}$.

(c) $6y^{2/3} \cdot 2y^{1/2} = 12y^{2/3 + 1/2} = 12y^{7/6}$, where $y \geq 0$

(d) $\left(\dfrac{3m^{5/6}}{y^{3/4}}\right)^2 \cdot \left(\dfrac{8y^3}{m^6}\right)^{2/3} = \dfrac{9m^{5/3}}{y^{3/2}} \cdot \dfrac{4y^2}{m^4} = 36m^{5/3-4}y^{2-3/2} = \dfrac{36y^{1/2}}{m^{7/3}}$

$(m > 0, \; y > 0)$

(e) $m^{2/3}(m^{7/3} + 2m^{1/3}) = (m^{2/3+7/3} + 2m^{2/3+1/3}) = m^3 + 2m$ ✚

The next example shows how to factor with rational exponents.

Example 4 Factor out the lowest power of the variable. Assume that all variables represent positive real numbers.

(a) $4m^{1/2} + 3m^{3/2} = m^{1/2}(4 + 3m)$

To check this result, multiply $m^{1/2}$ by $4 + 3m$.

(b) $y^{-1/3} + y^{2/3} = y^{-1/3}(1 + y)$ ✚

1.7   Exercises

Simplify each of the following. Use a calculator for Exercises 15–18. Assume that all variables represent nonnegative real numbers. See Examples 1 and 2.

1. $4^{1/2}$

2. $25^{1/2}$

3. $8^{2/3}$

4. $81^{3/4}$

5. $27^{-2/3}$

6. $32^{-4/5}$

7. $\left(\dfrac{4}{9}\right)^{-3/2}$

8. $\left(\dfrac{1}{8}\right)^{-5/3}$

9. $125^{-2/3}$

10. $64^{5/6}$

11. $\left(\dfrac{64}{27}\right)^{2/3}$

12. $\left(\dfrac{8}{125}\right)^{4/3}$

13. $\left(\dfrac{27}{64}\right)^{-4/3}$

14. $\left(\dfrac{121}{100}\right)^{-3/2}$

15. $3^{.4}$

16. $7^{-3.1}$

17. $1.48^{-2.9}$

18. $2^{5.41}$

19. $(16p^4)^{1/2}$

20. $(36r^6)^{1/2}$

21. $(27x^6)^{2/3}$

22. $(64a^{12})^{5/6}$

23. $(125y^9z^{12})^{1/3}$

24. $(256z^{16}r^8)^{1/4}$

Perform the indicated operations. Write answers using only positive exponents. Assume that all variables represent positive real numbers and that variables used as exponents represent rational numbers. See Example 3.

25. $2^{1/2} \cdot 2^{3/2}$

26. $5^{3/8} \cdot 5^{5/8}$

27. $27^{2/3} \cdot 27^{-1/3}$

28. $9^{-3/4} \cdot 9^{1/4}$

29. $\dfrac{4^{2/3} \cdot 4^{5/3}}{4^{1/3}}$

30. $\dfrac{3^{-5/2} \cdot 3^{3/2}}{3^{7/2} \cdot 3^{-9/2}}$

31. $(m^{2/3})(m^{5/3})$

32. $(x^{4/5})(x^{2/5})$

33. $(1 + n)^{1/2}(1 + n)^{3/4}$

34. $(m + 7)^{-1/6}(m + 7)^{-2/3}$

35. $(2y^{3/4}z)(3y^{-2}z^{-1/3})$

36. $(4a^{-1}b^{2/3})(a^{3/2}b^{-3})$

37. $(4a^{-2}b^7)^{1/2} \cdot (2a^{1/4}b^3)^5$

38. $(x^{-2}y^{1/3})^5 \cdot (8x^2y^{-2})^{-1/3}$

39. $\dfrac{7^{-1/3} \cdot 7r^{-3}}{7^{2/3} \cdot (r^{-2})^2}$

40. $\dfrac{12^{3/4} \cdot 12^{5/4} \cdot y^{-2}}{12^{-1} \cdot (y^{-3})^{-2}}$

41. $\dfrac{6k^{-4} \cdot (3k^{-1})^{-2}}{2^3 \cdot k^{1/2}}$

42. $\dfrac{8p^{-3} \cdot (4p^2)^{-2}}{p^{-5}}$

43. $\dfrac{a^{4/3} \cdot b^{1/2}}{a^{2/3} \cdot b^{-3/2}}$

44. $\dfrac{x^{1/3} \cdot y^{2/3} \cdot z^{1/4}}{x^{5/3} \cdot y^{-1/3} \cdot z^{3/4}}$

45. $\dfrac{k^{-3/5} \cdot h^{-1/3} \cdot t^{2/5}}{k^{-1/5} \cdot h^{-2/3} \cdot t^{1/5}}$

46. $\dfrac{m^{7/3} \cdot n^{-2/5} \cdot p^{3/8}}{m^{-2/3} \cdot n^{3/5} \cdot p^{-5/8}}$

47. $\dfrac{k^{3/2} \cdot k^{-1/2}}{k^{1/4} \cdot k^{3/4}}$

48. $\dfrac{m^{2/5} \cdot m^{3/5} \cdot m^{-4/5}}{m^{1/5} \cdot m^{-6/5}}$

49. $\dfrac{x^{-3} \cdot x^{4/3}}{x^{-2} \cdot x^{-3/4}}$

50. $\dfrac{-4a^{-1} \cdot a^{2/3}}{a^{-2}}$

51. $\dfrac{8y^{2/3}y^{-1}}{2^{-1}y^{3/4} \cdot y^{-1/6}}$

52. $\dfrac{9 \cdot k^{1/3} \cdot k^{-1/2} \cdot k^{-1/6}}{k^{-2/3}}$

53. $\dfrac{(k + 5)^{1/2}(k + 5)^{-1/4}}{(k + 5)^{3/4}}$

54. $\dfrac{(x + y)^{-5/8}(x + y)^{3/8}}{(x + y)^{1/8}(x + y)^{-1/8}}$

55. $\left(\dfrac{x^4y^3z}{16x^{-6}yz^5}\right)^{-1/2}$

56. $\left(\dfrac{p^3r^9}{27p^{-3}r^{-6}}\right)^{-1/3}$

Perform the indicated operations. Write answers without denominators. Assume that all variables used in denominators are not zero and all variables used as exponents represent rational numbers. See Example 3(d).

57. $(r^{3/p})^{2p}(r^{1/p})^{p^2}$

58. $(m^{2/x})^{x/3}(m^{x/4})^{2/x}$

59. $\dfrac{m^{1-a} \cdot m^a}{m^{-1/2}}$

60. $\dfrac{(y^{3-b})(y^{2b-1})}{y^{1/2}}$

61. $\dfrac{(x^{n/2})(x^{3n})^{1/2}}{x^{1/n}}$ **62.** $\dfrac{(a^{2/3})(a^{1/x})}{(a^{x/3})^{-2}}$ **63.** $\dfrac{(p^{1/n})(p^{1/m})}{p^{-m/n}}$ **64.** $\dfrac{(q^{2r/3})(q^{r})^{-1/3}}{(q^{4/3})^{1/r}}$

Find each of the following products. Assume that all variables represent positive real numbers. See Example 3(e).

65. $y^{5/8}(y^{3/8} - 10y^{11/8})$ **66.** $p^{11/5}(3p^{4/5} + 9p^{19/5})$

67. $-4k(k^{7/3} - 6k^{1/3})$ **68.** $-5y(3y^{9/10} + 4y^{3/10})$

69. $(x + x^{1/2})(x - x^{1/2})$ **70.** $(2z^{1/2} + z)(z^{1/2} - z)$

71. $(r^{1/2} - r^{-1/2})^2$ **72.** $(p^{1/2} - p^{-1/2})(p^{1/2} + p^{-1/2})$

73. $[(x^{1/2} - x^{-1/2})^2 + 4]^{1/2}$ **74.** $[(x^{1/2} + x^{-1/2})^2 - 4]^{1/2}$

Factor, using the given common factor. Assume all variables represent positive real numbers. See Example 4.

75. $4k^{7/4} + k^{3/4}$; $k^{3/4}$ **76.** $y^{9/2} - 3y^{5/2}$; $y^{5/2}$

77. $9z^{-1/2} + 2z^{1/2}$; $z^{-1/2}$ **78.** $3m^{2/3} - 4m^{-1/3}$; $m^{-1/3}$

79. $p^{-3/4} - 2p^{-7/4}$; $p^{-7/4}$ **80.** $6r^{-2/3} - 5r^{-5/3}$; $r^{-5/3}$

81. $(p + 4)^{-3/2} + (p + 4)^{-1/2} + (p + 4)^{1/2}$; $(p + 4)^{-3/2}$

82. $(3r + 1)^{-2/3} + (3r + 1)^{1/3} + (3r + 1)^{4/3}$; $(3r + 1)^{-2/3}$

One important application of mathematics to business and management concerns supply and demand. Usually, as the price of an item increases, the supply increases and the demand decreases. By studying past records of supply and demand at different prices, economists can construct an equation that describes (approximately) supply and demand for a given item. The next two exercises show examples of this.

83. The price of a certain type of solar heater is approximated by p, where

$$p = 2x^{1/2} + 3x^{2/3}$$

and x is the number of units supplied. Find the price when the supply is 64 units.

84. For a certain commodity the demand and the price are related by the equation

$$p = 1000 - 200x^{-2/3} \qquad (x > 0),$$

where x is the number of units of the product demanded. Find the price when the demand is 27.

In our system of government, the president is elected by the electoral college, and not by individual voters. Because of this, smaller states have a greater voice in the selection of a president than they otherwise would have. Two political scientists have studied the problems of campaigning for president under the current system and have concluded that candidates should allot their money according to the formula

$$\text{amount for large state} = \left(\dfrac{E_{\text{large}}}{E_{\text{small}}}\right)^{3/2} \times \text{amount for small state}.$$

Here E_{large} represents the electoral vote of the large state, and E_{small} represents the electoral vote of the small state. Find the amount that should be spent in each of the following larger states if $1,000,000$ is spent in the small state and the following statements are true.

85. The large state has 48 electoral votes, and the small state has 3.

86. The large state has 36 electoral votes, and the small state has 4.

87. 6 votes in a small state; 28 in a large

88. 9 votes in a small state; 32 in a large

(The next exercises require a calculator with a y^x key.) The Galapagos Islands are a chain of islands ranging in size from 2 to 2249 sq mi. A biologist has shown that the number of different land-plant species on an island in this chain is related to the size of the island approximately by

$$S = 28.6A^{.32},$$

where A is the area of the island in square miles and S is the number of different plant species on that island. Estimate S (rounding to the nearest whole number) for islands of the following areas.

89. 1 sq mi   **90.** 25 sq mi   **91.** 300 sq mi   **92.** 2000 sq mi

1.8   Radicals

In the last section the notation $a^{1/n}$ was used for the nth root of a for appropriate values of a and n. An alternative notation for $a^{1/n}$ uses *radical notation*.

Radical Notation for $a^{1/n}$

> If a is a real number, n is a positive integer, and $a^{1/n}$ exists, then
> $$\sqrt[n]{a} = a^{1/n}.$$

The symbol $\sqrt[n]{}$ is a **radical sign,** the number a is the **radicand,** and n is the **index** of the radical $\sqrt[n]{a}$. The familiar notation $\sqrt{a}$ is used instead of $\sqrt[2]{a}$.

Example 1   **(a)** $\sqrt[4]{16} = 16^{1/4} = 2$

(b) $\sqrt[5]{-32} = -2$

(c) $\sqrt[3]{1000} = 10$

(d) $\sqrt[6]{\dfrac{64}{729}} = \dfrac{2}{3}$ ✤

With $a^{1/n}$ written as $\sqrt[n]{a}$, $a^{m/n}$ also can be written using radicals.

Radical Notation for $a^{m/n}$

> If a is a real number, m is an integer, n is a positive integer, and $\sqrt[n]{a}$ exists, then
> $$a^{m/n} = (\sqrt[n]{a})^m = \sqrt[n]{a^m}.$$

Example 2   Write in radical form.

(a) $x^{5/6}$

$$x^{5/6} = \sqrt[6]{x^5} \quad \text{or} \quad (\sqrt[6]{x})^5 \quad (x \geq 0)$$

(b) $3x^{2/3} = 3\sqrt[3]{x^2}$

(c) $2p^{-1/2} = \dfrac{2}{p^{1/2}} = \dfrac{2}{\sqrt{p}} \quad (p > 0)$

(d) $(3a + b)^{1/4} = \sqrt[4]{3a + b} \quad (3a + b \geq 0)$

It is not possible to "distribute" exponents over a *sum,* so

$$(3a + b)^{1/4} \neq (3a)^{1/4} + b^{1/4}. \quad \clubsuit$$

Example 3   Write in exponential form.

(a) $\sqrt[4]{x^5} = x^{5/4} \quad (x \geq 0)$

(b) $\sqrt{3y} = (3y)^{1/2} \quad (y \geq 0)$

(c) $10(\sqrt[5]{z})^2 = 10z^{2/5}$

(d) $5\sqrt[3]{(2x^4)^7} = 5(2x^4)^{7/3} = 5 \cdot 2^{7/3}x^{28/3} \quad \clubsuit$

By the definition of $\sqrt[n]{a}$, for any positive integer n, if $\sqrt[n]{a}$ exists, then

$$(\sqrt[n]{a})^n = a.$$

If a is positive, or if a is negative and n is an odd positive integer,

$$\sqrt[n]{a^n} = a.$$

Because of the conditions just given, we *cannot* simply write $\sqrt{x^2} = x$. For example, if $x = -5$,

$$\sqrt{x^2} = \sqrt{(-5)^2} = \sqrt{25} = 5 \neq x.$$

To take care of the fact that a negative value of x can produce a positive result for the square root, use the following rule, which involves absolute value.

> For any real number a:
> $$\sqrt{a^2} = |a|$$

For example,

$$\sqrt{(-9)^2} = |-9| = 9, \quad \text{and} \quad \sqrt{13^2} = |13| = 13.$$

Example 4   Use absolute value to simplify the following expressions.

(a) $\sqrt{p^4} = |p^2| = p^2$

(b) $\sqrt{16m^8r^6} = |4m^4r^3| = 4m^4 |r^3|$

(c) $\sqrt{(2k + 3)^2} = |2k + 3|$

(d) $\sqrt{x^2 - 4x + 4} = \sqrt{(x - 2)^2} = |x - 2|$ ✦

To avoid difficulties when working with variable radicands, we will assume that all variables in radicands represent only nonnegative real numbers.

Three key rules for working with radicals are given below.

Rules for Radicals	For all real numbers a and b and positive integers m and n for which the indicated roots exist: $$\sqrt[n]{a} \cdot \sqrt[n]{b} = \sqrt[n]{ab} \qquad \sqrt[n]{\frac{a}{b}} = \frac{\sqrt[n]{a}}{\sqrt[n]{b}} \quad (b \neq 0) \qquad \sqrt[m]{\sqrt[n]{a}} = \sqrt[mn]{a}$$

Example 5   Use the rules for radicals to simplify each of the following.

(a) $\sqrt{6} \cdot \sqrt{54} = \sqrt{6 \cdot 54} = \sqrt{324} = 18$

(b) $\sqrt{\dfrac{7}{64}} = \dfrac{\sqrt{7}}{\sqrt{64}} = \dfrac{\sqrt{7}}{8}$

(c) $\sqrt[7]{\sqrt[3]{2}} = \sqrt[21]{2}$

(d) $\sqrt[4]{\sqrt{3}} = \sqrt[8]{3}$ ✦

Simplifying Radicals   In working with numbers, it is customary to write a number in its simplest form. For example, 10/2 is written as 5, $-9/6$ is written as $-3/2$, and 4/16 is written as 1/4. Similarly, expressions with radicals should be written in their simplest forms.

Simplified Radicals	An expression with radicals is simplifed when all of the following conditions are satisfied.
	1. The radicand has no factor raised to a power greater than or equal to the index.
	2. The radicand has no fractions.
	3. No denominator contains a radical.
	4. Exponents in the radicand and the index of the radical have no common factor.
	5. All indicated operations have been performed (if possible).

Example 6   Simplify each of the following. Assume that all variables represent nonnegative real numbers.

(a) $\sqrt{175} = \sqrt{25 \cdot 7} = \sqrt{25} \cdot \sqrt{7} = 5\sqrt{7}$

(b) $-3\sqrt[5]{32} = -3\sqrt[5]{2^5} = -3 \cdot 2 = -6$

(c) $\sqrt[3]{81x^5y^7z^6} = \sqrt[3]{27 \cdot 3 \cdot x^3 \cdot x^2 \cdot y^6 \cdot y \cdot z^6}$

$\qquad\qquad = \sqrt[3]{(27x^3y^6z^6)(3x^2y)}$

$\qquad\qquad = 3xy^2z^2\sqrt[3]{3x^2y}$ ✚

Radicals with the same radicand and the same index, such as $3\sqrt[4]{11pq}$ and $-7\sqrt[4]{11pq}$, are called **like radicals.** Only like radicals can be added or subtracted. As shown in parts (b) and (c) of the next example, it is sometimes necessary to simplify radicals before adding or subtracting.

Example 7   (a) $3\sqrt[4]{11pq} + (-7\sqrt[4]{11pq}) = -4\sqrt[4]{11pq}$

(b) $\sqrt{98x^3y} + 3x\sqrt{32xy}$

First remove all perfect square factors from under the radical. Then use the distributive property, as follows.

$$\sqrt{98x^3y} + 3x\sqrt{32xy} = \sqrt{49 \cdot 2 \cdot x^2 \cdot x \cdot y} + 3x\sqrt{16 \cdot 2 \cdot x \cdot y}$$

$$= 7x\sqrt{2xy} + (3x)(4)\sqrt{2xy}$$

$$= 7x\sqrt{2xy} + 12x\sqrt{2xy}$$

$$= 19x\sqrt{2xy}$$

(c) $\sqrt[3]{64m^4n^5} - \sqrt[3]{-27m^{10}n^{14}} = \sqrt[3]{(64m^3n^3)(mn^2)} - \sqrt[3]{(-27m^9n^{12})(mn^2)}$

$$= 4mn\sqrt[3]{mn^2} - (-3)m^3n^4\sqrt[3]{mn^2}$$

$$= 4mn\sqrt[3]{mn^2} + 3m^3n^4\sqrt[3]{mn^2}$$

$$= (4 + 3m^2n^3)mn\sqrt[3]{mn^2}$$ ✚

Example 8   Simplify the following radicals by first rewriting with rational exponents.

(a) $\sqrt[6]{3^2} = 3^{2/6} = 3^{1/3} = \sqrt[3]{3}$

(b) $\sqrt[6]{x^{12}y^3} = (x^{12}y^3)^{1/6} = x^2y^{3/6} = x^2y^{1/2} = x^2\sqrt{y}$

(c) $\sqrt[9]{\sqrt{6^3}} = \sqrt[9]{6^{3/2}} = (6^{3/2})^{1/9} = 6^{1/6} = \sqrt[6]{6}$ ✚

In Example 6(a), we simplified $\sqrt[6]{3^2}$ as $\sqrt[3]{3}$. However, to simplify $(\sqrt[6]{x})^2$, the variable x must be nonnegative. For example,

$$(-8)^{2/6} = [(-8)^{1/6}]^2.$$

This result is not a real number, since $(-8)^{1/6}$ is not a real number. On the other hand,

$$(-8)^{1/3} = -2.$$

Here, even though $2/6 = 1/3$,

$$(\sqrt[6]{x})^2 \neq \sqrt[3]{x}.$$

If a is nonnegative, then it is always true that $a^{m/n} = a^{mp/np}$. Reducing rational exponents on negative bases must be considered case by case.

Multiplying radical expressions is much like multiplying polynomials.

Example 9   **(a)** $(\sqrt{2} + 3)(\sqrt{8} - 5) = \sqrt{2}(\sqrt{8}) - \sqrt{2}(5) + 3\sqrt{8} - 3(5)$

$$= \sqrt{16} - 5\sqrt{2} + 3(2\sqrt{2}) - 15$$

$$= 4 - 5\sqrt{2} + 6\sqrt{2} - 15$$

$$= -11 + \sqrt{2}$$

(b) $(\sqrt{7} - \sqrt{10})(\sqrt{7} + \sqrt{10}) = (\sqrt{7})^2 - (\sqrt{10})^2$

$$= 7 - 10$$

$$= -3 \quad \clubsuit$$

Rationalizing the Denominator   The process of rationalizing the denominator, mentioned in condition 3 of the rules for simplifying radicals described above, is explained in the next few examples.

Example 10   Rationalize each denominator.

(a) $\dfrac{4}{\sqrt{3}}$

To rationalize the denominator, multiply by $\sqrt{3}/\sqrt{3}$ (which equals 1) so that the denominator of the product is a rational number.

$$\frac{4}{\sqrt{3}} \cdot \frac{\sqrt{3}}{\sqrt{3}} = \frac{4\sqrt{3}}{3}$$

(b) $\sqrt[4]{\dfrac{3}{5}}$

Start by using the fact that the radical of a quotient can be written as the quotient of radicals.

$$\sqrt[4]{\frac{3}{5}} = \frac{\sqrt[4]{3}}{\sqrt[4]{5}}$$

The denominator will be a rational number if it equals $\sqrt[4]{5^4}$. That is, four factors of 5 are needed under the radical. Since $\sqrt[4]{5}$ has just one factor of 5, three additional factors are needed, so

$$\frac{\sqrt[4]{3}}{\sqrt[4]{5}} = \frac{\sqrt[4]{3} \cdot \sqrt[4]{5^3}}{\sqrt[4]{5} \cdot \sqrt[4]{5^3}} = \frac{\sqrt[4]{3 \cdot 5^3}}{\sqrt[4]{5^4}} = \frac{\sqrt[4]{375}}{5}. \quad \clubsuit$$

Example 11   Simplify. Assume all variables represent positive real numbers.

(a) $\dfrac{\sqrt[4]{ab^3} \cdot \sqrt[4]{ab}}{\sqrt[4]{a^3b^3}}$

To rationalize the denominator, multiply by $\sqrt[4]{ab}/\sqrt[4]{ab}$, since $\sqrt[4]{a^3b^3} \cdot \sqrt[4]{ab} = \sqrt[4]{a^4b^4} = ab$.

$$\frac{\sqrt[4]{ab^3} \cdot \sqrt[4]{ab}}{\sqrt[4]{a^3b^3}} = \frac{\sqrt[4]{ab^3} \cdot \sqrt[4]{ab} \cdot \sqrt[4]{ab}}{\sqrt[4]{a^3b^3} \cdot \sqrt[4]{ab}}$$

$$= \frac{\sqrt[4]{ab^3 \cdot ab \cdot ab}}{\sqrt[4]{a^3b^3 \cdot ab}}$$

$$= \frac{\sqrt[4]{a^3b^5}}{\sqrt[4]{a^4b^4}}$$

$$= \frac{b\sqrt[4]{a^3b}}{ab} \qquad\qquad \sqrt[4]{b^5} = b\sqrt[4]{b}$$

$$= \frac{\sqrt[4]{a^3b}}{a} \qquad\qquad \text{Lowest terms}$$

(b) $\dfrac{-1}{\sqrt[3]{16}} - \dfrac{5}{\sqrt[3]{128}} + \dfrac{4}{\sqrt[3]{2}}$

Begin by simplifying all radicals:

$$\sqrt[3]{16} = \sqrt[3]{8 \cdot 2} = 2\sqrt[3]{2}$$

and

$$\sqrt[3]{128} = \sqrt[3]{64 \cdot 2} = 4\sqrt[3]{2}.$$

Now proceed as follows.

$$\frac{-1}{\sqrt[3]{16}} - \frac{5}{\sqrt[3]{128}} + \frac{4}{\sqrt[3]{2}} = \frac{-1}{2\sqrt[3]{2}} - \frac{5}{4\sqrt[3]{2}} + \frac{4}{\sqrt[3]{2}}$$

$$= \frac{-2}{4\sqrt[3]{2}} - \frac{5}{4\sqrt[3]{2}} + \frac{16}{4\sqrt[3]{2}} \qquad \text{Common denominator}$$

$$= \frac{9}{4\sqrt[3]{2}} \qquad\qquad \text{Combine terms}$$

$$= \frac{9\sqrt[3]{4}}{4\sqrt[3]{2} \cdot \sqrt[3]{4}} \qquad \text{Rationalize the denominator}$$

$$= \frac{9\sqrt[3]{4}}{4 \cdot \sqrt[3]{8}}$$

$$= \frac{9\sqrt[3]{4}}{4 \cdot 2}$$

$$= \frac{9\sqrt[3]{4}}{8} \quad \clubsuit$$

Example 12   Rationalize the denominator of $\dfrac{1}{1 - \sqrt{2}}$.

The best approach here is to multiply both numerator and denominator by the *conjugate** of the denominator, in this case $1 + \sqrt{2}$.

$$\frac{1}{1 - \sqrt{2}} = \frac{1(1 + \sqrt{2})}{(1 - \sqrt{2})(1 + \sqrt{2})} = \frac{1 + \sqrt{2}}{1 - 2} = -1 - \sqrt{2} \quad \clubsuit$$

1.8   Exercises

Write in radical form. Assume all variables are positive real numbers. See Example 2.

1. $m^{2/3}$ **2.** $p^{5/4}$ **3.** $5m^{4/5}$ **4.** $-2k^{1/6}$

5. $-4z^{-1/3}$ **6.** $10a^{-3/4}$ **7.** $(2m + p)^{2/3}$ **8.** $(5r + 3t)^{4/7}$

Write in exponential form. Assume all variables are nonnegative real numbers. See Example 3.

9. $\sqrt[5]{k^2}$ **10.** $-\sqrt[4]{z^5}$ **11.** $-\sqrt[3]{a^2}$ **12.** $2\sqrt{m^9}$

13. $-3\sqrt{5p^3}$ **14.** $m\sqrt{2y^5}$ **15.** $18\sqrt{m^2n^3p}$ **16.** $-12\sqrt[5]{x^3y^2z^4}$

Simplify each of the following. Assume that all variables represent positive real numbers. See Examples 1, 5, 6, 8, 10, and 11(a).

17. $\sqrt[3]{125}$ **18.** $\sqrt[4]{81}$ **19.** $\sqrt[4]{1296}$ **20.** $\sqrt[3]{343}$

21. $\sqrt[5]{-3125}$ **22.** $\sqrt[7]{-128}$ **23.** $\sqrt{50}$ **24.** $\sqrt{45}$

25. $\sqrt{2000}$ **26.** $\sqrt{3200}$ **27.** $\sqrt{5000}$ **28.** $\sqrt{27,000}$

29. $\sqrt[3]{81}$ **30.** $\sqrt[3]{16}$ **31.** $\sqrt[3]{250}$ **32.** $\sqrt[3]{128}$

33. $-\sqrt[4]{32}$ **34.** $-\sqrt[4]{243}$ **35.** $-\sqrt{\dfrac{9}{5}}$ **36.** $\sqrt{\dfrac{3}{8}}$

37. $-\sqrt[3]{\dfrac{3}{2}}$ **38.** $-\sqrt[3]{\dfrac{4}{5}}$ **39.** $\sqrt[4]{\dfrac{3}{2}}$ **40.** $\sqrt[4]{\dfrac{32}{81}}$

---

*The conjugate of $a\sqrt{m} + b\sqrt{n}$ is $a\sqrt{m} - b\sqrt{n}$.

41. $-\sqrt{24 \cdot 3^2 \cdot 2^4}$ **42.** $-\sqrt{18 \cdot 2^3 \cdot 4^2}$ **43.** $\sqrt[3]{16(-2)^4(2)^8}$ **44.** $\sqrt[3]{25(3)^4(5)^3}$

45. $\sqrt{8x^5z^8}$ **46.** $\sqrt{24m^6n^5}$ **47.** $\sqrt[3]{16z^5x^8y^4}$ **48.** $\sqrt[3]{81p^5x^2y^6}$

49. $\sqrt[4]{m^2n^7p^8}$ **50.** $\sqrt[4]{x^8y^6z^{10}}$ **51.** $\sqrt[6]{5^9r^8z^{10}}$ **52.** $\sqrt[4]{3^6m^6n^{14}}$

53. $\sqrt{\dfrac{2}{3x}}$ **54.** $\sqrt{\dfrac{5}{3p}}$ **55.** $\sqrt{\dfrac{x^5y^3}{z^2}}$ **56.** $\sqrt{\dfrac{g^3h^5}{r^3}}$

57. $\sqrt[3]{\dfrac{8}{x^2}}$ **58.** $\sqrt[3]{\dfrac{9}{16p^4}}$ **59.** $-\sqrt[3]{\dfrac{k^5m^3r^2}{r^8}}$ **60.** $-\sqrt[3]{\dfrac{9x^5y^6}{z^5w^2}}$

61. $\sqrt[4]{\dfrac{g^3h^5}{9r^6}}$ **62.** $\sqrt[4]{\dfrac{32x^{4/5}}{y^5}}$ **63.** $\dfrac{\sqrt[3]{mn} \cdot \sqrt[3]{m^2}}{\sqrt[3]{n^2}}$ **64.** $\dfrac{\sqrt[3]{8m^2n^3} \cdot \sqrt[3]{2m^2}}{\sqrt[3]{32m^4n^3}}$

65. $\dfrac{\sqrt[4]{32x^5y} \cdot \sqrt[4]{2xy^4}}{\sqrt[4]{4x^3y^2}}$ **66.** $\dfrac{\sqrt[4]{rs^2t^3} \cdot \sqrt[4]{r^3s^2t}}{\sqrt[4]{r^2t^3}}$ **67.** $\sqrt[3]{\sqrt{4}}$ **68.** $\sqrt[4]{\sqrt[3]{2}}$

69. $\sqrt[6]{\sqrt[3]{x}}$ **70.** $\sqrt[8]{\sqrt[4]{y}}$

Simplify each of the following, assuming that all variables represent nonnegative numbers. See Examples 7, 9, 10, and 11.

71. $4\sqrt{3} - 5\sqrt{12} + 3\sqrt{75}$ **72.** $2\sqrt{5} - 3\sqrt{20} + 2\sqrt{45}$ **73.** $\sqrt{50} - 8\sqrt{8} + 4\sqrt{18}$

74. $6\sqrt{27} - 3\sqrt{12} + 5\sqrt{48}$ **75.** $3\sqrt{28p} - 4\sqrt{63p} + \sqrt{112p}$ **76.** $9\sqrt{8k} + 3\sqrt{18k} - \sqrt{32k}$

77. $3\sqrt[3]{16} - 4\sqrt[3]{2}$ **78.** $\sqrt[3]{2} - \sqrt[3]{16} + 2\sqrt[3]{54}$ **79.** $2\sqrt[3]{3} + 4\sqrt[3]{24} - \sqrt[3]{81}$

80. $\sqrt[3]{32} - 5\sqrt[3]{4} + 2\sqrt[3]{108}$ **81.** $\dfrac{1}{\sqrt{3}} - \dfrac{2}{\sqrt{12}} + 2\sqrt{3}$ **82.** $\dfrac{1}{\sqrt{2}} + \dfrac{3}{\sqrt{8}} + \dfrac{1}{\sqrt{32}}$

83. $\dfrac{5}{\sqrt[3]{2}} - \dfrac{2}{\sqrt[3]{16}} + \dfrac{1}{\sqrt[3]{54}}$ **84.** $\dfrac{-4}{\sqrt[3]{3}} + \dfrac{1}{\sqrt[3]{24}} - \dfrac{2}{\sqrt[3]{81}}$ **85.** $\sqrt{a^3b^5} - 2\sqrt{a^7b^3} + \sqrt{a^3b^9}$

86. $\sqrt{p^7q^3} - \sqrt{p^5q^9} + \sqrt{p^9q}$ **87.** $(\sqrt{2} + 3)(\sqrt{2} - 3)$ **88.** $(\sqrt{5} + \sqrt{2})(\sqrt{5} - \sqrt{2})$

89. $(\sqrt[3]{11} - 1)(\sqrt[3]{11^2} + \sqrt[3]{11} + 1)$ **90.** $(\sqrt[3]{7} + 3)(\sqrt[3]{7^2} - 3\sqrt[3]{7} + 9)$

91. $(\sqrt{3} + \sqrt{8})^2$ **92.** $(\sqrt{2} - 1)^2$

93. $(3\sqrt{2} + \sqrt{3})(2\sqrt{3} - \sqrt{2})$ **94.** $(4\sqrt{5} - 1)(3\sqrt{5} + 2)$

95. $(2\sqrt[3]{3} + 1)(\sqrt[3]{3} - 4)$ **96.** $(\sqrt[3]{4} + 3)(5\sqrt[3]{4} + 1)$

Rationalize the denominator of each of the following. Assume that all variables represent nonnegative numbers and that no denominators are zero. See Example 12.

97. $\dfrac{3}{1 - \sqrt{2}}$ **98.** $\dfrac{2}{1 + \sqrt{5}}$ **99.** $\dfrac{\sqrt{3}}{\sqrt{5} + \sqrt{3}}$

100. $\dfrac{\sqrt{7}}{\sqrt{3} - \sqrt{7}}$ **101.** $\dfrac{-5}{2\sqrt{3} - 3\sqrt{2}}$ **102.** $\dfrac{6}{5\sqrt{2} + 2\sqrt{3}}$

103. $\dfrac{1 + \sqrt{3}}{3\sqrt{5} + 2\sqrt{3}}$

104. $\dfrac{\sqrt{7} - 1}{2\sqrt{7} + 4\sqrt{2}}$

105. $\dfrac{p}{\sqrt{p} + 2}$

106. $\dfrac{\sqrt{r}}{3 - \sqrt{r}}$

107. $\dfrac{a}{\sqrt{a + b} - 1}$

108. $\dfrac{3m}{2 + \sqrt{m + n}}$

109. $\dfrac{\sqrt{x} + \sqrt{x + 1}}{\sqrt{x} - \sqrt{x + 1}}$

110. $\dfrac{\sqrt{p} + \sqrt{p^2 - 1}}{\sqrt{p} - \sqrt{p^2 - 1}}$

In advanced mathematics it is sometimes useful to write a radical expression with a rational numerator. The procedure is similar to rationalizing the denominator. Rationalize the numerator of each of the following expressions.

111. $\dfrac{1 + \sqrt{2}}{2}$

112. $\dfrac{1 - \sqrt{3}}{3}$

113. $\dfrac{\sqrt{x}}{1 + \sqrt{x}}$

114. $\dfrac{\sqrt{p}}{1 - \sqrt{p}}$

115. $\dfrac{\sqrt{x} + \sqrt{x + 1}}{\sqrt{x} - \sqrt{x + 1}}$

116. $\dfrac{\sqrt{p} + \sqrt{p^2 - 1}}{\sqrt{p} - \sqrt{p^2 - 1}}$

Use a calculator to evaluate each of the following. Write all answers in scientific notation with three significant digits. For fourth roots, use the square root key twice.

117. $\sqrt{2.876 \times 10^7}$

118. $\sqrt{5.432 \times 10^9}$

119. $\sqrt[4]{3.87 \times 10^{-4}}$

120. $\sqrt[4]{5.913 \times 10^{-8}}$

121. $\dfrac{2.04 \times 10^{-3}}{\sqrt{5.97 \times 10^{-5}}}$

122. $\dfrac{3.86 \times 10^{-5}}{\sqrt{4.82 \times 10^{-5}}}$

123. $\sqrt{(4.721)^2 + (8.963)^2 - 2(4.721)(8.963)(.0468)}$

124. $\sqrt{(157.3)^2 + (184.7)^2 - 2(157.3)(184.7)(.9082)}$

Use absolute value to write the following without radicals. See Example 4.

125. $\sqrt{(m + n)^2}$

126. $\sqrt[3]{(a + 2b)^3}$

127. $\sqrt{z^2 - 6zx + 9x^2}$

128. $\sqrt[3]{(r + 2s)(r^2 + 4rs + 4s^2)}$

129. Find an approximate value for $\sqrt{5 + 2\sqrt{6}}$.

130. By squaring both sides, show that $\sqrt{5 + 2\sqrt{6}} = \sqrt{2} + \sqrt{3}$.

Chapter 1 *Summary*

To understand the concepts presented in this chapter, you should be familiar with the meanings of the following words, definitions, rules, and theorems. For easy reference, the section in which an item was first used is given next to the item.

Key Words

1.1 exponent
  base
1.2 coordinate
  inequalities
  absolute value
1.4 term
  coefficient
  like terms
  polynomial
  degree
  monomial
  binomial
  trinomial

1.5 factoring
  factored form
  prime polynomial
  irreducible polynomial
1.6 rational expression
  complex fraction
1.8 radical
  radicand
  rationalizing the
    denominator
  conjugate

Key Ideas

1.1 Sets of Numbers

Real numbers	$\{x \mid x$ corresponds to a point on a number line$\}$
Integers	$\{\ldots, -3, -2, -1, 0, 1, 2, 3, \ldots\}$
Rational numbers	$\left\{\dfrac{p}{q} \middle\mid p \text{ and } q \text{ are integers and } q \neq 0\right\}$
Irrational numbers	$\{x \mid x$ is real but not rational$\}$
Whole numbers	$\{0, 1, 2, 3, 4, \ldots\}$
Natural numbers	$\{1, 2, 3, 4, \ldots\}$

1.1 Properties of the Real Numbers

For all real numbers a, b, and c:

Commutative properties
$$a + b = b + a$$
$$ab = ba$$

Associative properties
$$(a + b) + c = a + (b + c)$$
$$(ab)c = a(bc)$$

Identity properties

There exists a unique real number 0 such that
$$a + 0 = a \quad \text{and} \quad 0 + a = a.$$
There exists a unique real number 1 such that
$$a \cdot 1 = a \quad \text{and} \quad 1 \cdot a = a.$$

Inverse properties	There exists a unique real number $-a$ such that

$$a + (-a) = 0 \quad \text{and} \quad (-a) + a = 0.$$

If $a \neq 0$, there exists a unique real number $1/a$ such that

$$a \cdot \frac{1}{a} = 1 \quad \text{and} \quad \frac{1}{a} \cdot a = 1.$$

Distributive property	$a(b + c) = ab + ac$

1.4 Special Products and 1.5 Results of Factoring

Difference of two squares	$x^2 - y^2 = (x + y)(x - y)$
Perfect square trinomials (squares of binomials)	$x^2 + 2xy + y^2 = (x + y)^2$ $x^2 - 2xy + y^2 = (x - y)^2$
Cubes of binomials	$(x + y)^3 = x^3 + 3x^2y + 3xy^2 + y^3$ $(x - y)^3 = x^3 - 3x^2y + 3xy^2 - y^3$
Difference of two cubes	$x^3 - y^3 = (x - y)(x^2 + xy + y^2)$
Sum of two cubes	$x^3 + y^3 = (x + y)(x^2 - xy + y^2)$

1.7 Rules for Exponents

Let r and s be rational numbers. The results below are valid for all real numbers a and b for which all indicated powers exist, and as long as no denominators are 0.

$$a^r \cdot a^s = a^{r+s} \qquad (ab)^r = a^r \cdot b^r \qquad (a^r)^s = a^{rs}$$

$$\frac{a^r}{a^s} = a^{r-s} \qquad \left(\frac{a}{b}\right)^r = \frac{a^r}{b^r} \qquad a^{-r} = \frac{1}{a^r}$$

1.8 Radical Notation

If a is a real number, n is a positive integer, and $a^{1/n}$ exists, then

$$\sqrt[n]{a} = a^{1/n}.$$

If m is an integer, n is a positive integer, and a is a real number for which $\sqrt[n]{a}$ exists, then

$$a^{m/n} = (\sqrt[n]{a})^m = \sqrt[n]{a^m}.$$

Chapter 1   *Review Exercises*

Let set $K = \{-12, -6, -9/10, -\sqrt{7}, -\sqrt{4}, 0, 1/8, \pi/4, 6, \sqrt{11}\}$. *List the elements of K that are members of the following sets.*

1. Natural numbers

2. Whole numbers

3. Integers

4. Rational numbers

5. Irrational numbers

6. Real numbers

For Exercises 7–12 choose all words from the following list that apply.

   (a) *natural numbers*   **(b)** *whole numbers*   **(c)** *integers*   **(d)** *rational numbers*

   (e) *irrational numbers*   **(f)** *real numbers*   **(g)** *meaningless*

7. 22

8. 0

9. $-\sqrt{25}$

10. 5/8

11. $-6/0$

12. $3\pi/4$

Use the order of operations to simplify each of the following.

13. $-4 - [2 - (-9)]$

14. $[8 - (-5)] - 4$

15. $2(-9) - 4(-6)$

16. $-7(-5) - 3(-9)$

17. $(-4 - 1)(-3 - 5) - 8$

18. $(6 - 9)(-2 - 7) - (-4)$

19. $\left(-\dfrac{5}{9} - \dfrac{2}{3}\right) - \dfrac{5}{6}$

20. $\left(-\dfrac{8}{5} - \dfrac{3}{4}\right) - \left(-\dfrac{1}{2}\right)$

21. $\dfrac{6(-4) - 9(-3)}{-5[-2 - (-6)]}$

22. $\dfrac{(-7)(-3) - (-8)(-5)}{(-4 - 2)(-1 - 6)}$

Evaluate each of the following if $a = -1$, $b = -2$, *and* $c = 4$.

23. $9a - 5b + 4c$

24. $6b - 11a - 3c$

25. $-4(2a - 5b)$

26. $6(3b + 2c)$

27. $\dfrac{9a + 2b}{a + b + c}$

28. $\dfrac{4a - 5c}{2a + 3b}$

Identify the properties illustrated in each of the following. Some may require more than one property.

29. $8(5 + 9) = (5 + 9)8$

30. $3 + (-3) = 0$

31. If $x \neq 0$, then $x \cdot \dfrac{1}{x} = 1$.

32. $4 \cdot 6 + 4 \cdot 12 = 4(6 + 12)$

33. $3 \cdot (4 \cdot 2) = (3 \cdot 4) \cdot 2$

34. $-(r - 2) = -r + 2$

35. $(9 + p) + 0 = 9 + p$

36. $[6 \cdot 5 + 8 \cdot 5] \cdot 2 = [(6 + 8) \cdot 5] \cdot 2$

Use the distributive property to rewrite sums as products and products as sums.

37. $9(p + q)$

38. $-11(m - n)$

39. $-3(5 - r)$

40. $k(r + s - t)$

41. $ab + ac$

42. $mn - 9m$

Write the following numbers in numerical order from smallest to largest.

43. $-7, -3, 8, \pi, -2, 0$

44. $5/6, 1/2, -2/3, -5/4, -3/8$

45. $|6 - 4|, -|-2|, |8 + 1|, -|3 - (-2)|$

46. $\sqrt{7}, -\sqrt{8}, -|\sqrt{16}|, |-\sqrt{12}|$

Write without absolute value bars.

47. $-|-6| + |3|$

48. $|-5| + |-9|$

49. $7 - |-8|$

50. $|3 - \sqrt{7}|$

51. $|\sqrt{8} - 3|$

52. $|\sqrt{52} - 8|$

53. $|m - 3|$ if $m > 3$

54. $|-6 - x^2|$

55. $|\pi - 4|$

56. $|3 + 5k|$ if $k < -3/5$

Perform each of the following operations. Assume that all variables appearing as exponents represent integers.

57. $(-9m^2 + 11m - 7) + (2m^2 - 12m + 4)$

58. $(3q^3 - 9q^2 + 6) + (4q^3 - 8q + 3)$

59. $(-7z^3 + 8z^2 - z) - (z^3 + 4z^2 + 10z)$

60. $2(3y^6 - 9y^2 + 2y) - (5y^6 - 10y^2 - 4y)$

61. $(8y - 7)(2y + 7)$

62. $(2r + 11s)(4r - 9s)$

63. $(3k - 5m)^2$

64. $(4a - 3b)^2$

65. $(3w - 2)(5w^2 - 4w + 1)$

66. $(2k + 5)(3k^3 - 4k^2 + 8k - 2)$

67. $(a - b + c)(a + b - 2c)$

68. $(x + 2y - z)^2$

69. $(p^q + 1)(p^q - 3)$

70. $(a^y + 6)^2$

Perform each division.

71. $\dfrac{15m^4n^5 - 10m^3n^7 + 20m^7n^3}{10mn^4}$

72. $\dfrac{10r^6m^3 + 8r^4m - 3r^{10}m^2}{12r^3m^4}$

73. $\dfrac{6p^2 - 5p - 56}{2p - 7}$

74. $\dfrac{72r^2 + 59r + 12}{8r + 3}$

75. $\dfrac{5m^3 - 7m^2 - 10m + 14}{m^2 - 2}$

76. $\dfrac{3b^3 - 8b^2 + 12b - 30}{b^2 + 4}$

77. $\dfrac{k^5 - 6k^4 + 9k^3 + k^2 - 12k + 3}{k^2 - 4k + 1}$

78. $\dfrac{z^5 + 2z^4 + 4z^3 - 8z^2 - 15z - 32}{z^2 + 2z + 4}$

Factor as completely as possible.

79. $7z^2 - 9z^3 + z$

80. $3(m - n) + 4k(m - n)$

81. $12p^5 - 8p^4 + 20p^3$

82. $3(z - 4)^2 + 9(z - 4)^3$

83. $r^2 + rp - 42p^2$

84. $z^2 - 6zk - 16k^2$

85. $6m^2 - 13m - 5$

86. $4k^2 + 11k - 3$

87. $30m^5 - 35m^4n - 25m^3n^2$

88. $48a^8 - 12a^7b - 90a^6b^2$

89. $169y^4 - 1$

90. $49m^8 - 9n^2$

91. $8y^3 - 1000z^6$

92. $125a^6 + 216b^9$

93. $6(3r - 1)^2 + (3r - 1) - 35$

94. $6(2q - 5)^2 + 13(2q - 5) - 15$

95. $ar - 3as + 5rb - 15sb$

96. $15mp + 9mq - 10np - 6nq$

97. $(16m^2 - 56m + 49) - 25a^2$

98. $r^9 - 8(r^3 - 1)^3$

Perform each of the following operations.

99. $\dfrac{5x^2y}{x + y} \cdot \dfrac{3x + 3y}{30xy^2}$

100. $\dfrac{3m - 9}{8m} \cdot \dfrac{16m + 24}{15}$

101. $\dfrac{k^2 + k}{8k^3} \cdot \dfrac{4}{k^2 - 1}$

102. $\dfrac{3r^3 - 9r^2}{r^2 - 9} \div \dfrac{8r^3}{r + 3}$

103. $\dfrac{x^2 + x - 2}{x^2 + 5x + 6} \div \dfrac{x^2 + 3x - 4}{x^2 + 4x + 3}$

104. $\left(1 - \dfrac{3}{p}\right)\left(1 + \dfrac{3}{p}\right)$

105. $\dfrac{27m^3 - n^3}{3m - n} \div \dfrac{9m^2 + 3mn + n^2}{9m^2 - n^2}$

106. $\dfrac{p^2 - 36q^2}{(p - 6q)^2} \cdot \dfrac{p^2 - 5pq - 6q^2}{p^2 - 6pq + 36q^2} \div \dfrac{5p}{p^3 + 216q^3}$

107. $\dfrac{1}{4y} + \dfrac{8}{5y}$

108. $\dfrac{m}{4 - m} + \dfrac{3m}{m - 4}$

109. $\dfrac{3}{x^2 - 4x + 3} - \dfrac{2}{x^2 - 1}$

110. $\left[\dfrac{1}{(x + h)^2 + 16} - \dfrac{1}{x^2 + 16}\right] \div h$

111. $\dfrac{\dfrac{1}{p} + \dfrac{1}{q}}{1 - \dfrac{1}{pq}}$

112. $\dfrac{3 + \dfrac{2m}{m^2 - 4}}{\dfrac{5}{m - 2}}$

Simplify each of the following. Write results without negative exponents. Assume that all variables represent nonzero real numbers.

113. 2^{-6}

114. -3^{-2}

115. $\left(\dfrac{-5}{4}\right)^{-2}$

116. $3^{-1} - 4^{-1}$

117. $(5z^3)(-2z^5)$

118. $(-3a^4b^7)(6a^{-2}b^2)$

119. $(8p^2q^3)(-2p^5q^{-4})$

120. $(-6p^5w^4m^{12})^0$

121. $(-6x^2y^{-3}z^2)^{-2}$

122. $\dfrac{-8y^7p^{-2}}{y^{-4}p^{-3}}$

123. $\dfrac{a^{-6}(a^{-8})}{a^{-2}(a^{11})}$

124. $\dfrac{5^{-1}r^4m^{-2}}{5r^{-2}m^{-4}}$

125. $\dfrac{6r^3s^{-2}}{6^{-1}r^4s^{-3}}$

126. $\dfrac{(p + q)^4(p + q)^{-3}}{(p + q)^6}$

127. $\dfrac{[p^2(m + n)^3]^{-2}}{p^{-2}(m + n)^{-5}}$

128. $[(ab^{-2})^2]^{-1}$

129. $\left(\dfrac{r^{-1}s^2}{t^{-2}}\right)^{-2}$

130. $\dfrac{(2x^{-3})^2(3x^2)^{-2}}{6(x^2y^3)}$

131. $\dfrac{(-5m^3n^{-4})^{-1}}{(10m^{-1}n^{-5})^{-2}}$

132. $\dfrac{(8a^2b)^2(3a^{-1}b^{-2})}{24(a^{-2}b^2)^{-1}(a^{-2})}$

Simplify. Write with only positive exponents. Assume that all variables represent positive real numbers.

133. $\dfrac{p^4(p^{-2})}{p^{5/3}}$

134. $(2 + p)^{5/4}(2 + p)^{-9/4}$

135. $(7r^{1/2})(2r^{3/4})(-r^{1/6})$

136. $(a^{3/4}b^{2/3})(a^{5/8}b^{-5/6})$

137. $\dfrac{y^{5/3} \cdot y^{-2}}{y^{-5/6}}$

138. $\left(\dfrac{25m^3n^5}{m^{-2}n^6}\right)^{-1/2}$

Simplify. Assume that all variables represent positive real numbers. Write without denominators.

139. $\dfrac{k^{2+p} \cdot k^{-4p}}{k^{6p}}$

140. $\dfrac{(r^{3+2z})(r^{-5-z})}{(r^{-z/2})^{-4}}$

Find each product. Assume that all variables represent positive real numbers.

141. $2z^{1/3}(5z^2 - 2)$

142. $-m^{3/4}(8m^{1/2} + 4m^{-3/2})$

143. $(p + p^{1/2})(3p - 5)$

144. $(m^{1/2} - 4m^{-1/2})^2$

Simplify. Assume that all variables represent positive numbers.

145. $\sqrt{200}$

146. $\sqrt[3]{16}$

147. $\sqrt[4]{1250}$

148. $-\sqrt{\dfrac{16}{3}}$

149. $\sqrt{\dfrac{7}{3r}}$

150. $-\sqrt[3]{\dfrac{2}{5p^2}}$

151. $\sqrt{\dfrac{2^7 y^8}{m^3}}$

152. $-\sqrt[3]{\dfrac{r^6 h^5}{z^2}}$

153. $\sqrt[4]{\sqrt[3]{m}}$

154. $\dfrac{\sqrt[4]{8p^2 q^5} \cdot \sqrt[4]{2p^3 q}}{\sqrt[4]{p^5 q^2}}$

155. $(\sqrt[3]{2} + 4)(\sqrt[3]{2^2} - 4\sqrt[3]{2} + 16)$

156. $\dfrac{3}{\sqrt{5}} - \dfrac{2}{\sqrt{45}} + \dfrac{6}{\sqrt{80}}$

157. $\sqrt{18m^3} - 3m\sqrt{32m} + 5\sqrt{m^3}$

158. $\dfrac{2}{7 - \sqrt{3}}$

159. $\dfrac{6}{3 - \sqrt{2}}$

160. $\dfrac{z}{\sqrt{z + 1}}$

161. $\dfrac{k}{\sqrt{k} - 3}$

162. $\dfrac{\sqrt{x} - \sqrt{x - 2}}{\sqrt{x} + \sqrt{x - 2}}$

Chapter 1   *Test*

Let set $A = \left\{ -10, -\dfrac{9}{3}, 0, \dfrac{2}{3}, \dfrac{4}{\pi}, \sqrt{25} \right\}$. *List the elements of set A that are members of the following sets.*

1. Irrational numbers

2. Real numbers

3. Integers

4. Natural numbers

5. Use the order of operations to simplify the following.

$$\frac{5 + 6 \div 2 \cdot 3 + 4}{(-3 - 2)(-4 + 2) - (3 \cdot 3 - 8)}$$

6. Evaluate the following when $x = -2$, $y = -3$, and $z = -4$.

$$\frac{2x - 3y - z}{-x - y + z}$$

Identify the properties illustrated in each of the following.

7. $2(x + 3) = 2(3 + x)$

8. $3(x + 4) = 3x + 12$

9. $(x + 0) + y = x + y$

10. $\left(6 \cdot \dfrac{1}{6} \right) - p = 1 - p$

11. Write the following numbers in numerical order from smallest to largest.

$$-|-2|, \ \sqrt{9}, \ |4 - 9|, \ -\sqrt{2}, \ \pi, \ 0, \ \frac{21}{4}$$

Perform each of the following operations. Assume that all variables appearing as exponents represent integers.

12. $(a^2 - 3a + 1) - (a - 2a^2) + (-4 + 3a - a^2)$

13. $(2x^n - 3)^2$

14. $(3y^2 - y + 4)(y + 2)$

Perform each division.

15. $\dfrac{-3m^5n^2 + 18m^3n}{-3m^2n}$

16. $\dfrac{x^3 - 4x^2 + 3x - 1}{x - 3}$

Factor as completely as possible.

17. $x^4 - 16$

18. $24m^3 - 14m^2 - 24m$

19. $4x^4 - 6x^2y^4 + 9y^8$

20. $x^3y^2 - 9x^3 - 8y^2 + 72$

Perform each of the following operations.

21. $\dfrac{5x^2 - 9x - 2}{30x^3 + 6x^2} \cdot \dfrac{2x^8 + 6x^7 + 4x^6}{x^4 - 3x^2 - 4}$

22. $\dfrac{x}{x^2 + 3x + 2} + \dfrac{2x}{2x^2 - x - 3}$

23. $\dfrac{a + b}{2a - 3} - \dfrac{a - b}{3 - 2a}$

24. $\dfrac{y - 2}{y - \dfrac{4}{y}}$

Simplify each of the following. Write results without negative exponents. Assume that all variables represent positive numbers.

25. -2^{-4}

26. $\left(\dfrac{x^n y^{2n-1}}{y^n}\right)^2$

27. $\left(\dfrac{x^{-1}y^3}{2x^0y^{-5}}\right)^{-2}$

28. $\left(\dfrac{x^{-2}y^{-1/3}z}{x^{-5/3}y^{-2/3}z^{2/3}}\right)^3$

Simplify. Assume that all variables represent positive numbers.

29. $\sqrt{18x^5y^8}$

30. $\sqrt[5]{\dfrac{x}{4y^2}}$

31. $\sqrt{32x} + \sqrt{2x} - \sqrt{18x}$

32. $\dfrac{3}{\sqrt{2} - 1}$

33. $(\sqrt{x} - \sqrt{y})(\sqrt{x} + \sqrt{y})$

2

Equations and Inequalities

Applications of algebra in everyday life frequently require solving one or more *equations*. In fields such as business, many applications of algebra also require the use of *inequalities*. Methods of solving several different kinds of equations and inequalities are discussed in this chapter.

2.1 Linear Equations

An **equation** is a statement that two expressions are equal. Examples of equations include

$$x + 2 = 9, \qquad 11y = 5y + 6y, \qquad \text{and} \qquad x^2 - 2x - 1 = 0.$$

To **solve** an equation means to find all numbers that make the equation a true statement. Such numbers are called **solutions** of the equation. A number that is a solution of an equation is said to **satisfy** the equation, and the solutions of an equation make up its **solution set.**

An equation satisfied by every number that is a meaningful replacement for the variable is called an **identity.** Examples of identities are

$$3x + 4x = 7x \qquad \text{and} \qquad x^2 - 3x + 2 = (x - 2)(x - 1).$$

Equations that are satisfied by some numbers but not by others are called **conditional equations.** Examples of conditional equations are

$$2m + 3 = 7 \quad \text{and} \quad \frac{5r}{r - 1} = 7.$$

The equation

$$3(x + 1) = 5 + 3x$$

is neither an identity nor a conditional equation. Multiplying on the left side gives

$$3x + 3 = 5 + 3x,$$

which is false for every value of x. Such an equation is called a **contradiction.**

Example 1 Decide whether each of the following equations is an identity, a conditional equation, or a contradiction.

(a) $9p^2 - 25 = (3p + 5)(3p - 5)$

Since the product of $3p + 5$ and $3p - 5$ is $9p^2 - 25$, the given equation is true for *every* value of p and is an identity.

(b) $5y - 4 = 11$

Replacing y with 3 gives

$$5 \cdot 3 - 4 = 11$$
$$11 = 11,$$

a true statement. On the other hand, $y = 4$ leads to

$$5 \cdot 4 - 4 = 11$$
$$16 = 11,$$

a false statement. The equation $5y - 4 = 11$ is true for some values of y, but not all, and thus is a conditional equation. (By the way, the word *some* in mathematics means "at least one." We can therefore say that the statement $5y - 4 = 11$ is true for *some* replacements of y, even though it turns out to be true only for $y = 3$.)

(c) $(a - 2)^2 + 3 = a^2 - 4a + 2$

The left side can be rewritten as follows.

$$a^2 - 4a + 4 + 3 = a^2 - 4a + 2$$
$$a^2 - 4a + 7 = a^2 - 4a + 2$$

The final equation is false for every value of a, so this equation is a contradiction. ✥

Equations with the same solution set are called **equivalent equations.** For example, $x + 1 = 5$ and $6x + 3 = 27$ are equivalent equations because they have the same solution set, {4}.

Example 2   Decide which of the following pairs of equations are equivalent.

(a) $2x - 1 = 3$ and $12x + 7 = 31$

Each of these equations has solution set $\{2\}$. Since the solution sets are equal, the equations are equivalent.

(b) $x = 3$ and $x^2 = 9$

The solution set for $x = 3$ is $\{3\}$, while the solution set for the equation $x^2 = 9$ is $\{3, -3\}$. Since the solution sets are not equal, the equations are not equivalent. ✛

One way to solve an equation is to rewrite it as a series of simpler equivalent equations. These simpler equations often can be obtained with the *addition and multiplication properties of equality*.

Addition and Multiplication Properties of Equality

For real numbers a, b, and c:

$a = b$ and $a + c = b + c$ are equivalent. *(The same number may be added to both sides of an equation.)*

If $c \neq 0$, then $a = b$ and $ac = bc$ are equivalent. *(The same nonzero number may be multiplied on both sides of an equation.).*

Example 3   Solve $3(2x - 4) = 7 - (x + 5)$.

Use the distributive property and then collect like terms to get the following sequence of simpler equivalent equations.

$$3(2x - 4) = 7 - (x + 5)$$
$$6x - 12 = 7 - x - 5$$
$$6x - 12 = 2 - x$$

Now add the same quantity to each side of the equation.

$$x + 6x - 12 = x + 2 - x$$
$$7x - 12 = 2$$
$$12 + 7x - 12 = 12 + 2$$
$$7x = 14$$

Finally, multiplying each side by the same number, 1/7, produces

$$\frac{1}{7} \cdot 7x = \frac{1}{7} \cdot 14$$
$$x = 2.$$

To check, replace x with 2 in the original equation, getting

$$3(2x - 4) = 7 - (x + 5) \qquad \text{Original equation}$$
$$3(2 \cdot \mathbf{2} - 4) = 7 - (\mathbf{2} + 5) \qquad \text{Let } x = 2$$
$$3(4 - 4) = 7 - (7)$$
$$0 = 0. \qquad\qquad \text{True}$$

Since replacing x with 2 results in a true statement, 2 is the solution of the given equation. The solution set is therefore $\{2\}$. ✚

The equation in Example 3 is a *linear equation,* as are most of the equations in this section.

Definition of Linear Equation	A **linear equation** in one variable is an equation that can be written in the form $$ax + b = 0,$$ where $a \neq 0$.

Example 4 Solve each equation.

(a) $\dfrac{3p - 1}{3} - \dfrac{2p}{p - 1} = p$

This equation does not satisfy the definition of a linear equation. However, the equation can be written as a linear equation and solved as one if the restrictions on the denominator are noted. Multiply both sides of the equation by the common denominator, $3(p - 1)$, assuming $p \neq 1$. Doing this gives

$$3(p - 1)\left(\frac{3p - 1}{3}\right) - 3(p - 1)\left(\frac{2p}{p - 1}\right) = 3(p - 1)$$
$$(p - 1)(3p - 1) - 3(2p) = 3p(p - 1)$$
$$3p^2 - 4p + 1 - 6p = 3p^2 - 3p.$$

A simpler equivalent equation comes from combining terms and adding $-3p^2$ to both sides, producing

$$-10p + 1 = -3p.$$

Now add $10p$ to both sides to get

$$1 = 7p.$$

Finally, multiplying both sides by 1/7 gives

$$\frac{1}{7} = p.$$

Check 1/7 in the given equation to verify that the solution set is {1/7}. The restriction $p \neq 1$ does not affect the solution set here, since $1/7 \neq 1$. Since the original equation and the linear equation found by multiplying both sides by $3(p - 1)$ have the same solution set, {1/7}, they are equivalent equations.

(b) $\dfrac{x}{x - 2} = \dfrac{2}{x - 2} + 2$

Multiply both sides of the equation by $x - 2$, assuming that $x - 2 \neq 0$.

$$x = 2 + 2(x - 2)$$
$$x = 2 + 2x - 4$$
$$x = 2$$

It is necessary to assume $x - 2 \neq 0$ in order to be able to multiply both sides of the equation by $x - 2$. Since $x = 2$, however, $x - 2 = 0$, and the multiplication property of equations does not apply. The solution is Ø. (Substituting 2 for x in the original equation would result in a denominator of 0.) The original equation and the linear equation $x = 2 + 2(x - 2)$ are *not* equivalent. ✦

Example 5 Solve $\dfrac{2}{x - 1} - \dfrac{4}{3x} = \dfrac{1}{x^2 - x}$.

First find a common denominator. Since $x^2 - x$ can be factored as $x(x - 1)$, a common denominator is $3x(x - 1)$. Multiply both sides of the equation by $3x(x - 1)$.

$$3x(x - 1)\,\frac{2}{x - 1} - 3x(x - 1)\,\frac{4}{3x} = 3x(x - 1)\,\frac{1}{x(x - 1)}$$
$$6x - 4(x - 1) = 3$$
$$6x - 4x + 4 = 3$$
$$2x = -1$$
$$x = -\frac{1}{2}$$

Check the answer in the original equation to verify that the solution set is {−1/2}. ✦

Sometimes an equation with more than one letter must be *solved for a specified variable*. This process is shown in the next example. (As a general rule letters from the beginning of the alphabet, such as a, b, c, and so on, are used to represent constants, while letters such as x, y, and z represent variables.)

Example 6 Solve the equation $3(2x - 5a) + 4b = 4x - 2$ for x.

Using the distributive property gives

$$6x - 15a + 4b = 4x - 2.$$

Treat x as the variable and the other letters as constants. Get all terms with x on one side and all terms without x on the other side.

$$6x - 4x = 15a - 4b - 2$$
$$2x = 15a - 4b - 2$$
$$x = \frac{15a - 4b - 2}{2} \quad \clubsuit$$

2.1 Exercises

Decide whether each of the following equations is an identity, a conditional equation, or a contradiction. See Example 1.

1. $x^2 + 5x = x(x + 5)$

2. $3y + 4 = 5(y - 2)$

3. $2(x - 7) = 5x + 3 - x$

4. $2x - 4 = 2(x + 2)$

5. $\dfrac{m + 3}{m} = 1 + \dfrac{3}{m}$

6. $\dfrac{p}{2 - p} = \dfrac{2}{p} - 1$

7. $4q + 20q - 25 = 24q + 5$

8. $3(k + 2) - 5(k + 2) = -2k - 4$

Decide which of the following pairs of equations are equivalent. See Example 2.

9. $3x - 5 = 7$
$-6x + 10 = 14$

10. $-x = 2x + 3$
$-3x = 3$

11. $\dfrac{3x}{x - 1} = \dfrac{2}{x - 1}$
$3x = 2$

12. $\dfrac{x + 1}{12} = \dfrac{5}{12}$
$x + 1 = 5$

13. $\dfrac{x}{x - 2} = \dfrac{2}{x - 2}$
$x = 2$

14. $\dfrac{x + 3}{x + 1} = \dfrac{2}{x + 1}$
$x = -1$

15. $x = 4$
$x^2 = 16$

16. $z^2 = 9$
$z = 3$

Solve each of the following equations. See Examples 3–5.

17. $4x - 1 = 15$

18. $-3y + 2 = 5$

19. $.2m - .5 = .1m + .7$

20. $.01p + 3.1 = 2.03p - 2.96$

21. $\dfrac{5}{6}k - 2k + \dfrac{1}{3} = \dfrac{2}{3}$

22. $\dfrac{3}{4} + \dfrac{1}{5}r - \dfrac{1}{2} = \dfrac{4}{5}r$

23. $3r + 2 - 5(r + 1) = 6r + 4$

24. $5(a + 3) + 4a - 5 = -(2a - 4)$

25. $2[m - (4 + 2m) + 3] = 2m + 2$

26. $4[2p - (3 - p) + 5] = -7p - 2$

27. $\dfrac{3x - 2}{7} = \dfrac{x + 2}{5}$

28. $\dfrac{2p + 5}{5} = \dfrac{p + 2}{3}$

29. $\dfrac{3k - 1}{4} = \dfrac{5k + 2}{8}$

30. $\dfrac{9x - 1}{6} = \dfrac{2x + 7}{3}$

31. $\dfrac{1}{4p} + \dfrac{2}{p} = 3$

32. $\dfrac{2}{t} + 6 = \dfrac{5}{2t}$

33. $\dfrac{m}{2} - \dfrac{1}{m} = \dfrac{6m + 5}{12}$

34. $\dfrac{-3k}{2} + \dfrac{9k - 5}{6} = \dfrac{11k + 8}{k}$

35. $\dfrac{2r}{r-1} = 5 + \dfrac{2}{r-1}$

36. $\dfrac{3x}{x+2} = \dfrac{1}{x+2} - 4$

37. $\dfrac{5}{2a+3} + \dfrac{1}{a-6} = 0$

38. $\dfrac{2}{x+1} = \dfrac{3}{5x+5}$

39. $\dfrac{4}{x-3} - \dfrac{8}{2x+5} + \dfrac{3}{x-3} = 0$

40. $\dfrac{5}{2p+3} - \dfrac{3}{p-2} = \dfrac{4}{2p+3}$

41. $\dfrac{2p}{p-2} = 3 + \dfrac{4}{p-2}$

42. $\dfrac{5k}{k+4} = 3 - \dfrac{20}{k+4}$

43. $\dfrac{3}{y-2} + \dfrac{1}{y+1} = \dfrac{1}{y^2 - y - 2}$

44. $\dfrac{2}{p+3} - \dfrac{5}{p-1} = \dfrac{1}{3 - 2p - p^2}$

45. $\dfrac{5}{x^2 - 2x - 3} = \dfrac{4}{x^2 + 5x + 4}$

46. $\dfrac{-2}{x^2 + 3x - 10} = \dfrac{7}{2 + x - x^2}$

47. $2(m+1)(m-1) = (2m+3)(m-2)$

48. $(2y-1)(3y+2) = 6(y+2)^2$

49. $(3x-4)^2 - 5 = 3(x+5)(3x+2)$

50. $(2x+5)^2 = 3x^2 + (x+3)^2$

Solve each of the following equations for x. See Example 6.

51. $2(x-a) + b = 3x + a$

52. $5x - (2a+c) = a(x+1)$

53. $ax + b = 3(x-a)$

54. $4a - ax = 3b + bx$

55. $\dfrac{x}{a-1} = ax + 3$

56. $\dfrac{2a}{x-1} = a - b$

57. $a^2x + 3x = 2a^2$

58. $ax + b^2 = bx - a^2$

59. $3x = (2x-1)(m+4)$

60. $-x = (5x+3)(3k+1)$

In the metric system of weights and measures, temperature is measured in degrees Celsius (°C) instead of degrees Fahrenheit (°F). To convert back and forth between the two systems, use the equations

$$C = \frac{5(F-32)}{9} \quad \text{and} \quad F = \frac{9}{5}C + 32.$$

In each of the following exercises, convert to the other system. Round answers to the nearest tenth of a degree if necessary.

61. 20°C   **62.** 100°C   **63.** 59°F   **64.** 86°F

65. 100°F   **66.** 350°F   **67.** 40°C   **68.** 85°C

When a consumer borrows money, the lender must tell the consumer the true annual interest rate of the loan. The method of finding the exact true annual interest rate requires special tables available from the government, but a quick approximate rate can be found by using the equation

$$A = \frac{2pf}{b(q+1)},$$

where p is the number of payments made in one year, f is the finance charge, b is the balance owed on the loan, and q is the total number of payments. Find the value of the variables not given in Exercises 69–74. Round A to the nearest percent and round other variables to the nearest whole number. (This formula is not accurate enough for the requirements of federal law.)

69. $p = 12$, $f = \$800$, $b = \$4000$, $q = 36$; find A

70. $p = 12$, $f = \$60$, $b = \$740$, $q = 12$; find A

71. $A = 14\%$ (or .14), $p = 12$, $b = \$2000$, $q = 36$; find f

72. $A = 11\%$, $p = 12$, $b = \$1500$, $q = 24$; find f

73. $A = 16\%$, $p = 12$, $f = \$370$, $q = 36$; find b

74. $A = 10\%$, $p = 12$, $f = \$490$, $q = 48$; find b

When a loan is paid off early, a portion of the finance charge must be returned to the borrower. By one method of calculating finance charge (called the rule of 78), the amount of unearned interest (finance charge to be returned) is given by

$$u = f \cdot \frac{n(n + 1)}{q(q + 1)},$$

where u represents unearned interest, f is the original finance charge, n is the number of payments remaining when the loan is paid off, and q is the original number of payments. Find the amount of the unearned interest in each of the following.

75. Original finance charge = $800, loan scheduled to run 36 months, paid off with 18 payments remaining

76. Original finance charge = $1400, loan scheduled to run 48 months, paid off with 12 payments remaining

77. Original finance charge = $950, loan scheduled to run 24 months, paid off with 6 payments remaining

78. Original finance charge = $175, loan scheduled to run 12 months, paid off with 3 payments remaining

79. Find the error in the following.

$$x^2 + 2x - 15 = x^2 - 3x$$
$$(x + 5)(x - 3) = x(x - 3)$$
$$x + 5 = x$$
$$5 = 0$$

Find the value of k that will make each equation equivalent to $x = 2$.

80. $9x - 7 = k$

81. $-5x + 11x - 2 = k + 4$

82. $\dfrac{8}{k + x} = 4$

83. $\sqrt{x + k} = 0$

84. $\sqrt{3x - 2k} = 4$

Solve the following equations. Round each solution to the nearest hundredth.

85. $9.06x + 3.59(8x - 5) = 12.07x + .5612$

86. $-5.74(3.1 - 2.7p) = 1.09p + 5.2588$

87. $\dfrac{2.5x - 7.8}{3.2} + \dfrac{1.2x + 11.5}{5.8} = 6$

88. $\dfrac{4.19x + 2.42}{.05} - \dfrac{5.03x - 9.74}{.02} = 1$

89. $\dfrac{2.63r - 8.99}{1.25} - \dfrac{3.90r - 1.77}{2.45} = r$

90. $\dfrac{8.19m + 2.55}{4.34} - \dfrac{8.17m - 9.94}{1.04} = 4m$

2.2 Applications of Linear Equations

Mathematics is an important problem-solving tool. Many times the solution of a problem depends on the use of a formula that expresses a relationship among several variables. For example, the formula

$$A = \frac{1}{2}(b_1 + b_2)h$$

gives the area of a trapezoid having bases of lengths b_1 and b_2 and height h. To solve the equation for b_1, get b_1 alone, with all other variables or numbers on the other side of the equals sign. To solve for b_1, first multiply both sides of the equation by 2.

$$2 \cdot A = 2 \cdot \frac{1}{2}(b_1 + b_2)h \qquad \text{Multiply by 2}$$

$$2A = (b_1 + b_2)h$$

$$\frac{2A}{h} = b_1 + b_2 \qquad \text{Multiply by } \frac{1}{h}$$

$$\frac{2A}{h} - b_2 = b_1 \qquad \text{Subtract } b_2$$

$$b_1 = \frac{2A}{h} - b_2$$

Recall from Section 2.1 that this process is called **solving for a specified variable.** Using the distributive property on the right side of the equation after the second step above would give

$$2A = (b_1 + b_2)h$$

$$2A = b_1 h + b_2 h$$

$$2A - b_2 h = b_1 h$$

$$\frac{2A - b_2 h}{h} = b_1,$$

which is equivalent to the result found by the first method.

Example 1   Solve $J\left(\dfrac{x}{k} + a\right) = x$ for x.

To get all terms with x on one side of the equation and all terms without x on the other, first use the distributive property.

$$J\left(\frac{x}{k}\right) + Ja = x$$

Eliminate the denominator, k, by assuming $k \neq 0$ and multiplying both sides by k.

$$k\,J\!\left(\frac{x}{k}\right) + k\,Ja = k\,x$$

$$Jx + kJa = kx$$

Then add $-Jx$ to both sides to get the two terms with x together.

$$kJa = kx - Jx$$

$$kJa = x(k - J) \qquad \text{Factor the right side}$$

Assuming $k \neq J$, multiply both sides by $1/(k - J)$ to find the solution.

$$x = \frac{kJa}{k - J} \quad \clubsuit$$

One of the main reasons for learning mathematics is to be able to use it in solving practical problems. For most students, however, learning how to apply mathematical skills to real situations is the most difficult task they face. In the rest of this section a few hints are given that may help with applications.

A common difficulty with "word problems" is trying to do everything at once. It is usually best to attack the problem in stages.

Solving Word Problems

1. Decide on an unknown, and name it with some variable that you *write down*. Most students try to skip this step. They are eager to get on with writing an equation. But this is an important step. If you don't know what "x" represents, how can you write a meaningful equation or interpret a result?

2. Draw a sketch or make a chart, if appropriate, showing the information given in the problem.

3. Decide on a variable expression to represent any other unknowns in the problem. For example, if W represents the width of a rectangle, L represents the length, and you know that the length is one more than twice the width, *write down* $L = 1 + 2W$.

4. Use the results of Steps 1 and 3 to write an equation.

5. Solve the equation.

6. Check the solution in the words of the original problem.

Notice how each of the steps listed above is carried out in the following examples.

Example 2   If the length of a side of a square is increased by 3 cm, the perimeter of the new square is 40 cm more than twice the length of a side of the original square. Find the dimensions of the original square.

First, decide what the variable should represent. Since the length of a side of the original square is to be found, let the variable represent that.

$$x = \text{length of side of the original square}$$

Now draw a figure using the given information, as in Figure 2.1.

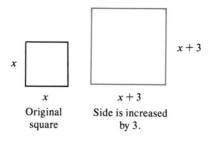

x

x
Original
square

$x + 3$

$x + 3$
Side is increased
by 3.

Figure 2.1

The length of a side of the new square is 3 cm more than the length of a side of the old square. Write a variable expression for that relationship.

$$x + 3 = \text{length of side of the new square}$$

Now write a variable expression for the perimeter of the new square. Since the perimeter of a square is 4 times the length of a side,

$$4(x + 3) = \text{perimeter of the new square.}$$

Now you can use the information given in the problem to write an equation. The perimeter of the new square is 40 cm more than twice the length of a side of the original square, so the equation is written as follows.

The new perimeter	is	40	more than	twice the side of the original square.
$4(x + 3)$	$=$	40	$+$	$2x$

Solve the equation.

$$4(x + 3) = 40 + 2x$$
$$4x + 12 = 40 + 2x$$
$$2x = 28$$
$$x = 14$$

This solution should be checked using the words of the original problem. The length of a side of the new square would be $14 + 3 = 17$ cm; its perimeter would be $4(17) = 68$ cm. Twice the length of a side of the original square is $2(14) = 28$ cm. Since $40 + 28 = 68$, the solution satisfies the problem.   ✤

If a problem asks for one unknown quantity, then only one condition is needed—a condition that tells you how to write the equation. If there is more than one unknown, then there must be at least as many conditions given in the problem as there are unknown quantities. In such cases, some conditions tell how the variables are related, while others tell how to write the necessary equations. In Example 2, one condition gave the length of the side of the new square, another gave the perimeter of the new square, and a third gave the wording that led to an equation.

Example 3   Chuck travels 80 km in the same time that Mary travels 180 km. Mary travels 50 km/hr faster than Chuck. Find the rate at which each person travels.

Let x represent Chuck's rate. Since Mary travels 50 km/hr faster,

$$x + 50 = \text{rate for Mary.}$$

Constant velocity problems of this kind are solved with the formula $d = rt$, where d is distance traveled in time t at a constant rate r. A chart is helpful for organizing the information given in this problem and the necessary variable expressions.

For Chuck, $d = 80$ and $r = x$. The formula $d = rt$ gives $t = d/r$, so that for Chuck, $t = 80/x$. For Mary, $d = 180$, $r = x + 50$, and $t = 180/(x + 50)$. This information is shown in the following chart.

	d	r	t
Chuck	80	x	$\dfrac{80}{x}$
Mary	180	$x + 50$	$\dfrac{180}{x + 50}$

Since they both travel for the same time,

$$\frac{80}{x} = \frac{180}{x + 50}.$$

Multiply both sides of this equation by $x(x + 50)$, getting

$$x(x + 50) \cdot \frac{80}{x} = x(x + 50) \cdot \frac{180}{x + 50}$$

$$80(x + 50) = 180x$$

$$80x + 4000 = 180x$$

$$4000 = 100x$$

$$40 = x.$$

Chuck's rate, which is represented by x, is 40 km/hr. Mary's rate is $x + 50$, or $40 + 50 = 90$ km/hr.   ✦

Example 4   Elizabeth Thornton is a chemist. She needs a 20% solution of potassium permanganate. She has a 15% solution on hand, as well as a 30% solution. How many liters of the 15% solution should she add to 3 liters of the 30% solution to get her 20% solution?

Figure 2.2

Let x be the number of liters of the 15% solution to be added. See Figure 2.2. Arrange the information of the problem in a table.

Strength	Liters of solution	Liters of pure potassium permanganate
15%	x	$.15x$
30%	3	$.30(3)$
20%	$3 + x$	$.20(3 + x)$

Since the number of liters of pure potassium permanganate in the 15% solution plus the number of liters in the 30% solution must equal the number of liters in the final 20% solution,

Liters in 15%		Liters in 30%		Liters in 20%
$.15x$	$+$	$.30(3)$	$=$	$.20(3 + x).$

Solve this equation as follows.

$$.15x + .90 = .60 + .20x$$
$$.30 = .05x$$
$$6 = x$$

By this result, 6 liters of the 15% solution should be mixed with 3 liters of the 30% solution, giving $6 + 3 = 9$ liters of 20% solution.   ✛

Example 5   One computer can do a job twice as fast as another. Working together, both computers can do the job in 2 hr. How long would it take the faster computer, working alone, to do the job?

Let x be the number of hours it would take the faster computer, working alone, to do the job. Then the slower computer will do the job alone in $2x$ hr.

In 1 hr the faster computer will do $1/x$ of the job, while in 1 hr the slower computer will do $1/(2x)$ of the job. Since the whole job can be done by the two machines working together in 2 hr, then half of the job will be done in 1 hr. This information leads to the equation below.

Portion done by faster computer in 1 hr		Portion done by slower computer in 1 hr		Portion done by the two computers in 1 hr
$\dfrac{1}{x}$	$+$	$\dfrac{1}{2x}$	$=$	$\dfrac{1}{2}$

Multiply both sides of the equation by $2x$.

$$2x\left(\frac{1}{x}\right) + 2x\left(\frac{1}{2x}\right) = 2x\left(\frac{1}{2}\right)$$
$$2 + 1 = x$$
$$3 = x$$

The faster computer could do the entire job, working alone, in 3 hr. The slower computer would need $2(3) = 6$ hr. ✚

The next example shows a business application of equations.

Example 6   A financial manager has \$14,000 to invest for her company. She plans to invest part of the money in tax-free bonds at 6% interest and the remainder at 9%. She wants to earn \$1005 per year in interest from the investments. Find the amount she should invest at each rate.

Let x represent the amount to be invested at 6%, so that $14,000 - x$ is the amount to be invested at 9%. Interest is given by the product of principal, rate, and time in years $(i = prt)$. Summarize this information in a chart.

Amount invested	Interest rate	Interest earned in 1 yr
x	$6\% = .06$	$.06x$
$14,000 - x$	$9\% = .09$	$.09(14,000 - x)$

Since the total interest is to be \$1005,

$$.06x + .09(14,000 - x) = 1005.$$

Solve this equation.

$$.06x + 1260 - .09x = 1005$$
$$-.03x = -255$$
$$x = 8500$$

The manager should invest \$8500 at 6%, and \$14,000 − \$8500 = \$5500 at 9%. ✤

2.2 Exercises

Solve each equation for the indicated variable. Assume that all denominators are nonzero. See Example 1.

1. $PV = k$ for V

2. $F = whA$ for h

3. $V = lwh$ for l

4. $i = prt$ for p

5. $V = V_0 + gt$ for g

6. $s = s_0 + gt^2 + k$ for g

7. $s = \dfrac{1}{2}gt^2$ for g

8. $A = \dfrac{1}{2}(B + b)h$ for h

9. $A = \dfrac{1}{2}(B + b)h$ for B

10. $C = \dfrac{5}{9}(F - 32)$ for F

11. $S = 2\pi(r_1 + r_2)h$ for r_1

12. $A = P\left(1 + \dfrac{i}{m}\right)$ for m

13. $g = \dfrac{4\pi^2 l}{t^2}$ for l

14. $P = \dfrac{E^2 R}{r + R}$ for R

15. $S = 2\pi rh + 2\pi r^2$ for h

16. $u = f \cdot \dfrac{k(k + 1)}{n(n + 1)}$ for f

17. $A = \dfrac{24f}{B(p + 1)}$ for f

18. $A = \dfrac{24f}{B(p + 1)}$ for B

19. $\dfrac{1}{R} = \dfrac{1}{r_1} + \dfrac{1}{r_2}$ for R

20. $m = \dfrac{Ft}{v_1 - v_2}$ for v_2

Solve each of the following problems that have solutions. See Examples 2–6.

21. A triangle has a perimeter of 30 cm. Two sides of the triangle are each twice as long as the shortest side. Find the length of the shortest side.

22. The length of a rectangle is 3 cm less than twice the width. The perimeter is 54 cm. Find the width.

23. A box contains 11 nickels. How many quarters must be added so that the box will contain \$2.30?

24. A cash drawer contains 15 twenty-dollar bills. How many five-dollar bills must be added to bring the value of the money up to \$390?

25. A pharmacist wishes to strengthen a mixture that is 10% alcohol to one that is 30% alcohol. How much pure alcohol should be added to 7 liters of the 10% mixture?

26. A student needs 10% hydrochloric acid for a chemistry experiment. How much 5% acid should be mixed with 60 ml of 20% acid to get a 10% solution?

27. The Old Time Goodies Store sells mixed nuts. Cashews sell for \$4 per quarter kg, hazelnuts for \$3 per quarter kg, and peanuts for \$1 per quarter kg. How many kg of peanuts should be added to 10 kg of cashews and 8 kg of hazelnuts to make a mixture that will sell for \$2.50 per quarter kg?

28. The Old Time Goodies Store also sells candy. The manager wants to prepare 200 kg of a mixture for a special Halloween promotion to sell at \$4.84/kg. How much \$4/kg candy should be combined with \$5.20/kg candy for the required mix?

Exercises 29 and 30 depend on the idea of the octane rating of gasoline, a measure of its antiknock qualities. In one measure of octane, a standard fuel is made with only two ingredients: heptane and isooctane. For this fuel, the octane rating is the percent of *isooctane. An actual gasoline blend is then compared to a standard fuel. For example, a gasoline with an octane rating of 98 has the same antiknock properties as a standard fuel that is 98% isooctane.*

29. How many liters of 94-octane gasoline should be mixed with 200 liters of 99-octane gasoline to get a mixture that is 97-octane?

30. A service station has 92-octane and 98-octane gasoline. How many liters of each should be mixed to provide 12 liters of 96-octane gasoline needed for chemical research?

31. On a vacation trip, Jose averaged 50 mph traveling from Amarillo to Flagstaff. Returning by a different route that covered the same number of miles, he averaged 55 mph. What is the distance between the two cities if his total traveling time was 32 hr?

32. Cindy left by plane to visit her mother in Hartford, 420 km away. Fifteen minutes later, her mother left to meet her at the airport. She drove the 20 km to the airport at 40 km/hr, arriving just as the plane taxied in. What was the speed of the plane?

33. Russ and Janet are running in the Apple Hill Fun Run. Russ runs at 7 mph, Janet at 5 mph. If they start at the same time, how long will it be before they are 1/2 mi apart?

34. If the run in Exercise 33 has a staggered start, and Janet starts first, with Russ starting 10 min later, how long will it be before he catches up with her?

35. Joann took 20 min to drive her boat upstream to water-ski at her favorite spot. Coming back later in the day, at the same boat speed, took her 15 min. If the current in that part of the river is 5 km/hr, what was her boat speed?

36. Joe traveled against the wind in a small plane for 3 hr. The return trip with the wind took 2.8 hr. Find the speed of the wind if the speed of the plane in still air is 180 mph.

37. Ron gets to work in 20 min when he drives his car. Riding his bike (by the same route) takes him 45 min. His average driving speed is 4.5 mph greater than his average speed on his bike. How far does he travel to work?

38. In the morning, Marge drove to a business appointment at 50 mph. Her average speed on the return trip in the afternoon was 40 mph. The return trip took 1/4 hr longer because of heavy traffic. How far did she travel to the appointment?

39. Mark can clean the house in 9 hr, while Wendy needs 6 hr. How long will it take them to clean the house if they work together?

40. Helen can paint a room in 5 hr. Jay can paint the same room in 4 hr. How long will it take them to paint the room together?

41. Two chemical plants are polluting a river. If plant A produces a predetermined maximum amount of pollutant twice as fast as plant B, and together they produce the maximum pollutant in 26 hr, how long will it take plant B alone?

42. A sewage treatment plant has two inlet pipes to its settling pond. One can fill the pond in 10 hr, the other in 12 hr. If the first pipe is open for 5 hr and then the second pipe is opened, how long will it take to fill the pond?

43. An inlet pipe can fill Dominic's pool in 5 hr, while an outlet pipe can empty it in 8 hr. In his haste to watch television, Dominic left both pipes open. How long did it take to fill the pool?

44. Suppose Dominic discovered his error (see Exercise 43) after an hour-long program. If he then closed the outlet pipe, how much more time would be needed to fill the pool?

45. A clock-radio is on sale for $49. If the sale price is 15% less than the regular price, what was the regular price?

46. A shopkeeper prices his items 20% over their wholesale price. If a lamp sells for $74, what is its wholesale price?

47. Jim Marshall invests $20,000 received from an insurance settlement in two ways, some at 10% and some at 6%. Altogether, he makes $1560 interest per year. How much is invested at each rate?

48. Candy Bowen received $52,000 profit from the sale of some land. She invested part at 7.5% interest and the rest at 8.5% interest. She earned a total of $4210 interest per year. How much did she invest at each rate?

49. Bill Cornett won $100,000 in a state lottery. He first paid income tax of 40% on the winnings. Of the rest, he invested some at 8.5% and some at 7%, making $5010 interest per year. How much is invested at each rate?

50. Marjorie Williams earned $48,000 from royalties on her cookbook. She paid a 40% income tax on these royalties. The balance was invested in two ways, some of it at 6.5% and some at 9%. The investments produced $2267 interest per year. Find the amount invested at each rate.

51. Karen Harr bought two plots of land for a total of $120,000. When she sold the first plot, she made a profit of 15%. When she sold the second, she lost 10%. Her total profit was $5500. How much did she pay for each piece of land?

52. Suppose $10,000 is invested at 6%. How much additional money must be invested at 8% to produce a yield of 7.2% on the entire amount invested?

53. Kathryn Johnson earns take-home pay of $298 a week. If her deductions for taxes, retirement, union dues, and medical plan amount to 26% of her wages, what is her weekly pay before deductions?

54. Barbara Burnett gives 10% of her net income to her church. This amounts to $167.20 per month. In addition, her paycheck deductions are 24% of her gross monthly income. What is her gross monthly income?

55. A bank pays 5% interest on passbook accounts and 7% interest on long-term deposits. Suppose a depositor divides $20,000 among the two types of deposits. Find the amount deposited at each rate if the total annual income from interest is $1340.

56. Miriam Cross wishes to sell a piece of property for $125,000. She wishes the money to be paid off in two ways—a short-term note at 12% and a long-term note at 10%. Find the amount of each note if the total annual interest income is $2000.

57. In planning his retirement, John Young deposits some money at 9% with twice as much deposited at 10%. Find the amount deposited at each rate if the total annual interest income is $5800.

58. A church building fund has invested $75,000 in two ways: part of the money at 7% and four times as much at 11%. Find the amount invested at each rate if the total annual income from interest is $7650.

2.3 Complex Numbers

In the next section we will solve *quadratic equations*, which have a term raised to the second power (for example, $x^2 - 4x + 3 = 0$). Solutions of quadratic equations may not be real numbers. For example, there are no real-number solutions to the quadratic equation

$$x^2 + 1 = 0.$$

A set of numbers is needed that permits the solution of *all* quadratic equations. To get such a set of numbers, the number i is defined as follows.

Definition of i

$$i = \sqrt{-1} \quad \text{or} \quad i^2 = -1.$$

Numbers of the form $a + bi$, where a and b are real numbers, are called **complex numbers.** Each real number is a complex number, since a real number a may be thought of as the complex number $a + 0i$. A complex number of the form $a + bi$, where b is nonzero, is called an **imaginary number.** Both the set of real numbers and the set of imaginary numbers are subsets of the set of complex numbers. (See Figure 2.3, which is an extension of Figure 1.6 in Section 1.1.) A complex number that is written in the form $a + bi$ or $a + ib$ is in **standard form.** (The form $a + ib$ is used to simplify certain symbols such as $i\sqrt{5}$, since $\sqrt{5}i$ could be too easily mistaken for $\sqrt{5i}$.)

Example 1 The following statements identify different kinds of complex numbers.

(a) -8, $\sqrt{7}$, and π are real numbers and complex numbers.

(b) $3i$, $-11i$, $i\sqrt{14}$, and $5 + i$ are imaginary numbers and complex numbers. ✚

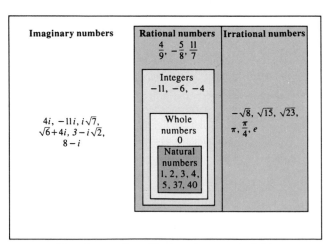

Complex numbers (Real numbers are shaded.)

Figure 2.3

Example 2   The list below shows several numbers, along with the standard form of each number.

Number	Standard form
$6i$	$0 + 6i$
-9	$-9 + 0i$
0	$0 + 0i$
$-i + 2$	$2 - i$
$8 + i\sqrt{3}$	$8 + i\sqrt{3}$ ✚

Many of the solutions to quadratic equations in the next section will involve expressions such as $\sqrt{-a}$, for a positive real number a, defined as follows.

Definition of
$\sqrt{-a}$

If $a \geq 0$, then
$$\sqrt{-a} = i\sqrt{a}.$$

Example 3   **(a)** $\sqrt{-16} = i\sqrt{16} = 4i$

(b) $\sqrt{-70} = i\sqrt{70}$ ✚

Products or quotients with negative radicands are simplified by first rewriting $\sqrt{-a}$ as $i\sqrt{a}$ for nonnegative numbers a. Then the properties of real numbers can be applied, together with the fact that $i^2 = -1$. When working with negative radicands, be sure to use the definition $\sqrt{-a} = i\sqrt{a}$ *before* using any of the other rules for radicals. In particular, the rule $\sqrt{c} \cdot \sqrt{d} = \sqrt{cd}$ is valid only when c and d are *not* both negative. For example,

$$\sqrt{(-4)(-9)} = \sqrt{36} = 6,$$

while

$$\sqrt{-4} \cdot \sqrt{-9} = 2i(3i) = 6i^2 = -6,$$

so

$$\sqrt{(-4)(-9)} \neq \sqrt{-4} \cdot \sqrt{-9}.$$

Example 4   **(a)** $\sqrt{-7} \cdot \sqrt{-7} = i\sqrt{7} \cdot i\sqrt{7}$

$$= i^2 \cdot (\sqrt{7})^2$$
$$= (-1) \cdot 7$$
$$= -7$$

(b) $\sqrt{-6} \cdot \sqrt{-10} = i\sqrt{6} \cdot i\sqrt{10}$

$$= i^2 \cdot \sqrt{60}$$
$$= -1 \cdot 2\sqrt{15}$$
$$= -2\sqrt{15}$$

(c) $\dfrac{\sqrt{-20}}{\sqrt{-2}} = \dfrac{i\sqrt{20}}{i\sqrt{2}} = \sqrt{10}$

(d) $\dfrac{\sqrt{-48}}{\sqrt{24}} = \dfrac{i\sqrt{48}}{\sqrt{24}} = i\sqrt{2}$ ✚

Operations on Complex Numbers   Complex numbers may be added, subtracted, multiplied, and divided using the properties of real numbers, as shown by the following definitions and examples.

The *sum* of two complex numbers $a + bi$ and $c + di$ is defined as follows.

Definition of Addition of Complex Numbers

$$(a + bi) + (c + di) = (a + c) + (b + d)i$$

Example 5   Find each sum.

(a) $(3 - 4i) + (-2 + 6i) = [3 + (-2)] + [-4 + 6]i$
$$= 1 + 2i$$

(b) $(-9 + 7i) + (3 - 15i) = -6 - 8i$ ✚

Since $(a + bi) + (0 + 0i) = a + bi$ for all complex numbers $a + bi$, the number $0 + 0i$ is called the **additive identity** for complex numbers. The sum of $a + bi$ and $-a - bi$ is $0 + 0i$, so the number $-a - bi$ is called the **negative** or **additive inverse** of $a + bi$.

Using this definition of additive inverse, *subtraction* of complex numbers $a + bi$ and $c + di$ is defined as

$$(a + bi) - (c + di) = (a + bi) + (-c - di) = (a - c) + (b - d)i.$$

Definition of Subtraction of Complex Numbers

$$(a + bi) - (c + di) = (a - c) + (b - d)i$$

Example 6   Subtract as indicated.

(a) $(-4 + 3i) - (6 - 7i) = (-4 - 6) + [3 - (-7)]i$
$$= -10 + 10i$$

(b) $(12 - 5i) - (8 - 3i) = 4 - 2i$ ✚

The product of two complex numbers can be found by multiplying as if the numbers were binomials and using the fact that $i^2 = -1$, as follows.

$$(a + bi)(c + di) = ac + adi + bic + bidi$$
$$= ac + adi + bci + bdi^2$$
$$= ac + (ad + bc)i + bd(-1)$$
$$(a + bi)(c + di) = (ac - bd) + (ad + bc)i$$

Based on this result, the product of the complex numbers $a + bi$ and $c + di$ is defined in the following way.

Definition of Multiplication of Complex Numbers	$(a + bi)(c + di) = (ac - bd) + (ad + bc)i$

This definition is hard to remember. To find a given product, it is easier just to multiply as with binomials.

Example 7   Find each of the following products.

(a) $(2 - 3i)(3 + 4i) = 2(3) + 2(4i) - 3i(3) - 3i(4i)$
$$= 6 + 8i - 9i - 12i^2$$
$$= 6 - i - 12(-1)$$
$$= 18 - i$$

(b) $(5 - 4i)(7 - 2i) = 5(7) + 5(-2i) - 4i(7) - 4i(-2i)$
$$= 35 - 10i - 28i + 8i^2$$
$$= 35 - 38i + 8(-1)$$
$$= 27 - 38i$$

(c) $(6 + 5i)(6 - 5i) = 6 \cdot 6 - 6 \cdot 5i + 6 \cdot 5i - (5i)^2$
$$= 36 - 25i^2$$
$$= 36 - 25(-1)$$
$$= 36 + 25$$
$$= 61 \quad \text{or} \quad 61 + 0i \text{ in standard form}$$

(d) i^{15}

Since $i^2 = -1$, the value of a power of i is found by writing the given power as a product involving i^2 or i^4. For example, $i^3 = i^2 \cdot i = (-1) \cdot i = -i$. Also, $i^4 = i^2 \cdot i^2 = (-1)(-1) = 1$. Using i^4 to rewrite i^{15} gives

$$i^{15} = i^{12} \cdot i^3 = (i^4)^3 \cdot i^3 = (1)^3(-i) = -i.$$

(e) $i^{-3} = i^{-4} \cdot i = (i^4)^{-1} \cdot i = (1)^{-1} \cdot i = i$ ✚

We can use methods similar to those of parts (d) and (e) of the last example to construct a table of powers of i.

Powers of i			
	$i^1 = i$	$i^5 = i$	$i^9 = i$
	$i^2 = -1$	$i^6 = -1$	$i^{10} = -1$
	$i^3 = -i$	$i^7 = -i$	$i^{11} = -i$
	$i^4 = 1$	$i^8 = 1$	$i^{12} = 1$, and so on.

Example 7(c) showed that $(6 + 5i)(6 - 5i) = 61$. The numbers $6 + 5i$ and $6 - 5i$ differ only in their middle signs; for this reason these numbers are called **conjugates** of each other. The product of a complex number and its conjugate is always a real number.

Property of Complex Conjugates

For real numbers a and b:

$$(a + bi)(a - bi) = a^2 + b^2$$

Example 8 The following list shows several pairs of conjugates, together with their products.

Number	Conjugate	Product
$3 - i$	$3 + i$	$(3 - i)(3 + i) = 9 + 1 = 10$
$2 + 7i$	$2 - 7i$	$(2 + 7i)(2 - 7i) = 53$
$-6i$	$6i$	$(-6i)(6i) = 36$ ✚

The conjugate of the divisor is used to find the *quotient* of two complex numbers. The quotient is found by multiplying both the numerator and the denominator by the conjugate of the denominator. The result should be written in standard form.

Example 9 **(a)** Find $\dfrac{3 + 2i}{5 - i}$.

Multiply numerator and denominator by the conjugate of $5 - i$.

$$\frac{3 + 2i}{5 - i} = \frac{(3 + 2i)(5 + i)}{(5 - i)(5 + i)}$$

$$= \frac{15 + 3i + 10i + 2i^2}{25 - i^2}$$

$$= \frac{13 + 13i}{26}$$

$$= \frac{13}{26} + \frac{13i}{26}$$

$$= \frac{1}{2} + \frac{1}{2}i$$

To check this answer, show that

$$(5 - i)\left(\frac{1}{2} + \frac{1}{2}i\right) = 3 + 2i.$$

(b) $\dfrac{3}{i} = \dfrac{3(-i)}{i(-i)}$ $-i$ is the conjugate of i

$\phantom{(b)\ \dfrac{3}{i}} = \dfrac{-3i}{-i^2}$

$\phantom{(b)\ \dfrac{3}{i}} = \dfrac{-3i}{1}$ $-i^2 = -(-1) = 1$

$\phantom{(b)\ \dfrac{3}{i}} = -3i$ $0 - 3i$ in standard form ✚

2.3 Exercises

Identify each complex number as real or imaginary. See Example 1.

1. $-9i$ **2.** 6 **3.** π **4.** $-\sqrt{7}$

5. $i\sqrt{6}$ **6.** $-3i$ **7.** $2 + 5i$ **8.** $-7 - 6i$

Write each of the following without negative radicands. See Examples 3 and 4.

9. $\sqrt{-100}$ **10.** $\sqrt{-169}$ **11.** $-\sqrt{-400}$ **12.** $-\sqrt{-225}$

13. $-\sqrt{-39}$ **14.** $-\sqrt{-95}$ **15.** $5 + \sqrt{-4}$ **16.** $-7 + \sqrt{-100}$

17. $-6 - \sqrt{-196}$ **18.** $13 + \sqrt{-16}$ **19.** $9 - \sqrt{-50}$ **20.** $-11 - \sqrt{-24}$

21. $\sqrt{-5} \cdot \sqrt{-5}$ **22.** $\sqrt{-20} \cdot \sqrt{-20}$ **23.** $\sqrt{-8} \cdot \sqrt{-2}$ **24.** $\sqrt{-27} \cdot \sqrt{-3}$

25. $\dfrac{\sqrt{-40}}{\sqrt{-10}}$ **26.** $\dfrac{\sqrt{-190}}{\sqrt{-19}}$ **27.** $\dfrac{\sqrt{-6} \cdot \sqrt{-2}}{\sqrt{3}}$ **28.** $\dfrac{\sqrt{-12} \cdot \sqrt{-6}}{\sqrt{8}}$

Add or subtract. Write each result in standard form. See Examples 5 and 6.

29. $(3 + 2i) + (4 - 3i)$ **30.** $(4 - i) + (2 + 5i)$

31. $(-2 + 3i) - (-4 + 3i)$ **32.** $(-3 + 5i) - (-4 + 3i)$

33. $(2 - 5i) - (3 + 4i) - (-2 + i)$ **34.** $(-4 - i) - (2 + 3i) + (-4 + 5i)$

35. $-i - 2 - (3 - 4i) - (5 - 2i)$ **36.** $3 - (4 - i) - 4i + (-2 + 5i)$

Multiply. Write each result in standard form. See Example 7.

37. $(2 + i)(3 - 2i)$ **38.** $(-2 + 3i)(4 - 2i)$ **39.** $(2 + 4i)(-1 + 3i)$

40. $(1 + 3i)(2 - 5i)$ **41.** $(-3 + 2i)^2$ **42.** $(2 + i)^2$

43. $(3 + i)(-3 - i)$ **44.** $(-5 - i)(5 + i)$ **45.** $(2 + 3i)(2 - 3i)$

46. $(6 - 4i)(6 + 4i)$ **47.** $(\sqrt{6} + i)(\sqrt{6} - i)$ **48.** $(\sqrt{2} - 4i)(\sqrt{2} - 4i)$

49. $i(3 - 4i)(3 + 4i)$ **50.** $i(2 + 7i)(2 - 7i)$ **51.** $3i(2 - i)^2$

52. $-5i(4 - 3i)^2$

Divide. Write each result in standard form. See Example 9.

53. $\dfrac{1 + i}{1 - i}$

54. $\dfrac{2 - i}{2 + i}$

55. $\dfrac{4 - 3i}{4 + 3i}$

56. $\dfrac{5 - 2i}{6 - i}$

57. $\dfrac{3 - 4i}{2 - 5i}$

58. $\dfrac{1 - 3i}{1 + i}$

59. $\dfrac{-3 + 4i}{2 - i}$

60. $\dfrac{5 + 6i}{5 - 6i}$

61. $\dfrac{2}{i}$

62. $\dfrac{-7}{3i}$

63. $\dfrac{1 - \sqrt{-5}}{3 + \sqrt{-4}}$

64. $\dfrac{2 + \sqrt{-3}}{1 - \sqrt{-9}}$

Perform the indicated operations. Write answers in standard form.

65. $\dfrac{2 + i}{3 - i} \cdot \dfrac{5 + 2i}{1 + i}$

66. $\dfrac{1 - i}{2 + i} \cdot \dfrac{4 + 3i}{1 + i}$

67. $\dfrac{6 + 2i}{5 - i} \cdot \dfrac{1 - 3i}{2 + 6i}$

68. $\dfrac{5 - 3i}{1 + 2i} \cdot \dfrac{2 - 4i}{1 + i}$

69. $\dfrac{5 - i}{3 + i} + \dfrac{2 + 7i}{3 + i}$

70. $\dfrac{4 - 3i}{2 + 5i} + \dfrac{8 - i}{2 + 5i}$

Find each of the following powers of i. See parts (d) and (e) of Example 7.

71. i^5

72. i^8

73. i^9

74. i^{11}

75. i^{12}

76. i^{25}

77. i^{43}

78. $1/i^9$

79. $1/i^{12}$

80. i^{-6}

81. i^{-15}

82. i^{-49}

Find all complex numbers $a + bi$ such that the square $(a + bi)^2$ is

83. real;

84. imaginary.

85. Show that $\dfrac{\sqrt{2}}{2} + \dfrac{\sqrt{2}}{2}i$ is a square root of i.

86. Show that $\dfrac{\sqrt{3}}{2} + \dfrac{1}{2}i$ is a cube root of i.

87. Prove the property of complex conjugates.

2.4 Quadratic Equations

As mentioned earlier, an equation of the form $ax + b = 0$ is a linear equation, while an equation with a term raised to the second power is a *quadratic equation.*

Definition of Quadratic Equation	An equation that can be written in the form $$ax^2 + bx + c = 0,$$ where *a, b,* and *c* are real numbers with $a \neq 0$, is a **quadratic equation.**

(Why is the restriction $a \neq 0$ necessary?) A quadratic equation written in the form $ax^2 + bx + c = 0$ is in *standard form.*

The simplest method of solving a quadratic equation, but one that is not always easily applied, is by factoring. This method depends on the following property.

Zero-Factor Property	If a and b are complex numbers, with $ab = 0$, then $a = 0$ or $b = 0$ or both.

The next example shows how the zero-factor property is used to solve a quadratic equation.

Example 1   Solve $6r^2 + 7r = 3$.

First write the equation in standard form as

$$6r^2 + 7r - 3 = 0.$$

Now factor $6r^2 + 7r - 3$ to get

$$(3r - 1)(2r + 3) = 0.$$

By the zero-factor property, the product $(3r - 1)(2r + 3)$ can equal 0 only if

$$3r - 1 = 0 \qquad \text{or} \qquad 2r + 3 = 0.$$

Solve each of these linear equations separately to find that the solutions of the original equation are $1/3$ and $-3/2$. Check these solutions by substituting in the original equation. The solution set is $\{1/3, -3/2\}$. ✦

A quadratic equation of the form $x^2 = k$ can be solved by factoring with the following sequence of equivalent equations.

$$x^2 = k$$
$$x^2 - k = 0$$
$$(x - \sqrt{k})(x + \sqrt{k}) = 0$$
$$x - \sqrt{k} = 0 \qquad \text{or} \qquad x + \sqrt{k} = 0$$
$$x = \sqrt{k} \qquad \text{or} \qquad x = -\sqrt{k}$$

This proves the following statement, which is sometimes called the **square root property.**

Square Root Property	The solution set of $x^2 = k$ is $\{\sqrt{k}, -\sqrt{k}\}$.

This solution set is often abbreviated as $\{\pm\sqrt{k}\}$. Both solutions are real if $k > 0$ and imaginary if $k < 0$. (If $k = 0$, there is only one solution.)

Example 2   Solve each equation.

(a) $z^2 = 17$

The solution set is $\{\pm\sqrt{17}\}$.

(b) $m^2 = -25$

Since $\sqrt{-25} = 5i$, the solution set of $m^2 = -25$ is $\{\pm 5i\}$.

(c) $(y - 4)^2 = 12$

Use a generalization of the square root property, working as follows.

$$(y - 4)^2 = 12$$
$$y - 4 = \pm\sqrt{12}$$
$$y = 4 \pm \sqrt{12}$$
$$y = 4 \pm 2\sqrt{3}$$

The solution set is $\{4 \pm 2\sqrt{3}\}$. ✦

Completing the Square   As suggested by Example 2(c), any quadratic equation can be solved using the square root property if it is first written in the form $(x + n)^2 = k$ for suitable numbers n and k. The next example shows how to write a quadratic equation in this form.

Example 3   Solve $x^2 - 4x = 8$

To write $x^2 - 4x = 8$ in the form $(x + n)^2 = k$, we must find a number that can be added to the left side of the equation to get a perfect square. The equation $(x + n)^2 = k$ can be written as $x^2 + 2xn + n^2 = k$. Comparing this expression with $x^2 - 4x = 8$ shows that

$$2xn = -4x$$
$$n = -2.$$

If $n = -2$, then $n^2 = 4$. Adding 4 to both sides of $x^2 - 4x = 8$ and factoring on the left gives

$$x^2 - 4x + 4 = 8 + 4$$
$$(x - 2)^2 = 12.$$

Now the square root property can be used as follows.

$$x - 2 = \pm\sqrt{12}$$
$$x = 2 \pm 2\sqrt{3}$$

The solution set is $\{2 \pm 2\sqrt{3}\}$. ✦

Example 4   Solve $9z^2 - 12z - 1 = 0$.

To get the coefficient of z^2 equal to 1 in the given equation, multiply both sides by 1/9 and then add 1/9 to both sides.

$$z^2 - \frac{4}{3}z = \frac{1}{9}$$

The term $-4z/3$ is twice the product of z and n, the number needed to make the left side a perfect square. That is, n, satisfies the equation

$$2zn = -\frac{4}{3}z$$

$$n = -\frac{2}{3}.$$

If $n = -2/3$, then $n^2 = 4/9$, which is added to both sides, getting

$$z^2 - \frac{4}{3}z + \frac{4}{9} = \frac{1}{9} + \frac{4}{9}.$$

Factoring on the left gives

$$\left(z - \frac{2}{3}\right)^2 = \frac{5}{9}.$$

Now use the square root property and the quotient property for radicals to get

$$z - \frac{2}{3} = \pm\sqrt{\frac{5}{9}}$$

$$z - \frac{2}{3} = \pm\frac{\sqrt{5}}{3}$$

$$z = \frac{2}{3} \pm \frac{\sqrt{5}}{3}.$$

These two solutions can be written as

$$\frac{2 \pm \sqrt{5}}{3},$$

with the solution set abbreviated as $\{(2 \pm \sqrt{5})/3\}$.   ❖

The process of changing $9z^2 - 12z - 1 = 0$ into the equivalent equation $(z - 2/3)^2 = 5/9$ in Example 4 is called *completing the square*.

Completing the Square	To solve $ax^2 + bx + c = 0$, $a \neq 0$, by completing the square: **1.** If $a \neq 1$, first multiply both sides by $1/a$, then rewrite the equation so that the constant is alone on one side of the equals sign. **2.** Square half the coefficient of x, and add the square to both sides. **3.** Factor, and use the square root property.

Quadratic Formula   The method of completing the square can be used to solve any quadratic equation. However, in the long run it is better to start with the general quadratic equation,

$$ax^2 + bx + c = 0, \quad a \neq 0,$$

and use the method of completing the square to solve this equation for x in terms of the constants a, b, and c. The result will be a general formula for solving any quadratic equation. First, use the fact that $a \neq 0$ and multiply both sides by $1/a$ to get

$$x^2 + \frac{b}{a}x + \frac{c}{a} = 0.$$

Add $-c/a$ to both sides.

$$x^2 + \frac{b}{a}x = -\frac{c}{a}$$

Now take half of b/a, and square the result:

$$\frac{1}{2} \cdot \frac{b}{a} = \frac{b}{2a} \quad \text{and} \quad \left(\frac{b}{2a}\right)^2 = \frac{b^2}{4a^2}.$$

Add the square to both sides, producing

$$x^2 + \frac{b}{a}x + \frac{b^2}{4a^2} = \frac{b^2}{4a^2} - \frac{c}{a}.$$

The expression on the left side of the equals sign can be written as the square of a binomial, while the expression on the right can be simplified.

$$\left(x + \frac{b}{2a}\right)^2 = \frac{b^2 - 4ac}{4a^2}$$

By the square root property, this last statement leads to

$$x + \frac{b}{2a} = \sqrt{\frac{b^2 - 4ac}{4a^2}} \quad \text{or} \quad x + \frac{b}{2a} = -\sqrt{\frac{b^2 - 4ac}{4a^2}}.$$

Since $4a^2 = (2a)^2$, or $4a^2 = (-2a)^2$,

$$x + \frac{b}{2a} = \frac{\sqrt{b^2 - 4ac}}{2a} \quad \text{or} \quad x + \frac{b}{2a} = \frac{-\sqrt{b^2 - 4ac}}{2a}.$$

Adding $-b/(2a)$ to both sides of each result gives

$$x = \frac{-b + \sqrt{b^2 - 4ac}}{2a} \quad \text{or} \quad x = \frac{-b - \sqrt{b^2 - 4ac}}{2a}.$$

A more compact form of this result, called the *quadratic formula*, is given on the following page.

Quadratic Formula	The solutions of the quadratic equation $ax^2 + bx + c = 0$, where $a \neq 0$, are $$\frac{-b \pm \sqrt{b^2 - 4ac}}{2a}.$$

Example 5 Solve $x^2 - 4x + 1 = 0$.

Here $a = 1$, $b = -4$, and $c = 1$. Substitute these values into the quadratic formula to get

$$x = \frac{-b \pm \sqrt{b^2 - 4ac}}{2a}$$

$$= \frac{-(-4) \pm \sqrt{(-4)^2 - 4(1)(1)}}{2(1)}$$

$$= \frac{4 \pm \sqrt{16 - 4}}{2}$$

$$= \frac{4 \pm 2\sqrt{3}}{2}$$

$$= \frac{2(2 \pm \sqrt{3})}{2}$$

$$= 2 \pm \sqrt{3}.$$

The solution set is $\{2 + \sqrt{3}, 2 - \sqrt{3}\}$, abbreviated as $\{2 \pm \sqrt{3}\}$. ❖

Example 6 Solve $2y^2 = y - 4$.

To find the values of a, b, and c, first rewrite the equation in standard form as $2y^2 - y + 4 = 0$. Then $a = 2$, $b = -1$, and $c = 4$. By the quadratic formula,

$$y = \frac{-(-1) \pm \sqrt{(-1)^2 - 4(2)(4)}}{2(2)}$$

$$= \frac{1 \pm \sqrt{1 - 32}}{4}$$

$$= \frac{1 \pm \sqrt{-31}}{4}$$

$$= \frac{1 \pm i\sqrt{31}}{4}.$$

The solution set is $\{(1 \pm i\sqrt{31})/4\}$. ❖

Example 7   Solve $\sqrt{2}m^2 + 5m - 3\sqrt{2} = 0$.

Here $a = \sqrt{2}$, $b = 5$, and $c = -3\sqrt{2}$, with

$$m = \frac{-5 \pm \sqrt{25 - 4(\sqrt{2})(-3\sqrt{2})}}{2\sqrt{2}}$$

$$= \frac{-5 \pm \sqrt{25 + 24}}{2\sqrt{2}}$$

$$= \frac{-5 \pm 7}{2\sqrt{2}}$$

Rationalize each denominator.

$$m = \frac{-12}{2\sqrt{2}} = \frac{-6}{\sqrt{2}} \cdot \frac{\sqrt{2}}{\sqrt{2}} = \frac{-6\sqrt{2}}{2} = -3\sqrt{2}$$

or $$m = \frac{2}{2\sqrt{2}} = \frac{1}{\sqrt{2}} \cdot \frac{\sqrt{2}}{\sqrt{2}} = \frac{\sqrt{2}}{2}$$

The solution set is $\{-3\sqrt{2}, \sqrt{2}/2\}$.   ✤

Example 8   Solve $x^3 + 8 = 0$.

Factor on the left side, and then set each factor equal to zero.

$$x^3 + 8 = 0$$
$$(x + 2)(x^2 - 2x + 4) = 0$$
$$x + 2 = 0 \quad \text{or} \quad x^2 - 2x + 4 = 0$$

The solution of $x + 2 = 0$ is $x = -2$. Now use the quadratic formula to solve $x^2 - 2x + 4 = 0$.

$$x^2 - 2x + 4 = 0$$

$$x = \frac{2 \pm \sqrt{4 - 16}}{2}$$

$$x = \frac{2 \pm \sqrt{-12}}{2}$$

$$x = \frac{2 \pm 2i\sqrt{3}}{2}$$

$$x = 1 \pm i\sqrt{3}$$

The solution set is $\{-2, 1 \pm i\sqrt{3}\}$.   ✤

Recall from Section 1.6 that a rational expression is not defined when its denominator is 0. Restrictions on the variable are found by determining the value or values that cause the expression in the denominator to equal 0.

Example 9   For each of the following, give restrictions on the variable.

(a) $\dfrac{2x - 5}{2x^2 - 9x - 5}$

Set the denominator equal to 0 and solve.

$$2x^2 - 9x - 5 = 0$$
$$(2x + 1)(x - 5) = 0$$

$2x + 1 = 0$ or $x - 5 = 0$

$x = -\dfrac{1}{2}$ or $x = 5$

The restrictions on the variable are $x \neq -1/2$ and $x \neq 5$.

(b) $\dfrac{1}{3x^2 - x + 4}$

Solve $3x^2 - x + 4 = 0$. Since the polynomial does not factor, use the quadratic formula.

$$x = \frac{-(-1) \pm \sqrt{(-1)^2 - 4(3)(4)}}{2(3)} = \frac{1 \pm \sqrt{-47}}{6}$$

Both solutions are imaginary numbers, so there are no real numbers that make the denominator equal to zero. Thus there are no real number restrictions on x.   ✤

The Discriminant   The quantity under the radical in the quadratic formula, $b^2 - 4ac$, is called the **discriminant.** When the numbers a, b, and c are *integers* (but not necessarily otherwise), the value of the discriminant can be used to determine whether the solutions will be rational, irrational, or imaginary numbers. If the discriminant is 0, there will be only one solution. (Why?)

The discriminant of a quadratic equation gives the following information about the solutions of the equation.

Discriminant

Discriminant	*Number of solutions*	*Kind of solutions*
Positive, perfect square	Two	Rational
Positive, but not a perfect square	Two	Irrational
Zero	One	Rational
Negative	Two	Imaginary

Example 10   Use the discriminant to determine whether the solutions of $5x^2 + 2x - 4 = 0$ are rational, irrational, or imaginary.

The discriminant is

$$b^2 - 4ac = (2)^2 - (4)(5)(-4) = 84.$$

Because the discriminant is positive, there are two real-number solutions. Since 84 is not a perfect square, the solutions will be irrational numbers. ✚

Example 11   Find a value of k so that the equation

$$16p^2 + kp + 25 = 0$$

has exactly one solution.

A quadratic equation with real coefficients will have exactly one solution if the discriminant is zero. Here, $a = 16$, $b = k$, and $c = 25$, giving the discriminant

$$b^2 - 4ac = k^2 - 4(16)(25) = k^2 - 1600.$$

The discriminant is 0 if

$$k^2 - 1600 = 0$$
$$k^2 = 1600,$$

from which $k = \pm 40$. ✚

2.4   Exercises

Solve the following equations by factoring or by using the square root property. See Examples 1 and 2.

1. $p^2 = 16$ **2.** $k^2 = 25$ **3.** $x^2 = 27$ **4.** $(p + 2)^2 = 7$

5. $(3k - 1)^2 = 12$ **6.** $(4t + 1)^2 = 20$ **7.** $p^2 - 5p + 6 = 0$ **8.** $q^2 + 2q - 8 = 0$

9. $6z^2 - 5z - 50 = 0$ **10.** $21p^2 = 10 - 29p$ **11.** $8k^2 + 14k + 3 = 0$ **12.** $18r^2 - 9r - 2 = 0$

Solve the following equations by completing the square. See Examples 3 and 4.

13. $p^2 - 8p + 15 = 0$ **14.** $m^2 + 5m = 6$ **15.** $x^2 - 2x - 4 = 0$

16. $r^2 + 8r + 13 = 0$ **17.** $2p^2 + 2p + 1 = 0$ **18.** $9z^2 - 12z + 8 = 0$

Solve the following equations by using the quadratic formula. See Examples 5 and 6.

19. $m^2 - m - 1 = 0$ **20.** $y^2 - 3y - 2 = 0$ **21.** $x^2 - 6x + 7 = 0$

22. $11p^2 - 7p + 1 = 0$ **23.** $4z^2 - 12z + 11 = 0$ **24.** $x^2 = 2x - 5$

25. $3k^2 + 2 = k$ **26.** $4 - \dfrac{11}{x} - \dfrac{3}{x^2} = 0$ **27.** $3 - \dfrac{4}{p} = \dfrac{2}{p^2}$

Use a calculator to give the solutions of the following equations to the nearest thousandth.

28. $5n^2 - 4 = 3n$ **29.** $9p^2 - 30p = 25$ **30.** $2m^2 = m + 2$

Use the quadratic formula to solve the following equations. See Example 7.

31. $m^2 - \sqrt{2}m - 1 = 0$ **32.** $z^2 - \sqrt{3}z - 2 = 0$

33. $\sqrt{2}p^2 - 3p + \sqrt{2} = 0$ **34.** $-\sqrt{6}k^2 - 2k + \sqrt{6} = 0$

35. $x^2 + \sqrt{5}x + 1 = 0$ **36.** $3\sqrt{5}m^2 - 2m = \sqrt{5}$

37. $(1 + \sqrt{2})x^2 - x + (1 - \sqrt{2}) = 2$ **38.** $(2 + \sqrt{5})r^2 - 3r + (2 - \sqrt{5}) = 0$

Solve each of the following equations by factoring first. See Example 8.

39. $x^3 - 1 = 0$ **40.** $x^3 + 64 = 0$ **41.** $x^3 + 27 = 0$

42. $x^3 + 1 = 0$ **43.** $x^3 - 64 = 0$ **44.** $x^3 - 27 = 0$

Solve the following equations by any method.

45. $8p^3 + 125 = 0$ **46.** $2 - \dfrac{5}{k} + \dfrac{2}{k^2} = 0$ **47.** $(m - 3)^2 = 5$ **48.** $t^2 - t = 3$

49. $\sqrt{3}p^2 - 2p + \sqrt{3} = 0$ **50.** $64r^3 - 343 = 0$ **51.** $2s^2 + 2s = 3$

52. $2x^2 - \sqrt{17}x + 1 = 0$ **53.** $2 - \dfrac{5}{r} + \dfrac{3}{r^2} = 0$ **54.** $y^2 = 24$

Give the restrictions on the variable for each of the following rational expressions. See Example 9.

55. $\dfrac{3 - 2x}{3x^2 - 19x - 14}$ **56.** $\dfrac{7 - 3x}{2x^2 + 7x - 15}$ **57.** $\dfrac{-3}{16y^2 + 8y + 1}$

58. $\dfrac{18}{25y^2 - 30y + 9}$ **59.** $\dfrac{4 + k^2}{4k^2 + k - 1}$ **60.** $\dfrac{2p^2 + 1}{-2p^2 + p + 4}$

61. $\dfrac{8 + 7x}{x^2 + x + 1}$ **62.** $\dfrac{-9x + 4}{7x^2 + 2x + 2}$

Identify the values of a, b, and c for each of the following: then evaluate the discriminant $b^2 - 4ac$ and use it to predict the type of solutions. Do not solve the equations. See Example 10.

63. $x^2 + 8x + 16 = 0$ **64.** $x^2 - 5x + 4 = 0$ **65.** $3m^2 - 5m + 2 = 0$

66. $8y^2 = 14y - 3$ **67.** $4p^2 = 6p + 3$ **68.** $2r^2 - 4r + 1 = 0$

69. $9k^2 + 11k + 4 = 0$ **70.** $3z^2 = 4z - 5$

Find all values of k for which each of the following equations has exactly one solution. See Example 11.

71. $9x^2 + kx + 4 = 0$ **72.** $25m^2 - 10m + k = 0$ **73.** $y^2 + 11y + k = 0$

74. $z^2 - 5z + k = 0$ **75.** $kr^2 + (2k + 6)r + 16 = 0$ **76.** $ky^2 + 2(k + 4)y + 25 = 0$

Use the methods shown earlier, along with the quadratic formula as necessary, to solve each equation for the indicated variable. Assume that no denominators are zero.

77. $s = \dfrac{1}{2}gt^2$ for t **78.** $A = \pi r^2$ for r **79.** $L = \dfrac{d^4 k}{h^2}$ for h

80. $F = \dfrac{kMv^4}{r}$ for v **81.** $s = s_0 + gt^2 + k$ for t **82.** $g = \dfrac{4\pi^2}{t^2}$ for t

83. $P = \dfrac{E^2 R}{(r + R)^2}$ for R **84.** $S = 2\pi rh + 2\pi r^2$ for r

For each of the following equations, (a) solve for x in terms of y, and (b) solve for y in terms of x.

85. $4x^2 - 2xy + 3y^2 = 2$ **86.** $3y^2 + 4xy - 9x^2 = -1$

Let r_1 and r_2 be the solutions of the quadratic equation $ax^2 + bx + c = 0$. Show that the following statements are correct.

87. $r_1 + r_2 = -b/a$ **88.** $r_1 \cdot r_2 = c/a$

2.5 Applications of Quadratic Equations

Many applications of mathematics lead to quadratic equations. Several typical examples are given in this section.

Example 1 A landscape contractor wants to make an exposed gravel border of uniform width around a rectangular pool in a garden. The pool is 10 ft long and 6 ft wide. There is enough material to cover 36 sq ft. How wide should the border be?

A sketch of the pool with border is shown in Figure 2.4. Let x represent the width of the border. Then the width of the larger rectangle is $6 + 2x$ and its length is $10 + 2x$. The area of the larger rectangle is $(6 + 2x)(10 + 2x)$. The area of the pool is $6 \cdot 10 = 60$. The area of the border is found by subtracting the area of the pool from the area of the larger rectangle. This difference should be 36 sq ft.

$$(6 + 2x)(10 + 2x) - 60 = 36$$

Now solve this equation.

$$60 + 32x + 4x^2 - 60 = 36$$
$$4x^2 + 32x - 36 = 0$$
$$x^2 + 8x - 9 = 0$$
$$(x + 9)(x - 1) = 0$$

The solutions are -9 and 1. The width of the border cannot be negative, so the border should be 1 ft wide. ✤

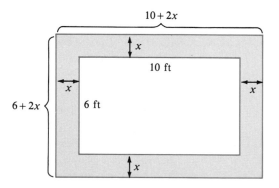

Figure 2.4

In work with applications, it is important to consider whether or not the solutions to the equation make sense in the application. Usually negative solutions are not acceptable.

Example 2   Pat and Mike clean the offices in a downtown office building each night. Working alone, Pat takes 1 hr less time than Mike to complete the job. Working together, they can finish the job in 6 hr. One night Pat calls in sick. How long should it take Mike to do the job alone?

Let

$$x = \text{the time for Mike to do the job,}$$

and $\quad x - 1 = \text{the time for Pat to do the job.}$

The part of the job that Mike can do in 1 hr is $1/x$, the part of the job that Pat can do in 1 hr is $1/(x - 1)$, and the part of the job they can do in 1 hr working together is $1/6$. This gives the equation

$$\frac{1}{x} + \frac{1}{x - 1} = \frac{1}{6}.$$

Multiply by the least common denominator, $6x(x - 1)$, and solve the resulting equation.

$$6x(x - 1)\left(\frac{1}{x}\right) + 6x(x - 1)\left(\frac{1}{x - 1}\right) = 6x(x - 1)\left(\frac{1}{6}\right)$$

$$6x - 6 + 6x = x^2 - x$$

$$0 = x^2 - 13x + 6$$

$$x = \frac{13 \pm \sqrt{(-13)^2 - 4(1)(6)}}{2(1)}$$

$$x = \frac{13 \pm \sqrt{169 - 24}}{2}$$

$$x = \frac{13 \pm \sqrt{145}}{2}$$

$$x \approx \frac{13 \pm 12.04}{2}$$

To the nearest tenth, $x = 12.5$ or $x = .5$. The solution $x = .5$ does not satisfy the conditions of the problem, since then Pat takes $x - 1 = -.5$ hr to complete the work. It will take Mike 12.5 hr to do the job alone. ✤

Example 3   A river excursion boat traveled upstream from Galt to Isleton, a distance of 12 miles. On the return trip downstream, the boat traveled 3 mph faster. If the return trip took 8 min less time, how fast did the boat travel upstream?

The chart below summarizes the information in the problem.

	d	r	t
Upstream	12	x	$\dfrac{12}{x}$
Downstream	12	$x + 3$	$\dfrac{12}{x + 3}$

The entries in the column for time are found from solving the distance formula, $d = rt$, for t in each case. Since rates are given in miles per hour, convert 8 min to hours as follows, letting H represent the equivalent number of hours.

$$\frac{H \text{ hr}}{1 \text{ hr}} = \frac{8 \text{ min}}{60 \text{ min}}$$

$$H = \frac{8}{60} = \frac{2}{15}$$

Now write an equation using the fact that the time for the return trip (downstream) was 8 min or 2/15 hr less than the time upstream.

$$\frac{12}{x + 3} = \frac{12}{x} - \frac{2}{15}$$

Solve the equation, first multiplying on both sides by the common denominator, $15x(x + 3)$, to get

$$
\begin{aligned}
12(15x) &= 12\,(15)(x + 3) - 2x(x + 3) \\
180x &= 180x + 540 - 2x^2 - 6x \\
0 &= 540 - 2x^2 - 6x \\
2x^2 + 6x - 540 &= 0 \\
2(x^2 + 3x - 270) &= 0 \\
2(x + 18)(x - 15) &= 0 \\
x = -18 \quad &\text{or} \quad x = 15.
\end{aligned}
$$

Reject the negative solution. The boat traveled 15 mph upstream.   ❖

Example 4 A lot is in the shape of a right triangle. The longer leg of the triangle is 20 m longer than twice the length of the shorter leg. The hypotenuse is 10 m longer than the longer leg. Find the lengths of the three sides of the lot.

Let s = length of the shorter leg. Then $2s + 20$ represents the length of the longer leg, and $(2s + 20) + 10 = 2s + 30$ represents the length of the hypotenuse. Application of the Pythagorean theorem gives the equation

$$s^2 + (2s + 20)^2 = (2s + 30)^2$$
$$s^2 + 4s^2 + 80s + 400 = 4s^2 + 120s + 900$$
$$s^2 - 40s - 500 = 0$$
$$(s - 50)(s + 10) = 0$$
$$s = 50 \quad \text{or} \quad s = -10.$$

Since s represents a length, the value -10 is not reasonable. The shorter leg is 50 m long, the longer leg 120 m long, and the hypotenuse 130 m long. ✦

Example 5   The manager of an 80-unit apartment complex knows from experience that at a rent of \$300, all the units will be full. On the average, one additional unit will remain vacant for each \$20 increase in rent over \$300. Currently, the revenue from the complex is \$35,000. How many apartments are rented?

Let x = number of \$20 increases over \$300. Then $80 - x$ apartments are rented at $300 + 20x$ dollars each (there are x increases of \$20 each in rent), and the revenue is

$$(80 - x)(300 + 20x) = 35,000.$$

To solve the equation, first multiply on the left side.

$$24,000 + 1600x - 300x - 20x^2 = 35,000$$
$$-20x^2 + 1300x - 11,000 = 0$$
$$x^2 - 65x + 550 = 0$$
$$(x - 55)(x - 10) = 0$$
$$x = 55 \quad \text{or} \quad x = 10.$$

If $x = 55$, 25 apartments will be rented at \$1400 per month for total revenue of \$35,000. (Although this is possible, it is probably unlikely.) When $x = 10$, 70 apartments are rented at \$500 per month for revenue of \$35,000. Both answers are usable, although one may not be reasonable in the actual situation. ✦

2.5   Exercises

Solve the following problems. See Examples 1–5.

1. Find two consecutive even integers whose product is 288.

2. The sum of the squares of two consecutive integers is 481. Find the integers.

3. Two integers have a sum of 10. The sum of the squares of the integers is 148. Find the integers.

4. The difference of two integers is 12. Their product is 493. Find the integers.

5. A shopping center has a rectangular area of 40,000 sq yd enclosed on three sides for a parking lot. The length is 200 yd more than twice the width. What are the dimensions of the lot?

6. An ecology center wants to set up an experimental garden using 300 m of fencing to enclose a rectangular area of 5000 sq m. Find the dimensions of the rectangle.

7. A manufacturing firm wants to package its product in a can 3 inches high with a surface area of 8π sq in. What should the radius of the circular top and bottom of the can be? (*Hint:* The surface area of a circular can is $S = 2\pi rh + 2\pi r^2$, where r is the radius of the top (or bottom) and h is the height.)

8. In Exercise 7, what radius would produce a can with a volume of π times the radius?

9. Pat Kelley went into a frame-it-yourself shop. He wanted a frame 3 inches longer than it was wide. The frame he chose extended 1.5 inches beyond the picture on each side. Find the outside dimensions of the frame if the area of the unframed picture is 70 sq inches.

10. Kathy Monaghan wants to buy a rug for a room that is 12 ft wide and 15 ft long. She wants to leave a uniform strip of floor around the rug. She can afford to buy 108 sq ft of carpeting. What dimensions should the rug have?

11. A rectangular page in a book is to have an 18 cm by 23 cm illustration in the center with equal margins on all four sides. How wide should the margins be if the page has an area of 594 sq cm?

12. A landscape architect has included a rectangular flower bed measuring 9 ft by 5 ft in her plans for a new building. She wants to use two colors of flowers in the bed, one in the center and the other for a border of the same width on all four sides. If she has enough plants to cover 24 sq ft for the border, how wide can the border be?

13. Bob Jensen can clean a garage in 9 hr less time than his brother Paul. Working together, they can do the job in 20 hr. How long would it take each one to do the job alone?

14. An experienced roofer can do a complete roof in a housing development in 2 hr less time than an inexperienced roofer. If the two work together on a roof, they complete the job in 180/19 hr. How long would it take the experienced roofer to do a roof by himself?

15. Two typists are working on a special project. The experienced typist could complete the project in 2 hr less time than the new typist. Together they complete the project in 2.4 hr. How long would it have taken the experienced typist working alone?

16. It takes two copy machines 5/6 hr to make the copies for a company newsletter. One copy machine would take 1 hr longer than the other to do the job alone. How long would it take the faster machine to complete the job alone?

17. Paula drives 10 mph faster than Steve. Both start at the same time for Atlanta from Chattanooga, a distance of about 100 mi. It takes Steve 1/3 hr longer than Paula to make the trip. What is Steve's average speed?

18. Marjorie Seachrist walks 1 mph faster than her daughter Sibyl. In a walk for charity, both walked the full distance of 24 mi. Sibyl took 2 hr longer than Marjorie. What was Sibyl's average speed?

19. A plane flew 1000 mi. It later took off and flew 2025 mi at an average speed of 50 mph faster than the first trip. The second trip took 2 hr more time than the first. Find the time for the first trip.

20. The Branson family traveled 100 mi to a lake for their vacation. On the return trip their average speed was 50 mph faster. The total time for the round trip was 11/3 hr. What was the family's average speed on their trip to the lake?

21. The manager of a cherry orchard is trying to decide when to schedule the annual harvest. If the cherries are picked now, the average yield per tree will be 100 lb, and the cherries can be sold for 40 cents per pound. Past experience shows that the yield per tree will increase about 5 lb per week, while the price will decrease about 2 cents per pound per week. How many weeks should the manager wait to get an average revenue of $40 per tree?

22. A local group of scouts has been collecting old aluminum cans for recycling. The group has already collected 12,000 lb of cans, for which they could currently receive $4 per hundred pounds. The group can continue to collect cans at the rate of 400 lb per day. However, a glut in the old-can market has caused the recycling company to announce that it will lower its price, starting immediately, by $.10 per hundred pounds per day. The scouts can make only one trip to the recycling center. How many days should they wait in order to get $490 for their cans?

23. A local club is arranging a charter flight to Miami. The cost of the trip is $225 each for 75 passengers, with a refund of $5 per passenger for each passenger in excess of 75. How many passengers must take the flight to produce revenue of $16,000?

24. An art film theatre has 80 customers when tickets cost $4. For each additional 25-cent charge the theatre has one less customer. How many additional 25-cent charges would produce revenue of $455?

25. A kite is flying on 50 ft of string. How high is it above the ground if its height is 10 ft more than the horizontal distance from the person flying it? Assume the string is being released at ground level.

26. A boat is being pulled into a dock with a rope attached at water level. When the boat is 12 ft from the dock, the length of the rope from the boat to the dock is 3 ft longer than twice the height of the dock above the water. Find the height of the dock.

Find the solution for each of the following problems. Round to the nearest hundredth.

27. One leg of a right triangle is 4 cm longer than the other leg. The hypotenuse is 10 cm longer than the shorter leg. Find each of the three sides of the triangle.

28. A rectangle has a diagonal of 16 cm. The width of the rectangle is 3.2 cm less than the length. Find the length and width of the rectangle.

2.6 Equations Quadratic in Form

The equation $12m^4 - 11m^2 + 2 = 0$ is not a quadratic equation because of the m^4 term. However, with the substitutions

$$x = m^2 \quad \text{and} \quad x^2 = m^4$$

the given equation becomes

$$12x^2 - 11x + 2 = 0,$$

which is a quadratic equation. This quadratic equation can be solved to find x, and then $x = m^2$ can be used to find the values of m, the solutions to the original equation.

An equation is said to be **quadratic in form** if it can be written as

$$au^2 + bu + c = 0,$$

where $a \neq 0$ and u is some algebraic expression.

Example 1   Solve $12m^4 - 11m^2 + 2 = 0$.

As mentioned above, this equation is quadratic in form. By making the substitution $x = m^2$, the equation becomes

$$12x^2 - 11x + 2 = 0,$$

which can be solved by factoring in the following way.

$$12x^2 - 11x + 2 = 0$$
$$(3x - 2)(4x - 1) = 0$$
$$x = \frac{2}{3} \quad \text{or} \quad x = \frac{1}{4}$$

The original equation contains the variable m. To find m, use the fact that $x = m^2$ and replace x with m^2, getting

$$m^2 = \frac{2}{3} \qquad \text{or} \qquad m^2 = \frac{1}{4}$$

$$m = \pm\sqrt{\frac{2}{3}} \qquad \text{or} \qquad m = \pm\sqrt{\frac{1}{4}}$$

$$m = \pm\frac{\sqrt{2}}{\sqrt{3}}$$

$$m = \frac{\pm\sqrt{6}}{3} \qquad \text{or} \qquad m = \pm\frac{1}{2}.$$

These four solutions of the given equation $12m^4 - 11m^2 + 2 = 0$ make up the solution set $\{\sqrt{6}/3, -\sqrt{6}/3, 1/2, -1/2\}$, abbreviated as $\{\pm\sqrt{6}/3, \pm 1/2\}$. Of course, it is possible to factor $12m^4 - 11m^2 + 2 = 0$ directly (without using substitution) as $(3m^2 - 2)(4m^2 - 1) = 0$. This approach is much quicker and produces the same results.   ✛

Example 2   Solve $6p^{-2} + p^{-1} = 2$.

Let $u = p^{-1}$ so that $u^2 = p^{-2}$. Then rearrange terms to get

$$6u^2 + u - 2 = 0.$$

Factor on the left, and then place each factor equal to 0, giving

$$(3u + 2)(2u - 1) = 0$$
$$3u + 2 = 0 \qquad \text{or} \qquad 2u - 1 = 0$$
$$u = -\frac{2}{3} \qquad \text{or} \qquad u = \frac{1}{2}.$$

Since $u = p^{-1}$,

$$p^{-1} = -\frac{2}{3} \quad \text{or} \quad p^{-1} = \frac{1}{2},$$

from which

$$p = -\frac{3}{2} \quad \text{or} \quad p = 2.$$

The solution set of $6p^{-2} + p^{-1} = 2$ is $\{-3/2, 2\}$.   ✛

Example 3   Solve $4p^4 - 16p^2 + 13 = 0$.
   Let $u = p^2$ to get

$$4u^2 - 16u + 13 = 0.$$

To solve this equation, use the quadratic formula with $a = 4$, $b = -16$, and $c = 13$. Substituting these values into the quadratic formula gives

$$u = \frac{-(-16) \pm \sqrt{(-16)^2 - 4(4)(13)}}{2(4)}$$

$$= \frac{16 \pm \sqrt{48}}{8}$$

$$= \frac{16 \pm 4\sqrt{3}}{8}$$

$$= \frac{4 \pm \sqrt{3}}{2}.$$

Since $p^2 = u$,

$$p^2 = \frac{4 + \sqrt{3}}{2} \quad \text{or} \quad p^2 = \frac{4 - \sqrt{3}}{2}.$$

Finally,

$$p = \pm\sqrt{\frac{4 + \sqrt{3}}{2}} \quad \text{or} \quad p = \pm\sqrt{\frac{4 - \sqrt{3}}{2}}.$$

Rationalizing the denominators gives the solution set

$$\left\{ \frac{\pm\sqrt{8 + 2\sqrt{3}}}{2}, \frac{\pm\sqrt{8 - 2\sqrt{3}}}{2} \right\}.$$ ✛

   To solve equations containing radicals or rational exponents, such as $x = \sqrt{15 - 2x}$, or $(x + 1)^{1/2} = x$, use the following result.

> If P and Q are algebraic expressions, then every solution of the equation $P = Q$ is also a solution of the equation $(P)^n = (Q)^n$, for any positive integer n.

Be very careful when using this result. It does *not* say that the equations $P = Q$ and $(P)^n = (Q)^n$ are equivalent; it says only that each solution of the original equation $P = Q$ is also a solution of the new equation $(P)^n = (Q)^n$. The new equation may have *more* solutions than the original equation, however. For example, the solution set of the equation $x = -2$ is $\{-2\}$. If we square both sides of the equation $x = -2$, we get the new equation $x^2 = 4$, which has solution set $\{-2, 2\}$. Since the solution sets are not equal, the equations are not equivalent. Because of this,

when an equation contains radicals or rational exponents, it is *essential* to check all proposed solutions in the original equation.

Example 4   Solve $x = \sqrt{15 - 2x}$.

The equation $x = \sqrt{15 - 2x}$ can be solved by squaring both sides as follows.

$$x^2 = (\sqrt{15 - 2x})^2$$
$$x^2 = 15 - 2x$$
$$x^2 + 2x - 15 = 0$$
$$(x + 5)(x - 3) = 0$$
$$x = -5 \quad \text{or} \quad x = 3$$

Now the proposed solutions must be checked in the original equation, $x = \sqrt{15 - 2x}$.

If $x = -5$, does $x = \sqrt{15 - 2x}$?

$$-5 = \sqrt{15 + 10}$$
$$-5 = 5$$

False

If $x = 3$, does $x = \sqrt{15 - 2x}$?

$$3 = \sqrt{15 - 6}$$
$$3 = 3$$

True

As this check shows, only 3 is a solution, giving the solution set $\{3\}$. ✛

Example 5   Solve $\sqrt{2x + 3} - \sqrt{x + 1} = 1$.

Separate the radicals by writing the equation as

$$\sqrt{2x + 3} = 1 + \sqrt{x + 1}.$$

Now square both sides. Be very careful when squaring on the right side of this equation. Recall that $(a + b)^2 = a^2 + 2ab + b^2$; replace a with 1 and b with $\sqrt{x + 1}$ to get the next equation, the result of squaring both sides of $\sqrt{2x + 3} = 1 + \sqrt{x + 1}$.

$$2x + 3 = 1 + 2\sqrt{x + 1} + x + 1$$
$$x + 1 = 2\sqrt{x + 1}$$

One side of the equation still contains a radical; to eliminate it, square both sides again.

$$x^2 + 2x + 1 = 4(x + 1)$$
$$x^2 - 2x - 3 = 0$$
$$(x - 3)(x + 1) = 0$$
$$x = 3 \quad \text{or} \quad x = -1$$

Check these proposed solutions in the original equation.

If $x = 3$, does $\sqrt{2x + 3} - \sqrt{x + 1} = 1$?
$$\sqrt{9} - \sqrt{4} = 1$$
$$3 - 2 = 1 \qquad \text{True}$$

If $x = -1$, does $\sqrt{2x + 3} - \sqrt{x + 1} = 1$?
$$\sqrt{1} - \sqrt{0} = 1$$
$$1 - 0 = 1 \qquad \text{True}$$

Both proposed solutions 3 and -1 are solutions of the original equation, giving $\{3, -1\}$ as the solution set. ✤

Example 6   Solve $(5x^2 - 6)^{1/4} = x$.

Since the equation involves a fourth root, begin by raising both sides to the fourth power.

$$[(5x^2 - 6)^{1/4}]^4 = x^4$$
$$5x^2 - 6 = x^4$$
$$x^4 - 5x^2 + 6 = 0$$

Now substitute y for x^2.

$$y^2 - 5y + 6 = 0$$
$$(y - 3)(y - 2) = 0$$
$$y = 3 \quad \text{or} \quad y = 2$$

Since $y = x^2$,

$$x^2 = 3 \qquad \text{or} \qquad x^2 = 2$$
$$x = \pm\sqrt{3} \qquad \text{or} \qquad x = \pm\sqrt{2}.$$

Checking the four proposed solutions, $\sqrt{3}$, $-\sqrt{3}$, $\sqrt{2}$, and $-\sqrt{2}$, in the original equation shows that only $\sqrt{3}$ and $\sqrt{2}$ are solutions, so the solution set is $\{\sqrt{3}, \sqrt{2}\}$. ✤

2.6 Exercises

Solve the following equations by first rewriting in quadratic form. See Examples 1–3.

1. $m^4 + 2m^2 - 15 = 0$

2. $3k^4 + 10k^2 - 25 = 0$

3. $2r^4 - 7r^2 + 5 = 0$

4. $4x^4 - 8x^2 + 3 = 0$

5. $(g - 2)^2 - 6(g - 2) + 8 = 0$

6. $(p + 2)^2 - 2(p + 2) - 15 = 0$

7. $-(r + 1)^2 - 3(r + 1) + 3 = 0$

8. $-2(z - 4)^2 + 2(z - 4) + 3 = 0$

9. $6(k + 2)^4 - 11(k + 2)^2 + 4 = 0$

10. $8(m - 4)^4 - 10(m - 4)^2 + 3 = 0$

11. $7p^{-2} + 19p^{-1} = 6$

12. $5k^{-2} - 43k^{-1} = 18$

13. $(r - 1)^{2/3} + (r - 1)^{1/3} = 12$

14. $(y + 3)^{2/3} - 2(y + 3)^{1/3} - 3 = 0$

15. $5r = r^{-1}$

16. $a^{-1} = a^{-2}$

17. $3 + \dfrac{5}{p^2 + 1} = \dfrac{2}{(p^2 + 1)^2}$

18. $6 = \dfrac{7}{2y - 3} + \dfrac{3}{(2y - 3)^2}$

Solve the following equations. See Examples 4–6.

19. $\sqrt{3z + 7} = 3z + 5$

20. $\sqrt{4r + 13} = 2r - 1$

21. $\sqrt{4k + 5} - 2 = 2k - 7$

22. $\sqrt{6m + 7} - 1 = m + 1$

23. $\sqrt{4x} - x + 3 = 0$

24. $\sqrt{2t} - t + 4 = 0$

25. $\sqrt{y} = \sqrt{y - 5} + 1$

26. $\sqrt{2m} = \sqrt{m + 7} - 1$

27. $\sqrt{m + 7} + 3 = \sqrt{m - 4}$

28. $\sqrt{r + 5} - 2 = \sqrt{r - 1}$

29. $\sqrt{y + 2} = \sqrt{2y + 5} - 1$

30. $\sqrt{4x + 1} = \sqrt{x - 1} + 2$

31. $\sqrt{2z} = \sqrt{3z + 12} - 2$

32. $\sqrt{5k + 1} - \sqrt{3k} = 1$

33. $\sqrt{r + 2} = 1 - \sqrt{3r + 7}$

34. $\sqrt{2p - 5} - 2 = \sqrt{p - 2}$

35. $\sqrt{2\sqrt{7x + 2}} = \sqrt{3x + 2}$

36. $\sqrt{3\sqrt{2m + 3}} = \sqrt{5m - 6}$

37. $3 - \sqrt{x} = \sqrt{2\sqrt{x} - 3}$

38. $\sqrt{x} + 2 = \sqrt{4 + 7\sqrt{x}}$

39. $\sqrt[3]{4n + 3} = \sqrt[3]{2n - 1}$

40. $\sqrt[3]{2z} = \sqrt[3]{5z + 2}$

41. $\sqrt[3]{t^2 + 2t - 1} = \sqrt[3]{t^2 + 3}$

42. $\sqrt[3]{2x^2 - 5x + 4} = \sqrt[3]{2x^2}$

43. $(2r + 5)^{1/3} = (6r - 1)^{1/3}$

44. $(3m + 7)^{1/3} = (4m + 2)^{1/3}$

45. $\sqrt[4]{q - 15} = 2$

46. $\sqrt[4]{3x + 1} = 1$

47. $\sqrt[4]{y^2 + 2y} = \sqrt[4]{3}$

48. $\sqrt[4]{k^2 + 6k} = 2$

49. $(z^2 + 24z)^{1/4} = 3$

50. $(3t^2 + 52t)^{1/4} = 4$

51. $(2r - 1)^{2/3} = r^{1/3}$

52. $(z - 3)^{2/5} = (4z)^{1/5}$

53. $k^{2/3} = 2k^{1/3}$

54. $3m^{3/4} = m^{1/2}$

Find all solutions for each of the following, rounded to the nearest thousandth.

55. $6p^4 - 41p^2 + 63 = 0$

56. $20z^4 - 67z^2 + 56 = 0$

57. $3k^2 - \sqrt{4k^2 + 3} = 0$

58. $2r^2 - \sqrt{r^2 + 1} = 0$

Solve each equation for the indicated variable. Assume that all denominators are nonzero.

59. $d = k\sqrt{h}$ for h

60. $v = \dfrac{k}{\sqrt{d}}$ for d

61. $P = 2\sqrt{\dfrac{L}{g}}$ for L

62. $c = \sqrt{a^2 + b^2}$ for a

63. $x^{2/3} + y^{2/3} = a^{2/3}$ for y

64. $m^{3/4} + n^{3/4} = 1$ for m

2.7   Inequalities

An equation says that two expressions are equal, while an **inequality** says that one expression is greater than, greater than or equal to, less than, or less than or equal to, another. As with equations, a value of the variable for which the inequality is true is a solution of the inequality, and the set of all such solutions is the solution set of the inequality. Two inequalities with the same solution set are equivalent.

Inequalities are solved with the following properties of inequalities.

Properties of Inequalities

> For real numbers a, b, and c:
>
> **(a)** $a < b$ and $a + c < b + c$ are equivalent.
>   *(The same number may be added to both sides of an inequality.)*
> **(b)** If $c > 0$, then $a < b$ and $ac < bc$ are equivalent.
>   *(Both sides of an inequality may be multiplied by the same positive number.)*
> **(c)** If $c < 0$, then $a < b$ and $ac > bc$ are equivalent.
>   *(Both sides of an inequality may be multiplied by the same negative number, as long as the direction of the inequality symbol is reversed.)*
> Replacing $<$ with $>$, $\le$, or $\ge$ results in equivalent properties.

Pay careful attention to part (c): if both sides of an inequality are multiplied by a negative number, the direction of the inequality symbol must be reversed. For example, starting with the true statement $-3 < 5$ and multiplying both sides by the positive number 2 gives

$$-3 \cdot 2 < 5 \cdot 2$$
$$-6 < 10,$$

still a true statement. On the other hand, starting with $-3 < 5$ and multiplying both sides of the *negative* number -2 gives a true result only if the direction of the inequality symbol is reversed.

$$-3(-2) > 5(-2)$$
$$6 > -10$$

Example 1   Solve the inequality $-3x + 5 > -7$.

Use the properties of inequalities. Adding -5 on both sides gives

$$-3x + 5 + (-5) > -7 + (-5)$$
$$-3x > -12.$$

Now multiply both sides by $-1/3$. Since $-1/3 < 0$, reverse the direction of the inequality symbol.

$$-\frac{1}{3}(-3x) < -\frac{1}{3}(-12)$$

$$x < 4$$

The original inequality is satisfied by any real number less than 4. The solution set can be written $\{x \mid x < 4\}$. A graph of the solution set is shown in Figure 2.5, where the parenthesis is used to show that 4 itself does not belong to the solution set.   ✛

Figure 2.5

The set $\{x \mid x < 4\}$, the solution set for the inequality in Example 1, is an example of an **interval.** A simplified notation, called **interval notation,** is used for writing intervals. With this notation, the interval in Example 1 can be written as $(-\infty, 4)$. The symbol $-\infty$ is not a real number; it is used to show that the interval includes all real numbers less than 4. The interval $(-\infty, 4)$ is an example of an **open interval,** since the endpoint, 4, is not part of the interval. Examples of other sets written in interval notation are shown below. A square bracket is used to show that a number *is* part of the graph, and a parenthesis is used to indicate that a number is *not* part of the graph. Whenever two real numbers a and b are used to write an interval, it is assumed that $a < b$.

Type of interval	*Set*	*Interval notation*	*Graph*
Open interval	$\{x \mid a < x\}$	$(a, +\infty)$	
	$\{x \mid a < x < b\}$	(a, b)	
	$\{x \mid x < b\}$	$(-\infty, b)$	
Half-open interval	$\{x \mid a \leq x\}$	$[a, +\infty)$	
	$\{x \mid a < x \leq b\}$	$(a, b]$	
	$\{x \mid a \leq x < b\}$	$[a, b)$	
	$\{x \mid x \leq b\}$	$(-\infty, b]$	
Closed interval	$\{x \mid a \leq x \leq b\}$	$[a, b]$	

Example 2 Solve $4 - 3y \leq 7 + 2y$. Write the solution in interval notation and graph the solution on a number line.

Write the following series of equivalent inequalities.

$$4 - 3y \leq 7 + 2y$$
$$-4 - 2y + 4 - 3y \leq -4 - 2y + 7 + 2y$$
$$-5y \leq 3$$
$$(-1/5)(-5y) \geq (-1/5)(3)$$
$$y \geq -3/5$$

In set-builder notation, the solution set is $\{y | y \geq -3/5\}$, while in interval notation the solution set is $[-3/5, +\infty)$. See Figure 2.6 for the graph of the solution set.   ✛

Figure 2.6

From now on, the solutions of all inequalities will be written with interval notation.

The inequality $-2 < 5 + 3m < 20$ in the next example says that $5 + 3m$ is between -2 and 20. This inequality can be solved using an extension of the properties of inequalities given above.

Example 3 Solve $-2 < 5 + 3m < 20$.

Write equivalent inequalities as follows.

$$-2 < 5 + 3m < 20$$
$$-7 < 3m < 15 \qquad \text{Add } -5$$
$$-7/3 < m < 5 \qquad \text{Multiply by } 1/3$$

The solution, graphed in Figure 2.7, is the interval $(-7/3, 5)$.   ✛

Figure 2.7

A product will break even, or begin to produce a profit, only if the revenue from selling the product at least equals the cost of producing it. If R represents revenue and C is cost, then the **break-even point** is the point where $R \geq C$.

Example 4   Find the break-even point if the revenue and cost of a certain product are given by $R = 4x$ and $C = 2x + 1000$, where x is the number of units produced.
Set $R \geq C$ and solve for x.

$$R \geq C$$
$$4x \geq 2x + 1000$$
$$2x \geq 1000$$
$$x \geq 500$$

The break-even point is at $x = 500$. This product will at least break even only if the number of units produced is in the interval $[500, +\infty)$. ✚

Quadratic Inequalities   Section 2.4 introduced quadratic equations. The solution of *quadratic inequalities* depends on the solution of quadratic equations.

Definition of Quadratic Inequality	A **quadratic inequality** is an inequality that can be written in the form $$ax^2 + bx + c < 0,$$ for real numbers $a \neq 0$, b, and c. (The symbol $<$ can be replaced with $>$, $\leq$, or $\geq$.)

Example 5   Solve the quadratic inequality $x^2 - x - 12 < 0$.
Begin by finding the values of x that satisfy $x^2 - x - 12 = 0$.

$$x^2 - x - 12 = 0$$
$$(x + 3)(x - 4) = 0$$
$$x = -3 \quad \text{or} \quad x = 4$$

These two points, -3 and 4, divide a number line into the three regions shown in Figure 2.8. If a point in region *A*, for example, makes the polynomial $x^2 - x - 12$ negative, then all points in region *A* will make that polynomial negative.

Figure 2.8

To find the regions that make $x^2 - x - 12$ negative (<0), draw a number line that shows where factors are positive or negative, as in Figure 2.9. First decide on the sign of the factor $x + 3$ in each of the three regions; then do the same thing for the factor $x - 4$. The results are shown in Figure 2.9.

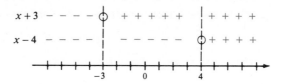

Figure 2.9

Now consider the sign of the product of the two factors in each region. As Figure 2.9 shows, both factors are negative in the interval $(-\infty, -3)$; therefore their product is positive in that interval. For the interval $(-3, 4)$, one factor is positive and the other is negative, giving a negative product. In the last region, $(4, +\infty)$, both factors are positive, so their product is positive. The polynomial $x^2 - x - 12$ is negative (what the original inequality calls for) when the product of its factors is negative, that is, for the interval $(-3, 4)$. The graph of this solution set is shown in Figure 2.10.   ✚

Figure 2.10

Figure 2.9 is an example of a **sign graph,** a graph that shows the values of the variable that make the factors of a quadratic inequality positive or negative.

Example 6   Solve the inequality $2x^2 + 5x - 12 \geq 0$.
Begin by finding the values of x that satisfy $2x^2 + 5x - 12 = 0$.

$$2x^2 + 5x - 12 = 0$$
$$(2x - 3)(x + 4) = 0$$
$$x = 3/2 \quad \text{or} \quad x = -4$$

These two points divide the number line into the three regions shown in the sign graph in Figure 2.11. Since both factors are negative in the first interval, their product, $2x^2 + 5x - 12$, is positive there. In the second interval, the factors have opposite signs, and therefore their product is negative. Both factors are positive

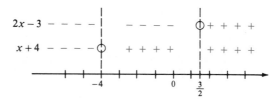

Figure 2.11

in the third interval, and their product also is positive there. Thus, the polynomial $2x^2 + 5x - 12$ is positive or zero in the interval $(-\infty, -4]$ and also in the interval $[3/2, +\infty)$. Since both of these intervals belong to the solution set, the result can be written as the *union** of the two intervals,

$$(-\infty, -4] \cup [3/2, +\infty).$$

The graph of the solution set is shown in Figure 2.12.   ✛

Figure 2.12

Rational Inequalities   The inequalities discussed in the remainder of this section involve quotients of algebraic expressions, and for this reason they are called **rational inequalities.** These inequalities can be solved with a sign graph in much the same way as quadratic inequalities.

Example 7   Solve the inequality $\dfrac{5}{x + 4} \geq 1$.

Multiplying both sides of the inequality by $x + 4$ would get rid of the denominators, but then it would be necessary to consider whether $x + 4$ is positive or negative. Instead, subtract 1 from both sides of the inequality, getting

$$\frac{5}{x + 4} - 1 \geq 0.$$

Writing the left side as a single fraction gives

$$\frac{5 - (x + 4)}{x + 4} \geq 0$$

or

$$\frac{1 - x}{x + 4} \geq 0.$$

The quotient can change sign only when the denominator is 0 or when the numerator is 0. This occurs at

$$1 - x = 0 \quad \text{or} \quad x + 4 = 0$$
$$x = 1 \quad \text{or} \quad x = -4.$$

---

*The **union** of sets A and B, written $A \cup B$, is defined as $A \cup B = \{x \mid x$ is an element of A or x is an element of $B\}$.

Make a sign graph as before. This time, consider the sign of the quotient of the two quantities rather than the sign of their product. See Figure 2.13.

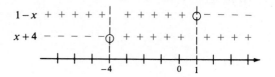

Figure 2.13

The quotient of two numbers is positive if both numbers are positive or if both numbers are negative. On the other hand, the quotient is negative if the two numbers have opposite signs. The sign graph in Figure 2.13 shows that values in the interval $(-4, 1)$ give a positive quotient and are part of the solution. With a quotient, the endpoints must be considered separately to make sure that no denominator is 0. Here, -4 gives a 0 denominator but 1 satisfies the given inequality. In interval notation, the solution set is $(-4, 1]$.   ✤

As suggested by Example 7, be very careful with the endpoints of the intervals in the solution of rational inequalities.

Example 8   Solve $\dfrac{2x - 1}{3x + 4} < 5.$

Begin by subtracting 5 on both sides and combining the terms on the left into a single fraction.

$$\frac{2x - 1}{3x + 4} < 5$$

$$\frac{2x - 1}{3x + 4} - 5 < 0$$

$$\frac{2x - 1 - 5(3x + 4)}{3x + 4} < 0$$

$$\frac{-13x - 21}{3x + 4} < 0$$

To draw a sign graph, first solve the equations

$$-13x - 21 = 0 \qquad \text{and} \qquad 3x + 4 = 0,$$

getting the solutions

$$x = -\frac{21}{13} \qquad \text{and} \qquad x = -\frac{4}{3}.$$

Use the values $-21/13$ and $-4/3$ to divide the number line into three intervals. Now complete a sign graph and find the intervals where the quotient is negative. See Figure 2.14.

Figure 2.14

From the sign graph, values of x in the two intervals $(-\infty, -21/13)$ and $(-4/3, +\infty)$ make the quotient negative, as required. Neither endpoint satisfies the given inequality, so the solution set should be written $(-\infty, -21/13) \cup (-4/3, +\infty)$. ✚

2.7 Exercises

Write each of the following in interval notation. Graph each interval.

1. $-1 < x < 4$ **2.** $x \geq -3$ **3.** $x < 0$ **4.** $8 > x > 3$

5. $2 > x \geq 1$ **6.** $-4 \geq x > -5$ **7.** $-9 > x$ **8.** $6 \leq x$

Using the variable x, write each of the following intervals as an inequality.

9. $(-4, 3)$

10. $[2, 7)$

11. $(-\infty, -1]$

12. $(3, +\infty)$

13.

14.

15.

16.

Solve the following inequalities. Write the solutions in interval notation. Graph each solution. See Examples 1–3.

17. $2x + 1 \leq 9$

18. $3y - 2 \leq 10$

19. $-3p - 2 \leq 1$

20. $-5r + 3 \geq -2$

21. $2(m + 5) - 3m + 1 \geq 5$

22. $6m - (2m + 3) \geq 4m - 5$

23. $8k - 3k + 2 < 2(k + 7)$

24. $2 - 4x + 5(x - 1) < -6(x - 2)$

25. $\dfrac{4x + 7}{-3} \leq 2x + 5$

26. $\dfrac{2z - 5}{-8} \geq 1 - z$

27. $2 \leq y + 1 \leq 5$

28. $-3 \leq 2t \leq 6$

29. $-10 > 3r + 2 > -16$

30. $4 > 6a + 5 > -1$

31. $-3 \leq \dfrac{x - 4}{-5} < 4$

32. $1 < \dfrac{4m - 5}{-2} < 9$

Solve the following quadratic inequalities. Write the solutions in interval notation. Graph each solution. (Hint: In Exercises 43–46, use the quadratic formula.) See Examples 5 and 6.

33. $x^2 \le 9$

34. $p^2 > 16$

35. $r^2 + 4r + 6 \ge 3$

36. $z^2 + 6z + 16 < 8$

37. $x^2 - x \le 6$

38. $r^2 + r < 12$

39. $2k^2 - 9k > -4$

40. $3n^2 \le -10 - 13n$

41. $x^2 > 0$

42. $p^2 < -1$

43. $x^2 + 5x - 2 < 0$

44. $4x^2 + 3x + 1 \le 0$

45. $m^2 - 2m \le 1$

46. $p^2 + 4p > -1$

The following inequalities are not quadratic inequalities, but they still may be solved in much the same way (with four regions). Solve each inequality. Write solutions in interval notation.

47. $(2r - 3)(r + 2)(r - 3) \ge 0$

48. $(y + 5)(3y - 4)(y + 2) > 0$

49. $x^3 - 4x \le 0$

50. $r^3 - 9r \ge 0$

51. $4m^3 + 7m^2 - 2m > 0$

52. $6p^3 - 11p^2 + 3p > 0$

Work each of the following problems.

53. Rick has 420 points so far in his math class. He must have at least 80% of a total of 700 points to get a B in the class. If the final is worth 200 points, what is the lowest score he can get on the final to get a B in the class?

54. On a trip to the market, Kevin bought twice as many apples as oranges. If he bought at least a dozen pieces of fruit, what is the smallest number of oranges he bought?

Find all intervals where the following products will at least break even. See Example 4.

55. The cost to produce x units of wire is $C = 50x + 5000$, while the revenue is $R = 60x$.

56. The cost to produce x units of squash is $C = 100x + 6000$, while the revenue is $R = 500x$.

57. $C = 85x + 900$; $R = 105x$

58. $C = 70x + 500$; $R = 60x$

Use the ideas introduced in this section to solve each of the following.

59. The commodity market is very unstable; money can be made or lost quickly when investing in soybeans, wheat, pork bellies, and the like. Suppose that an investor kept track of her total profit, P, at time t, measured in months, after she began investing, and found that

$$P = 4t^2 - 29t + 30.$$

Find the time intervals where she has been ahead. (*Hint:* $t > 0$ in this case.)

60. Suppose the velocity of an object is given by

$$v = 2t^2 - 5t - 12,$$

where t is time in seconds. (Here t can be positive or negative.) Find the intervals where the velocity is negative.

61. An analyst has found that his company's profits, in hundreds of thousands of dollars, are given by

$$P = 3x^2 - 35x + 50,$$

where x is the amount, in hundreds of dollars, spent on advertising. For what values of x does the company make a profit?

62. The manager of a large apartment complex has found that the profit is given by

$$P = -x^2 + 250x - 15,000,$$

where x is the number of units rented. For what values of x does the complex produce a profit?

63. The formula for converting from Celsius to Fahrenheit temperature is

$$F = \frac{9}{5}C + 32.$$

What temperature range in °F corresponds to 0°C to 30°C?

64. A projectile is fired from ground level. After t sec its height above the ground is $220t - 16t^2$ ft. For what time period is the projectile at least 624 ft above the ground?

Solve the following inequalities. Give answers in interval notation. See Examples 7 and 8.

65. $\dfrac{m - 3}{m + 5} \le 0$

66. $\dfrac{r + 1}{r - 4} > 0$

67. $\dfrac{k - 1}{k + 2} > 1$

68. $\dfrac{a - 6}{a + 2} < -1$

69. $\dfrac{3}{x - 6} \le 2$

70. $\dfrac{1}{k - 2} < \dfrac{1}{3}$

71. $\dfrac{1}{m - 1} < \dfrac{5}{4}$

72. $\dfrac{6}{5 - 3x} \le 2$

73. $\dfrac{10}{3 + 2x} \le 5$

74. $\dfrac{1}{x + 2} \ge 3$

75. $\dfrac{7}{k + 2} \ge \dfrac{1}{k + 2}$

76. $\dfrac{5}{p + 1} > \dfrac{12}{p + 1}$

77. $\dfrac{3}{2r - 1} > \dfrac{-4}{r}$

78. $\dfrac{-5}{3h + 2} \ge \dfrac{5}{h}$

79. $\dfrac{4}{y - 2} \le \dfrac{3}{y - 1}$

80. $\dfrac{4}{n + 1} < \dfrac{2}{n + 3}$

81. $\dfrac{y + 3}{y - 5} \le 1$

82. $\dfrac{a + 2}{3 + 2a} \le 5$

Solve the following rational inequalities using methods similar to those used above.

83. $\dfrac{2x - 3}{x^2 + 1} \ge 0$

84. $\dfrac{9x - 8}{4x^2 + 25} < 0$

85. $\dfrac{(3x - 5)^2}{(2x - 5)^3} > 0$

86. $\dfrac{(5x - 3)^3}{(8x - 25)^2} \le 0$

87. $\dfrac{(2x - 3)(3x + 8)}{(x - 6)^3} \ge 0$

88. $\dfrac{(9x - 11)(2x + 7)}{(3x - 8)^3} > 0$

Use the discriminant to find the values of k where the following equations have real solutions.

89. $x^2 - kx + 8 = 0$

90. $x^2 + kx - 5 = 0$

91. $x^2 + kx + 2k = 0$

92. $kx^2 + 4x + k = 0$

93. If $a > b > 0$, show that $1/a < 1/b$.

94. If $a > b$, is it always true that $1/a < 1/b$?

95. Suppose $a > b > 0$. Show that $a^2 > b^2$.

96. Let $b > 0$. When is $b^2 > b$?

2.8 Absolute Value Equations and Inequalities

In this section we describe methods of solving equations and inequalities involving absolute value. Recall from Chapter 1 that the absolute value of a number a, written $|a|$, gives the distance from a to 0 on a number line. By this definition, the absolute value equation $|x| = 3$ can be solved by finding all real numbers at a distance of 3 units from 0. As shown in Figure 2.15, there are two numbers satisfying this condition, 3 and -3, so that the solution set of the equation $|x| = 3$ is the set $\{3, -3\}$.

Figure 2.15

Example 1   Solve $|p - 4| = 5$.

The expression $|p - 4|$ represents the distance between p and 4. The equation $|p - 4| = 5$ can be solved by finding all real numbers that are 5 units from 4. As shown in Figure 2.16, these numbers are -1 and 9. The solution set is $\{-1, 9\}$.   ✚

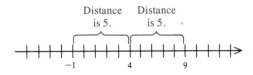

Figure 2.16

The definition of absolute value leads to the following properties of absolute value that can be used to solve absolute value equations algebraically.

For any positive number b:

(1) $|a| = b$ if and only if $a = b$ or $a = -b$

(2) $|a| = |b|$ if and only if $a = b$ or $a = -b$

Example 2 Solve each equation.

(a) $|5 - 3m| = 12$

Use property (1) above, with $a = 5 - 3m$, to write

$$5 - 3m = 12 \quad \text{or} \quad 5 - 3m = -12.$$

Solve each equation.

$$
\begin{array}{ll}
5 - 3m = 12 & \qquad 5 - 3m = -12 \\
-3m = 7 & \qquad -3m = -17 \\
m = -\dfrac{7}{3} & \qquad m = \dfrac{17}{3}
\end{array}
$$

The solution set is $\{-7/3,\ 17/3\}$.

(b) $|4m - 3| = |m + 6|$

By property (2) above, this equation will be true if

$$4m - 3 = m + 6 \quad \text{or} \quad 4m - 3 = -(m + 6).$$

Solve each equation.

$$
\begin{array}{ll}
4m - 3 = m + 6 & \qquad 4m - 3 = -(m + 6) \\
3m = 9 & \qquad 4m - 3 = -m - 6 \\
m = 3 & \qquad 5m = -3 \\
 & \qquad m = -\dfrac{3}{5}
\end{array}
$$

The solution set of $|4m - 3| = |m + 6|$ is thus $\{3,\ -3/5\}$. ✥

Absolute Value Inequalities The method used to solve absolute value equations can be generalized to solve inequalities with absolute value.

Example 3 Solve $|x| < 5$.

Since absolute value gives the distance between a number and 0, the inequality $|x| < 5$ is satisfied by all real numbers whose distance from 0 is less than 5. As shown in Figure 2.17, the solution includes all numbers from -5 to 5, or $-5 < x < 5$. In interval notation, the solution is written as the open interval $(-5, 5)$. A graph of the solution set is shown in Figure 2.17. ✥

Figure 2.17

In a similar way, the solution of $|x| > 5$ consists of all real numbers whose distance from 0 is greater than 5. This includes those numbers greater than 5 or those less than -5, or

$$x < -5 \quad \text{or} \quad x > 5.$$

In interval notation, the solution is written $(-\infty, -5) \cup (5, +\infty)$. The solution set is shown in Figure 2.18.

The following properties of absolute value, which can be obtained from the definition of absolute value, are used to solve absolute value inequalities.

Figure 2.18

For any positive number b:

(1) $|a| < b$ if and only if $-b < a < b;$

(2) $|a| > b$ if and only if $a < -b$ or $a > b$.

Example 4   Solve $|x - 2| < 5$.

This inequality is satisfied by all real numbers whose distance from 2 is less than 5. As shown in Figure 2.19, the solution set is the interval $(-3, 7)$. Property (1) above can be used to solve the inequality as follows. Let $a = x - 2$ and $b = 5$, so that $|x - 2| < 5$ if and only if

$$-5 < x - 2 < 5.$$

Adding 2 to each portion of this inequality produces

$$-3 < x < 7,$$

giving the interval solution $(-3, 7)$.   ✤

Figure 2.19

Example 5   Solve $|x - 8| \geq 1$.

All numbers whose distance from 8 is greater than or equal to 1 are solutions. As shown in Figure 2.20, the solution set is $(-\infty, 7] \cup [9, +\infty)$. To find the solution using property (2) above, let $a = x - 8$ and $b = 1$ so that $|x - 8| \geq 1$ if and only if

$$x - 8 \leq -1 \quad \text{or} \quad x - 8 \geq 1.$$

Solve each inequality separately to get the solution set found from the figure, $(-\infty, 7] \cup [9, +\infty)$, as mentioned above.   ✚

Figure 2.20

The properties given above for solving absolute value inequalities require that the absolute value expression be *alone* on one side of the inequality. Example 6 shows how to meet this requirement.

Example 6   Solve $|2 - 7m| - 1 > 4$.

In order to use the properties of absolute value given above, first add 1 to both sides; this gives

$$|2 - 7m| > 5.$$

Now use property (2) above. By this property, $|2 - 7m| > 5$ if and only if

$$2 - 7m < -5 \quad \text{or} \quad 2 - 7m > 5.$$

Solve each of these inequalities separately to get the solution set $(-\infty, -3/7) \cup (1, +\infty)$.   ✚

Example 7   Solve $|2 - 5x| \geq -4$.

Since the absolute value of a number is always nonnegative, $|2 - 5x| \geq -4$ is always true. The solution set includes all real numbers. In interval notation, this is written $(-\infty, +\infty)$.   ✚

2.8 Exercises

Solve each of the following equations. See Examples 1 and 2.

1. $|a - 2| = 1$ **2.** $|x - 3| = 2$ **3.** $|3m - 1| = 2$ **4.** $|4p + 2| = 5$

5. $|5 - 3x| = 3$ **6.** $|-3a + 7| = 3$ **7.** $\left|\dfrac{z - 4}{2}\right| = 5$ **8.** $\left|\dfrac{m + 2}{2}\right| = 7$

9. $\left|\dfrac{5}{r-3}\right| = 10$ **10.** $\left|\dfrac{3}{2h-1}\right| = 4$ **11.** $\left|\dfrac{6y+1}{y-1}\right| = 3$ **12.** $\left|\dfrac{3a-4}{2a+3}\right| = 1$

13. $|2k-3| = |5k+4|$ **14.** $|p+1| = |3p-1|$ **15.** $|4-3y| = |7+2y|$

16. $|2+5a| = |4-6a|$ **17.** $|x+2| = |x-1|$ **18.** $|y-5| = |y+3|$

Solve each of the following inequalities. Give the solution in interval notation.
See Examples 3–7.

19. $|x| \le 3$ **20.** $|y| \le 10$ **21.** $|m| > 1$ **22.** $|z| > 5$

23. $|a| < -2$ **24.** $|b| > -5$ **25.** $|x| - 3 \le 7$ **26.** $|r| + 3 \le 10$

27. $|2x+5| < 3$ **28.** $\left|x - \dfrac{1}{2}\right| < 2$ **29.** $|3m-2| > 4$ **30.** $|4x-6| > 10$

31. $|3z+1| \ge 7$ **32.** $|8b+5| \ge 7$ **33.** $\left|5x + \dfrac{1}{2}\right| - 2 < 5$ **34.** $\left|x + \dfrac{2}{3}\right| + 1 < 4$

Write each of the following as an absolute value statement. For example, write
"k is at least 4 units from 1" as $|k-1| \ge 4$.

35. x is within 4 units of 2   **36.** m is no more than 8 units from 9

37. z is no less than 2 units from 12   **38.** p is at least 5 units from 9

39. k is 6 units from 1   **40.** r is 5 units from 3

41. If x is within .0004 units of 2, then y is within .00001 units of 7.

42. y is within 10^{-6} units of 10 whenever x is within 2×10^{-4} units of 5.

43. If $|x-2| < 3$, find values of m and n such that $m < 3x + 5 < n$.

44. If $|x+8| < 16$, find values of p and q such that $p < 2x - 1 < q$.

Chapter 2 *Summary*

Key Words

2.1 equation
solution
solution set
identity
conditional equation
contradiction
equivalent equations
linear equation
2.2 solving for a specified
variable
2.3 complex number
imaginary number
conjugates

2.4 quadratic equation
completing the square
quadratic formula
discriminant
2.6 quadratic in form
2.7 inequality
interval notation
open interval
half-open interval
closed interval

Key Ideas

2.3 Definition of i

$i = \sqrt{-1}$ or $i^2 = -1$

2.4 Zero-Factor Property

If a and b are complex numbers, with $ab = 0$, then $a = 0$ or $b = 0$ or both.

2.4 Square Root Property

The solution set of $x^2 = k$ is $\{\sqrt{k}, -\sqrt{k}\}$.

2.4 Quadratic Formula

The solutions of the quadratic equation $ax^2 + bx + c = 0$, where $a \neq 0$, are

$$x = \frac{-b \pm \sqrt{b^2 - 4ac}}{2a}.$$

2.8 Properties of Absolute Value

If b is a positive number, then

$$|a| = b \quad \text{if and only if} \quad a = b \text{ or } a = -b;$$
$$|a| = |b| \quad \text{if and only if} \quad a = b \text{ or } a = -b;$$
$$|a| < b \quad \text{if and only if} \quad -b < a < b;$$
$$|a| > b \quad \text{if and only if} \quad a < -b \text{ or } a > b.$$

Chapter 2   *Review Exercises*

Solve each of the following equations.

1. $2m + 7 = 3m + 1$

2. $4k - 2(k - 1) = 12$

3. $5y - 2(y + 4) = 3(2y + 1)$

4. $\dfrac{2}{x - 3} = \dfrac{3}{x + 3}$

5. $\dfrac{10}{4z - 4} = \dfrac{1}{1 - z}$

6. $\dfrac{2}{p} - \dfrac{4}{3p} = 8 + \dfrac{3}{p}$

7. $\dfrac{5}{3r} - 10 = \dfrac{3}{2r}$

8. $\dfrac{p}{p - 2} - \dfrac{3}{5} = \dfrac{2}{p - 2}$

9. $5(x - 3)(2x + 1) = 10(x + 2)(x - 4)$

10. $(3k + 1)^2 = 6k^2 + 3(k - 1)^2$

Solve for x.

11. $3(x + 2b) + a = 2x - 6$

12. $9x - 11(k + p) = x(a - 1)$

13. $\dfrac{x}{m - 2} = kx - 3$

14. $r^2x - 5x = 3r^2$

Solve each of the following for the indicated variable.

15. $2a + ay = 4y - 4a$ for y

16. $\dfrac{3m}{m - x} = 2m + x$ for x

17. $F = \dfrac{9}{5}C + 32$ for C

18. $A = P + Pi$ for P

19. $A = I\left(1 - \dfrac{j}{n}\right)$ for j

20. $A = \dfrac{24f}{b(p + 1)}$ for f

21. $\dfrac{1}{k} = \dfrac{1}{r_1} + \dfrac{1}{r_2}$ for r_1

22. $m = \dfrac{Ft}{\sqrt{I} - \sqrt{2}}$ for t

23. $V = \pi r^2 L$ for L

24. $P(r + R)^2 = E^2R$ for P

25. $\dfrac{xy^2 - 5xy + 4}{3x} = 2p$ for x

26. $\dfrac{zx^2 - 5x + z}{z + 1} = 9$ for z

Solve each of the following problems.

27. A stereo is on sale for 15% off. The sale price is $425. What was the original price?

28. To make a special candy mix for Valentine's Day, the owner of a candy store wants to combine chocolate hearts that sell for $5 per pound with candy kisses that sell for $3.50 per pound. How many pounds of each kind should be used to get 30 lb of a mix which can be sold for $4.50 per pound?

29. Alison can ride her bike to the university library in 20 min. The trip home, which is all uphill, takes her 30 min. If her rate is 8 mph slower on the return trip, how far does she live from the library?

30. Ed and Susan are stuffing envelopes for a political campaign. Working together, they can stuff 5000 envelopes in 4 hr. When Susan worked at the job alone, it took her 6 hr to stuff the 5000 envelopes. How long would it take Ed working alone to stuff 5000 envelopes?

31. A realtor borrowed $90,000 to develop some property. He was able to borrow part of the money at 11.5% interest and the rest at 12%. The annual interest on the two loans amounts to $10,525. How much was borrowed at each rate?

32. An excursion boat travels upriver to a landing and then returns to its starting point. The trip upriver takes 1.2 hr, and the trip back takes .9 hr. If the average speed on the return trip is 5 mph faster than on the trip upriver, what is the boat's speed upriver?

Perform each operation. Write each result in standard form.

33. $(6 - i) + (4 - 2i)$

34. $(-11 + 2i) - (5 - 7i)$

35. $15i - (3 + 2i) - 5$

36. $-6 + 4i - (5i - 2)$

37. $(5 - i)(3 + 4i)$ 　　 **38.** $(8 - 2i)(1 - i)$ 　　 **39.** $(5 - 11i)(5 + 11i)$ 　　 **40.** $(7 + 6i)(7 - 6i)$

41. $(4 - 3i)^2$ 　　 **42.** $(1 + 7i)^2$ 　　 **43.** $-5i(3 - i)^2$ 　　 **44.** $4i(2 + 5i)(2 - i)$

45. $\dfrac{6 + i}{1 - i}$ 　　 **46.** $\dfrac{5 + 2i}{5 - 2i}$ 　　 **47.** $\dfrac{4 + 3i}{2 + i}$ 　　 **48.** $\dfrac{7 - i}{1 + i}$

Find each of the following powers of i.

49. i^7 　　　　　　 **50.** i^{15} 　　　　　　 **51.** i^{123}

52. i^{507} 　　　　　 **53.** i^{-18} 　　　　　 **54.** i^{-23}

Solve each equation.

55. $(b + 7)^2 = 5$ 　　 **56.** $(3y - 2)^2 = 8$ 　　 **57.** $2a^2 + a - 15 = 0$ 　　 **58.** $12x^2 = 8x - 1$

59. $2q^2 - 11q = 21$ 　　 **60.** $3x^2 + 2x = 16$ 　　 **61.** $2 - \dfrac{5}{p} = \dfrac{3}{p^2}$ 　　 **62.** $\dfrac{4}{m^2} = 2 + \dfrac{7}{m}$

63. $\sqrt{2}x^2 - 4x + \sqrt{2} = 0$ 　　　　　　 **64.** $p^2 - \sqrt{5}p - 1 = 0$

Evaluate the discriminant for each of the following, and then use it to predict the type of solutions for the equation.

65. $8y^2 = 2y - 6$ 　　 **66.** $6k^2 - 2k = 3$ 　　 **67.** $16r^2 + 3 = 26r$

68. $8p^2 + 10p = 7$ 　　 **69.** $25z^2 - 110z + 121 = 0$ 　　 **70.** $4y^2 - 8y + 17 = 0$

Work each word problem.

71. Rick plans to replace the vinyl floor covering in his 10- by 12-ft kitchen. He wants to have a border of even width of a special material. He can afford only 21 sq ft of this material. How wide a border can he have?

72. Calvin wants to fence off a rectangular playground beside an apartment building. The building forms one boundary, so he needs to fence only the other three sides. The area of the playground is to be 11,250 sq m. He has enough material to build 325 m of fence. Find the length and width of the playground.

73. It takes two gardeners 3 hr (working together) to mow the lawns in a city park. One gardener could do the entire job in 1 hr less time than the other. How long would it take the slower gardener to complete the work alone? Give the answer to the nearest tenth.

74. Steve and Paula sell pies. It takes Paula 1 hr longer than Steve to bake a day's supply of pies. Working together, it takes them 1 1/5 hr to bake the pies. How long would it take Steve working alone?

75. In a marathon (a 26-mile run), the winner finished 2/5 hr before his friend. If the difference in their rates was 4/3 mph, what was the winner's average speed in the race?

76. An Ironman race consists of three parts; a one-mile swim, a ten-mile run, and a twenty-mile bicycle ride. Russ's times for the run and the bicycle ride were the same. His speed swimming was 2.5 mph and he completed the entire race in 4.4 hr. Find his speed on the bicycle.

Solve each equation.

77. $4a^4 + 3a^2 - 1 = 0$

78. $2x^4 - x^2 = 0$

79. $(r + 1)^2 - 3(r + 1) = 4$

80. $4(y - 2)^2 - 9(y - 2) + 2 = 0$

81. $(2z + 3)^{2/3} + (2z + 3)^{1/3} = 6$

82. $5\sqrt{m} = \sqrt{3m + 2}$

83. $\sqrt{4y - 2} = \sqrt{3y + 1}$

84. $\sqrt{2x + 3} = x + 2$

85. $\sqrt{p + 2} = 2 + p$

86. $\sqrt{k} = \sqrt{k + 3} - 1$

87. $\sqrt{x + 3} - \sqrt{3x + 10} = 1$

88. $\sqrt{5x - 15} - \sqrt{x + 1} = 2$

89. $\sqrt{x^2 + 3x} - 2 = 0$

90. $\sqrt[3]{2r} = \sqrt[3]{3r + 2}$

91. $\sqrt[3]{6y + 2} = \sqrt[3]{4y}$

92. $(x - 2)^{2/3} = x^{1/3}$

Solve the following inequalities. Write answers in interval notation.

93. $-9x < 4x + 7$

94. $11y \geq 2y - 8$

95. $-5z - 4 \geq 3(2z - 5)$

96. $-(4a + 6) < 3a - 2$

97. $3r - 4 + r > 2(r - 1)$

98. $7p - 2(p - 3) \leq 5(2 - p)$

99. $5 \leq 2x - 3 \leq 7$

100. $-8 > 3a - 5 > -12$

101. $x^2 + 3x - 4 \leq 0$

102. $p^2 + 4p > 21$

103. $6m^2 - 11m - 10 < 0$

104. $k^2 - 3k - 5 \geq 0$

105. $z^3 - 16z \leq 0$

106. $2r^3 - 3r^2 - 5r < 0$

107. $\dfrac{3a - 2}{a} > 4$

108. $\dfrac{5p + 2}{p} < -1$

109. $\dfrac{3}{r - 1} \leq \dfrac{5}{r + 3}$

110. $\dfrac{3}{x + 2} > \dfrac{2}{x - 4}$

Work the following word problems.

111. Steve and Paula (Exercise 74) have found that the profit from their pie shop is given by

$$P = -x^2 + 28x + 60,$$

where x is the number of units of pies sold daily. For what values of x is the profit positive?

112. A company produces videotapes. The revenue from the sale of x units of tapes is $R = 8x$. The cost to produce x units of tapes is $C = 3x + 1500$. In what interval will the company at least break even?

Solve each equation.

113. $|a + 4| = 7$

114. $|-y + 2| = 3$

115. $\left| \dfrac{7}{2 - 3a} \right| = 9$

116. $\left|\dfrac{8r - 1}{3r + 2}\right| = 7$ **117.** $|5r - 1| = |2r + 3|$ **118.** $|k + 7| = |k - 8|$

Solve each inequality. Write solutions with interval notation.

119. $|m| \leq 7$ **120.** $|r| < 2$ **121.** $|p| > 3$ **122.** $|z| > -1$

123. $|2z + 9| \leq 3$ **124.** $|5m - 8| \leq 2$ **125.** $|7k - 3| < 5$ **126.** $|2p - 1| > 2$

127. $|3r + 7| - 5 > 0$ **128.** $|1 - y| - 2 \geq 4$

Chapter 2 *Test*

Solve each of the following equations.

1. $x - (2x + 1) = 7 - 3(x + 1)$

2. $\dfrac{x}{x - 3} = \dfrac{3}{x - 3} + 4$

Solve each of the following for the indicated variable.

3. $k^2x = 2x + 2b$ for x

4. $\dfrac{1}{c} - \dfrac{1}{a} = \dfrac{1}{b}$ for c

Solve each of the following problems.

5. How many quarts of a 60% alcohol solution must be added to 40 qt of a 20% alcohol solution to obtain a mixture that is 30% alcohol?

6. Fred and Wilma start from the same point and travel on a straight road. Fred travels at 30 mph, while Wilma travels at 50 mph. If Wilma starts 3 hr after Fred, find the distance they travel before meeting.

Perform each operation. Write each result in standard form.

7. $(7 - i) - (6 - 10i)$ **8.** $(4 + 3i)(-5 + 2i)$ **9.** $\dfrac{10i}{1 - 3i}$

10. Is i^{297} equal to i, -1, $-i$, or 1?

Solve each equation.

11. $6x(2 - x) = 7$

12. $\dfrac{3x - 2}{3x + 2} = \dfrac{2x + 3}{4x - 1}$

13. Evaluate the discriminant, and use it to predict the type of solutions for the equation $3x^2 = 5x - 2$.

14. Tony Lally has a rectangular-shaped flower box that measures 4 ft by 6 ft. He wants to double the available area by increasing the length and width by the same amount. What should the new dimensions be?

Solve each equation.

15. $\sqrt{5 + 2x} = x + 1$

16. $\sqrt{2x + 1} - \sqrt{x} = 1$

17. $x^4 - 3x^2 - 10 = 0$

18. $2 - \sqrt[3]{2x + x^2} = 0$

Solve each of the following inequalities. Write each answer in interval notation.

19. $-2(x - 1) - 10 \leq 2(2 + x)$

20. $4 \geq 3 + \dfrac{x}{2} \geq -2$

21. $2x^2 - x - 3 \geq 0$

22. $\dfrac{6}{2x - 5} \leq 2$

23. Solve $|-5 - 3x| = 4$.

Solve each inequality. Write solutions in interval notation.

24. $|2x - 5| < 9$

25. $|2x + 1| - 11 \geq 0$

3

Relations

A major goal of this text is to make a careful study of *relations* and *functions* (a special kind of relation discussed in the next chapter). In this chapter we discuss several important first- and second-degree relations.

3.1 Graphing Relations

Many things in daily life are related. For example, a student's grade in a course usually is related to the amount of time spent studying, while the number of miles per gallon of gas used on a car trip depends on the speed of the car. Driving at 55 mph might give 31 miles per gallon, while speeding at 65 mph might reduce the gas mileage to 28 miles per gallon.

Pairs of related numbers, such as 55 and 31 or 65 and 28 in the gas mileage illustration, can be written as *ordered pairs*. An **ordered pair** of numbers consists of two numbers, written inside parentheses, in which the sequence of the numbers is important. For example, (4, 2) and (2, 4) are different ordered pairs because the order of the numbers is different. Notation such as (3, 4) has already been used in this book to show an interval on the number line. Now the same notation is used to indicate an ordered pair of numbers. In virtually every case, the intended use will be clear from the context of the discussion.

A set of ordered pairs is called a **relation.** The **domain** of a relation is the set of first elements in the ordered pairs, and the **range** of the relation is the set of all possible second elements. In the driving example above, the domain is the set of all possible speeds and the range is the set of resulting miles per gallon.

Ordered pairs can be used to express the solutions of equations in two variables. For example, we say that (1, 2) is a solution of $2x - y = 0$ since substituting 1 for x and 2 for y in the equation gives

$$2(1) - 2 = 0$$
$$0 = 0,$$

a true statement. When an ordered pair represents the solution of an equation with the variables x and y, the x-value is written first.

Although any set of ordered pairs is a relation, in mathematics we are most interested in those relations that are solution sets of equations. We may say that an equation *defines a relation,* or that it is the *equation of the relation.* For simplicity, we often refer to equations such as

$$y = 3x + 5 \qquad \text{or} \qquad x^2 + y^2 = 16$$

as relations, although technically the solution set of the equation is the relation.

Example 1 For each relation defined below, give three ordered pairs that belong to the relation, and state the domain and range of the relation.

(a) $\{(2, 5), (7, -1), (10, 3), (-4, 0), (0, 5)\}$

Three ordered pairs from the relation are any three of the five ordered pairs in the set. The domain is the set of first elements,

$$\{2, 7, 10, -4, 0\},$$

and the range is the set of second elements,

$$\{5, -1, 3, 0\}.$$

(b) $y = 4x - 1$

To find an ordered pair of the relation, choose any number for x or y and substitute in the equation to get the corresponding value. For example, let $x = -2$. Then

$$y = 4(-2) - 1 = -9,$$

giving the ordered pair $(-2, -9)$. If $y = 3$, then

$$3 = 4x - 1$$
$$4 = 4x$$
$$1 = x,$$

and the ordered pair is (1, 3). Verify that (0, -1) also belongs to the relation. Since x and y can take any real-number values, both the domain and range are $(-\infty, +\infty)$.

(c) $x = \sqrt{y - 1}$

Verify that the ordered pairs (1, 2), (0, 1), and (2, 5) belong to the relation. Since x equals the positive square root of $y - 1$, the domain is restricted to $[0, +\infty)$. Also, only nonnegative numbers have a square root, so the range is determined by the inequality

$$y - 1 \geq 0$$
$$y \geq 1,$$

giving $[1, +\infty)$ as the range.   ✥

The Rectangular Coordinate System   Since the study of relations often involves looking at their *graphs,* this section begins with a brief review of the coordinate plane. As mentioned in Chapter 1, each real number corresponds to a point on a number line. This correspondence is set up by establishing a coordinate system for the line. This idea is extended to the two dimensions of a plane by drawing two perpendicular lines, one horizontal and one vertical. These lines intersect at a point O called the **origin.** The horizontal line is called the **x-axis,** and the vertical line is called the **y-axis.**

Starting at the origin, the x-axis can be made into a number line by placing positive numbers to the right and negative numbers to the left. The y-axis can be made into a number line with positive numbers going up and negative numbers going down.

The x-axis and y-axis together make up a **rectangular coordinate system,** or **Cartesian coordinate system** (named for one of its co-inventors, René Descartes; the other co-inventor was Pierre de Fermat). The plane into which the coordinate system is introduced is the **coordinate plane,** or **xy-plane.** The x-axis and y-axis divide the plane into four regions, or **quadrants,** labeled as shown in Figure 3.1. The points on the x-axis and y-axis belong to no quadrant.

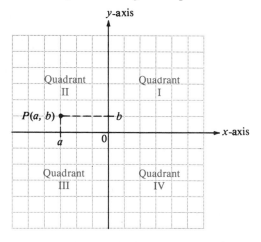

Figure 3.1

Each point P in the xy-plane corresponds to a unique ordered pair (a, b) of real numbers. The numbers a and b are the **coordinates** of point P. To locate on the xy-plane the point corresponding to the ordered pair $(3, 4)$, for example, draw a vertical line through 3 on the x-axis and a horizontal line through 4 on the y-axis. These two lines cross at point A in Figure 3.2. Point A corresponds to the ordered pair $(3, 4)$. Also in Figure 3.2, B corresponds to the ordered pair $(-5, 6)$, C to $(-2, -4)$, D to $(4, -3)$, and E to $(-3, 0)$. The point P corresponding to the ordered pair (a, b) often is written as $P(a, b)$ as in Figure 3.1 and referred to as "the point (a, b)."

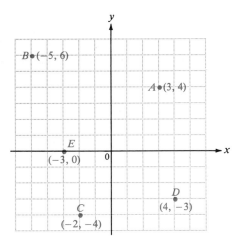

Figure 3.2

The Distance Formula   The *distance formula,* to be developed next, can be used to find the equation of a relation based on its definition as a set of ordered pairs (or, equivalently, as a set of points). Figure 3.3 shows the points $(-4, 3)$ and $(8, -2)$.

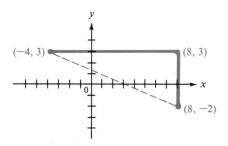

Figure 3.3

To find the distance between these two points, complete a right triangle as shown in the figure. This right triangle has its 90° angle at (8, 3). The horizontal side of the triangle has length

$$|8 - (-4)| = 12,$$

where absolute value is used to make sure that the distance is not negative. The vertical side of the triangle has length

$$|3 - (-2)| = 5.$$

By the Pythagorean theorem, the length of the remaining side of the triangle is

$$\sqrt{12^2 + 5^2} = \sqrt{144 + 25} = \sqrt{169} = 13.$$

The distance between $(-4, 3)$ and $(8, -2)$ is 13.

To obtain a general formula for the distance between two points on a coordinate plane, let $P(x_1, y_1)$ and $R(x_2, y_2)$ be any two distinct points in a plane, as shown in Figure 3.4. Complete a right triangle by locating point Q with coordinates (x_2, y_1). Using the Pythagorean theorem gives the distance between P and R, written $d(P, R)$, as

$$d(P, R) = \sqrt{(x_2 - x_1)^2 + (y_2 - y_1)^2}.$$

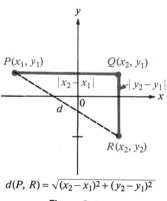

$$d(P, R) = \sqrt{(x_2 - x_1)^2 + (y_2 - y_1)^2}$$

Figure 3.4

The distance formula can be summarized as follows.

Distance Formula

Suppose that $P(x_1, y_1)$ and $R(x_2, y_2)$ are two points in a coordinate plane. Then the distance between P and R, written $d(P, R)$, is given by the **distance formula,**

$$d(P, R) = \sqrt{(x_2 - x_1)^2 + (y_2 - y_1)^2}.$$

Although the proof of the distance formula assumes that P and R are not on a horizontal or vertical line, the result is true for any two points.

Example 2 Find the distance between $P(-8, 4)$ and $Q(3, -2)$.
According to the distance formula,

$$\begin{aligned} d(P, Q) &= \sqrt{[3 - (-8)]^2 + (-2 - 4)^2} \\ &= \sqrt{11^2 + (-6)^2} \\ &= \sqrt{121 + 36} \\ &= \sqrt{157}. \end{aligned}$$

Using a calculator with a square root key gives 12.530 as an approximate value for $\sqrt{157}$. ✛

Example 3 Are the points $M(-2, 5)$, $N(12, 3)$, and $Q(10, -11)$ the vertices of a right triangle?
To decide whether or not this triangle is a right triangle, use the converse of the Pythagorean theorem: if the sides a, b, and c of a triangle satisfy $a^2 + b^2 = c^2$, then the triangle is a right triangle. A triangle with the three given points as vertices is shown in Figure 3.5. This triangle is a right triangle if the square of the length of the longest side equals the sum of the squares of the lengths of the other two sides. Use the distance formula to find the length of each side of the triangle.

$$\begin{aligned} d(M, N) &= \sqrt{[12 - (-2)]^2 + (3 - 5)^2} = \sqrt{196 + 4} = \sqrt{200} \\ d(M, Q) &= \sqrt{[10 - (-2)]^2 + (-11 - 5)^2} = \sqrt{144 + 256} = \sqrt{400} = 20 \\ d(N, Q) &= \sqrt{(10 - 12)^2 + (-11 - 3)^2} = \sqrt{4 + 196} = \sqrt{200} \end{aligned}$$

By these results,

$$[d(M, Q)]^2 = [d(M, N)]^2 + [d(N, Q)]^2,$$

proving that the triangle is a right triangle with hypotenuse connecting M and Q. ✛

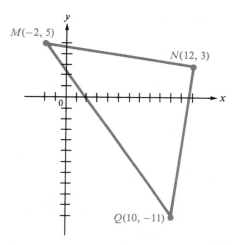

Figure 3.5

The Midpoint Formula   The midpoint formula is used to find the coordinates of the midpoint of a line segment. (Recall that the midpoint of a line segment is equidistant from the endpoints of the segment.) To develop the midpoint formula, let (x_1, y_1) and (x_2, y_2) be any two distinct points in a plane. (Although Figure 3.6 shows $x_1 < x_2$, no particular order is required.) Assume that the two points are not on a horizontal or vertical line. Let (x, y) be the midpoint of the segment connecting (x_1, y_1) and (x_2, y_2). Draw vertical lines from each of the three points to the x-axis, as shown in Figure 3.6.

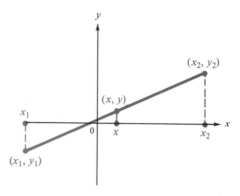

Figure 3.6

Since (x, y) is the midpoint of the line segment connecting (x_1, y_1) and (x_2, y_2), the distance between x and x_1 equals the distance between x and x_2, so that

$$x_2 - x = x - x_1$$
$$x_2 + x_1 = 2x$$
$$x = \frac{x_1 + x_2}{2}.$$

By this result, the x-coordinate of the midpoint is the average of the x-coordinates of the endpoints of the segment. In a similar manner, the y-coordinate of the midpoint is $(y_1 + y_2)/2$, proving the following statement.

Midpoint Formula	The midpoint of the line segment with endpoints (x_1, y_1) and (x_2, y_2) is $$\left(\frac{x_1 + x_2}{2}, \frac{y_1 + y_2}{2}\right).$$

In other words, the midpoint formula says that the coordinates of the midpoint of a segment are found by calculating the *average* of the x-coordinates and the average of the y-coordinates of the endpoints of the segment.

Example 4   Find the midpoint M of the segment with endpoints $(8, -4)$ and $(-9, 6)$.
Use the midpoint formula to find that the coordinates of M are

$$\left(\frac{8 + (-9)}{2}, \frac{-4 + 6}{2}\right) = \left(-\frac{1}{2}, 1\right). \quad \clubsuit$$

Example 5 A line segment has an endpoint at $(2, -8)$ and a midpoint at $(-1, -3)$. Find the other endpoint of the segment.

The formula for the x-coordinate of the midpoint is $(x_1 + x_2)/2$. Here the x-coordinate of the midpoint is -1. Letting $x_1 = 2$ gives

$$-1 = \frac{2 + x_2}{2}$$
$$-2 = 2 + x_2$$
$$-4 = x_2.$$

In the same way, $y_2 = 2$ and the endpoint is $(-4, 2)$. $\quad \clubsuit$

Circles By definition, a **circle** is the set of all points in a plane that lie a given distance from a given point. The given distance is the **radius** of the circle and the given point is the **center.** Since a circle is a set of points, it is a relation. The equation of a circle can be found from its definition by using the distance formula.

Figure 3.7 shows a circle of radius 3 with center at the origin. To find the equation of this circle, let (x, y) be any point on the circle. The distance between (x, y) and the center of the circle, $(0, 0)$, is given by

$$\sqrt{(x - 0)^2 + (y - 0)^2}.$$

Since this distance equals the radius, 3,

$$\sqrt{(x - 0)^2 + (y - 0)^2} = 3$$
$$\sqrt{x^2 + y^2} = 3$$
$$x^2 + y^2 = 9.$$

As suggested by Figure 3.7, the domain of the relation is $[-3, 3]$, while the range of the relation is $[-3, 3]$.

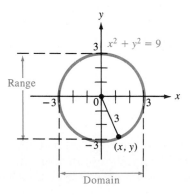

Figure 3.7

Example 6   Find an equation for the circle having radius 6 and center at $(-3, 4)$.

This circle is shown in Figure 3.8. Its equation can be found by using the distance formula. Let (x, y) be any point on the circle. The distance from (x, y) to $(-3, 4)$ is given by

$$\sqrt{[x - (-3)]^2 + (y - 4)^2} = \sqrt{(x + 3)^2 + (y - 4)^2}.$$

This same distance is given by the radius, 6. Therefore,

$$\sqrt{(x + 3)^2 + (y - 4)^2} = 6$$

or

$$(x + 3)^2 + (y - 4)^2 = 36.$$

The domain is $[-9, 3]$, and the range is $[-2, 10]$, as seen in Figure 3.8.   ✤

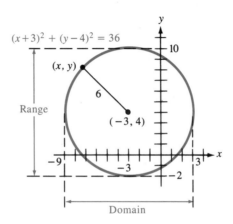

Figure 3.8

Generalizing from the work in Example 6 gives the following result.

Center-Radius Form of the Equation of a Circle

> The circle with center (h, k) and radius r has equation
> $$(x - h)^2 + (y - k)^2 = r^2,$$
> the **center-radius form** of the equation of a circle.

Starting with the center-radius form of the equation of a circle, $(x - h)^2 + (y - k)^2 = r^2$, and squaring $x - h$ and $y - k$ gives an equation of the form

$$x^2 + y^2 + cx + dy + e = 0, \qquad (*)$$

where c, d, and e are real numbers. This result is the **general form of the equation of a circle.** Also, starting with an equation in the form of (*), the process of completing the square can be used to get an equation of the form

$$(x - h)^2 + (y - k)^2 = m$$

for some number m. If $m > 0$, then $r^2 = m$, and the equation represents a circle with radius $\sqrt{m}$. If $m = 0$, then the equation represents the single point (h, k). If $m < 0$, no points satisfy the equation.

Example 7 Find the center and radius for $x^2 - 6x + y^2 + 10y + 25 = 0$.

Since this equation has the form of equation (*) above, it represents either a circle, a single point, or no points at all. To decide which, complete the square on x and y separately. Start with

$$(x^2 - 6x \qquad) + (y^2 + 10y \qquad) = -25.$$

Half of -6 is -3, and $(-3)^2 = 9$. Also, half of 10 is 5, and $5^2 = 25$. Add 9 and 25 on the left, and to compensate, add 9 and 25 on the right:

$$(x^2 - 6x + 9) + (y^2 + 10y + 25) = -25 + 9 + 25$$
$$(x - 3)^2 + (y + 5)^2 = 9.$$

Since $9 > 0$, the equation represents a circle with center at $(3, -5)$ and radius 3. ✚

For any set of ordered pairs of real numbers, there is a corresponding set of points in a coordinate plane. If (x, y) is one of the ordered pairs in the set, then the corresponding point (x, y) can be located on the coordinate plane. This set of points is the **graph** of the set of ordered pairs.

Example 8 Graph each relation.

(a) $(x - 3)^2 + (y + 2)^2 = 16$

By the center-radius form of the equation of a circle, this relation is a circle with center at $(3, -2)$ and radius 4. The graph is shown in Figure 3.9.

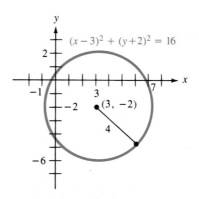

Figure 3.9

(b) $x^2 + 10x + y^2 - 4y + 33 = 0$.

Completing the square as above gives

$$(x^2 + 10x + 25) + (y^2 - 4y + 4) = -33 + 25 + 4$$
$$(x + 5)^2 + (y - 2)^2 = -4.$$

Since $-4 < 0$, there are no ordered pairs of real numbers (x, y) that satisfy the equation. Thus, there is no graph. ✤

Symmetry Looking back at the graph in Figure 3.7, notice that the graph could be cut in half by the y-axis with each half the mirror image of the other half. Such a graph is said to be *symmetric with respect to the y-axis*. The idea of symmetry is helpful in drawing graphs.

Figure 3.10(a) shows a graph that is symmetric with respect to the y-axis. As suggested by Figure 3.10(a), for a graph to be symmetric with respect to the y-axis, the point $(-x, y)$ must be on the graph whenever (x, y) is on the graph.

Similarly, if the graph in Figure 3.10(b) were folded in half along the x-axis, the portion from the top would exactly match the portion from the bottom. Such a graph is *symmetric with respect to the x-axis*. As the graph suggests, symmetry with respect to the x-axis means that the point $(x, -y)$ must be on the graph whenever the point (x, y) is on the graph.

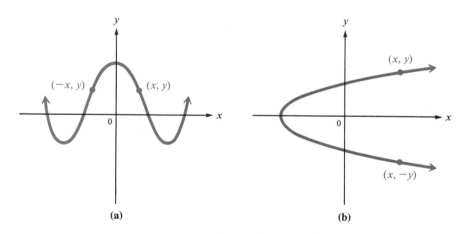

Figure 3.10

The following tests tell when a graph is symmetric with respect to the x-axis or y-axis.

Symmetry with Respect to an Axis	The graph of an equation is **symmetric with respect to the y-axis** if the replacement of x with $-x$ results in an equivalent equation.
	The graph of an equation is **symmetric with respect to the x-axis** if the replacement of y with $-y$ results in an equivalent equation.

Example 9   Test for symmetry with respect to the *x*-axis or *y*-axis.

(a) $y = x^2 + 4$

Replace *x* with $-x$:

$$y = x^2 + 4 \qquad \text{becomes} \qquad y = (-x)^2 + 4 = x^2 + 4.$$

The result is the same as the original equation, so the graph (shown in Figure 3.11) is symmetric with respect to the *y*-axis. The graph is *not* symmetric with respect to the *x*-axis, since replacing *y* with $-y$ gives

$$-y = x^2 + 4,$$

which is not equivalent to the original equation.

Figure 3.11

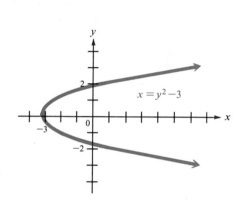

Figure 3.12

(b) $x = y^2 - 3$

Replace *y* with $-y$ to get $x = (-y)^2 - 3 = y^2 - 3$, the same as the original equation. The graph is symmetric with respect to the *x*-axis, as shown in Figure 3.12. Is the graph symmetric with respect to the *y*-axis?

(c) $2x + y = 4$

Replace *x* with $-x$ and then replace *y* with $-y$; in neither case does an equivalent equation result. This graph is symmetric neither with respect to the *x*-axis nor the *y*-axis.   ✜

Another kind of symmetry is found when it is possible to rotate a graph 180° about the origin and have the result coincide exactly with the original graph. Symmetry of this type is called *symmetry with respect to the origin*. It can be shown that rotating a graph 180° is equivalent to saying that the point $(-x, -y)$ is on the graph whenever (x, y) is on the graph. Figure 3.13 on page 152 shows

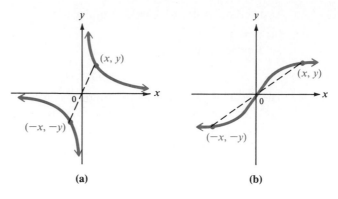

Figure 3.13

two graphs that are symmetric with respect to the origin. A test for this type of symmetry is given below.

Symmetry with Respect to the Origin	The graph of an equation is **symmetric with respect to the origin** if the replacement of both x with $-x$ and y with $-y$ results in an equivalent equation.

The tests for symmetry are listed below.

Tests for Symmetry

	Symmetric with respect to		
	x-axis	*y-axis*	*origin*
Test	Replace y with $-y$.	Replace x with $-x$.	Replace x with $-x$ and replace y with $-y$.
Example	(1)	(2)	(3)

3.1 Exercises

Find the distance d(P, Q) and the midpoint of segment PQ. See Examples 2 and 4.

1. $P(5, 7)$, $Q(13, -1)$

2. $P(-2, 5)$, $Q(4, -3)$

3. $P(-8, -2)$, $Q(-3, -5)$

4. $P(-6, -10)$, $Q(6, 5)$

5. $P(\sqrt{2}, -\sqrt{5})$, $Q(3\sqrt{2}, 4\sqrt{5})$

6. $P(5\sqrt{7}, -\sqrt{3})$, $Q(-\sqrt{7}, 8\sqrt{3})$

Give the distance between the following points rounded to the nearest thousandth. See Example 2.

7. $(3, -7)$, $(-5, 19)$

8. $(-4, 6)$, $(8, -5)$

Find the other endpoint of each segment having endpoint and midpoint as given. See Example 5.

9. Endpoint $(-3, 6)$, midpoint $(5, 8)$

10. Endpoint $(-5, 3)$, midpoint $(-7, 6)$

Decide whether or not the following points are the vertices of a right triangle. See Example 3.

11. $(-4, 0)$, $(1, 3)$, $(-6, -2)$

12. $(-8, 2)$, $(5, -7)$, $(3, -9)$

13. $(\sqrt{3}, 2\sqrt{3} + 3)$, $(\sqrt{3} + 4, -\sqrt{3} + 3)$, $(2\sqrt{3}, 2\sqrt{3} + 4)$

14. $(4 - \sqrt{3}, -2\sqrt{3})$, $(2 - \sqrt{3}, -\sqrt{3})$, $(3 - \sqrt{3}, -2\sqrt{3})$

If three points lie on a straight line, then the sum of the distances between two pairs of points will equal the distance between the other pair. Use the distance formula to decide whether or not the points in Exercises 15–18 lie on a straight line.

15. $(0, 7)$, $(3, -5)$, $(-2, 15)$

16. $(1, -4)$, $(2, 1)$, $(-1, -14)$

17. $(0, -9)$, $(3, 7)$, $(-2, -19)$

18. $(1, 3)$, $(5, -12)$, $(-1, 11)$

Graph each of the following circles. See Examples 7 and 8.

19. $x^2 + y^2 = 36$

20. $x^2 + y^2 = 81$

21. $(x - 2)^2 + y^2 = 36$

22. $x^2 + (y + 3)^2 = 49$

23. $(x + 2)^2 + (y - 5)^2 = 12$

24. $(x - 4)^2 + (y - 3)^2 = 8$

Find an equation for each of the following circles. See Example 6.

25. Center $(1, 4)$, radius 3

26. Center $(-2, 5)$, radius 4

27. Center $(-1, 2)$, passing through $(2, 6)$

28. Center $(2, -7)$, passing through $(-2, -4)$

29. Center $(-3, -2)$, tangent to the x-axis

30. Center $(5, -1)$, tangent to the y-axis

31. Endpoints of a diameter at $(4, 3)$ and $(-4, -3)$

32. Endpoints of a diameter at $(-1, 3)$ and $(4, -9)$

Plot the following points, and then use the same axes to plot the points that are symmetric to the given point with respect to the following elements: (a) x-axis, (b) y-axis, (c) origin.

33. $(5, -3)$ **34.** $(-6, 1)$ **35.** $(-4, -2)$

36. $(-8, 3)$ **37.** $(-8, 0)$ **38.** $(0, -3)$

In Exercises 39–48, use the tests for symmetry to decide whether the graph of each relation is symmetric with respect to the x-axis, the y-axis, or the origin. See Example 9.

39. $x^2 + y^2 = 5$ **40.** $y^2 = 4 - x^2$ **41.** $y = x^2 - 8x$ **42.** $y = 4x - x^2$

43. $y = x^3$ **44.** $y = -x^3$ **45.** $y = \dfrac{1}{1 + x^2}$ **46.** $y = \dfrac{-1}{x^2 + 9}$

47. $xy = 2$ **48.** $xy = -6$

Give all quadrants in which the set of points satisfying the following conditions for ordered pairs (x, y) are located.

49. $x > 0$ **50.** $y < 0$ **51.** $xy < 0$ **52.** $\dfrac{x}{y} > 0$

53. $|x| < 3, y < -2$ **54.** $|y| < 2, x > 1$

55. Show that the points $(-2, 2)$, $(13, 10)$, $(21, -5)$, and $(6, -13)$ are the vertices of a rhombus (all sides equal in length).

56. Are the points $A(1, 1)$, $B(5, 2)$, $C(3, 4)$, $D(-1, 3)$ the vertices of a parallelogram? Of a rhombus (all sides equal in length)?

57. Use the distance formula to write an equation for all points that are 5 units from $(0, 0)$. Sketch a graph showing these points.

58. Write an equation for all points 3 units from $(-5, 6)$. Sketch a graph showing these points.

59. Find all points (x, y) with $x = y$ that are 4 units from $(1, 3)$.

60. Find all points satisfying $x + y = 0$ that are 8 units from $(-2, 3)$.

61. Let F be some algebraic expression involving x as the only variable. Suppose the equation $y = F$ is changed to $y = -F$. What is the effect on the graph of $y = F$?

Sketch examples of graphs that satisfy the following conditions.

62. Symmetric with respect to the x-axis but not to the y-axis

63. Symmetric with respect to the y-axis but not to the x-axis

64. Symmetric with respect to the origin but to neither the x-axis nor the y-axis

65. Find the equation of the circle having the points $(3, -5)$ and $(-7, 2)$ as endpoints of a diameter.

66. Suppose a circle is tangent to both axes, is in the third quadrant, and has a radius of $\sqrt{2}$. Find its equation.

67. One circle has center at (3, 4) and radius 5. A second circle has center at $(-1, -3)$ and radius 4. Do the circles cross?

68. Does a circle with radius 6 and center at (0, 5) cross a circle with center at $(-5, -4)$ and radius 4?

3.2   Linear Relations

Every *linear relation* has a graph that is a straight line. Examples of linear relations are $2x + 3y = 5$, $x + 5 = 0$, and $y = -2$.

Definition of Linear Relation	A **linear relation** in two variables is a relation that can be written in the form $$Ax + By = C,$$ where A, and B, and C are real numbers with A and B not both 0.

A linear relation written in the form $Ax + By = C$ is in **standard form.**

In the equation $Ax + By = C$, any number can be used for x or y, so the domain and range of a linear relation in which neither A nor B is 0 are both the set of real numbers $(-\infty, +\infty)$.

The graph of a linear relation can be found by plotting points. Since the graph is a line, only two points are needed. Two points that are especially useful for sketching the graph of a line are its *intercepts*. An **x-intercept** is an x-value at which a graph crosses the x-axis. A **y-intercept** is a y-value at which a graph crosses the y-axis. Since $y = 0$ on the x-axis, an x-intercept is found by setting y equal to 0 in the equation and solving for x. Similarly, a y-intercept is found by setting $x = 0$ in the equation and solving for y.

Example 1   Graph $3x + 2y = 6$.

Use the intercepts. The y-intercept is found by letting $x = 0$.

$$3 \cdot 0 + 2y = 6$$
$$2y = 6$$
$$y = 3$$

For the x-intercept, let $y = 0$, getting

$$3x + 2 \cdot \quad = 6$$
$$3x = 6$$
$$x = 2.$$

Plotting $(0, 3)$ and $(2, 0)$ gives the graph in Figure 3.14. A third point could be found as a check if desired. ✚

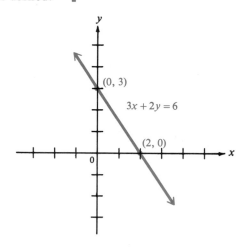

Figure 3.14

Example 2   Graph $y = -3$.

 Since y always equals -3, the value of y can never be 0. This means that the graph has no x-intercept. The only way a straight line can have no x-intercept is for it to be parallel to the x-axis, as shown in Figure 3.15. Notice that the domain of this linear relation is $(-\infty, +\infty)$, but the range is $\{-3\}$. ✚

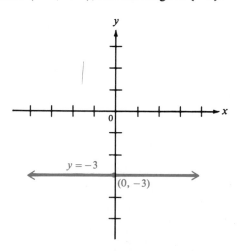

Figure 3.15

Example 3   Graph $4x - 5y = 0$.

Find the intercepts. If $x = 0$, then

$$4(0) - 5y = 0$$
$$-5y = 0$$
$$y = 0.$$

Letting $y = 0$ leads to the same ordered pair, $(0, 0)$. The graph of this relation has just one intercept—at the origin. Find another point by choosing a different value for x (or y). Choosing $x = 5$ gives

$$4(5) - 5y = 0$$
$$20 - 5y = 0$$
$$20 = 5y$$
$$4 = y,$$

which leads to the ordered pair $(5, 4)$. Complete the graph using the two points $(0, 0)$ and $(5, 4)$, with a third point as a check. See Figure 3.16.   ✚

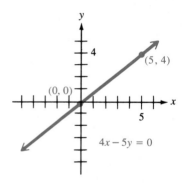

Figure 3.16

Slope   An important characteristic of a straight line is its *slope,* a numerical measure of the steepness of the line. To find this measure, start with the line through the two distinct points (x_1, y_1) and (x_2, y_2), as shown in Figure 3.17 on the following page, where $x_1 \neq x_2$. The difference

$$x_2 - x_1$$

is called the **change in x** and denoted by Δx (read "delta *x*"), where Δ is the Greek letter *delta.* In the same way, the **change in y** can be written

$$\Delta y = y_2 - y_1.$$

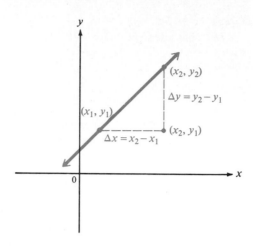

Figure 3.17

The *slope* of a nonvertical line is defined as the quotient of the change in y and the change in x, as follows.

Definition of Slope

The slope m of the line through the points (x_1, y_1) and (x_2, y_2) is

$$m = \frac{\Delta y}{\Delta x} = \frac{y_2 - y_1}{x_2 - x_1},$$

where $\Delta x \neq 0$.

The slope of a line can be found only if the line is nonvertical. This guarantees that $x_2 \neq x_1$ so that the denominator $x_2 - x_1 \neq 0$. It is not possible to define the slope of a vertical line.

> **The slope of a vertical line is undefined.**

Example 4 Find the slope of the line through each of the following pairs of points.

(a) $(-4, 8), (2, -3)$

Let $x_1 = -4$, $y_1 = 8$, and $x_2 = 2$, $y_2 = -3$. Then

$$\Delta y = -3 - 8 = -11$$

and
$$\Delta x = 2 - (-4) = 6.$$

The slope is

$$m = \frac{\Delta y}{\Delta x} = -\frac{11}{6}.$$

(b) $(2, 7)$, $(2, -4)$

A sketch shows that the line through $(2, 7)$ and $(2, -4)$ is vertical. As mentioned above, the slope of a vertical line is not defined. (An attempt to use the definition of slope here would produce a zero denominator.)

(c) $(5, -3)$ and $(-2, -3)$

By definition of slope,

$$m = \frac{-3 - (-3)}{-2 - 5} = \frac{0}{-7} = 0. \quad \clubsuit$$

Drawing a graph through the points in Example 4(c) produces a line that is horizontal, which suggests the following generalization.

The slope of a horizontal line is 0.

Figure 3.18 shows lines of various slopes. As the figure shows, a line with a positive slope goes up from left to right, but a line with a negative slope goes down.

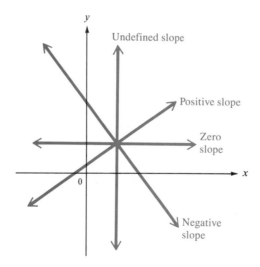

Figure 3.18

It can be shown, using theorems for similar triangles, that the slope is independent of the choice of points on the line. That is, the slope of a line is the same no matter which pair of distinct points on the line are used to find it.

Example 5   Find the slope of the line with equation $4x - 3y = 12$ by using two different pairs of points.

The intercepts are $(0, -4)$ and $(3, 0)$. Using these points gives

$$m = \frac{0 - (-4)}{3 - 0} = \frac{4}{3}.$$

Two other ordered pairs that satisfy the equation are $(2, -4/3)$ and $(15/4, 1)$. With these values, the slope formula gives

$$m = \frac{1 - \left(-\dfrac{4}{3}\right)}{\dfrac{15}{4} - 2} = \frac{\dfrac{7}{3}}{\dfrac{7}{4}} = \frac{7}{3} \cdot \frac{4}{7} = \frac{4}{3},$$

the same slope found with the intercepts. ✤

Example 6   Graph the line passing through $(-1, 5)$ and having slope $-5/3$.

First locate the point $(-1, 5)$ as shown in Figure 3.19. Since the slope of this line is $-5/3$, a change of 3 units horizontally produces a change of -5 units vertically. This gives a second point, $(2, 0)$, which can then be used to complete the graph. ✤

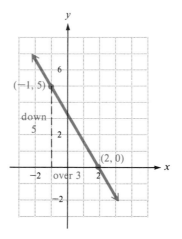

Figure 3.19

Equations of a Line   Figure 3.20 shows the line passing through the fixed point (x_1, y_1) and having slope m. (Assuming that the line has a slope guarantees that it is not vertical.) Let (x, y) be any other point on the line. By the definition of slope, the slope of the line is

$$\frac{y - y_1}{x - x_1}.$$

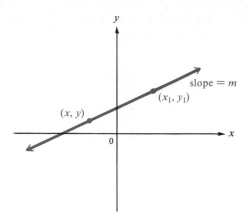

Figure 3.20

Since the slope of the line is m,

$$\frac{y - y_1}{x - x_1} = m.$$

Multiplying both sides by $x - x_1$ gives

$$y - y_1 = m(x - x_1).$$

This result, called the *point-slope form* of the equation of a line, identifies points on a given line: a point (x, y) lies on the line through (x_1, y_1) with slope m if and only if

$$y - y_1 = m(x - x_1).$$

Point-Slope Form

> The line with slope m passing through the point (x_1, y_1) has an equation
> $$y - y_1 = m(x - x_1),$$
> the **point-slope form** of the equation of a line.

Example 7   Write an equation of the line through $(-4, 1)$ with slope -3.

Here $x_1 = -4$, $y_1 = 1$, and $m = -3$. Use the point-slope form of the equation of a line to get

$$y - 1 = -3[x - (-4)]$$
$$y - 1 = -3(x + 4)$$
$$y - 1 = -3x - 12$$

or

$$3x + y = -11,$$

in standard form.   ✤

Example 8   Find an equation of the line through $(-3, 2)$ and $(2, -4)$.
Find the slope first. By the definition of slope,

$$m = \frac{-4 - 2}{2 - (-3)}$$

$$= \frac{-6}{5}.$$

Either $(-3, 2)$ or $(2, -4)$ can be used for (x_1, y_1). Choosing $x_1 = -3$ and $y_1 = 2$ in the point-slope form gives

$$y - 2 = \frac{-6}{5}[x - (-3)]$$

$$5(y - 2) = -6(x + 3)$$

$$5y - 10 = -6x - 18$$

$$6x + 5y = -8.$$

Verify that the same equation results if $(2, -4)$ is used instead of $(-3, 2)$ in the point-slope form.   ✛

As a special case of the point-slope form of the equation of a line, suppose that a line passes through the point $(0, b)$, so the line has y-intercept b. If the line has slope m, then using the point-slope form with $x_1 = 0$ and $y_1 = b$ gives

$$y - y_1 = m(x - x_1)$$

$$y - b = m(x - 0)$$

$$y = mx + b$$

as an equation of the line. Since this result shows the slope of the line and the y-intercept, it is called the *slope-intercept form* of the equation of a line.

Slope-Intercept Form

The line with slope m and y-intercept b has an equation

$$y = mx + b,$$

the **slope-intercept** form of the equation of a line.

Example 9   Find the slope and y-intercept of $3x - y = 2$. Graph the line.
First write $3x - y = 2$ in the slope-intercept form, $y = mx + b$. Do this by solving for y, getting $y = 3x - 2$. This result shows that the slope is $m = 3$ and the y-intercept is $b = -2$. To draw the graph, first locate the y-intercept. See Figure 3.21 on the facing page. Then, as in the previous section, use the slope of 3, or 3/1, to get a second point on the graph. The line through these two points is the graph of $3x - y = 2$.   ✛

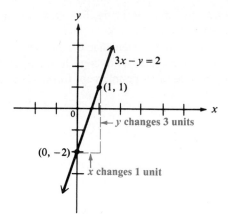

Figure 3.21

In the discussion above, it was assumed that the given line had a slope. The only lines having undefined slope are vertical lines. The vertical line through the point (a, b) passes through all points of the form (a, y), for any value of y. This fact determines the equation of a vertical line.

Equation of a Vertical Line	An equation of the vertical line through the point (a, b) is $x = a$.

For example, the vertical line through $(-4, 9)$ has equation $x = -4$, while the vertical line through $(0, 1/4)$ has equation $x = 0$.

The horizontal line through the point (a, b) passes through all points of the form (x, b), for any value of x. Therefore, the equation of a horizontal line involves only the variable y.

Equation of a Horizontal Line	An equation of the horizontal line through the point (a, b) is $y = b$.

For example, the horizontal line through $(1, -3)$ has the equation $y = -3$. See Figure 3.15 for the graph of this equation.

Parallel and Perpendicular Lines   Slopes can be used to decide whether or not two lines are parallel. Since two parallel lines are equally "steep," they should have the same slope. Also, two different lines with the same "steepness" are parallel. The following result summarizes this discussion.

Parallel Lines	Two different nonvertical lines are parallel if and only if they have the same slope.

Example 10   Find the equation of the line that goes through the point (3, 5) and is parallel to the line $2x + 5y = 4$.

Since it is given that the point (3, 5) is on the line, we need only find the slope to use the point-slope form. Find the slope by writing the equation of the given line in slope-intercept form. (That is, solve for y.)

$$2x + 5y = 4$$

$$y = -\frac{2}{5}x + \frac{4}{5}$$

The slope is $-2/5$. Since the lines are parallel, $-2/5$ is also the slope of the line whose equation is to be found. Substituting $m = -2/5$, $x_1 = 3$, and $y_1 = 5$ into the point-slope form gives

$$y - y_1 = m(x - x_1)$$

$$y - 5 = -\frac{2}{5}(x - 3)$$

$$5(y - 5) = -2(x - 3)$$

$$5y - 25 = -2x + 6$$

$$2x + 5y = 31. \quad \clubsuit$$

Slopes are also used to determine if two lines are perpendicular. Whenever two lines have slopes with a product of -1, the lines are perpendicular.

Perpendicular Lines	Two lines, neither of which is vertical, are perpendicular if and only if their slopes have a product of -1.

Exercises 71–74 outline a proof of this statement.

Example 11   Find the equation of the line L through (3, 0) and perpendicular to $5x - y = 4$.

Use the point-slope form to get the equation of line L. To find the slope of line L, first find the slope of $5x - y = 4$ by solving for y.

$$5x - y = 4$$

$$y = 5x - 4$$

This line has a slope of 5. Since the lines are perpendicular, if line L has slope m, then

$$5m = -1$$

$$m = \frac{-1}{5}.$$

Thus, the required equation is

$$y - 0 = \frac{-1}{5}(x - 3)$$

$$5y = -x + 3$$

$$x + 5y = 3. \quad \clubsuit$$

All the lines discussed above have equations that could be written in the form $Ax + By = C$ for real numbers A, B, and C. As mentioned earlier, the equation $Ax + By = C$ is the standard form of the equation of a line. The various forms of linear equations are listed below.

Linear Equations

General equation	Type of equation
$Ax + By = C$	Standard form (if $A \neq 0$ and $B \neq 0$), x-intercept C/A, y-intercept C/B, slope $-A/B$
$x = k$	Vertical line, x-intercept k, no y-intercept, undefined slope
$y = k$	Horizontal line, y-intercept k, no x-intercept, slope 0
$y = mx + b$	Slope-intercept form, y-intercept b, slope m
$y - y_1 = m(x - x_1)$	Point-slope form, slope m, through (x_1, y_1)

A straight line is often the best approximation of a set of data points that result from a real situation. One way to find the equation of such a straight line is to use two typical data points and the point-slope form of the equation of a line.

Example 12   Consumer prices in the United States at the beginning of each year, measured as a percent of the 1967 average, have produced a graph that is approximately linear. In 1984 the consumer price index was 300%, and in 1987 it was 333%. Let y represent the consumer price index in year x, where $x = 0$ corresponds to 1980. Find the equation of the line through the given two data points.

Since $x = 0$ corresponds to 1980, 1984 is equivalent to $x = 4$ and 1987 is equivalent to $x = 7$. The two data points are $(4, 300)$ and $(7, 333)$. The slope of the line through these points is

$$m = \frac{333 - 300}{7 - 4} = \frac{33}{3} = 11.$$

Choosing (4, 300) as (x_1, y_1) in the point-slope form of the equation of a line gives

$$y - y_1 = m(x - x_1)$$
$$y - 300 = 11(x - 4)$$
$$y - 300 = 11x - 44$$
$$y = 11x + 256$$

as an equation of the line through the given points.

Once the equation is known, it can be used to estimate the consumer price index in future years. For example, for 1989, $x = 9$ and the consumer price index will be

$$y = 11(9) + 256 = 355,$$

or 355% of the 1967 average, assuming no change in the pattern of increases. It is not wise to use an equation like this to make predictions very far into the future, however. ✛

3.2 Exercises

Graph each of the following. See Examples 1–3.

1. $x - y = 4$ **2.** $x + y = 4$ **3.** $3x - y = 6$ **4.** $2x - 3y = 6$

5. $2x + 5y = 10$ **6.** $4x - 3y = 9$ **7.** $x = 2$ **8.** $y = -3$

9. $y = 3x$ **10.** $x = -2y$ **11.** $x = -5y$ **12.** $y = x$

Find the slope of each of the following lines. (Hint: In Exercises 19–24, find two points on the line.) See Examples 4 and 5.

13. Through $(-2, 1)$ and $(3, 2)$ **14.** Through $(-2, 3)$ and $(-1, 2)$ **15.** Through $(8, 4)$ and $(-1, -3)$

16. Through $(-4, -3)$ and $(5, 0)$ **17.** Through $(1, 5)$ and $(-2, 5)$ **18.** Through $(2, 3)$ and $(2, 7)$

19. $y = 3x - 2$ **20.** $y = -2x + 1$ **21.** $3x + 4y = 6$

22. $2x + y = 8$ **23.** $y = 4$ **24.** $x = -6$

25. Through $(-1.978, 4.806)$ and $(3.759, 8.125)$

26. Through $(11.72, 9.811)$ and $(-12.67, -5.009)$

Graph the line passing through the given point and having the indicated slope. See Example 6.

27. Through $(-1, 3)$, $m = 3/2$ **28.** Through $(-2, 8)$, $m = -1$

29. Through $(3, -4)$, $m = -1/3$ **30.** Through $(-2, -3)$, $m = -3/4$

31. Through $(-1, 4)$, $m = 0$ **32.** Through $(9/4, 2)$, undefined slope

Write an equation in standard form for each of the following lines.
See Examples 7–8.

33. Through $(1, 3)$, $m = -2$

34. Through $(2, 4)$, $m = -1$

35. Through $(-5, 4)$, $m = -3/2$

36. Through $(-4, 3)$, $m = 3/4$

37. Through $(-8, 1)$, undefined slope

38. Through $(6, 1)$, $m = 0$

39. Through $(-1, 3)$ and $(3, 4)$

40. Through $(8, -1)$ and $(4, 3)$

41. Through $(0, 3)$ and $(4, 0)$

42. Through $(-3, 0)$ and $(0, -5)$

43. x-intercept 3, y-intercept -2

44. x-intercept -2, y-intercept 4

45. Vertical, through $(-6, 5)$

46. Horizontal, through $(8, 7)$

47. Through $(-1.76, 4.25)$, slope -5.081

48. Through $(5.469, 11.08)$, slope 4.723

Give the slope and y-intercept of each of the following lines. See Example 9.

49. $y = 3x - 1$

50. $y = -2x + 7$

51. $4x - y = 7$

52. $2x + 3y = 16$

53. $4y = -3x$

54. $2y - x = 0$

Write an equation in standard form for each of the following lines.
See Examples 10 and 11.

55. Through $(-1, 4)$, parallel to $x + 3y = 5$

56. Through $(3, -2)$, parallel to $2x - y = 5$

57. Through $(1, 6)$, perpendicular to $3x + 5y = 1$

58. Through $(-2, 0)$, perpendicular to $8x - 3y = 7$

59. Through $(-5, 7)$, perpendicular to $y = -2$

60. Through $(1, -4)$, perpendicular to $x = 4$

In Exercises 61–66, use slopes to decide whether or not the given points lie
on a straight line. (Hint: In each exercise, first find the slope of the line through
the first and second points and then the slope of the line through the second and third
points. If these slopes are the same, the points lie on a straight line.)

61. $M(1, -2)$, $N(3, -18)$, $P(-2, 22)$

62. $A(0, -2)$, $B(3, 7)$, $C(-4, -14)$

63. $X(1, 3)$, $Y(-4, 73)$, $Z(5, -50)$

64. $R(0, -7)$, $S(8, -11)$, $T(-9, -25)$

65. $P(4, 3)$, $Q(2, 0)$, $R(-18, -12)$

66. $D(4, -5)$, $E(3, -5/2)$, $F(-6, 18)$

67. Find k so that the line through $(4, -1)$ and $(k, 2)$ is
 (a) parallel to $3y + 2x = 6$;
 (b) perpendicular to $2y - 5x = 1$.

68. Find r so that the line through $(2, 6)$ and $(-4, r)$ is
 (a) parallel to $2x - 3y = 4$;
 (b) perpendicular to $x + 2y = 1$.

69. Use slopes to show that the quadrilateral with vertices at $(1, 3)$, $(-5/2, 2)$, $(-7/2, 4)$, and $(2, 1)$ is a parallelogram (has opposite sides parallel).

70. Use slopes to show that the square with vertices at $(-2, 5)$, $(4, 5)$, $(4, -1)$, and $(-2, -1)$ has diagonals that are perpendicular.

To prove that two perpendicular lines, neither of which is vertical, have slopes with a product of −1, go through the following steps. Let line L_1 have equation $y = m_1x + b_1$, and let line L_2 have equation $y = m_2x + b_2$. Assume that L_1 and L_2 are perpendicular, and complete right triangle MPN as shown in the figure.

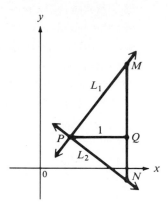

71. Show that MQ has length m_1.

72. Show that QN has length $-m_2$.

73. Show that triangles MPQ and PNQ are similar.

74. Show that $m_1/1 = 1/-m_2$ and that $m_1m_2 = -1$.

In each of the following problems, assume that the data can be approximated fairly closely by a straight line. Find the equation of the line. See Example 12.

75. A company finds that it can make a total of 20 solar heaters for $13,900, while 10 heaters cost $7500. Let y be the total cost to produce x solar heaters.

76. The sales of a small company were $27,000 in its second year of operation and $63,000 in its fifth year. Let y represent sales in year x.

77. When a certain industrial pollutant is introduced into a river, the reproduction of catfish declines. In a given period of time, dumping three tons of the pollutant results in a fish population of 37,000. Also, 12 tons of pollutant produce a fish population of 28,000. Let y be the fish population when x tons of pollutant are introduced into the river.

78. In the snake *Lampropelbis polyzona*, total length y is related to tail length x in the domain 30 mm ≤ x ≤ 200 mm by a linear function. Find such a linear function, if a snake 455 mm long has a 60 mm tail, and a 1050 mm snake has a 140 mm tail.

79. According to research done by the political scientist James March, if the Democrats win 45% of the two-party vote for the House of Representatives, they win 42.5% of the seats. If the Democrats win 55% of the vote, they win 67.5% of the seats. Let y be the percent of seats won, and x the percent of the two-party vote.

80. If the Republicans win 45% of the two-party vote, they win 32.5% of the seats (see Exercise 79). If they win 60% of the vote, they get 70% of the seats. Let y represent the percent of the seats, and x the percent of the vote.

3.3   Parabolas

In this section we discuss relations of the form $y = ax^2 + bx + c$ and $x = ay^2 + by + c$, where $a \neq 0$. The simplest equation of the form $y = ax^2 + bx + c$ is $y = x^2$, with $a = 1$, $b = 0$, and $c = 0$, so this relation is graphed first.

Example 1   Graph $y = x^2$.

Set x equal to 0 in $y = x^2$ to get $y = 0$, which shows that the only intercept is (0, 0). Replace x with $-x$ in the equation to see that the graph is symmetric with respect to the y-axis. Verify that this is the only symmetry. A table giving a few ordered pairs with positive x-values that satisfy $y = x^2$ is shown with the graph in Figure 3.22. Plot these points and the points in quadrant II obtained by symmetry, and draw a smooth curve through the points to complete the graph.   ✛

The graph in Figure 3.22 is called a **parabola.** The line of symmetry for a parabola, the y-axis here, is the **axis** of the parabola. The lowest point on this parabola, the point (0, 0), is called the **vertex** of the parabola. Since any value may be used for x, the domain is $(-\infty, +\infty)$. Here, $y = x^2$, and x^2 is always greater than or equal to zero, so the range is $[0, +\infty)$.

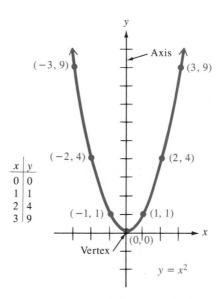

x	y
0	0
1	1
2	4
3	9

Figure 3.22

Starting with $y = x^2$, there are several possible ways to get a more general expression:

$$y = ax^2 \qquad\qquad \text{Multiply by a positive or negative coefficient.}$$
$$y = x^2 + k \qquad\qquad \text{Add a positive or negative constant.}$$
$$y = (x - h)^2 \qquad\qquad \text{Replace } x \text{ with } x - h, \text{ where } h \text{ is a constant.}$$
$$y = a(x - h)^2 + k \qquad \text{Do all of the above.}$$

The graph of each of these relations is still a parabola, but it is modified from that of $y = x^2$. The next few examples show how these changes modify the parabola. The first example shows the result of changing $y = x^2$ to $y = ax^2$.

Example 2   Graph each relation.

(a) $y = 2x^2$

A table of selected ordered pairs is given with the graph in Figure 3.23. The y-values of the ordered pairs of this relation are twice as large as the corresponding y-values for the graph of $y = x^2$. This makes the graph rise more rapidly, so the parabola is narrower than the parabola for $y = x^2$.

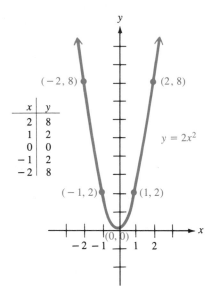

x	y
2	8
1	2
0	0
-1	2
-2	8

Figure 3.23

(b) $y = -\dfrac{1}{2}x^2$

The coefficient $-1/2$ causes the y-values to be closer to the x-axis than for $y = x^2$, making the graph broader than that of $y = x^2$. Because the y-values are negative for each nonzero x-value, this graph opens downward. Again the axis is the line $x = 0$, and the vertex, the *highest* point on the graph, is (0, 0). See Figure 3.24.   ❖

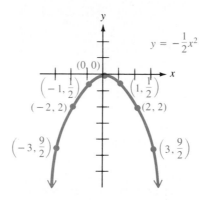

Figure 3.24

As Example 2 suggests, $|a|$ in $y = ax^2$ determines the width of a parabola, so that it is narrower than the graph of $y = x^2$ when $|a| > 1$ and broader than the graph of $y = x^2$ when $|a| < 1$.

The next two examples show how changing $y = x^2$ to $y = x^2 + k$, or to $y = (x - h)^2$, respectively, affects the graph of a parabola.

Example 3   Graph $y = x^2 - 4$.

Each value of y will be 4 less than the corresponding value of y in $y = x^2$. This means that $y = x^2 - 4$ has the same shape as $y = x^2$ but is shifted 4 units down. See Figure 3.25. The vertex of the parabola (on this parabola, the lowest point) is at (0, −4). The axis of the parabola is the vertical line $x = 0$. When

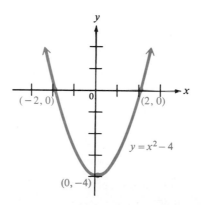

Figure 3.25

the vertex of a parabola is shifted, the intercepts are usually good choices for additional points to plot. Here, the y-intercept is -4, which is the y-value of the vertex. The x-intercepts are found by setting $y = 0$:

$$y = x^2 - 4$$
$$0 = x^2 - 4$$
$$4 = x^2$$
$$x = 2 \quad \text{or} \quad x = -2.$$

The x-intercepts are $(-2, 0)$ and $(2, 0)$. ✤

The vertical shift of the graph in Example 3 is called a **translation.** Example 4 below shows a horizontal translation, which is a shift to the right or left.

Example 4 Graph $y = (x - 4)^2$.

Comparing the two tables of ordered pairs shown in Figure 3.26, for $y = x^2$ and $y = (x - 4)^2$, indicates that this parabola is translated 4 units to the right as compared to the graph of $y = x^2$. For example, the vertex is $(4, 0)$ instead of $(0, 0)$, and $x = -2$ in $y = x^2$ corresponds to the same y-value as $x = 2$ in $y = x^2 - 4$, a difference of $2 - (-2) = 4$ units. The axis of $y = (x - 4)^2$ is the vertical line $x = 4$. See Figure 3.26. ✤

$y = x^2$

x	y
-2	4
-1	1
0	0
1	1
2	4

$y = (x - 4)^2$

x	y
2	4
3	1
4	0
5	1
6	4

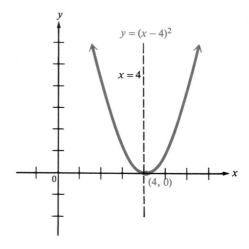

Figure 3.26

A combination of all the changes mentioned above produces the results shown in Example 5.

Example 5   Graph $y = -(x + 3)^2 + 1$.

   This parabola is translated 3 units to the left and 1 unit up. Because of the negative sign, it opens downward, so that the vertex, the point $(-3, 1)$, is the *highest* point on the graph. The axis is the line $x = -3$. The y-intercept is $y = -(0 + 3)^2 + 1 = -8$. By symmetry about the axis $x = -3$, the point $(-6, -8)$ also is on the graph. The x-intercepts are found by solving the equation

$$0 = -(x + 3)^2 + 1$$
$$0 = -(x^2 + 6x + 9) + 1$$
$$0 = -x^2 - 6x - 9 + 1$$
$$0 = -x^2 - 6x - 8$$
$$0 = x^2 + 6x + 8$$
$$0 = (x + 2)(x + 4),$$

from which $x = -2$ or $x = -4$. The graph is shown in Figure 3.27.   ✤

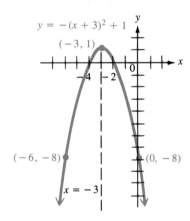

Figure 3.27

Examples 2–5 suggest the following generalizations.

Graph of a Parabola

The graph of

$$y = a(x - h)^2 + k, \quad \text{where } a \neq 0,$$

(a) is a parabola with vertex (h, k), and the vertical line $x = h$ as axis;
(b) opens upward if $a > 0$ and downward if $a < 0$;
(c) is broader than $y = x^2$ if $0 < |a| < 1$ and narrower than $y = x^2$ if $|a| > 1$.

Given the expression $y = ax^2 + bx + c$, where a, b, and c are real numbers and $a \neq 0$, the process of *completing the square* (discussed in Chapter 2) can be used to change $ax^2 + bx + c$ to the form $a(x - h)^2 + k$, so that the vertex and axis may be identified. Follow the steps given in the next example.

Example 6   Graph $y = -3x^2 - 2x + 1$.

To complete the square, first factor out -3 to get

$$y = -3\left(x^2 + \frac{2}{3}x \qquad\right) + 1.$$

(This is necessary to make the coefficient of x^2 equal 1.) As shown earlier, a number must be added to get a perfect square trinomial. But in this case, to avoid changing the left side of the equation, *subtract* the same number *inside* the parentheses, in effect adding 0. As before, add half the coefficient of x, 1/3, squared: $(1/3)^2 = 1/9$. Add and subtract 1/9 inside the parentheses as follows:

$$y = -3\left(x^2 + \frac{2}{3}x + \frac{1}{9} - \frac{1}{9}\right) + 1.$$

Using the distributive property and simplifying gives

$$y = -3\left(x^2 + \frac{2}{3}x + \frac{1}{9}\right) - 3\left(-\frac{1}{9}\right) + 1$$

$$y = -3\left(x^2 + \frac{2}{3}x + \frac{1}{9}\right) + \frac{1}{3} + 1$$

$$y = -3\left(x^2 + \frac{2}{3}x + \frac{1}{9}\right) + \frac{4}{3}.$$

Factor to get

$$y = -3\left(x + \frac{1}{3}\right)^2 + \frac{4}{3}.$$

Now the equation of the parabola is written in the form $y = a(x - h)^2 + k$, and this rewritten equation shows that the axis of the parabola is the vertical line

$$x + \frac{1}{3} = 0 \qquad \text{or} \qquad x = -\frac{1}{3},$$

and that the vertex is $(-1/3, 4/3)$. Use these results, together with the intercepts, symmetry, and additional ordered pairs as needed to get the graph in Figure 3.28. ✛

$y = -3x^2 - 2x + 1$

$y = -3\left(x + \dfrac{1}{3}\right)^2 + \dfrac{4}{3}$

Figure 3.28

In Example 5, the x-value of the vertex, -3, was halfway between the x-intercepts, -4 and -2. This happens whenever the parabola has two x-intercepts because the parabola is symmetric about its axis, which goes through the vertex. See Figure 3.27. In the general equation of a parabola, $y = ax^2 + bx + c$, where $a \neq 0$, the x-intercepts are found by solving the quadratic equation $ax^2 + bx + c = 0$. By the quadratic formula, these intercepts are

$$x = \frac{-b + \sqrt{b^2 - 4ac}}{2a} \quad \text{and} \quad x = \frac{-b - \sqrt{b^2 - 4ac}}{2a}.$$

Then, by the midpoint formula, the x-value of the vertex is

$$\frac{1}{2}\left(\frac{-b + \sqrt{b^2 - 4ac}}{2a} + \frac{-b - \sqrt{b^2 - 4ac}}{2a}\right) = \frac{1}{2}\left(\frac{-b + \sqrt{b^2 - 4ac} - b - \sqrt{b^2 - 4ac}}{2a}\right)$$

$$= \frac{1}{2}\left(\frac{-2b}{2a}\right)$$

$$= \frac{-b}{2a}.$$

This gives a formula for the x-value of the vertex of a parabola.

Vertex of a Parabola

> The x-value of the vertex of the parabola
>
> $$y = ax^2 + bx + c, \text{ where } a \neq 0, \text{ is } -\frac{b}{2a}.$$

Example 7   Use the formula above to find the vertex of the parabola $y = 2x^2 - 4x + 3$.

In this equation, $a = 2$, $b = -4$, and $c = 3$. By the formula given above, the x-value of the vertex of parabola is

$$x = -\frac{b}{2a} = -\frac{-4}{2(2)} = 1.$$

The y-value is found by substituting 1 for x into the equation $y = 2x^2 - 4x + 3$ to get

$$y = 2(1)^2 - 4(1) + 3 = 1.$$

The vertex is $(1, 1)$. ✥

Geometric Definition of a Parabola   Geometrically, a parabola is defined as the set of all points in a plane that are equally distant from a fixed point and a fixed line not containing the point. The point is called the **focus** and the line is the **directrix.** The line through the focus and perpendicular to the directrix is the axis of the parabola. The point on the axis that is equally distant from the focus and the directrix is the vertex of the parabola.

The parabola in Figure 3.29 has the point $(0, p)$ as focus and the line $y = -p$ as directrix. The vertex is $(0, 0)$. Let (x, y) be any point on the parabola. The distance from (x, y) to the directrix is $|y - (-p)|$, while the distance from (x, y) to $(0, p)$ is $\sqrt{(x - 0)^2 + (y - p)^2}$. Since (x, y) is equally distant from the directrix and the focus,

$$|y - (-p)| = \sqrt{(x - 0)^2 + (y - p)^2}.$$

Square both sides, getting

$$(y + p)^2 = x^2 + (y - p)^2$$
$$y^2 + 2py + p^2 = x^2 + y^2 - 2py + p^2,$$
$$4py = x^2,$$

the equation of the parabola with focus $(0, p)$ and directrix $y = -p$. Solving $4py = x^2$ for y gives

$$y = \frac{1}{4p}x^2,$$

so that $1/(4p) = a$ when the equation is written in the form $y = ax^2 + bx + c$.

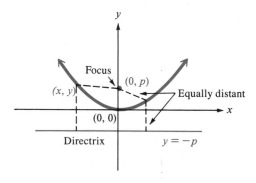

Figure 3.29

This result could be extended to a parabola with vertex at (h, k), focus p units above (h, k), and directrix p units below (h, k), or to a parabola with vertex at (h, k), focus p units below (h, k), and directrix p units above (h, k).

Horizontal Parabolas   The directrix of a parabola could be the *vertical* line $x = -p$, where $p > 0$, with focus on the x-axis at $(p, 0)$, producing a parabola opening to the right. This parabola would have equation $y^2 = 4px$ or $x = [1/(4p)]y^2$. The next examples show the graphs of horizontal parabolas with equations of the form $x = ay^2 + by + c$.

Example 8   Graph $x = y^2$.

The equation $x = y^2$ can be obtained from $y = x^2$ by exchanging x and y. Choosing values of y and finding the corresponding values of x gives the parabola in Figure 3.30. The graph of $x = y^2$, shown in color, is symmetric with respect to the line $y = 0$ and has vertex at $(0, 0)$. For comparison, the graph of $y = x^2$ is shown in black. These graphs are symmetric to each other with respect to the line $y = x$.   ✚

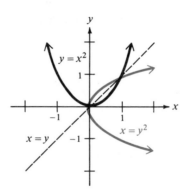

Figure 3.30

Example 9   Graph $x = 2y^2 + 6y + 5$.

To write this equation in the form $x = a(y - k)^2 + h$, complete the square on y as follows:

$$x = 2(y^2 + 3y \qquad) + 5$$
$$= 2\left(y^2 + 3y + \frac{9}{4} - \frac{9}{4}\right) + 5$$
$$= 2\left(y^2 + 3y + \frac{9}{4}\right) + 2\left(-\frac{9}{4}\right) + 5$$
$$x = 2\left(y + \frac{3}{2}\right)^2 + \frac{1}{2}.$$

As this result shows, the vertex of the parabola is the point (1/2, −3/2). The axis is the horizontal line

$$y + \frac{3}{2} = 0 \quad \text{or} \quad y = -\frac{3}{2}.$$

There is no y-intercept, since the vertex is on the right of the y-axis and the graph opens to the right. However, the x-intercept is

$$x = 2(0)^2 + 6(0) + 5 = 5.$$

Using the vertex, the axis of symmetry, and the x-intercept, and plotting a few additional points gives the graph in Figure 3.31. ❖

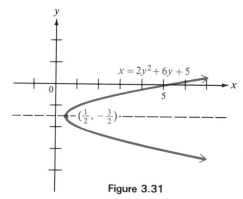

Figure 3.31

The types of parabolas discussed in this section are summarized below.

Graphs of Parabolas

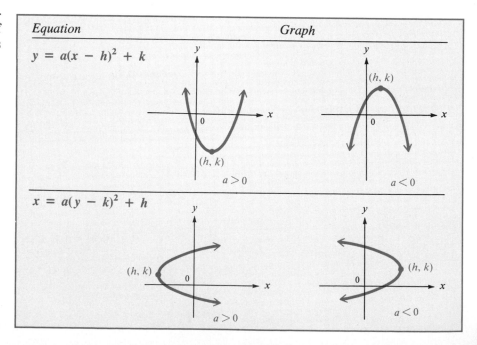

Applications of Parabolas   Parabolas have many practical applications. For example, if a light source is placed at the focus of a parabolic reflector, as in Figure 3.32, light rays reflect parallel to the axis, making a spotlight or flashlight. The process also works in reverse. Light rays from a distant source come in parallel to the axis and are reflected to a point at the focus. (If such a reflector is aimed at the sun, a temperature of several thousand degrees may be obtained.) This use of parabolic reflection is seen in the satellite dishes used to pick up signals from cable television stations.

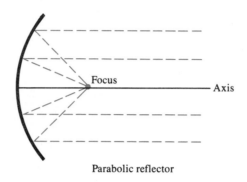

Parabolic reflector

Figure 3.32

The fact that the vertex of a vertical parabola is the highest or lowest point on the graph makes equations of the form $y = ax^2 + bx + c$ important in problems where the maximum or minimum value of some quantity is to be found. When $a < 0$, the y-value of the vertex gives the maximum value of y and the x-value tells where it occurs. Similarly, when $a > 0$, the y-value of the vertex gives the minimum y-value.

Example 10   Ms. Whitney owns and operates Aunt Emma's Pie Shop. She has hired a consultant to analyze her business operations. The consultant tells her that her profit P is given by

$$P = 120x - x^2,$$

where x is the number of units of pies that she makes. How many units of pies should be made in order to maximize the profit? What is the maximum possible profit?

The profit relation can be rewritten as $P = -x^2 + 120x + 0$. Use the methods of this section to find that the vertex of this parabola is (60, 3600). Since $a < 0$, the vertex is the highest point on the graph and produces a *maximum* rather than a minimum. Figure 3.33 on the following page shows the portion of the profit graph located in quadrant I. (Why is quadrant I the only one of interest here?) The maximum profit of $3600 occurs when 60 units of pies are made. In this case, profit increases as more and more pies are made up to 60 units and then decreases as more pies are made past this point. ✛

Figure 3.33

3.3 Exercises

1. Graph the following on the same coordinate system.

(a) $y = 2x^2$ (b) $y = 3x^2$ (c) $y = \frac{1}{2}x^2$ (d) $y = \frac{1}{3}x^2$

(e) How does the coefficient affect the shape of the graph?

2. Graph the following on the same coordinate system.

(a) $y = x^2 + 2$ (b) $y = x^2 - 1$ (c) $y = x^2 + 1$ (d) $y = x^2 - 2$

(e) How do these graphs differ from the graph of $y = x^2$?

3. Graph the following on the same coordinate system.

(a) $y = (x - 2)^2$ (b) $y = (x + 1)^2$ (c) $y = (x + 3)^2$ (d) $y = (x - 4)^2$

(e) How do these graphs differ from the graph of $y = x^2$?

4. Use the quadratic formula to find the values of x when $y = 0$. Use the two values you get to locate the x value of the vertex. Find the axis of each parabola.

(a) $y = x^2 + 8x + 13$ (b) $y = x^2 - 12x + 30$

(c) $y = 3x^2 - 2x + 6$ (d) $y = 5x^2 + 6x - 3$

Graph each of the following parabolas. Give the vertex and axis of each. See Examples 1–7.

5. $y = (x - 2)^2$ **6.** $y = (x + 4)^2$ **7.** $y = (x + 3)^2 - 4$

8. $y = (x - 5)^2 - 4$ **9.** $y = -2(x + 3)^2 + 2$ **10.** $y = -3(x - 2)^2 + 1$

11. $y = -\frac{1}{2}(x + 1)^2 - 3$ **12.** $y = \frac{2}{3}(x - 2)^2 - 1$ **13.** $y = x^2 - 2x + 3$

14. $y = x^2 + 6x + 5$ **15.** $y = -x^2 - 4x + 2$ **16.** $y = -x^2 + 6x - 6$

17. $y = 2x^2 - 4x + 5$ **18.** $y = -3x^2 + 24x - 46$

Find several points satisfying each of the following and then sketch the graph.

19. $y = .14x^2 + .56x - .3$ **20.** $y = .82x^2 + 3.24x - .4$

21. $y = -.09x^2 - 1.8x + .5$ **22.** $y = -.35x^2 + 2.8x - .3$

Graph each horizontal parabola. See Examples 8 and 9.

23. $x = y^2 + 2$ **24.** $x = -y^2$ **25.** $x = (y + 1)^2$

26. $x = (y - 3)^2$ **27.** $x = (y + 2)^2 - 1$ **28.** $x = (y - 4)^2 + 2$

29. $x = -2(y + 3)^2$ **30.** $x = -3(y - 1)^2 + 2$ **31.** $x = \dfrac{1}{2}(y + 1)^2 + 3$

32. $x = \dfrac{2}{3}(y - 3)^2 + 2$ **33.** $x = y^2 + 2y - 8$ **34.** $x = y^2 - 6y + 7$

35. $x = -2y^2 + 2y - 3$ **36.** $x = -4y^2 - 4y - 3$

Work the following word problems. See Example 10.

37. Glenview Community College wants to construct a rectangular parking lot on land bordered on one side by a highway. It has 320 ft of fencing with which to fence off the other three sides. What should be the dimensions of the lot if the enclosed area is to be a maximum? [*Hint:* Let x represent the width of the lot and let $320 - 2x$ represent the length. Graph the area parabola, $A = x(320 - 2x)$, and investigate the vertex.]

38. What would be the maximum area that could be enclosed by the college's 320 ft of fencing if it decided to close the entrance by enclosing all four sides of the lot? (See Exercise 37.)

39. George runs a sandwich shop. By studying data concerning his past costs, he has found that the cost of operating his shop is given by

$$C = 2x^2 - 1200x + 180,100,$$

where C is the daily cost to make x sandwiches. Find the number of sandwiches George must sell to minimize the cost. What is the minimum cost?

40. The revenue of a charter bus company depends on the number of unsold seats. If the revenue R is given by

$$R = 5000 + 50x - x^2,$$

where x is the number of unsold seats, find the maximum revenue and the number of unsold seats which produce maximum revenue.

41. The number of mosquitoes M, in millions, in a certain area of Kentucky depends on the June rainfall x, in inches, approximately as follows.

$$M = 10x - x^2$$

Find the rainfall that will produce the maximum number of mosquitoes.

42. If an object is thrown upward with an initial velocity of 32 ft per sec, then its height after t sec is given by

$$h = 32t - 16t^2.$$

Find the maximum height attained by the object. Find the number of seconds it takes the object to hit the ground.

43. Find two numbers whose sum is 20 and whose product is a maximum. (*Hint:* Let x and $20 - x$ be the two numbers, and write an equation for the product.)

44. A charter flight charges a fare of $200 per person plus $4 per person for each unsold seat on the plane. If the plane holds 100 passengers, and if x represents the number of unsold seats, find the following.
  (a) An expression for the total revenue received for the flight (*Hint:* Multiply the number of people flying, $100 - x$, by the price per ticket.)
  (b) The graph for the expression of part (a)
  (c) The number of unsold seats that will produce the maximum revenue
  (d) The maximum revenue

45. The demand for a certain type of cosmetic is given by

$$p = 500 - x,$$

where p is the price when x units are demanded.
  (a) Find the revenue R that would be obtained at a price of x. (*Hint:* Revenue = Demand × Price.)
  (b) Graph the revenue relation R.
  (c) From the graph of the revenue relation, estimate the price that will produce maximum revenue.
  (d) What is the maximum revenue?

46. For the months of June through October, the percent of maximum possible chlorophyll production in a leaf is approximated by C, where

$$C = 10x + 50.$$

Here x is time in months with $x = 1$ representing June. From October through December, C is approximated by

$$C = -20(x - 5)^2 + 100,$$

with x as above. Find the percent of maximum possible chlorophyll production in each of the following months: **(a)** June **(b)** July **(c)** September **(d)** October **(e)** November **(f)** December.

47. Use your results from Exercise 46 to sketch a graph of $y = C$ from June through December. In what month is chlorophyll production at a maximum?

48. An arch is shaped like a parabola. It is 30 m wide at the base and 15 m high. How wide is the arch 10 m from the ground?

49. A culvert is shaped like a parabola, 18 cm across the top and 12 cm deep. How wide is the culvert 8 cm from the top?

50. (a) Graph the parabola $y = 2x^2 + 5x - 3$.
  (b) Use the graph to find the solution of the quadratic inequality $2x^2 + 5x - 3 < 0$.
  (c) Use the graph to find the solution of the quadratic inequality $2x^2 + 5x - 3 > 0$.

51. Find a value of c so that $y = x^2 - 10x + c$ has exactly one x-intercept.

52. Find b so that $y = x^2 + bx + 9$ has exactly one x-intercept.

Use the geometric definition of parabola given in the text to find the equation of each parabola having the given point as focus.

53. $(0, 3)$, directrix $y = -3$ **54.** $(0, -5)$, directrix $y = 5$ **55.** $(-2, 0)$, directrix $x = 2$

56. $(8, 0)$, directrix $x = -8$ **57.** $(3, 6)$, vertex $(3, 4)$ **58.** $(-5, 2)$, vertex $(-5, 5)$

59. $(-4, 1)$, directrix $x = -7$ **60.** $(3, -4)$, directrix $x = 9$

3.4 Ellipses and Hyperbolas

This section completes our discussion of second-degree relations by introducing the *ellipse* and the *hyperbola*.

Definition of Ellipse	An **ellipse** is the set of all points in a plane the sum of whose distances from two fixed points is constant. The two fixed points are called the **foci** of the ellipse.

For example, the ellipse in Figure 3.34 has foci at points F and F'. By the definition, the ellipse is made up of all points P such that the sum $d(P, F) + d(P, F')$ is a constant. The ellipse in Figure 3.34 has its **center** at the origin. Points V and V' are the **vertices** of the ellipse, and the line segment connecting V and V' is the **major axis.** The foci always lie on the major axis. The line segment from B to B' is the **minor axis.**

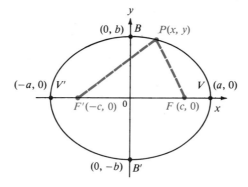

Figure 3.34

If the foci are chosen to be on the x-axis (or y-axis), with the center of the ellipse at the origin, then the distance formula and the definition of an ellipse can be used to obtain the following result. (See Exercise 57.)

Equation of an Ellipse	The ellipse centered at the origin with x-intercepts a and $-a$, and y-intercepts b and $-b$, has equation $$\frac{x^2}{a^2} + \frac{y^2}{b^2} = 1,$$ where $a \neq b$.

In an equation of an ellipse, the coefficients of x^2 and y^2 must be different positive numbers. (What happens if the coefficients are equal?)

An ellipse is the graph of a relation. As suggested by the graph in Figure 3.34, if the ellipse has equation $(x^2/a^2) + (y^2/b^2) = 1$, the domain is $[-a, a]$ and the range is $[-b, b]$. Notice that the ellipse in Figure 3.34 is symmetric with respect to the x-axis, the y-axis, and the origin. More generally, every ellipse is symmetric with respect to its major axis, its minor axis, and its center.

Ellipses have many useful applications. As the earth makes its year-long journey around the sun, it traces an ellipse. Spacecraft travel around the earth in elliptical orbits, and planets make elliptical orbits around the sun. An interesting recent application is the use of an elliptical tub in the nonsurgical removal of kidney stones.

Example 1   Graph $4x^2 + 9y^2 = 36$.

To get the form of the equation of an ellipse, divide both sides by 36.

$$\frac{x^2}{9} + \frac{y^2}{4} = 1$$

This ellipse is centered at the origin, with x-intercepts 3 and -3, and y-intercepts 2 and -2. Additional ordered pairs that satisfy the equation of the ellipse may be found and plotted as needed (a calculator with a square root key will be helpful). The domain of this relation is $[-3, 3]$, and the range is $[-2, 2]$. The graph is shown in Figure 3.35.   ❖

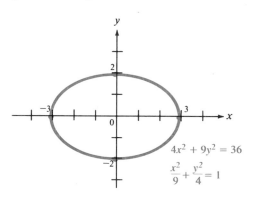

Figure 3.35

Example 2   Give the equation of the ellipse with center at the origin, a vertex at $(5, 0)$, and minor axis of length 6.

The equation will have the form $(x^2/a^2) + (y^2/b^2) = 1$. One vertex is at $(5, 0)$, so $a = 5$. The minor axis equals $2b$, so

$$2b = 6$$
$$b = 3.$$

The equation is

$$\frac{x^2}{25} + \frac{y^2}{9} = 1. \quad ✚$$

Just as a circle need not have its center at the origin, an ellipse may also be translated away from the origin.

Example 3   Graph $\dfrac{(x - 2)^2}{9} + \dfrac{(y + 1)^2}{16} = 1$.

This relation represents an ellipse centered at $(2, -1)$. Graph the ellipse using the fact that $a = 3$ and $b = 4$. Start at $(2, -1)$ and locate two points each 3 units away from $(2, -1)$ on a horizontal line, one to the right of $(2, -1)$ and one to the left. Locate two other points on a vertical line through $(2, -1)$, one 4 units up and one 4 units down. Since $b > a$, the vertices are on the vertical line through the center. The vertices are $(2, 3)$ and $(2, -5)$. Find additional points as necessary. The final graph is shown in Figure 3.36. As the graph suggests, the domain is $[-1, 5]$, and the range is $[-5, 3]$.   ✚

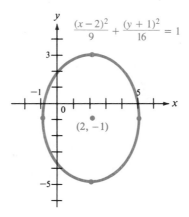

Figure 3.36

Hyperbolas   An ellipse was defined as the set of all points in a plane the sum of whose distances from two fixed points is constant. A **hyperbola** is the set of all points in a plane the *difference* of whose distances from two fixed points (called **foci**) is a positive constant. Some applications of hyperbolas are given in the exercises.

As with ellipses, the equation of a hyperbola can be found from the distance formula and the definition of a hyperbola. (See Exercise 58.)

Equations of Hyperbolas

A hyperbola centered at the origin, with x-intercepts a and $-a$, has an equation of the form

$$\frac{x^2}{a^2} - \frac{y^2}{b^2} = 1,$$

while a hyperbola centered at the origin, with y-intercepts b and $-b$, has an equation of the form

$$\frac{y^2}{b^2} - \frac{x^2}{a^2} = 1.$$

Some texts use $y^2/a^2 - x^2/b^2 = 1$ for this last equation. For a brief introduction such as this, the form given is commonly used.

The x-intercepts are the **vertices** of a hyperbola with the equation $(x^2/a^2) - (y^2/b^2) = 1$, and the y-intercepts are the vertices of a hyperbola with the equation $(y^2/b^2) - (x^2/a^2) = 1$. The line segment between the vertices is the **transverse axis** of the hyperbola, and the midpoint of the transverse axis is the **center** of the hyperbola.

Example 4   Graph the hyperbola $\dfrac{x^2}{16} - \dfrac{y^2}{9} = 1$.

By the first equation of a hyperbola given earlier, the hyperbola is centered at the origin and has x-intercepts 4 and -4. However, if $x = 0$,

$$-\frac{y^2}{9} = 1 \qquad \text{or} \qquad y^2 = -9,$$

which has no real solutions. For this reason, the graph has no y-intercepts. To complete the graph, find some other ordered pairs that belong to it. For example, if $x = 6$,

$$\frac{6^2}{16} - \frac{y^2}{9} = 1$$

$$-\frac{y^2}{9} = 1 - \frac{36}{16}$$

$$\frac{y^2}{9} = \frac{20}{16}$$

$$y^2 = \frac{180}{16} = \frac{45}{4}$$

$$y = \pm\frac{3\sqrt{5}}{2} \approx \pm 3.4.$$

The graph includes the points (6, 3.4) and (6, −3.4). If $x = -6$, $y \approx \pm 3.4$, so the points (−6, 3.4) and (−6, −3.4) also are on the graph. These points, along with other points on the graph, were used to help sketch the final graph shown in Figure 3.37. The graph suggests that the domain of this hyperbola is $(-\infty, 4] \cup [4, +\infty)$, while the range is $(-\infty, +\infty)$. Using the tests for symmetry would show that this hyperbola is symmetric with respect to the x-axis, the y-axis, and the origin.   ✚

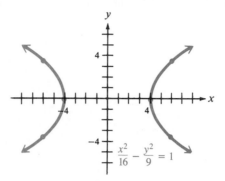

Figure 3.37

Starting with

$$\frac{x^2}{a^2} - \frac{y^2}{b^2} = 1$$

and solving for y gives

$$\frac{x^2}{a^2} - 1 = \frac{y^2}{b^2}$$

$$\frac{x^2 - a^2}{a^2} = \frac{y^2}{b^2}$$

$$y = \pm \frac{b}{a}\sqrt{x^2 - a^2}. \qquad (1)$$

If x^2 is very large in comparison to a^2, the difference $x^2 - a^2$ would be very close to x^2. If this happens, then the points satisfying equation (1) above would be very close to one of the lines

$$y = \pm \frac{b}{a}x.$$

Thus, as $|x|$ gets larger and larger, the points of the hyperbola $x^2/a^2 - y^2/b^2 = 1$ come closer and closer to the lines $y = (\pm b/a)x$. These lines, called the *asymptotes* of the hyperbola, are very helpful when graphing the hyperbola. An **asymptote** is a line that the graph of a relation approaches but never reaches. Asymptotes are discussed again in Section 5.6.

Example 5 Graph $\dfrac{x^2}{25} - \dfrac{y^2}{49} = 1$.

For this hyperbola, $a = 5$ and $b = 7$. With these values, $y = (\pm b/a)x$ becomes $y = (\pm 7/5)x$. Use the x-intercepts: if $x = 5$, then $y = (\pm 7/5)(5) = \pm 7$, and if $x = -5$, $y = \pm 7$. These four points, $(5, 7)$, $(5, -7)$, $(-5, 7)$, and $(-5, -7)$, lead to the rectangle shown in Figure 3.38. The extended diagonals of this rectangle are the asymptotes of the hyperbola. The hyperbola has x-intercepts 5 and -5. The final graph is shown in Figure 3.38. ✤

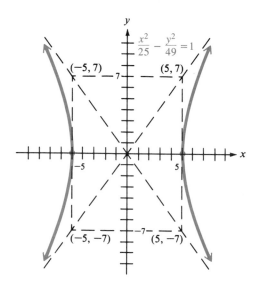

Figure 3.38

The rectangle used to graph the hyperbola in Example 5 is called the *fundamental rectangle*. For a given hyperbola, the equations of the asymptotes can be found by replacing the 1 with 0 in either of the general equations of a hyperbola.

Asymptotes of a Hyperbola	The **asymptotes** of the hyperbola with equation

$$\dfrac{x^2}{a^2} - \dfrac{y^2}{b^2} = 1 \quad \text{or} \quad \dfrac{y^2}{b^2} - \dfrac{x^2}{a^2} = 1$$

are the extended diagonals of the **fundamental rectangle** with vertices at (a, b), $(a, -b)$, $(-a, b)$, and $(-a, -b)$. The equations of the asymptotes can be found by replacing 1 with 0 in the equation of the hyperbola and solving for y.

Example 6   Graph $25y^2 - 4x^2 = 100$.

Divide each side by 100 to get

$$\frac{y^2}{4} - \frac{x^2}{25} = 1.$$

This hyperbola is centered at the origin, has foci on the y-axis, and has y-intercepts 2 and -2. To find the equations of the asymptotes, replace 1 with 0, getting

$$\frac{y^2}{4} - \frac{x^2}{25} = 0$$

$$\frac{y^2}{4} = \frac{x^2}{25}$$

$$y^2 = \frac{4x^2}{25}$$

$$y = \pm\frac{2}{5}x.$$

As indicated above, use the points $(5, 2)$, $(5, -2)$, $(-5, 2)$, and $(-5, -2)$ to get the fundamental rectangle shown in Figure 3.39. Use the diagonals of this rectangle to determine the asymptotes for the graph, as shown in Figure 3.39. ✛

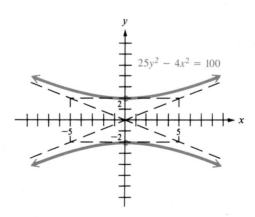

Figure 3.39

Example 7   Graph $\dfrac{(y + 2)^2}{9} - \dfrac{(x + 3)^2}{4} = 1$.

This equation represents a hyperbola that is translated away from the origin, with center at $(-3, -2)$. See Figure 3.40 on the following page.   ✛

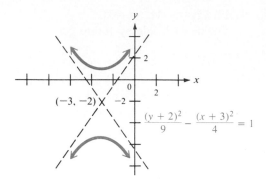

$$\frac{(y + 2)^2}{9} - \frac{(x + 3)^2}{4} = 1$$

Figure 3.40

Example 8 Write the equation of the hyperbola centered at $(-2, 1)$, with a vertex at $(-2, 3)$, and with a equal to half of b.

Since the center is at $(-2, 1)$ and $a < b$, the equation will have the form

$$\frac{(y - 1)^2}{b^2} - \frac{(x + 2)^2}{a^2} = 1.$$

The distance from the center to the given vertex at $(-2, 3)$ gives b for this hyperbola. Using the distance formula,

$$b = \sqrt{(-2 + 2)^2 + (1 - 3)^2} = 2.$$

From the information given, $a = (1/2)b = 1$, so the equation is

$$\frac{(y - 1)^2}{4} - \frac{(x + 2)^2}{1} = 1 \quad \clubsuit$$

Conic Sections The graphs of the second-degree equations studied in this chapter, parabolas, hyperbolas, ellipses, and circles, are called **conic sections** since each can be obtained by cutting a cone with a plane, as shown in Figure 3.41.

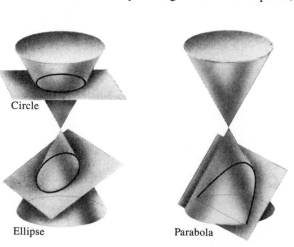

Circle

Ellipse

Parabola

Hyperbola

Figure 3.41

The following chart summarizes our work with conic sections.

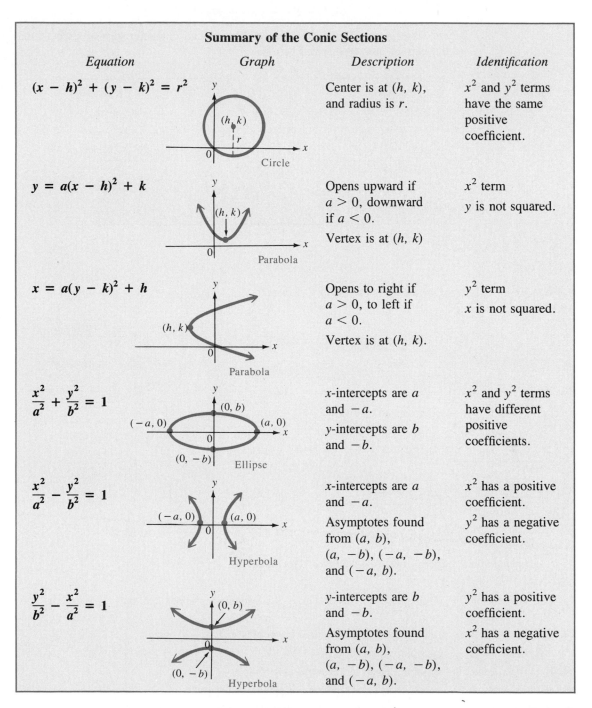

Summary of the Conic Sections

Equation	Graph	Description	Identification
$(x - h)^2 + (y - k)^2 = r^2$	Circle	Center is at (h, k), and radius is r.	x^2 and y^2 terms have the same positive coefficient.
$y = a(x - h)^2 + k$	Parabola	Opens upward if $a > 0$, downward if $a < 0$. Vertex is at (h, k)	x^2 term y is not squared.
$x = a(y - k)^2 + h$	Parabola	Opens to right if $a > 0$, to left if $a < 0$. Vertex is at (h, k).	y^2 term x is not squared.
$\dfrac{x^2}{a^2} + \dfrac{y^2}{b^2} = 1$	Ellipse	x-intercepts are a and $-a$. y-intercepts are b and $-b$.	x^2 and y^2 terms have different positive coefficients.
$\dfrac{x^2}{a^2} - \dfrac{y^2}{b^2} = 1$	Hyperbola	x-intercepts are a and $-a$. Asymptotes found from (a, b), $(a, -b)$, $(-a, -b)$, and $(-a, b)$.	x^2 has a positive coefficient. y^2 has a negative coefficient.
$\dfrac{y^2}{b^2} - \dfrac{x^2}{a^2} = 1$	Hyperbola	y-intercepts are b and $-b$. Asymptotes found from (a, b), $(a, -b)$, $(-a, -b)$, and $(-a, b)$.	y^2 has a positive coefficient. x^2 has a negative coefficient.

lines, points included here

All conic sections presented in this chapter have equations of the form

$$Ax^2 + Bx + Cy^2 + Dy + E = 0,$$

where either A or C must be nonzero. The special characteristics of each of the conic sections are summarized below.

Equations of Conic Sections

Conic section	Characteristic	Example
Parabola	Either $A = 0$ or $C = 0$, but not both.	$x^2 = y + 4$ $(y - 2)^2 = -(x + 3)$
Circle	$A = C \neq 0$	$x^2 + y^2 = 16$
Ellipse	$A \neq C, AC > 0$	$\dfrac{x^2}{16} + \dfrac{y^2}{25} = 1$
Hyperbola	$AC < 0$	$x^2 - y^2 = 1$

3.4   Exercises

Sketch the graph of each of the following ellipses or hyperbolas.
See Examples 1 and 3–7.

1. $\dfrac{x^2}{9} + \dfrac{y^2}{4} = 1$
2. $\dfrac{x^2}{16} + \dfrac{y^2}{36} = 1$
3. $\dfrac{x^2}{6} + \dfrac{y^2}{9} = 1$
4. $\dfrac{x^2}{8} - \dfrac{y^2}{12} = 1$

5. $x^2 + 4y^2 = 16$
6. $25x^2 + 9y^2 = 225$
7. $x^2 = 9 + y^2$
8. $y^2 = 16 + x^2$

9. $25x^2 - 4y^2 = -100$
10. $4x^2 - y^2 = -16$
11. $\dfrac{x^2}{1/9} + \dfrac{y^2}{1/16} = 1$
12. $\dfrac{x^2}{4/25} + \dfrac{y^2}{9/49} = 1$

13. $\dfrac{64x^2}{9} + \dfrac{25y^2}{36} = 1$
14. $\dfrac{121x^2}{25} + \dfrac{16y^2}{9} = 1$

15. $\dfrac{(x - 1)^2}{9} + \dfrac{(y + 3)^2}{25} = 1$
16. $\dfrac{(x + 3)^2}{16} + \dfrac{(y - 2)^2}{36} = 1$

17. $\dfrac{(x - 3)^2}{16} - \dfrac{(y + 2)^2}{49} = 1$
18. $\dfrac{(y - 5)^2}{4} - \dfrac{(x + 1)^2}{9} = 1$

19. $\dfrac{(y + 1)^2}{25} - \dfrac{(x - 3)^2}{36} = 1$
20. $\dfrac{(x + 2)^2}{16} - \dfrac{(y + 2)^2}{25} = 1$

Write an equation for each of the following ellipses or hyperbolas.
See Examples 2 and 8.

21. Ellipse, center at the origin, length of major axis $= 12$, endpoint of minor axis at $(0, 4)$

22. Ellipse, center at the origin, vertex $(6, 0)$, length of minor axis $= 8$

23. Hyperbola, center at the origin, vertex $(0, 2)$, $a = 2b$

24. Hyperbola, center at the origin, length of transverse axis $= 8$, $b = 3a$

25. Ellipse, center at $(2, -2)$, $a = 4$, $b = 3$, major axis parallel to the x-axis

26. Hyperbola, center at $(-3, 1)$, $a = 4$, $b = 2$, transverse axis parallel to the y-axis

27. Hyperbola, center at (4, 3), vertex (1, 3), $b = 2$

28. Ellipse, center at $(-1, -2)$, length of minor axis = 4, vertex $(-1, 1)$

Identify each of the following conic sections. Draw a graph of each that has a graph.

29. $x^2 = 25 + y^2$

30. $x^2 = 25 - y^2$

31. $9x^2 + 36y^2 = 36$

32. $x^2 = 4y - 8$

33. $\dfrac{(x + 3)^2}{16} + \dfrac{(y - 2)^2}{16} = 1$

34. $\dfrac{(x - 4)}{8} - \dfrac{(y + 1)^2}{2} = 0$

35. $y^2 - 4y = x + 4$

36. $11 - 3x = 2y^2 - 8y$

37. $(x + 7)^2 + (y - 5)^2 + 4 = 0$

38. $4(x - 3)^2 + 3(y + 4)^2 = 0$

39. $3x^2 + 6x + 3y^2 - 12y = 12$

40. $2x^2 - 8x + 5y^2 + 20y = 12$

41. $x^2 - 6x + y = 0$

42. $x - 4y^2 - 8y = 0$

43. $4x^2 - 8x - y^2 - 6y = 6$

44. $x^2 + 2x = y^2 - 4y - 2$

45. $4x^2 - 8x + 9y^2 + 54y = -84$

46. $3x^2 + 12x + 3y^2 = -11$

47. $6x^2 - 12x + 6y^2 - 18y + 25 = 0$

48. $4x^2 - 24x + 5y^2 + 10y + 41 = 0$

49. $x^2 + y^2 = 2y + 48$

50. $x^2 + 4x + y^2 = 21$

51. $x^2 - 2.84x + y^2 + 1.4y + 1.8664 = 0$

52. $x^2 + 7.4x + y^2 - 3.8y + 16.09 = 0$

Work the following word problems.

53. Draftspeople often use the method shown in the sketch to draw an ellipse. Explain why the method works.

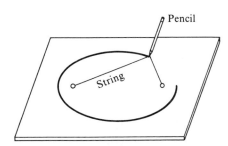

54. Ships and planes often use a location finding system called LORAN. With this system, a radio transmitter at M on the figure sends out a series of pulses. When each pulse is received at transmitter S, it then sends out a pulse. A ship at P receives pulses from both M and S. A receiver on the ship measures the difference in the arrival times of the pulses. The navigator then consults a special map, showing certain curves according to the differences in arrival times. In this way, the ship can be located as lying on a portion of which curve?

55. Microphones are placed at points $(-c, 0)$ and $(c, 0)$. An explosion occurs at point $P(x, y)$ having positive x-coordinate. (See the figure.) The sound is detected at the closer microphone t sec before being detected at the farther microphone. Assume that sound travels at a speed of 330 m per sec, and show that P must be on the hyperbola

$$\frac{x^2}{330^2 t^2} - \frac{y^2}{4c^2 - 330^2 t^2} = \frac{1}{4}.$$

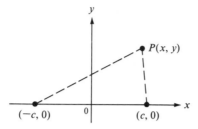

56. The orbit of Venus is an ellipse with the sun at one focus. An approximate equation for its orbit is

$$\frac{x^2}{5013} + \frac{y^2}{4970} = 1,$$

where x and y are measured in millions of miles. Find the widest possible distance across the ellipse.

57. Suppose that $(c, 0)$ and $(-c, 0)$ are the foci of an ellipse. Suppose that the sum of the distances from any point (x, y) of the ellipse to the two foci is $2a$. See the accompanying figure.

(a) Use the distance formula to show that the equation of the resulting ellipse is

$$\frac{x^2}{a^2} + \frac{y^2}{a^2 - c^2} = 1.$$

(b) Show that a and $-a$ are the x-intercepts.
(c) Let $b^2 = a^2 - c^2$, and show that b and $-b$ are the y-intercepts.

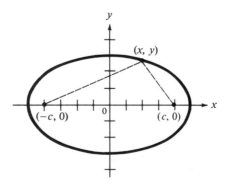

58. Suppose a hyperbola has center at the origin, foci at $F'(-c, 0)$ and $F(c, 0)$, and the value $d(P, F') - d(P, F) = 2a$. Let $b^2 = c^2 - a^2$, and show that the equation of the hyperbola is

$$\frac{x^2}{a^2} - \frac{y^2}{b^2} = 1.$$

3.5 Inequalities

Many mathematical descriptions of real situations are best expressed as inequalities rather than equations. For example, a firm might be able to use a machine *no more* than 12 hours a day, while production of *at least* 500 cases of a certain product might be required to meet a contract. The simplest way to see the solution of an inequality in two variables is to draw its graph.

A line divides a plane into three sets of points: the points of the line itself and the points belonging to the two regions determined by the line. Each of these two regions is called a **half-plane.** In Figure 3.42, line r divides the plane into three different sets of points: line r, half-plane P, and half-plane Q. The points on r belong neither to P nor to Q. Line r is the boundary of each half-plane.

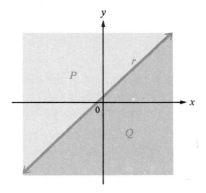

Figure 3.42

A **linear inequality in two variables** is an inequality of the form

$$Ax + By \le C,$$

where A, B, and C are real numbers, with A and B not both equal to 0. (The symbol $\le$ could be replaced with $\ge$, $<$, or $>$.) The graph of a linear inequality is a half-plane, perhaps with its boundary. For example, to graph the linear inequality $3x - 2y \leq 6$, first graph the boundary, $3x - 2y = 6$, as shown in Figure 3.43.

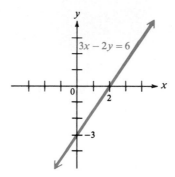

Figure 3.43

Since the points of the line $3x - 2y = 6$ satisfy $3x - 2y \leq 6$, this line is part of the solution. To decide which half-plane (the one above the line $3x - 2y = 6$ or the one below the line) is part of the solution, solve the original inequality for y.

$$3x - 2y \leq 6$$
$$-2y \leq -3x + 6$$
$$y \geq \frac{3}{2}x - 3 \qquad \text{Change} \leq \text{to} \geq$$

For a particular value of x, the inequality will be satisfied by all values of y that are *greater than* or equal to $(3/2)x - 3$. This means that the solution includes the half-plane *above* the line, as well as the line itself. See Figure 3.44.

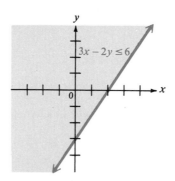

Figure 3.44

There is an alternative method for deciding which side of the boundary line to shade. Choose as a test point any point not on the graph of the equation. The origin, $(0, 0)$, is often a good test point (as long as it does not lie on the boundary line). Substituting 0 for x and 0 for y in the inequality $3x - 2y \leq 6$ gives

$$3(0) - 2(0) \leq 6$$
$$0 \leq 6,$$

a true statement. Since $(0, 0)$ leads to a true result, it is part of the solution of the inequality. Shade the side of the graph containing $(0, 0)$, as shown in Figure 3.44.

Example 1   Graph $x + 4y > 4$.

The boundary here is the straight line $x + 4y = 4$. Since the points on this line do not satisfy $x + 4y > 4$, it is customary to graph the boundary as a dashed line, as in Figure 3.45. To decide which half-plane satisfies the inequality, use a test point. Choosing $(0, 0)$ as a test point gives $0 + 4 \cdot 0 > 4$, or $0 > 4$, a false statement. Since $(0, 0)$ leads to a false statement, shade the side of the graph *not* containing $(0, 0)$, as in Figure 3.45.   ✚

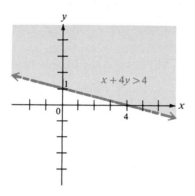

Figure 3.45

Example 2   Graph $y \leq 2x^2 - 3$.

First graph the boundary, the parabola $y = 2x^2 - 3$, as shown in Figure 3.46. Then select any test point not on the parabola, such as $(0, 0)$. Since $(0, 0)$ does not satisfy the original inequality, shade the portion of the graph that does not include $(0, 0)$, as shown in Figure 3.46. Because of the $=$ portion of $\leq$, the points on the parabola itself also belong to the graph.   ✚

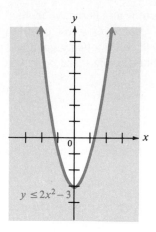

Figure 3.46

Example 3 Graph each inequality.

(a) $y^2 \leq 1 + 4x^2$

Write the boundary $y^2 = 1 + 4x^2$ as $y^2 - 4x^2 = 1$, a hyperbola with y-intercepts 1 and -1, as shown in Figure 3.47. Select any point not on the hyperbola and test it in the original inequality. Since $(0, 0)$ satisfies this inequality, shade the area between the two branches of the hyperbola, as shown in Figure 3.47. The points on the hyperbola are part of the solution.

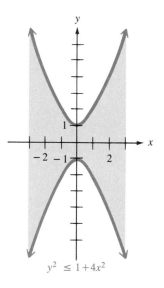

Figure 3.47

(b) $y^2 \geq 1 + 4x^2$

The boundary is the same as in part (a). Since (0, 0) makes this inequality false, the *two* regions above and below the branches of the hyperbola are shaded. Again, the points on the hyperbola itself are included in the solution. ✚

Example 4   Graph $2x - y \leq 4$, where $x < 3$.

Figure 3.48 shows the graphs of both inequalities. The heavily shaded area to the left of the line $x = 3$ and above the line $2x - y = 4$ represents the region that satisfies both inequalities. The point where the boundary lines intersect (3, 2), belongs to the first inequality but not to the second. For this reason, (3, 2) is not part of the solution. This is shown with an open dot on the graph at (3, 2). ✚

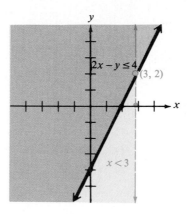

Figure 3.48

Example 5   Graph $16x^2 + y^2 \leq 16$, where $y \geq 0$.

The inequality $16x^2 + y^2 \leq 16$ has as its boundary the ellipse $16x^2 + y^2 = 16$, graphed in Figure 3.49. To decide which region to shade, choose any point off the ellipse, such as (0, 0). Substituting 0 for x and 0 for y in the inequality $16x^2 + y^2 \leq 16$ gives

$$16 \cdot 0^2 + 0^2 \leq 16 \quad \text{or} \quad 0 \leq 16,$$

a true statement. This true statement shows that the solution includes the region containing (0, 0), which is the region inside the ellipse. The inequality $y \geq 0$ includes all points on or above the x-axis. The final graph consists of the region inside the ellipse and above the x-axis, together with its boundary, as shown in Figure 3.49. ✚

Figure 3.49

3.5 Exercises

Graph each of the following inequalities. See Examples 1–3.

1. $x \le 3$ **2.** $y \le -2$ **3.** $x + 2y \le 6$ **4.** $x - y \ge 2$

5. $2x + 3y \ge 4$ **6.** $4y - 3x < 5$ **7.** $3x - 5y > 6$ **8.** $x < 3 + 2y$

9. $5x \le 4y - 2$ **10.** $2x > 3 - 4y$ **11.** $y < 3x^2 + 2$ **12.** $y \le x^2 - 4$

13. $y > (x - 1)^2 + 2$ **14.** $y > 2(x + 3)^2 - 1$ **15.** $x^2 + (y + 3)^2 \le 16$

16. $(x - 4)^2 + (y + 3)^2 \le 9$ **17.** $4x^2 \le 4 - y^2$ **18.** $x^2 + 9y^2 > 9$

19. $9x^2 - 16y^2 > 144$ **20.** $4x^2 \le 36 + 9y^2$

Graph each of the following inequalities. See Examples 4 and 5.

21. $x - 3y < 4$, where $x \le 0$ **22.** $2x + y > 5$, where $y \ge 0$

23. $3x + 2y \ge 6$, where $y \le 2$ **24.** $4x + 3y \le 6$, where $x \le -2$

25. $x + 2y < 4$, where $3x - y > 5$ **26.** $4x + 1 < 2y$, where $-x > y - 3$

27. $2x + 3y < 6$, where $x - 5y \ge 10$ **28.** $3x - 4y > 6$, where $2x + 3y > 4$

29. $y^2 \le x + 3$, where $y \ge 0$ **30.** $x^2 \le 1 - y$, where $x \ge 0$

31. $y^2 \le 49 - x^2$, where $y \le 0$ **32.** $x^2 > 16 - y^2$, where $x \le 0$

33. $\dfrac{x^2}{36} < 1 - \dfrac{y^2}{121}$, where $x \ge 0$ **34.** $\dfrac{y^2}{16} \le 1 + \dfrac{x^2}{25}$, where $y \ge 0$

Chapter 3 *Summary*

Key Words

3.1 ordered pair
relation
domain
range
origin
x-axis
y-axis
rectangular coordinate
  system
Cartesian coordinate
  system
coordinate plane
xy-plane
quadrants
coordinates
midpoint
circle
radius
center

3.2 linear relation
x-intercept
y-intercept
3.3 parabola
axis
vertex, vertices
translation
focus, foci
directrix
3.4 ellipse
major axis
minor axis
hyperbola
transverse axis
asymptotes
fundamental rectangle
conic sections
3.5 half-plane

Key Ideas

3.1 Distance Formula

Suppose $P(x_1, y_1)$ and $R(x_2, y_2)$ are two points in a coordinate plane. Then the distance between P and R, written $d(P, R)$, is

$$d(P, R) = \sqrt{(x_2 - x_1)^2 + (y_2 - y_1)^2}.$$

3.1 Midpoint Formula

The midpoint of the line segment with endpoints (x_1, y_1) and (x_2, y_2) is

$$\left(\frac{x_1 + x_2}{2}, \frac{y_1 + y_2}{2} \right).$$

3.2 Definition of Slope

The slope m of the line through the points (x_1, y_1) and (x_2, y_2) is

$$m = \frac{\Delta y}{\Delta x} = \frac{y_2 - y_1}{x_2 - x_1},$$

where $\Delta x \neq 0$.

3.2 Point-Slope Form

The line with slope m passing through the point (x_1, y_1) has an equation

$$y - y_1 = m(x - x_1),$$

the point-slope form of the equation of a line.

3.2 Slope-Intercept Form

The line with slope m and y-intercept b has an equation

$$y = mx + b,$$

the slope-intercept form of the equation of a line.

3.1 Equations
3.3 of Conic
3.4 Sections

Circle $(x - h)^2 + (y - k)^2 = r^2$

Parabola $y = a(x - h)^2 + k$ or $x = a(y - k)^2 + h$

Ellipse $\dfrac{x^2}{a^2} + \dfrac{y^2}{b^2} = 1$

Hyperbola $\dfrac{x^2}{a^2} - \dfrac{y^2}{b^2} = 1$ or $\dfrac{y^2}{b^2} - \dfrac{x^2}{a^2} = 1$

Chapter 3   *Review Exercises*

Find the distance between each of the following pairs of points, and state the midpoint of each pair.

1. $P(3, -1)$, $Q(-4, 5)$ **2.** $M(-8, 2)$, $N(3, -7)$ **3.** $A(-6, 3)$, $B(-6, 8)$

4. Are the points $(5, 7)$, $(3, 9)$, $(6, 8)$ the vertices of a right triangle?

5. Find all possible values of k so that $(-1, 2)$, $(-10, 5)$, and $(-4, k)$ are the vertices of a right triangle.

6. Find all possible values of x so that the distance between $(x, -9)$ and $(3, -5)$ is 6.

7. Find all points (x, y) with $x = 6$ so that (x, y) is 4 units from $(1, 3)$.

8. Find all points (x, y) with $x + y = 0$ so that (x, y) is 6 units from $(-2, 3)$.

Find equations for each of the circles below.

9. Center $(-2, 3)$, radius 5

10. Center $(\sqrt{5}, -\sqrt{7})$, radius $\sqrt{3}$

11. Center $(-8, 1)$, passing through $(0, 16)$

12. Center $(3, -6)$, tangent to the x-axis

Find the center and radius of each of the following that are circles.

13. $x^2 - 4x + y^2 + 6y + 12 = 0$

14. $x^2 - 6x + y^2 - 10y + 30 = 0$

15. $x^2 + 7x + y^2 + 3y + 1 = 0$

16. $x^2 + 11x + y^2 - 5y + 46 = 0$

Decide whether the equations below have graphs that are symmetric with respect to the x-axis, the y-axis, or the origin.

17. $3y^2 - 5x^2 = 15$ **18.** $x + y^2 = 8$ **19.** $y^3 = x + 1$ **20.** $x^2 = y^3$

21. $|y| = -x$ **22.** $|x + 2| = |y - 3|$ **23.** $|x| = |y|$ **24.** $xy = 8$

Find the slope for each of the following lines that has a slope.

25. Through $(8, 7)$ and $(1/2, -2)$

26. Through $(2, -2)$ and $(3, -4)$

27. Through $(5, 6)$ and $(5, -2)$

28. Through $(0, -7)$ and $(3, -7)$

29. $9x - 4y = 2$ **30.** $11x + 2y = 3$ **31.** $x - 5y = 0$

32. $x - 2 = 0$ **33.** $y + 6 = 0$ **34.** $y = x$

Graph each of the following equations.

35. $3x + 7y = 14$ **36.** $2x - 5y = 5$ **37.** $3y = x$

38. $y = 3$ **39.** $x = -5$ **40.** $y = x$

For each of the following lines, write the equation in standard form.

41. Through $(-2, 4)$ and $(1, 3)$

42. Through $(-2/3, -1)$ and $(0, 4)$

43. Through $(3, -5)$ with slope -2

44. Through $(-4, 4)$ with slope $3/2$

45. Through $(1/5, 1/3)$ with slope $-1/2$

46. x-intercept -3, y-intercept 5

47. No x-intercept, y-intercept $3/4$

48. Through $(2, -1)$, parallel to $3x - y = 1$

49. Through $(0, 5)$, perpendicular to $8x + 5y = 3$

50. Through $(2, -10)$, perpendicular to a line with undefined slope

51. Through $(3, -5)$, parallel to $y = 4$

52. Through $(-7, 4)$, perpendicular to $y = 8$

Graph each of the following lines.

53. Through $(2, -4)$, $m = 3/4$

54. Through $(0, 5)$, $m = -2/3$

55. Through $(-4, 1)$, $m = 3$

56. Through $(-3, -2)$, $m = -1$

Graph each of the following equations.

57. $y = x^2 - 4$

58. $y = 6 - x^2$

59. $y = 3(x + 1)^2 - 5$

60. $y = -\dfrac{1}{4}(x - 2)^2 + 3$

61. $y = x^2 - 4x + 2$

62. $y = -3x^2 - 12x - 1$

Give the vertex and axis of the graph of each of the following equations.

63. $y = -(x + 3)^2 - 9$

64. $y = (x - 7)^2 + 3$

65. $y = x^2 - 7x + 2$

66. $y = -x^2 - 4x + 1$

67. $y = -3x^2 - 6x + 1$

68. $y = 4x^2 - 4x + 3$

Use parabolas to solve each of the following problems.

69. Find two numbers whose sum is 11 and whose product is a maximum.

70. Find two numbers having a sum of 40 such that the sum of the square of one and twice the square of the other is a minimum.

71. Find the rectangular region of maximum area that can be enclosed with 180 m of fencing.

72. Find the rectangular region of maximum area that can be enclosed with 180 m of fencing if no fencing is needed along one side of the region.

Graph each parabola. Give the vertex and axis of each.

73. $x = y^2 - 2$

74. $x = (y + 3)^2$

75. $x = -(y + 1)^2 + 2$

76. $x = y^2 - 4y$

77. $x = -y^2 + 5y + 1$

78. $x = -3y^2 + 6y - 2$

Graph each of the following equations.

79. $\dfrac{x^2}{25} + \dfrac{y^2}{4} = 1$

80. $\dfrac{x^2}{3} + \dfrac{y^2}{16} = 1$

81. $\dfrac{x^2}{4} - \dfrac{y^2}{9} = 1$

82. $\dfrac{y^2}{100} - \dfrac{x^2}{25} = 1$

83. $x^2 = 16 + y^2$

84. $4x^2 + 25y^2 = 100$

85. $\dfrac{25x^2}{9} + \dfrac{4y^2}{25} = 1$

86. $\dfrac{100x^2}{49} + \dfrac{9y^2}{16} = 1$

87. $\dfrac{(x + 1)^2}{16} - \dfrac{(y - 2)^2}{4} = 1$

88. $\dfrac{(x - 2)^2}{9} + \dfrac{(y + 3)^2}{4} = 1$

Graph each inequality.

89. $5x - y \leq 20$

90. $3x - 5 > 10$

91. $y \leq (x - 4)^2$

92. $y < (x - 1)^2 + 3$

93. $25y^2 - 36x^2 > 900$

94. $(x - 2)^2 + (y + 3)^2 \geq 9$

Graph the following inequalities.

95. $2x - y \geq 4$, where $y \geq -2$

96. $3x + 7y < 21$, where $x < 4$

97. $y \leq x^2$, where $x \geq 0$

98. $25x^2 + 4y^2 \geq 100$, where $y \geq 0$

Chapter 3 *Test*

1. Find the distance between $P(-2, 1)$ and $Q(3, 4)$.

2. Find x so that the distance from the origin to the point $(x, -4)$ is 5.

3. Find the equation for a circle centered at $(5, -1)$ and passing through $(8, 3)$.

4. Find the center and radius of the circle with equation

$$x^2 + y^2 + 4x - 6y = -12.$$

5. Decide whether the equation $y = 2x^3 - x$ has a graph that is symmetric with respect to the x-axis, y-axis, or the origin.

Find the slope for each of the following lines.

6. Through $(-3, 4)$ and $\left(\dfrac{3}{2}, \dfrac{1}{2}\right)$

7. $5x - 4y = 0$

8. Graph the equation $4x - 5y = 10$.

For each of the following lines, write the equation in standard form.

9. Through $(-2, 3)$ and $(6, -1)$

10. Through $(2, 7)$, perpendicular to $x = 4$

11. Through $(-6, 2)$, parallel to $2x - y = 8$

12. Graph the line through $(2, -3)$ with slope $-2/3$.

13. Graph $y = 3x^2 - 12x + 9$.

14. Give the vertex and axis of the graph of $y = x^2 - 5x + 4$.

15. A developer has 80 ft of fencing. He wants to put a fence around a rectangular plot of land adjacent to a river. Use a parabola to find the maximum area he can enclose. (Note: the side next to the river will not be fenced.)

16. Graph the parabola $x = -(y + 2)^2 - 4$. Give its vertex and axis.

Graph each of the following equations.

17. $\dfrac{(x + 3)^2}{4} + \dfrac{(y - 2)^2}{9} = 1$

18. $\dfrac{y^2}{25} - \dfrac{x^2}{9} = 1$

19. Graph the inequality $(x - 4)^2 + (y + 3)^2 > 9$.

20. Graph the inequality $x > y^2$, where $y \leq 2$.

4

Functions

In this chapter we begin a careful study of functions by discussing some general topics. In the chapters that follow we will examine specific types of functions and their graphs in detail. The techniques of graphing used in the previous chapter will be very helpful in understanding the functions discussed in this chapter.

4.1 Functions

The graph in Figure 4.1(a) shows the velocity in miles per hour at time t in seconds for a Porsche 928 as it accelerates from rest. For example, the graph shows that 15 sec after starting, the car is going 90 mph, and 19 sec after starting, the speed is 100 mph. The graph in Figure 4.1(b) shows the variation in blood pressure for a typical person. (Systolic and diastolic pressures are the upper and lower limit in the periodic changes in pressure that produce the pulse. The length of time between peaks is called the period of the pulse.)

Each of the graphs in Figure 4.1 is an example of a *function,* which is a special kind of relation.

Definition of Function	A **function** is a relation in which each element in the domain corresponds to exactly one element in the range.

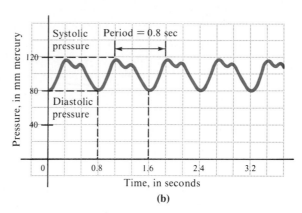

Figure 4.1

Suppose a group of students get together each Monday evening to study algebra (and perhaps watch football). A number giving the student's weight to the nearest kilogram can be associated with each member of this set of students. Since each student has only one weight at a given time, the relationship between students and their weights is a function. The domain is the set of all students in the group, while the range is the set of all possible weights of the students.

In most mathematical applications of functions, the correspondence between the domain and range elements is defined with an equation, like $y = 5x - 11$. The letter x here is called the **independent variable** because it represents the elements in the domain, and y is called the **dependent** variable because it represents the elements in the range. (These equations need not use only x and y as variables; any appropriate letters may be used. In physics, for example, t is often used to represent the independent variable *time*.)

Example 1   Decide whether the following sets are functions. Give the domain and range of each relation.

(a) {(1, 2), (3, 4), (5, 6), (7, 8), (9, 10)}

The domain is the set {1, 3, 5, 7, 9}, and the range is {2, 4, 6, 8, 10}. Since each element in the domain corresponds to just one element in the range, this set is a function.

Figure 4.1(a): from *Road and Track,* April 1978. Reprinted with permission of Road & Track.
Figure 4.1(b): from *Calculus for the Life Sciences* by Rodolfo De Sapio. Copyright © 1976, 1978 by W. H. Freeman and Company. Reprinted by permission.

(b) $\{(1, 1), (1, 2), (1, 3), (2, 4)\}$

The domain here is $\{1, 2\}$, and the range is $\{1, 2, 3, 4\}$. Since one element in the domain, 1, has been assigned three different elements from the range, this set is *not* a function.

(c) $\{(-5, 2), (-4, 2), (-3, 2), (-2, 2), (-1, 2)\}$

Here, the domain is $\{-5, -4, -3, -2, -1\}$, and the range is $\{2\}$. Although every element in the domain corresponds to the same range element, this is a function because each element in the domain has exactly one range element assigned to it.

(d) $\{(x, y) \mid y = x - 2, x$ any real number$\}$

Since y is always found by subtracting 2 from x, each x corresponds to just one y, so this relation is a function. Any number can be used for x, and each x will give a number 2 smaller for y; thus, both the domain and the range are the set of all real numbers, or in interval notation, $(-\infty, +\infty)$. ✤

It is common to use the letters f, g, and h to name functions. If f is a function and x is an element in the domain of f, then $f(x)$, read "f of x," is the corresponding element in the range. For example, if f is the function in which a value of x is squared to give the corresponding value in the range, f could be defined as

$$f(x) = x^2.$$

If $x = -5$ is an element from the domain of f, the corresponding element from the range is found by replacing x with -5:

$$f(-5) = (-5)^2 = 25.$$

The number -5 in the domain corresponds to 25 in the range.

This *function notation* can be summarized as follows.

$$y = \big(f(x)\big) = \overbrace{3x + 2}$$

Name of the function — Defining expression

Name of the independent variable

Value of the function

Example 2   Let $g(x) = 3\sqrt{x}$ and $h(x) = 1 + 4x$. Find each of the following.

(a) $g(16)$

To find $g(16)$, replace x in $g(x) = 3\sqrt{x}$ with 16, getting

$$g(16) = 3\sqrt{16} = 3 \cdot 4 = 12.$$

(b) $h(-3) = 1 + 4(-3) = -11$

(c) $g(-4)$ does not exist; -4 is not in the domain of g since $\sqrt{-4}$ is not a real number.

(d) $h(\pi) = 1 + 4\pi$

(e) $g(m) = 3\sqrt{m}$, if m represents a nonnegative real number. ✛

Example 3   Let $f(x) = 2x^2 - 3x$, and find the quotient

$$\frac{f(x + h) - f(x)}{h}, \quad \text{where } h \ne 0,$$

which is important in calculus.

To find $f(x + h)$, replace x in $f(x)$ with $x + h$, to get

$$f(x + h) = 2(x + h)^2 - 3(x + h).$$

Then

$$
\begin{aligned}
\frac{f(x + h) - f(x)}{h} &= \frac{2(x + h)^2 - 3(x + h) - (2x^2 - 3x)}{h} \\
&= \frac{2(x^2 + 2xh + h^2) - 3x - 3h - 2x^2 + 3x}{h} \\
&= \frac{2x^2 + 4xh + 2h^2 - 3x - 3h - 2x^2 + 3x}{h} \\
&= \frac{4xh + 2h^2 - 3h}{h} \\
&= \frac{h(4x + 2h - 3)}{h} \\
&= 4x + 2h - 3. \quad ✛
\end{aligned}
$$

Notice that $f(x + h)$ is not the same as $f(x) + h$. For $f(x) = 2x^2 - 3x$, as shown in Example 3,

$$f(x + h) = 2(x + h)^2 - 3(x + h) = 2x^2 + 4xh + 2h^2 - 3x - 3h$$

but

$$f(x) + h = (2x^2 - 3x) + h = 2x^2 - 3x + h.$$

Throughout this book, if the domain for a function specified by an algebraic formula is not given, it will be assumed to be the largest possible set of real numbers for which the formula is meaningful. For example, if

$$f(x) = \frac{-4x}{2x - 3},$$

then any real number can be used for x except $x = 3/2$, which makes the denominator equal to 0. Based on our assumption, the domain of this function must be $\{x \mid x \ne 3/2\}$.

Example 4   Give the domain and range of each of the following functions.

(a) $f(x) = 4 - 3x$

Here x can take on any value at all, as can $f(x)$, making both the domain and range the set of all real numbers. In interval notation, both the domain and range are $(-\infty, +\infty)$.

(b) $f(x) = x^2 + 4$

Since any real number can be squared, the domain is the set of all real numbers, $(-\infty, +\infty)$. The square of any number is nonnegative, so that $x^2 \geq 0$ and $x^2 + 4 \geq 4$. The range is the interval $[4, +\infty)$.

(c) $g(x) = \sqrt{x - 2}$

If $\sqrt{x - 2}$ is to be a real number, then

$$x - 2 \geq 0 \quad \text{or} \quad x \geq 2,$$

so the domain of the function is given by the interval $[2, +\infty)$. Since $\sqrt{x - 2}$ is nonnegative, the range is $[0, +\infty)$.

(d) $h(x) = -\sqrt{100 - x^2}$

For $-\sqrt{100 - x^2}$ to be a real number,

$$100 - x^2 \geq 0.$$

Since $100 - x^2$ factors as $(10 - x)(10 + x)$, use a sign graph to verify that

$$-10 \leq x \leq 10,$$

making the domain of the function $[-10, 10]$. For any value of x in the domain,

$$-10 \leq -\sqrt{100 - x^2} \leq 0,$$

so the range is $[-10, 0]$. ✚

For most of the functions discussed in this book, the domain can be found with the methods already presented. As Example 4 suggests, one reason for studying inequalities is to find domains. The range, however, often must be found by using graphing, more involved algebra, or calculus. For example, the graph in Figure 4.2 suggests that the function has domain $[-3, 3]$ and range $[-1, 3]$.

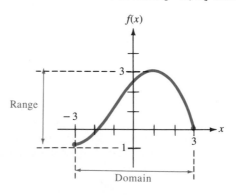

Figure 4.2

There is a quick way to tell whether a given graph is the graph of a function. Figure 4.3 shows two graphs. In the graph for part (a), each value of x leads to only one value of y, so that this is the graph of a function. On the other hand, the graph in part (b) is not the graph of a function. For example, if $x = x_1$, the vertical line through x, intersects the graph at two points, showing that there are two values of y that correspond to this x-value. This idea is known as the *vertical line test* for a function.

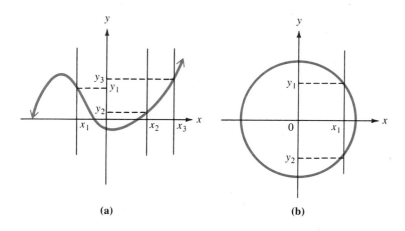

(a)   (b)

Figure 4.3

Vertical Line Test

If each vertical line cuts a graph in no more than one point, the graph is the graph of a function.

In manufacturing, the cost of making a product usually consists of two parts. One part is a *fixed cost* for designing the product, setting up a factory, training workers, and so on. Usually the fixed cost is constant for a particular product and does not change as more items are made. The other part of the cost is a *variable cost* per item for labor, materials, packaging, shipping, and so on. The variable cost is often the same per item, so that the total amount of variable cost increases as more items are produced. A *linear cost function* has the form $C(x) = mx + b$, where m represents the variable cost per item and b represents the fixed cost to produce x items. The revenue from selling a product depends on the price per item and the number of items sold, as given by the *revenue function*, $R(x) = px$, where p is the price per item and $R(x)$ is the revenue from the sale of x items. The profit is described by the *profit function* given by $P(x) = R(x) - C(x)$.

Example 5   **(a)** Assume that the cost to produce an item is a linear function. If the fixed cost is $1500 and the variable cost is $100, write a cost function for the product.

Since the cost function is linear it will have the form $C(x) = mx + b$, with $m = 100$ and $b = 1500$. That is,

$$C(x) = 100x + 1500.$$

(b) Find the revenue function if the item in part (a) sells for \$125.
The revenue function is

$$R(x) = px = 125x.$$

(c) Give the profit function for the item in part (a).
The profit function is given by

$$
\begin{aligned}
P(x) &= R(x) - C(x) \\
&= 125x - (100x + 1500) \\
&= 125x - 100x - 1500 \\
&= 25x - 1500. \quad \clubsuit
\end{aligned}
$$

4.1 Exercises

For each of the following, find the indicated function values: **(a)** $f(-2)$, **(b)** $f(0)$, **(c)** $f(1)$, *and* **(d)** $f(4)$.

1.

2.

3.

4.

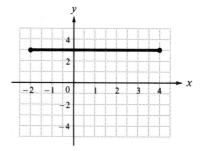

Let $f(x) = 3x - 1$ *and* $g(x) = x^2$. *Find each of the following function values.*
See Example 2.

5. $f(0)$ **6.** $f(-1)$ **7.** $f(-3)$ **8.** $f(4)$

9. $g(2)$ **10.** $g(0)$ **11.** $f(a)$ **12.** $g(b)$

13. $f(1) + g(1)$ **14.** $f(-2) - g(-2)$ **15.** $f(3) \cdot g(3)$ **16.** $\dfrac{g(-1)}{f(-1)}$

17. $f(-2m)$ **18.** $f(-11p)$ **19.** $f(5a - 2)$ **20.** $f(3 + 2k)$

21. $g(5p - 2)$ **22.** $g(-6k + 1)$

Let $f(x) = -4.6x^2 - 8.9x + 1.3$. Find each of the following function values.

23. $f(3)$ **24.** $f(-5)$ **25.** $f(-4.2)$ **26.** $f(-1.8)$

Give the domain and range of each function defined as follows. See Example 4.

27. $f(x) = 2x - 1$ **28.** $g(x) = 3x + 5$ **29.** $g(x) = x^4$ **30.** $h(x) = (x - 2)^2$

31. $f(x) = \sqrt{8 + x}$ **32.** $f(x) = -\sqrt{x + 6}$ **33.** $h(x) = \sqrt{16 - x^2}$ **34.** $m(x) = \sqrt{x^2 - 25}$

35.   **36.**   **37.**

38.   **39.**   **40.**

                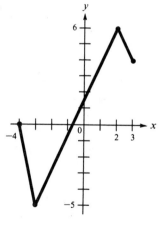

Let $f(x) = 2^x$ for all rational numbers x. Find each of the following function values. Assume that m and r represent rational numbers.

41. $f(1)$ **42.** $f(2)$ **43.** $f(4)$ **44.** $f(-1)$

45. $f(-2)$ **46.** $f(-4)$ **47.** $f(m)$ **48.** $f(-5r)$

49. $f(1/2)$ **50.** $f(1/4)$

For each of the functions defined as follows, find **(a)** $f(x + h)$,
(b) $f(x + h) - f(x)$, *and* **(c)** $\dfrac{f(x + h) - f(x)}{h}$. *See Example 3.*

51. $f(x) = x^2 - 4$ **52.** $f(x) = 8 - 3x^2$ **53.** $f(x) = 6x + 2$

54. $f(x) = 4x + 11$ **55.** $f(x) = 2x^3 + x^2$ **56.** $f(x) = -4x^3 - 8x$

Assume that each of the following can be expressed as a linear cost function. Find
(a) *the cost function,* **(b)** *the revenue function, and* **(c)** *the profit function.*
See Example 5.

	Fixed Cost	Variable Cost	Price of Item
57.	$500	$10	$35
58.	$180	$11	$20
59.	$2700	$150	$280
60.	$1650	$200	$305

Break-even analysis *is another application of functions. Suppose a firm that produces chicken feed finds that the total cost,* $C(x)$, *of producing x units is given by*

$$C(x) = 20x + 100,$$

while the revenue from the sale of x units is given by $R(x)$, *where*

$$R(x) = 24x.$$

The firm will break even *(no profit and no loss) as long as revenue just equals cost.*

61. Find x so that $R(x) = C(x)$. This value of x gives the *break-even point*.

62. Graph R and C on the same axis. Identify the break-even point. Identify any regions of profit or loss.

You are the manager of a firm. You are considering the manufacture of a new product, so you ask the accounting department to produce cost estimates and the sales department to produce sales estimates. After you receive the data, you must decide whether to go ahead with production of the new product. Analyze the following data (find a break-even point) and then decide what you would do. See Exercises 61 and 62.

63. $C(x) = 85x + 900$; $R(x) = 105x$; not more than 38 units can be sold

64. $C(x) = 105x + 6000$; $R(x) = 250x$; not more than 400 units can be sold

65. $C(x) = 70x + 500$; $R(x) = 60x$

66. $C(x) = 950x + 1250$; $R(x) = 950x$

4.2   Some Useful Functions

Many of the graphs discussed in Chapter 3 are the graphs of functions. By the vertical line test, any straight line that is not vertical is the graph of a function, as is the graph of any vertical parabola. Because of this, *linear* and quadratic functions can be defined as follows.

Definitions of Linear and Quadratic Functions	If a, b, and c are real numbers, then the function defined by $$f(x) = ax + b$$ is a **linear function,** and the function defined by $$f(x) = ax^2 + bx + c \quad (a \neq 0)$$ is a **quadratic function.**

Example 1   Graph each function defined as follows.

(a) $f(x) = -4x + 5$

This defines a linear function with x-intercept 5/4 and y-intercept 5. The graph is shown in Figure 4.4.

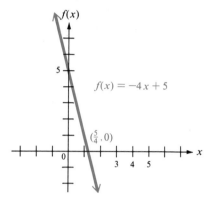

Figure 4.4

(b) $f(x) = -(x - 2)^2 + 5$

The graph of this quadratic function, a parabola with vertex at (2, 5) and opening downward, is shown in Figure 4.5.   ✛

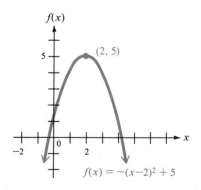

Figure 4.5

In the rest of this section and in the next two sections some useful functions that are neither linear nor quadratic are discussed.

One common function that is not linear and not quadratic is the **absolute value function,** defined by $f(x) = |x|$, or $y = |x|$. Since $|x|$ can be found for any real number x, the domain is $(-\infty, +\infty)$. Also, $|x| \geq 0$ for any real number x, so the range is $[0, +\infty)$. When $x \geq 0$, then $y = |x| = x$, so $y = x$ is graphed for nonnegative values of x. On the other hand, if $x < 0$, then $y = |x| = -x$, and $y = -x$ is graphed for negative values of x. The final graph is shown in Figure 4.6. By the vertical line test, the graph is that of a function. Notice that the graph of $f(x) = |x|$ is symmetric with respect to the y-axis. Like the parabola, the graphs of absolute value functions always have an axis of symmetry.

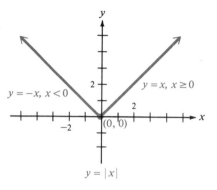

Figure 4.6

Example 2   Graph $y = |x - 2|$.

Again, the domain is $(-\infty, +\infty)$, and the range is $[0, +\infty)$. Since $y \geq 0$, the lowest point on the graph (the "vertex") is found by setting $y = 0$ to get $(2, 0)$. Graphing the point $(2, 0)$ and a few other selected points leads to the graph in Figure 4.7. This graph has the same shape as that of $y = |x|$, but the "vertex" point is translated 2 units to the right, from $(0, 0)$ to $(2, 0)$. The axis of symmetry is the vertical line $x = 2$, through $(2, 0)$.   ✛

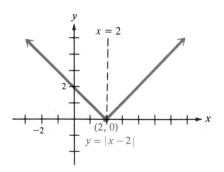

Figure 4.7

Example 3   Graph $y = |3x + 4| + 1$.

The value of y is always greater than or equal to 1, since $|3x + 4| \geq 0$. Thus, the y-value of the "vertex" is 1. Find the x-value of the "vertex" by substituting 1 for y in the equation.

$$y = |3x + 4| + 1$$
$$1 = |3x + 4| + 1$$
$$0 = |3x + 4|$$
$$x = -\frac{4}{3}$$

Plotting a few other ordered pairs leads to the graph in Figure 4.8. The graph shows the "vertex" translated to $(-4/3, 1)$. The axis of symmetry is $x = -4/3$. The coefficient of x, 3, determines the slopes of the two lines that form the graph. One has slope 3, and the other has slope -3. Because the absolute value of the slopes is greater than 1, the lines are steeper than the lines that form the graph of $y = |x|$.   ✛

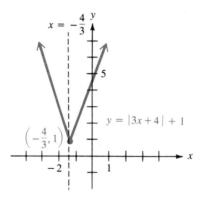

Figure 4.8

The graphs of the absolute value functions of Examples 2 and 3 are made up of portions of two different straight lines. Such functions, called **functions defined piecewise,** are often defined with different equations for different parts of the domain.

Example 4   Graph the function

$$y = \begin{cases} x + 1 & \text{if } x \leq 2 \\ -2x + 7 & \text{if } x > 2. \end{cases}$$

For $x \leq 2$, graph $y = x + 1$, and for $x > 2$, graph $y = -2x + 7$. The graph is shown in Figure 4.9 on the following page.   ✛

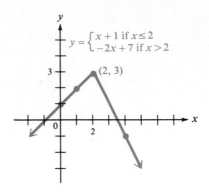

Figure 4.9

The **greatest integer function,** written $y = [\![x]\!]$, is defined as follows:

$[\![x]\!]$ is the greatest integer less than or equal to x.

For example, $[\![8]\!] = 8$, $[\![-5]\!] = -5$, $[\![\pi]\!] = 3$, $[\![12 \ 1/9]\!] = 12$, $[\![-2.001]\!] = -3$, and so on.

Example 5 Graph $y = [\![x]\!]$.

For any value of x in the interval $[0, 1)$, $[\![x]\!] = 0$. Also, for x in $[1, 2)$, $[\![x]\!] = 1$. This process continues; for x in $[2, 3)$, the value of $[\![x]\!]$ is 2. The values of y are constant between integers, but they jump at integer values of x. This makes the graph, shown in Figure 4.10, a series of line segments. In each case, the left endpoint of the segment is included, and the right endpoint is excluded. The domain of the function is $(-\infty, +\infty)$, while the range is the set of integers, $\{. \ . \ . \ , -2, -1, 0, 1, 2, . \ . \ .\}$. ❖

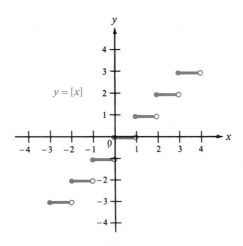

Figure 4.10

Each of the graphs in Examples 5, 6, and 7 is made up of a series of horizontal line segments. (See Figures 4.10, 4.11, and 4.12.) The functions producing these graphs are called **step functions.**

Example 6   Graph $y = \left[\!\left[\dfrac{1}{2}x + 1 \right]\!\right]$.

If x is in the interval $[0, 2)$, then $y = 1$. For x in $[2, 4)$, $y = 2$, and so on. The graph is shown in Figure 4.11.   ✚

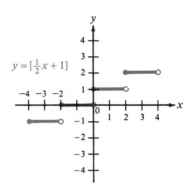

Figure 4.11

Example 7   The Terrific Taxi Company charges a base rate of $2 plus $1 per mile or part of a mile. Find the cost of a trip of 5 mi; of 4.5 mi; of 9.5 mi. Graph the ordered pairs (miles, cost). Is this graph the graph of a function?

The cost for a 5-mile trip is

$$2 + 5(1) = 7, \text{ or } \$7.$$

For a 4.5-mile trip, the cost will be the same as for a 5-mile trip, $7. A 9.5-mile trip costs the same as a 10-mile trip, $2 + 10(1) = 12$, or $12. The graph, a step function, is shown in Figure 4.12. The vertical line test shows that the graph is that of a function.   ✚

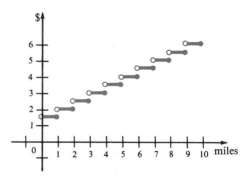

Figure 4.12

By the vertical line test, vertical parabolas are the only conic sections that are the graphs of functions. However, *half* of a conic section often represents a function. These functions are sometimes called **root functions.**

Example 8 Graph $f(x) = -\sqrt{1 + 4x^2}$

Let $y = f(x)$. Squaring both sides gives

$$y^2 = 1 + 4x^2$$

or

$$y^2 - 4x^2 = 1.$$

Use the fact that $4 = 1/(1/4)$ to write the equation as

$$\frac{y^2}{1} - \frac{x^2}{1/4} = 1,$$

the equation of a hyperbola with y-intercepts 1 and -1. Use the points $(1/2, 1)$, $(1/2, -1)$, $(-1/2, 1)$, and $(-1/2, -1)$ to sketch the asymptotes. Since the given equation $f(x) = -\sqrt{1 + 4x^2}$ restricts y to nonpositive values, the graph is the lower branch of the hyperbola, as shown in Figure 4.13. The domain of $f(x) = -\sqrt{1 + 4x^2}$ is the set of all x such that $1 + 4x^2 \ge 0$. This condition is satisfied for all x in the interval $(-\infty, +\infty)$. The range is $(-\infty, -1]$. ✛

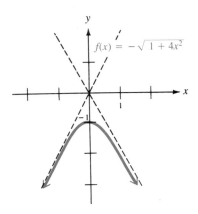

Figure 4.13

4.2 Exercises

Graph each of the following equations. Identify those that define functions, and give the domain and range of each relation. See Examples 1–3.

1. $y = 3x - 5$
2. $y = \dfrac{3}{4}x - 1$
3. $y = x^2 + 2x$
4. $y = (x - 3)^2 + 1$

5. $y = |x + 1|$
6. $y = |x - 1|$
7. $y = |2 - x|$
8. $y = |-3 - x|$

9. $y = |5x + 4|$
10. $3y = |x + 2|$
11. $y + 1 = |x - 2|$
12. $2y - 1 = |x + 3|$

13. $x = |y|$ **14.** $x = |y + 1|$ **15.** $y = |x| + 4$ **16.** $y = 2|x| - 1$

17. $2x = 3|y| - 1$ **18.** $x - 1 = |3y + 2|$ **19.** $y = |x| - x$ **20.** $x = |y| + y$

For each of the following, find **(a)** $f(-5)$, **(b)** $f(-1)$, **(c)** $f(0)$, **(d)** $f(3)$, *and* **(e)** $f(5)$.

21. $f(x) = \begin{cases} 2x & \text{if } x \le -1 \\ x - 1 & \text{if } x > -1 \end{cases}$ **22.** $f(x) = \begin{cases} x - 2 & \text{if } x < 3 \\ 4 - x & \text{if } x \ge 3 \end{cases}$

23. $f(x) = \begin{cases} 3x + 5 & \text{if } x \le 0 \\ 4 - 2x & \text{if } 0 < x < 2 \\ x & \text{if } x \ge 2 \end{cases}$ **24.** $f(x) = \begin{cases} 4x + 1 & \text{if } x < 2 \\ 3x & \text{if } 2 \le x \le 5 \\ 3 - 2x & \text{if } x > 5 \end{cases}$

Graph each of the functions defined as follows. See Example 4.

25. $f(x) = \begin{cases} x - 1 & \text{if } x \le 3 \\ 2 & \text{if } x > 3 \end{cases}$ **26.** $f(x) = \begin{cases} 6 - x & \text{if } x \le 3 \\ 3x - 6 & \text{if } x > 3 \end{cases}$

27. $f(x) = \begin{cases} 4 - x & \text{if } x < 2 \\ 1 + 2x & \text{if } x \ge 2 \end{cases}$ **28.** $f(x) = \begin{cases} -2 & \text{if } x \ge 1 \\ 2 & \text{if } x < 1 \end{cases}$

29. $f(x) = \begin{cases} 2x + 1 & \text{if } x \ge 0 \\ x & \text{if } x < 0 \end{cases}$ **30.** $f(x) = \begin{cases} 5x - 4 & \text{if } x \ge 1 \\ x & \text{if } x < 1 \end{cases}$

31. $f(x) = \begin{cases} 2 + x & \text{if } x < -4 \\ -x & \text{if } -4 \le x \le 5 \\ 3x & \text{if } x > 5 \end{cases}$ **32.** $f(x) = \begin{cases} -2x & \text{if } x < -3 \\ 3x - 1 & \text{if } -3 \le x \le 2 \\ -4x & \text{if } x > 2 \end{cases}$

33. $f(x) = \begin{cases} |x| & \text{if } x > -2 \\ x & \text{if } x \le -2 \end{cases}$ **34.** $f(x) = \begin{cases} |x| - 1 & \text{if } x > -1 \\ x - 1 & \text{if } x \le -1 \end{cases}$

Graph each of the following. See Examples 5 and 6.

35. $f(x) = [\![-x]\!]$ **36.** $f(x) = [\![2x]\!]$ **37.** $f(x) = [\![2x - 1]\!]$ **38.** $f(x) = [\![3x + 1]\!]$

39. $f(x) = [\![3x]\!]$ **40.** $f(x) = [\![3x]\!] + 1$ **41.** $f(x) = [\![3x]\!] - 1$ **42.** $f(x) = x - [\![x]\!]$

Graph each of the following, and give the domain and range. See Example 8.

43. $y = \sqrt{4 - x}$ **44.** $y = 3\sqrt{7 - 2x}$ **45.** $y = \sqrt{x^2 - 9}$ **46.** $y = -\sqrt{16 - x^2}$

47. $\dfrac{y}{3} = -\sqrt{1 + \dfrac{x^2}{16}}$ **48.** $\dfrac{y}{2} = \sqrt{1 - \dfrac{x^2}{25}}$ **49.** $y = \sqrt{1 - \dfrac{x^2}{64}}$ **50.** $y = \sqrt{1 + \dfrac{x^2}{36}}$

Work the following exercises. See Example 7.

51. Suppose a chain-saw rental firm charges a fixed $4 sharpening fee plus $7 per day or fraction of a day. Let $S(x)$ represent the cost of renting a saw for x days. Find the value of **(a)** $S(1)$, **(b)** $S(1.25)$, and **(c)** $S(3.5)$.
(d) Graph $y = S(x)$.
(e) Give the domain and range of S.

52. Assume that it costs 30¢ to mail a letter weighing one ounce or less, and then 27¢ for each additional ounce or fraction of an ounce. Let $L(x)$ be the cost of mailing a letter weighing x oz. Find **(a)** $L(.75)$, **(b)** $L(1.6)$, and **(c)** $L(4)$.
(d) Graph $y = L(x)$.
(e) Give the domain and range of L.

53. Use the greatest integer function and write an expression for the number of ounces for which postage will be charged on a letter weighing x oz (See Exercise 52).

54. Washington, D.C., taxi rates in a recent year were 90¢ for the first 1/9 mi and 10¢ for each additional 1/9 mi or fraction of 1/9. Let $C(x)$ be the cost for a taxi ride of x 1/9 mi. Find **(a)** $C(1)$, **(b)** $C(2.3)$, and **(c)** $C(8)$.
    (d) Graph $y = C(x)$.
    (e) Give the domain and range of C.

55. For a lift truck rental of no more than three days, the charge is $300. An additional charge of $75 is made for each day or portion of a day after three. Graph the ordered pairs (number of days, cost).

56. A car rental costs $37 for one day, which includes 50 free miles. Each additional 25 mi or portion costs $10. Graph the ordered pairs (miles, cost).

57. When a diabetic takes long-acting insulin, the insulin reaches its peak effect on the blood sugar level in about 3 hr. This effect remains fairly constant for 5 hr, then declines, and is very low until the next injection. In a typical patient, the level of insulin might be given by the following function.

$$i(t) = \begin{cases} 40t + 100 & \text{if } 0 \le t \le 3 \\ 220 & \text{if } 3 < t \le 8 \\ -80t + 860 & \text{if } 8 < t \le 10 \\ 60 & \text{if } 10 < t \le 24 \end{cases}$$

Here $i(t)$ is the blood sugar level, in appropriate units, at time t measured in hours from the time of the injection. Chuck takes his insulin at 6 A.M. Find the blood sugar level at each of the following times.

(a) 7 A.M. **(b)** 9 A.M. **(c)** 10 A.M **(d)** noon
(e) 2 P.M. **(f)** 5 P.M. **(g)** midnight **(h)** Graph $y = i(t)$.

58. To rent a midsized car from Avis costs $27 per day or fraction of a day. If you pick up the car in Lansing and drop it in West Lafayette, there is a fixed $25 dropoff charge. Let $C(x)$ represent the cost of renting the car for x days, taking it from Lansing to West Lafayette. Find each of the following.

(a) $C(3/4)$ **(b)** $C(9/10)$ **(c)** $C(1)$
(d) $C(1\ 5/8)$ **(e)** $C(2.4)$ **(f)** Graph $y = C(x)$.
(g) Is C a function? **(h)** Is C a linear function?

4.3 Algebra of Functions

When a company accountant sits down to estimate the firm's overhead, the first step might be to find functions representing the cost of materials, labor charges, equipment maintenance, and so on. The sum of these various functions could then be used to find the total overhead for the company. Methods of combining functions are discussed in this section.

Given two functions f and g, their *sum*, written $f + g$, is defined as

$$(f + g)(x) = f(x) + g(x),$$

for all x such that both $f(x)$ and $g(x)$ exist. Similar definitions can be given for the difference, $f - g$, product, fg, and quotient, f/g, of functions; however, the quotient,

$$\left(\frac{f}{g}\right)(x) = \frac{f(x)}{g(x)},$$

is defined only for those values of x where both $f(x)$ and $g(x)$ exist, and in addition, $g(x) \neq 0$. The various operations on functions are defined as follows.

Operations on Functions

If f and g are functions, then for all values of x for which both $f(x)$ and $g(x)$ exist,
the sum of f and g is

$$(f + g)(x) = f(x) + g(x),$$

the difference of f and g is

$$(f - g)(x) = f(x) - g(x),$$

the product of f and g is

$$(fg)(x) = f(x) \cdot g(x),$$

and the quotient of f and g is

$$\left(\frac{f}{g}\right)(x) = \frac{f(x)}{g(x)}, \quad \text{where } g(x) \neq 0.$$

Example 1   Let $f(x) = x^2 + 1$, and $g(x) = 3x + 5$. Find each of the following.
(a) $(f + g)(1)$
    Since $f(1) = 2$ and $g(1) = 8$, use the definition above to get

$$(f + g)(1) = f(1) + g(1) = 2 + 8 = 10.$$

(b) $(f - g)(-3) = f(-3) - g(-3) = 10 - (-4) = 14$
(c) $(fg)(5) = f(5) \cdot g(5) = 26 \cdot 20 = 520$
(d) $\left(\frac{f}{g}\right)(0) = \frac{f(0)}{g(0)} = \frac{1}{5}$ ✦

Example 2   Let $f(x) = 8x - 9$ and $g(x) = \sqrt{2x - 1}$.
(a) $(f + g)(x) = f(x) + g(x) = 8x - 9 + \sqrt{2x - 1}$
(b) $(f - g)(x) = f(x) - g(x) = 8x - 9 - \sqrt{2x - 1}$
(c) $(fg)(x) = f(x) \cdot g(x) = (8x - 9)\sqrt{2x - 1}$
(d) $\left(\frac{f}{g}\right)(x) = \frac{f(x)}{g(x)} = \frac{8x - 9}{\sqrt{2x - 1}}$

The domain of f is the set of all real numbers, while the domain of $g(x) = \sqrt{2x - 1}$ includes just those real numbers that make $2x - 1 \geq 0$; the domain of g is the interval $[1/2, +\infty)$. The domain of $f + g$, $f - g$, and fg is thus $[1/2, +\infty)$. With f/g, the denominator cannot be zero, so the value $1/2$ is excluded from the domain. The domain of f/g is $(1/2, +\infty)$. ✛

The domains of $f + g$, $f - g$, fg, and f/g are summarized below. (Recall that the intersection of two sets is the set of all elements belonging to *both* sets.)

Domains of $f + g$, $f - g$, fg, f/g	For functions f and g, the domains of $f + g$, $f - g$, and fg include all real numbers in the intersection of the domains of f and g, while the domain of f/g includes those real numbers in the intersection of the domains of f and g for which $g(x) \neq 0$.

Composition of Functions The sketch in Figure 4.14 shows a function f that assigns to each element x of set X some element y of set Y. Suppose also that a function g takes each element of set Y and assigns a value z of set Z. Using both f and g, then, an element x in X is assigned to an element z in Z. The result of this process is a new function h, which takes an element x in X and assigns an element z in Z. This function h is called the *composition* of functions g and f, written $g \circ f$, and defined as follows.

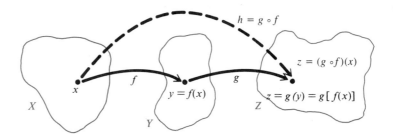

Figure 4.14

Definition of Composition of Functions	If f and g are functions, then the **composite function,** or **composition,** of g and f is $$(g \circ f)(x) = g[f(x)]$$ for all x in the domain of f such that $f(x)$ is in the domain of g.

Suppose an oil well off the California coast is leaking, with the leak spreading oil in a circular layer over the surface. At any time t, in minutes, after the beginning of the leak, the radius of the circular oil slick is $r(t) = 5t$ feet. Since

$A(r) = \pi r^2$ gives the area of a circle of radius r, the area can be expressed as a function of time by substituting $5t$ for r in $A(r) = \pi r^2$ to get

$$A(r) = \pi r^2$$
$$A[r(t)] = \pi(5t)^2 = 25\pi t^2$$

The function $A(r(t))$ is a composite function of the functions A and r.

Example 3   Let $f(x) = 4x + 1$ and $g(x) = 2x^2 + 5x$. Find each of the following.
(a) $(g \circ f)(x)$

By definition, $(g \circ f)(x) = g[f(x)]$. Using the given functions,

$$\begin{aligned}
(g \circ f)(x) = g[f(x)] &= g(4x + 1) \\
&= 2(4x + 1)^2 + 5(4x + 1) \\
&= 2(16x^2 + 8x + 1) + 20x + 5 \\
&= 32x^2 + 16x + 2 + 20x + 5 \\
&= 32x^2 + 36x + 7.
\end{aligned}$$

(b) $(f \circ g)(x)$

If we use the definition above with f and g interchanged, $(f \circ g)(x)$ becomes $f[g(x)]$, with

$$\begin{aligned}
(f \circ g)(x) = f[g(x)] \\
&= f(2x^2 + 5x) \\
&= 4(2x^2 + 5x) + 1 \\
&= 8x^2 + 20x + 1. \quad \clubsuit
\end{aligned}$$

As this example shows, it is not always true that $f \circ g = g \circ f$. In fact, two composite functions are equal only for a special class of functions, discussed in the next section. In Example 3, the domain of both composite functions is the set of all real numbers.

Example 4   Let $f(x) = 1/x$ and $g(x) = \sqrt{3 - x}$. Find $f \circ g$ and $g \circ f$. Give the domain of each.

First find $f \circ g$.

$$\begin{aligned}
(f \circ g)(x) = f[g(x)] \\
&= f(\sqrt{3 - x}) \\
&= \frac{1}{\sqrt{3 - x}}
\end{aligned}$$

The radical $\sqrt{3 - x}$ is a nonzero real number only when $3 - x > 0$ or $x < 3$, so that the domain of $f \circ g$ is the interval $(-\infty, 3)$.

Use the same functions to find $g \circ f$, as follows.

$$(g \circ f)(x) = g[f(x)]$$

$$= g\left(\frac{1}{x}\right)$$

$$= \sqrt{3 - \frac{1}{x}}$$

$$= \sqrt{\frac{3x - 1}{x}}$$

The domain of $g \circ f$ is the set of all real numbers x such that $x \neq 0$ and $3 - f(x) \geq 0$. By the methods of Section 2.7, the domain of $g \circ f$ is the set $(-\infty, 0) \cup [1/3, +\infty)$. ✤

4.3   Exercises

For each pair of functions defined as follows, find **(a)** $f + g$, **(b)** $f - g$, **(c)** fg, *and* **(d)** f/g. *Give the domain of each. See Example 2.*

1. $f(x) = 4x - 1$, $g(x) = 6x + 3$

2. $f(x) = 9 - 2x$, $g(x) = -5x + 2$

3. $f(x) = 3x^2 - 2x$, $g(x) = x^2 - 2x + 1$

4. $f(x) = 6x^2 - 11x$, $g(x) = x^2 - 4x - 5$

5. $f(x) = \sqrt{2x + 5}$, $g(x) = \sqrt{4x - 9}$

6. $f(x) = \sqrt{11x - 3}$, $g(x) = \sqrt{2x - 15}$

7. $f(x) = 4x^2 - 11x + 2$, $g(x) = x^2 + 5$

8. $f(x) = 15x^2 - 2x + 1$, $g(x) = 16 + x^2$

Let $f(x) = 4x^2 - 2x$ *and let* $g(x) = 8x + 1$. *Find each of the following.*
See Examples 1–3.

9. $(f + g)(3)$

10. $(f + g)(-5)$

11. $(fg)(4)$

12. $(fg)(-3)$

13. $\left(\dfrac{f}{g}\right)(-1)$

14. $\left(\dfrac{f}{g}\right)(4)$

15. $(f + g)(m)$

16. $(f - g)(2k)$

17. $(f \circ g)(2)$

18. $(f \circ g)(-5)$

19. $(g \circ f)(2)$

20. $(g \circ f)(-5)$

21. $(f \circ g)(k)$

22. $(g \circ f)(5z)$

Find $f \circ g$ *and* $g \circ f$ *and their domains for each pair of functions defined as follows.*
See Examples 3 and 4.

23. $f(x) = 8x + 12$, $g(x) = 3x - 1$

24. $f(x) = -6x + 9$, $g(x) = 5x + 7$

25. $f(x) = 5x + 3$, $g(x) = -x^2 + 4x + 3$

26. $f(x) = 4x^2 + 2x + 8$, $g(x) = x + 5$

27. $f(x) = -x^3 + 2$, $g(x) = 4x$

28. $f(x) = 2x$, $g(x) = 6x^2 - x^3$

29. $f(x) = \dfrac{1}{x}$, $g(x) = x^2$

30. $f(x) = \dfrac{2}{x^4}$, $g(x) = 2 - x$

31. $f(x) = \sqrt{x + 2}$, $g(x) = 8x - 6$

32. $f(x) = 9x - 11$, $g(x) = 2\sqrt{x + 2}$

33. $f(x) = \dfrac{1}{x - 5}$, $g(x) = \dfrac{2}{x}$

34. $f(x) = \dfrac{8}{x - 6}$, $g(x) = \dfrac{4}{3x}$

35. $f(x) = \sqrt{x + 1}$, $g(x) = \dfrac{-1}{x}$

36. $f(x) = \dfrac{8}{x}$, $g(x) = \sqrt{3 - x}$

For each pair of functions defined as follows, show that $(f \circ g)(x) = x$ and
$(g \circ f)(x) = x$.

37. $f(x) = 8x, \quad g(x) = \dfrac{1}{8}x$

38. $f(x) = \dfrac{3}{4}x, \quad g(x) = \dfrac{4}{3}x$

39. $f(x) = 8x - 11, \quad g(x) = \dfrac{x + 11}{8}$

40. $f(x) = \dfrac{x - 3}{4}, \quad g(x) = 4x + 3$

41. $f(x) = x^3 + 6, \quad g(x) = \sqrt[3]{x - 6}$

42. $f(x) = \sqrt[5]{x - 9}, \quad g(x) = x^5 + 9$

In each of the following exercises, a function h is given. Find functions f and g such that
$h(x) = (f \circ g)(x)$. Many such pairs of functions exist.

43. $h(x) = (6x - 2)^2$

44. $h(x) = (11x^2 + 12x)^2$

45. $h(x) = \sqrt{x^2 - 1}$

46. $h(x) = \dfrac{1}{x^2 + 2}$

47. $h(x) = \dfrac{(x - 2)^2 + 1}{5 - (x - 2)^2}$

48. $h(x) = (x + 2)^3 - 3(x + 2)^2$

49. Suppose the population P of a certain species of fish depends on the number x (in hundreds) of a smaller kind of fish which serves as its food supply, so that

$$P(x) = 2x^2 + 1.$$

Suppose, also, that the number x (in hundreds) of the smaller species of fish depends upon the amount a (in appropriate units) of its food supply, a kind of plankton. Suppose

$$x = f(a) = 3a + 2.$$

Find $(P \circ f)(a)$, the relationship between the population P of the large fish and the amount a of plankton available.

50. Suppose the demand for a certain brand of vacuum cleaner is given by

$$D(p) = \dfrac{-p^2}{100} + 500,$$

where p is the price in dollars. If the price, in terms of the cost, c, is expressed as

$$p(c) = 2c - 10,$$

find $D(c)$, the demand in terms of the cost.

51. When a thermal inversion layer is over a city (as happens often in Los Angeles), pollutants cannot rise vertically but are trapped below the layer and must disperse horizontally. Assume that a factory smokestack begins emitting a pollutant at 8 A.M. Assume that the pollutant disperses horizontally over a circular area. If t represents the time, in hours, since the factory began emitting pollutants ($t = 0$ represents 8 A.M.), assume that the radius of the circle of pollution is $r(t) = 2t$ mi. Let $A(r) = \pi r^2$ represent the area of a circle of radius r. Find and interpret $(A \circ r)(t)$.

52. An oil well off the Gulf Coast is leaking, with the leak spreading oil over the surface as a circle. At any time t, in minutes, after the beginning of the leak, the radius of the circular oil slick on the surface is $r(t) = 4t$ ft. Let $A(r) = \pi r^2$ represent the area of a circle of radius r. Find and interpret $(A \circ r)(t)$.

4.4 *Inverse Functions*

Addition and subtraction are inverse operations: starting with a number x, adding 5, and subtracting 5 gives x back as a result. Similarly, some functions are inverses of each other. For example, the functions

$$f(x) = 8x \qquad \text{and} \qquad g(x) = \frac{1}{8}x$$

are inverses of each other. This means that if a value of x such as $x = 12$ is chosen, so that

$$f(12) = 8 \cdot 12 = 96,$$

calculating $g(96)$ gives

$$g(96) = \frac{1}{8} \cdot 96 = 12.$$

Thus, $\qquad\qquad\qquad g[f(12)] = 12.$

Also, $f[g(12)] = 12$. For these functions f and g, it can be shown that

$$f[g(x)] = x \qquad \text{and} \qquad g[f(x)] = x$$

for any value of x.

This section will show how to start with a function such as $f(x) = 8x$ and obtain the inverse function $g(x) = (1/8)x$. Not all functions have inverse functions. The only functions that do have inverse functions are *one-to-one functions*.

One-to-One Functions   For the function $y = 5x - 8$, any two different values of x produce two different values of y. On the other hand, for the function $y = x^2$, two different values of x can lead to the *same* value of y; for example, both $x = 4$ and $x = -4$ give $y = 4^2 = (-4)^2 = 16$. A function such as $y = 5x - 8$, where different elements from the domain lead to different elements from the range, is called a *one-to-one function*.

Definition of One-to-One Function

A function f is a **one-to-one function** if, for elements a and b from the domain of f,

$$a \neq b \qquad \text{implies} \qquad f(a) \neq f(b).$$

Example 1   Decide whether each of the following functions is one-to-one.

(a) $f(x) = -4x + 12$

Suppose that $a \neq b$. Then $-4a \neq -4b$, and $-4a + 12 \neq -4b + 12$. Thus, the fact that $a \neq b$ implies that $f(a) \neq f(b)$, so f is one-to-one.

(b) $f(x) = \sqrt{25 - x^2}$

If $x = 3$, then $f(3) = \sqrt{25 - 3^2} = \sqrt{16} = 4$. For $x = -3$, we have $f(-3) = \sqrt{25 - (-3)^2} = \sqrt{16} = 4$. Both $x = 3$ and $x = -3$ lead to $y = 4$, so f is not one-to-one. ✤

As shown in part (b) of Example 1, a way to show that a function is *not* one-to-one is to produce a pair of unequal numbers that lead to the same function value.

In addition, there is a useful graphical test that tells whether or not a function is one-to-one. Figure 4.15(a) shows the graph of a function cut by a horizontal line. Each point where the horizontal line cuts the graph has the same value of y but a different value of x. Here, three different values of x lead to the same value of y; therefore, the function is not one-to-one. On the other hand, the graph of Figure 4.15(b) can be cut by any horizontal line in no more than one point, so it is the graph of a one-to-one function.

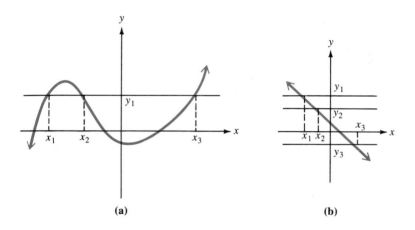

Figure 4.15

This *horizontal line test* for one-to-one functions can be summarized as follows.

Horizontal Line Test

If each horizontal line cuts the graph of a function in no more than one point, then the function is one-to-one.

Inverse Functions In the previous section the composition of two functions was discussed. For example, if

$$f(x) = 8 + 3x \qquad \text{and} \qquad g(x) = \frac{x - 8}{3},$$

then $\qquad (f \circ g)(x) = 8 + 3\left(\frac{x - 8}{3}\right) = 8 + x - 8 = x$

and $\qquad (g \circ f)(x) = \frac{(8 + 3x) - 8}{3} = \frac{3x}{3} = x.$

These two functions f and g have a special property:

$$(f \circ g)(x) = x \qquad \text{and} \qquad (g \circ f)(x) = x.$$

For this reason, f and g are *inverse functions* of each other.

Definition of Inverse Functions	Two one-to-one functions f and g are **inverses** of each other if $$f \circ g(x) = x \text{ for every } x \text{ in the domain of } g,$$ and $\qquad g \circ f(x) = x$ for every x in the domain of f.

Example 2   Let $f(x) = x^3 - 1$, and let $g(x) = \sqrt[3]{x + 1}$. Is g the inverse function of f? Use the definition to get

$$(f \circ g)(x) = f[g(x)] = (\sqrt[3]{x + 1})^3 - 1$$
$$= x + 1 - 1 = x$$
$$(g \circ f)(x) = g[f(x)] = \sqrt[3]{(x^3 - 1) + 1} = \sqrt[3]{x^3} = x.$$

The domain of f, $(-\infty, +\infty)$, is the range of g, and the range of f, $(-\infty, +\infty)$, is the domain of g. Function g is the inverse function of f. Also, f is the inverse function of g. ✚

A special notation is often used for inverse functions: if g is the inverse function of f, then g can be written as f^{-1} (read "f-inverse"). In Example 2,

$$f^{-1}(x) = \sqrt[3]{x + 1}.$$

Do not confuse the -1 in f^{-1} with a negative exponent. The symbol $f^{-1}(x)$ does not represent $1/f(x)$; it represents the inverse function of f. Keep in mind that a function f can have an inverse function f^{-1} if and only if f is one-to-one.

In the example at the beginning of this section, for the inverse functions f and g, $f(12) = 96$ and $g(96) = 12$; that is $(12, 96)$ belonged to f and $(96, 12)$ belonged to g. The inverse of any one-to-one function f can be found by exchanging the components of the ordered pairs of f. This means that the domain of f becomes the range of f^{-1} and the range of f becomes the domain of f^{-1}. See Figure 4.16.

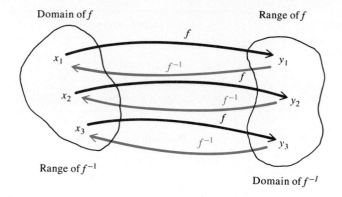

Figure 4.16

Given a one-to-one function f and a value of x in the domain of f, the corresponding value in the range of f can be found by means of the equation $y = f(x)$. By the inverse function f^{-1}, it is possible to take y and produce x: $x = f^{-1}(y)$. Therefore, to find the equation for f^{-1}, it is necessary only to solve $y = f(x)$ for x.

For example, let $f(x) = 7x - 2$. Then $y = 7x - 2$. This function is one-to-one, so that f^{-1} exists. Solve the equation for x, as follows:

$$y = 7x - 2$$
$$7x = y + 2$$
$$x = \frac{y + 2}{7},$$

or, since $x = f^{-1}(y)$,

$$f^{-1}(y) = \frac{y + 2}{7}.$$

It is common to use x to represent an element from the domain of a function. Replacing y with x gives

$$f^{-1}(x) = \frac{x + 2}{7}.$$

Check that $(f \circ f^{-1})(x) = x$ and $(f^{-1} \circ f)(x) = x$, making f^{-1} the inverse of f.

In summary, finding the equation of an inverse function involves the following steps.

Finding an
Equation
for f^{-1}

> 1. Let $y = f(x)$ be a one-to-one function.
> 2. Solve for x. Let $x = f^{-1}(y)$.
> 3. Exchange x and y to get $y = f^{-1}(x)$.
> 4. Check the domains and ranges: the domain of f and the range of f^{-1} are equal, as are the domain of f^{-1} and the range of f.

Example 3   For each of the following functions, find its inverse function where possible.

(a) $f(x) = \dfrac{4x + 6}{5}$

This function is one-to-one and thus has an inverse. Let $y = f(x)$, and solve for x, getting

$$y = \frac{4x + 6}{5}$$

$$5y = 4x + 6$$

$$5y - 6 = 4x$$

$$\frac{5y - 6}{4} = x.$$

Finally, exchange x and y, and let $y = f^{-1}(x)$, to get

$$\frac{5x - 6}{4} = y,$$

or

$$f^{-1}(x) = \frac{5x - 6}{4}.$$

Check that the domain of f is the range of f^{-1} and the range of f is the domain of f^{-1}. Since both f and f^{-1} are linear functions, their domains and ranges are $(-\infty, +\infty)$.

(b) $f(x) = x^3 - 1$

Two different values of x will produce two different values of $x^3 - 1$, so the function is one-to-one and has an inverse. To find the inverse, first solve $y = x^3 - 1$ for x, as follows.

$$y = x^3 - 1$$

$$y + 1 = x^3$$

$$\sqrt[3]{y + 1} = x$$

Exchange x and y, giving

$$\sqrt[3]{x + 1} = y,$$

or

$$f^{-1}(x) = \sqrt[3]{x + 1}.$$

In Example 2, we verified that $\sqrt[3]{x + 1}$ is the inverse of $x^3 - 1$.

(c) $f(x) = x^2$

We can find two different values of x that give the same value of y. For example, both $x = 4$ and $x = -4$ give $y = 16$. Therefore, the function is not one-to-one and thus has no inverse function. ✦

Suppose f and f^{-1} are inverse functions, and $f(a) = b$ for real numbers a and b. Then, by the definition of inverse, $f^{-1}(b) = a$. This shows that if a point (a, b) is on the graph of f, then (b, a) will belong to the graph of f^{-1}. As shown in Figure 4.17, the points (a, b) and (b, a) are symmetric with respect to the line $y = x$. Thus, the graph of f^{-1} can be obtained from the graph of f by reflecting the graph of f about the line $y = x$.

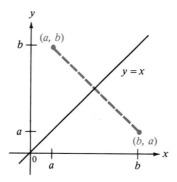

Figure 4.17

For example, Figure 4.18 shows the graph of $f(x) = x^3 - 1$ as a solid curve and the graph of $f^{-1}(x) = \sqrt[3]{x + 1}$ as a dashed curve. These graphs are symmetric with respect to the line $y = x$.

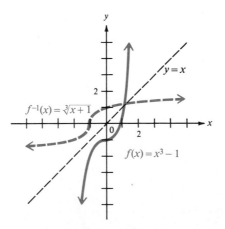

Figure 4.18

Example 4 Let $f(x) = \sqrt{x + 5}$ with domain $[-5, +\infty)$. Find $f^{-1}(x)$.

The function f is one-to-one and has an inverse function. To find this inverse function, start with

$$y = \sqrt{x + 5}$$

and solve for x, to get

$$y^2 = x + 5$$
$$y^2 - 5 = x.$$

Exchanging x and y gives

$$x^2 - 5 = y.$$

However, $x^2 - 5$ is not $f^{-1}(x)$. In the definition of f above, the domain was given as $[-5, +\infty)$. The range of f is $[0, +\infty)$. As mentioned above, the range of f equals the domain of f^{-1}, so f^{-1} must be given as

$$f^{-1}(x) = x^2 - 5, \qquad \text{domain } [0, +\infty).$$

As a check, the range of f^{-1}, $[-5, +\infty)$, equals the domain of f. Graphs of f and f^{-1} are shown in Figure 4.19. The line $y = x$ is included on the graph to show that the graphs of f and f^{-1} are mirror images with respect to this line. ✦

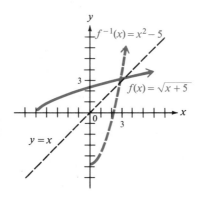

Figure 4.19

4.4 Exercises

Decide whether each of the following functions is one-to-one. See Example 1.

1.

2.

3.

4.

5.

6.

7. $y = 4x - 5$

8. $x + y = 0$

9. $y = 6 - x$

10. $y = -x^2$

11. $y = (x - 2)^2$

12. $y = -(x + 3)^2 - 8$

13. $y = \sqrt{36 - x^2}$

14. $y = -\sqrt{100 - x^2}$

15. $y = |25 - x^2|$

16. $y = -|16 - x^2|$

17. $y = x^3 - 1$

18. $y = -\sqrt[3]{x + 5}$

19. $y = (\sqrt{x} + 1)^2$

20. $y = (3 - 2\sqrt{x})^2$

21. $y = \dfrac{1}{x + 2}$

22. $y = \dfrac{-4}{x - 8}$

23. $y = 9$

24. $y = -4$

Decide whether the functions in each pair are inverses of each other. See Example 2.

25.

26.

27.

28.

29.

30.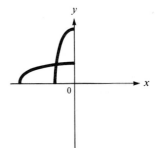

31. $f(x) = -\dfrac{3}{11}x, \quad g(x) = -\dfrac{11}{3}x$

32. $f(x) = 2x + 4, \quad g(x) = \dfrac{1}{2}x - 2$

33. $f(x) = 5x - 5, \quad g(x) = \dfrac{1}{5}x + 1$

34. $f(x) = 8x - 7, \quad g(x) = \dfrac{x + 8}{7}$

35. $f(x) = \dfrac{1}{x+1}$, $\quad g(x) = \dfrac{x-9}{12}$

36. $f(x) = \dfrac{1}{x+1}$, $\quad g(x) = \dfrac{1-x}{x}$

37. $f(x) = \dfrac{2}{x+6}$, $\quad g(x) = \dfrac{6x+2}{x}$

38. $f(x) = \dfrac{1}{x}$, $\quad g(x) = \dfrac{1}{x}$

39. $f(x) = 4x$, $\quad g(x) = -\dfrac{1}{4}x$

40. $f(x) = \sqrt{x+8}$, domain $[-8, +\infty)$, and $g(x) = x^2 - 8$, domain $[0, +\infty)$

41. $f(x) = x^2 + 3$, domain $[0, +\infty)$, and $g(x) = \sqrt{x-3}$, domain $[3, +\infty)$

42. $f(x) = |x-1|$, domain $[-1, +\infty)$, and $g(x) = |x+1|$, domain $[1, +\infty)$

43. $f(x) = -|x+5|$, domain $[-5, +\infty)$, and $g(x) = |x-5|$, domain $[5, +\infty)$

Graph the inverse of each one-to-one function.

44.

45.

46.

47.

48.

49.

For each function defined as follows that is one-to-one, write an equation for the inverse function in the form of $y = f^{-1}(x)$, and then graph f and f^{-1} on the same axes. See Examples 3 and 4.

50. $y = 3x - 4$

51. $y = 4x - 5$

52. $y = \dfrac{1}{3}x$

53. $y = -\dfrac{2}{5}x$

54. $y = x^3 + 1$

55. $y = -x^3 - 2$

56. $x + 4y = 12$

57. $3x + y = 9$

58. $y = x^2$

59. $y = -x^2 + 2$

60. $y = \dfrac{1}{x}$

61. $xy = 4$

62. $xy = 3$

63. $y = \dfrac{2}{x}$

64. $f(x) = 4 - x^2$, domain $(-\infty, 0]$

65. $f(x) = \sqrt{6+x}$, domain $[-6, +\infty)$

66. $f(x) = \sqrt{9 - x}$, domain $(-\infty, 9]$

67. $f(x) = \sqrt{x^2 - 9}$, domain $(-\infty, -3]$

68. $f(x) = -\sqrt{x^2 - 16}$, domain $[4, +\infty)$

69. $f(x) = (x - 1)^2$, domain $(-\infty, 1]$

70. Let $f(x) = ax + b$ and $g(x) = cx + d$. Find all values of a, b, c, and d for which $f^{-1} = g$.

4.5 Variation

In many applications of mathematics, it is necessary to express relationships between quantities. For example, in chemistry the ideal gas law shows how temperature, pressure, and volume are related. In physics, various formulas in optics show how the focal length of a lens and the size of an image are related. All of these relations are functions.

If the quotient of two variables is constant, then one variable *varies directly* or is *directly proportional* to the other. This idea can be stated in a different way as follows.

Definition of Direct Variation

> y **varies directly** as x, or y is **directly proportional** to x, if a nonzero real number k, called the **constant of variation**, exists such that
>
> $$y = kx.$$

If y varies directly as x, y is a function of x.

Example 1 Suppose the area of a certain rectangle varies directly as the length. If the area is 50 m^2 when the length is 10 m, find the area when the length is 25 m.

Since the area varies directly as the length,

$$A = kl,$$

where A represents the area of the rectangle, l is the length, and k is a nonzero constant that must be found. When $A = 50$ and $l = 10$, the equation $A = kl$ becomes

$$50 = 10k \quad \text{or} \quad k = 5.$$

Using this value of k, the relationship between the area and the length can be expressed as

$$A = 5l.$$

To find the area when the length is 25, replace l with 25 to get

$$A = 5l = 5(25) = 125.$$

The area of the rectangle is 125 m^2 when the length is 25 m. ✦

Sometimes y varies as a power of x. In this case y is a polynomial function of x.

Definition of Direct Variation as *n*th Power	Let n be a positive real number. Then y **varies directly as the *n*th power** of x, or y is **directly proportional to the *n*th power** of x, if a nonzero real number k exists such that $$y = kx^n.$$

The phrase "directly proportional" is sometimes abbreviated to just "proportional."

For example, the area of a square of side x is given by the formula $A = x^2$, so that the area varies directly as the square of the length of a side. Here $k = 1$.

The case where y increases as x decreases is an example of *inverse variation*, where y is a rational function of x.

Definition of Inverse Variation	Let n be a positive real number. Then y **varies inversely as the *n*th power** of x, or y is **inversely proportional** to the nth power of x, if a nonzero real number k exists such that $$y = \frac{k}{x^n}.$$ If $n = 1$, then $y = k/x$, and y **varies inversely** as x.

Example 2 In a certain manufacturing process, the cost of producing a single item varies inversely as the square of the number of items produced. If 100 items are produced, each costs \$2. Find the cost per item if 400 items are produced.

Let x represent the number of items produced and y the cost per item, and write

$$y = \frac{k}{x^2}$$

for some nonzero constant k. Since $y = 2$ when $x = 100$,

$$2 = \frac{k}{100^2} \quad \text{or} \quad k = 20{,}000.$$

Thus, the relationship between x and y is given by

$$y = \frac{20{,}000}{x^2}.$$

When 400 items are produced, the cost per item is

$$y = \frac{20{,}000}{400^2} = .125, \text{ or } 12.5\text{¢}. \quad \clubsuit$$

One variable may depend on more than one other variable. Such variation is called *combined variation*. More specifically, when a variable depends on the *product* of two or more other variables, it is referred to as *joint variation*.

Definition of Joint Variation

> Let m and n be real numbers. Then y **varies jointly** as the nth power of x and the mth power of z if a real number k exists such that
> $$y = kx^n z^m.$$

Example 3   The area of a triangle varies jointly as the lengths of the base and the height. A triangle with a base of 10 ft and a height of 4 ft has an area of 20 sq ft. Find the area of a triangle with a base of 3 cm and a height of 8 cm.

Let A represent the area, b the base, and h the height of the triangle. Then

$$A = kbh$$

for some number k. Since A is 20 when b is 10 and h is 4,

$$20 = k(10)(4)$$
$$\frac{1}{2} = k.$$

Then

$$A = \frac{1}{2}bh,$$

and, when $b = 3$ cm and $h = 8$ cm,

$$A = \frac{1}{2}(3)(8) = 12 \text{ sq cm.} \quad \clubsuit$$

The steps involved in solving a problem in variation are summarized here.

Solving Variation Problems

> 1. Write, in an algebraic form, the general relationship among the variables. Use the constant k.
> 2. Substitute given values of the variables and find the value of k.
> 3. Substitute this value of k into the formula from Step 1, thus obtaining a specific formula.
> 4. Solve for the required unknown.

The final example in this section shows combined variation.

Example 4   Suppose that m varies directly as the square of p and inversely as q. Suppose also that $m = 8$ when $p = 2$ and $q = 6$. Find m if $p = 6$ and $q = 10$.

Here m depends on two variables, p and q.

$$m = \frac{kp^2}{q}$$

Since $m = 8$ when $p = 2$ and $q = 6$,

$$8 = \frac{k \cdot 2^2}{6}$$

$$k = 12$$

and it follows that

$$m = \frac{12p^2}{q}.$$

For $p = 6$ and $q = 10$,

$$m = \frac{12 \cdot 6^2}{10}$$

$$m = \frac{216}{5}.$$

4.5 Exercises

Express each of the following statements as an equation.

1. a varies directly as b.

2. m is proportional to n.

3. x is inversely proportional to y.

4. p varies inversely as y.

5. r varies jointly as s and t.

6. R is proportional to m and p.

7. w is proportional to x^2 and inversely proportional to y.

8. c varies directly as d and inversely as f^2 and g.

Solve each of the following problems. See Example 4.

9. If m varies directly as x and y, and $m = 10$ when $x = 4$ and $y = 7$, find m when $x = 11$ and $y = 8$.

10. Suppose m varies directly as z and p. If $m = 10$ when $z = 3$ and $p = 5$, find m when $z = 5$ and $p = 7$.

11. Suppose r varies directly as the square of m, and inversely as s. If $r = 12$ when $m = 6$ and $s = 4$, find r when $m = 4$ and $s = 10$.

12. Suppose p varies directly as the square of z, and inversely as r. If $p = 32/5$ when $z = 4$ and $r = 10$, find p when $z = 2$ and $r = 16$.

13. Let a be proportional to m and n^2, and inversely proportional to y^3. If $a = 9$ when $m = 4$, $n = 9$, and $y = 3$, find a if $m = 6$, $n = 2$, and $y = 5$.

14. If y varies directly as x, and inversely as m^2 and r^2, and $y = 5/3$ when $x = 1$, $m = 2$, and $r = 3$, find y if $x = 3$, $m = 1$, and $r = 8$.

15. Let r vary directly as p^2 and q^3, and inversely as z^2 and x. If $r = 334.6$ when $p = 1.9$, $q = 2.8$, $z = .4$, and $x = 3.7$, find r when $p = 2.1$, $q = 4.8$, $z = .9$, and $x = 1.4$.

16. Let p vary directly as x and z^2, and inversely as m, n, and q. If $p = 9.25$ when $x = 3.2$, $z = 10.1$, $m = 2$, $n = 1.7$, and $q = 8.3$, find p when $x = 8.1$, $z = 4.3$, $m = 4$, $n = 2.1$, and $q = 6.2$.

Solve each of the following problems. See Examples 1–3.

17. The distance a body falls from rest varies directly as the square of the time it falls (disregarding air resistance). If an object falls 1024 ft in 8 sec, how far will it fall in 12 sec?

18. Hooke's law for an elastic spring states that the distance a spring stretches varies directly as the force applied. If a force of 15 lb stretches a certain spring 8 inches, how much will a force of 30 lb stretch the spring?

19. In electric current flow, it is found that the resistance (measured in units called ohms) offered by a fixed length of wire of a given material varies inversely as the square of the diameter of the wire. If a wire .01 inch in diameter has a resistance of .4 ohm, what is the resistance of a wire of the same length and material with a diameter of .03 inch?

20. The illumination produced by a light source varies inversely as the square of the distance from the source. The illumination of a light source at 5 m is 70 candela. What is the illumination 12 m from the source?

21. The pressure exerted by a certain liquid at a given point is proportional to the depth of the point below the surface of the liquid. If the pressure 20 m below the surface is 70 kg per cm^2, what pressure is exerted 40 m below the surface?

22. The distance that a person can see to the horizon from a point above the surface of the earth varies directly as the square root of the height of the point above the earth. A person on a hill 121 m high can see for 15 km to the horizon. How far is the horizon from a hill 900 m high?

23. Simple interest varies jointly as principal and time. If $1000 left at interest for 2 yr earned $110, find the amount of interest earned by $5000 for 5 yr.

24. The volume of a right circular cylinder is jointly proportional to the square of the radius of the circular base and to the height. If the volume is 300 cu cm when the height is 10.62 cm and the radius is 3 cm, find the volume for a cylinder with a radius of 4 cm and a height of 15.92 cm.

25. The Downtown Construction Company is designing a building whose roof rests on round concrete pillars. The company's engineers know that the maximum load a cylindrical column of circular cross section can hold varies directly as the fourth power of the diameter and inversely as the square of the height. If a 9 m column 1 m in diameter will support a load of 8 metric tons, how many metric tons will be supported by a column 12 m high and 2/3 m in diameter?

26. The maximum load of a horizontal beam that is supported at both ends varies directly as the width and square of the height and inversely as the length between supports. If a beam 8 m long, 12 cm wide, and 15 cm high can support a maximum of 400 kg, what is the maximum load of a beam of the same material 16 m long, 24 cm wide, and 8 cm high?

27. The force needed to keep a car from skidding on a curve varies inversely as the radius of the curve and jointly as the weight of the car and the square of the speed. It takes 3000 lb of force to keep a 2000-lb car from skidding on a curve of radius 500 ft at 30 mph. What force is needed to keep the same car from skidding on a curve of radius 800 ft at 60 mph?

28. The period of a pendulum varies directly as the square root of the length of the pendulum and inversely as the square root of the acceleration due to gravity. Find the period when the length is 121 cm and the acceleration due to gravity is 980 cm per sec squared, if the period is 6π sec when the length is 289 cm and the acceleration due to gravity is 980 cm per sec squared.

29. The pressure on a point in a liquid is directly proportional to the distance from the surface to the point. In a certain liquid the pressure at a depth of 4 m is 60 kg per m². Find the pressure at a depth of 10 m.

30. The volume V of a gas varies directly as the temperature T and inversely as the pressure P. If V is 10 when T is 280 and P is 6, find V if T is 300 and P is 10.

31. A sociologist has decided to rank the cities in a nation by population, with population varying inversely as the ranking. If a city with a population of 1,000,000 ranks 8th, find the population of a city that ranks 2nd.

32. Under certain conditions, the length of time that it takes for fruit to ripen during the growing season varies inversely as the average maximum temperature during the season. If it takes 25 days for fruit to ripen with an average maximum temperature of 80°, find the number of days it would take at 75°.

33. The number of long-distance phone calls between two cities in a certain time period varies directly as the populations p_1 and p_2 of the cities, and inversely as the distance between them. If 10,000 calls are made between two cities 500 mi apart, having populations of 50,000 and 125,000, find the number of calls between two cities 800 mi apart having populations of 20,000 and 80,000.

34. The horsepower needed to run a boat through water varies directly as the cube of the speed. If 80 horsepower are needed to go 15 km/hr in a certain boat, how many horsepower would be needed to go 30 km/hr?

35. According to Poiseuille's law, the resistance to flow of a blood vessel, R, is directly proportional to the length, l, and inversely proportional to the fourth power of the radius, r. If $R = 25$ when $l = 12$ and $r = .2$, find R as r increases to .3, while l is unchanged.

36. The Stefan-Boltzmann law says that the radiation of heat from an object is directly proportional to the fourth power of the Kelvin temperature of the object. For a certain object, $R = 213.73$ at room temperature (293° Kelvin). Find R if the temperature increases to 335° Kelvin.

37. Suppose a nuclear bomb is detonated at a certain site. The effects of the bomb will be felt over a distance from the point of detonation that is directly proportional to the cube root of the yield of the bomb. Suppose a 100-kiloton bomb has certain effects to

a radius of 3 km from the point of detonation. Find the distance that the effects would be felt for a 1500-kiloton bomb.

38. The maximum speed possible on a length of railroad track is directly proportional to the cube root of the amount of money spent on maintaining the track. Suppose that a maximum speed of 25 km/hr is possible on a stretch of track for which $450,000 was spent on maintenance. Find the maximum speed if the amount spent on maintenance is increased to $1,750,000.

39. Assume that a person's weight varies directly as the cube of his or her height. Find the weight of a 20-inch, 7-lb baby who grows up to be an adult 67 inches tall. (How reasonable does our assumption about weight and height seem?)

40. The cost of a pizza varies directly as the square of its radius. If a pizza with a radius of 15 cm costs $7, find the cost of a pizza having a radius of 22.5 cm. (You might want to do some research at a nearby pizza establishment and see if this assumption is reasonable.)

Chapter 4   *Summary*

Key Words

4.1 independent variable
dependent variable

4.2 linear function
quadratic function
absolute value function
function defined piecewise
greatest integer function
step function
root function

4.5 direct variation
constant of variation
inverse variation
joint variation

Key Ideas

4.1 Function

A function is a relation in which each element in the domain corresponds to exactly one element in the range.

4.1 Vertical Line Test

If each vertical line cuts a graph in no more than one point, the graph is the graph of a function.

4.3 Operations on Functions

If f and g are functions, then for all values of x for which both $f(x)$ and $g(x)$ exist, the sum of f and g is

$$(f + g)(x) = f(x) + g(x),$$

the difference of f and g, is

$$(f - g)(x) = f(x) - g(x),$$

the product of f and g is

$$(fg)(x) = f(x) \cdot g(x),$$

and the quotient of f and g is

$$\left(\frac{f}{g}\right)(x) = \frac{f(x)}{g(x)}, \quad \text{where } g(x) \neq 0.$$

4.3 Composition of Functions

If f and g are functions, then the composite function, or composition, of g and f is

$$(g \circ f)(x) = g[f(x)]$$

for all x in the domain of f such that $f(x)$ is in the domain of g.

4.4 One-to-One Function

A function f is a one-to-one function if, for elements a and b from the domain of f,

$$a \neq b \quad \text{implies} \quad f(a) \neq f(b).$$

4.4 Horizontal Line Test

If each horizontal line cuts the graph of a function in no more than one point, then the function is one-to-one.

4.4 Inverse Functions

Two one-to-one functions f and g are inverses of each other if

$$(f \circ g)(x) = x \text{ for every } x \text{ in the domain of } g,$$

and $\quad (g \circ f)(x) = x$ for every x in the domain of f.

Chapter 4   *Review Exercises*

Give the domain of each of the following.

1. $f(x) = -x$

2. $f(x) + 6 = -x$

3. $f(x) = |x| - 4$

4. $f(x) = -3|x - 4|$

5. $f(x) = -(x + 3)^2$

6. $f(x) = \sqrt{5 - x}$

7. $f(x) = \sqrt{49 - x^2}$

8. $f(x) = -\sqrt{x^2 - 4}$

9. $f(x) = \sqrt{\dfrac{1}{x^2 + 9}}$

10. $f(x) = \sqrt{x^2 + 3x - 10}$

11. $f(x) = \dfrac{x + 2}{x - 7}$

12. $f(x) = \dfrac{3x - 8}{4x - 7}$

Let $f(x) = -x^2 + 4x + 2$ and $g(x) = 3x + 5$. Find each of the following.

13. $f(-3)$

14. $f(4)$

15. $f(0)$

16. $g(3)$

17. $f[g(-1)]$

18. $g[f(-2)]$

19. $f(3 + z)$

20. $g(5y - 4)$

For each of the following, find $\dfrac{f(x + h) - f(x)}{h}$, where $h \neq 0$.

21. $f(x) = -2x^2 + 4x - 3$

22. $f(x) = 12x - 5$

23. $f(x) = -x^3 + 2x^2$

24. $f(x) = -3$

Graph each function defined as follows.

25. $f(x) = -|x|$

26. $f(x) = |x| - 3$

27. $f(x) = -|x| - 2$

28. $f(x) = -|x + 1| + 3$

29. $f(x) = 2|x - 3| - 4$

30. $f(x) = [\![x - 3]\!]$

31. $f(x) = \left[\!\!\left[\dfrac{1}{2}x - 2\right]\!\!\right]$

32. $f(x) = \begin{cases} -4x + 2 & \text{if } x \leq 1 \\ 3x - 5 & \text{if } x > 1 \end{cases}$

33. $f(x) = \begin{cases} 3x + 1 & \text{if } x < 2 \\ -x + 4 & \text{if } x \geq 2 \end{cases}$

34. $f(x) = \begin{cases} |x| & \text{if } x < 3 \\ 6 - x & \text{if } x \geq 3 \end{cases}$

35. $f(x) = \sqrt{x^2 - 16}$

36. $f(x) = -\sqrt{9 - x^2}$

37. $f(x) = \sqrt{9 - x}$

38. $f(x) = \sqrt{x - 1}$

39. A trailer hauling service charges $45, plus $2 per mile or part of a mile. Graph the ordered pairs (miles, cost).

40. Let f be a function that gives the cost to rent a floor polisher for x hr. The cost is a flat $3 for cleaning the polisher plus $4 per day or fraction of a day for using the polisher.

   (a) Graph f.

   (b) Give the domain and range of f.

Let $f(x) = 3x^2 - 4$ and $g(x) = x^2 - 3x - 4$. Find each of the following values.

41. $(f + g)(x)$

42. $(fg)(x)$

43. $(f - g)(4)$

44. $(f + g)(-4)$

45. $(f + g)(2k)$

46. $(fg)(1 + r)$

47. $(f/g)(3)$

48. $(f/g)(-1)$

49. Give the domain of $(fg)(x)$.

50. Give the domain of $(f/g)(x)$.

Let $f(x) = \sqrt{x - 2}$ and $g(x) = x^2$. Find each of the following.

51. $(f \circ g)(x)$

52. $(g \circ f)(x)$

53. $(f \circ g)(-6)$

54. $(f \circ g)(2)$

55. $(g \circ f)(3)$

56. $(g \circ f)(24)$

Which of the functions graphed or defined as follows are one-to-one?

57.

58.

59.

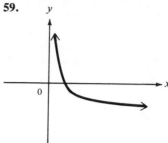

60. $f(x) = \dfrac{8x - 9}{5}$

61. $f(x) = -x^2 + 11$

62. $f(x) = \sqrt{5 - x}$

63. $f(x) = \sqrt{100 - x^2}$

64. $f(x) = -\sqrt{1 - \dfrac{x^2}{100}}$, domain $[0, 10]$

For each of the following that defines a one-to-one function, write an equation for the inverse function in the form $y = f^{-1}(x)$ and then graph f and f^{-1}.

65. $f(x) = 12x + 3$

66. $f(x) = \dfrac{2}{x - 9}$

67. $f(x) = x^3 - 3$

68. $f(x) = \sqrt{x^2 - 6}$, domain $(-\infty, -\sqrt{6}]$

69. $f(x) = \sqrt{25 - x^2}$, domain $[0, 5]$

70. $f(x) = -\sqrt{x - 3}$

Write each of the following statements as an equation.

71. m varies directly as the square of z.

72. y varies inversely as r and directly as the cube of p.

73. Y varies jointly as M and the square of N and inversely as the cube of X.

74. A varies jointly as the third power of t and the fourth power of s, and inversely as p and the square of h.

Solve each of the following problems.

75. Suppose r varies directly as x and inversely as the square of y. If r is 10 when x is 5 and y is 3, find r when x is 12 and y is 4.

76. Suppose m varies jointly as n and the square of p, and inversely as q. If m is 20 when n is 5, p is 6, and q is 18, find m when n is 7, p is 11, and q is 2.

77. Suppose Z varies jointly as the square of J and the cube of M, and inversely as the fourth power of W. If Z is 125 when J is 3, M is 5 and W is 1, find Z if J is 2, M is 7, and W is 3.

78. The power a windmill obtains from the wind varies directly as the cube of the wind velocity. If a 10-km/hr wind produces 10,000 units of power, how much power is produced by a wind of 15 km/hr?

79. Hooke's law for an elastic spring states that the distance a spring stretches varies directly as the force applied. If a force of 32 lb stretches a certain spring 48 inches, how much will a force of 24 lb stretch the spring?

80. The weight w of an object varies inversely as the square of the distance d between the object and the center of the earth. If a man weighs 90 kg on the surface of the earth, how much would he weigh 800 km above the surface? (The radius of the earth is about 6400 km.)

Chapter 4 *Test*

Give the domain of each of the following.

1. $f(x) = 3 + |x + 8|$

2. $f(x) = \sqrt{x^2 + 7x + 12}$

3. $f(x) = \dfrac{x - 1}{2x + 8}$

4. $f(x) = \sqrt{\dfrac{1}{9 - x^2}}$

Let $f(x) = 6 - 3x - x^2$ and $g(x) = 2x + 3$. Find each of the following.

5. $f(-2)$

6. $g[f(2)]$

7. $g\left(\dfrac{1}{2}y - 4\right)$

Graph each function defined as follows.

8. $f(x) = |x - 2| - 1$

9. $f(x) = [\![x + 1]\!]$

10. $f(x) = \begin{cases} 3 & \text{if } x < -2 \\ 2 - \dfrac{1}{2}x & \text{if } x \geq -2 \end{cases}$

11. $f(x) = \sqrt{x^2 - 4}$

12. $f(x) = \sqrt{3 - x}$

Let $f(x) = 3x^2 - 4x + 7$ and $g(x) = 2x + 1$. Find each of the following values.

13. $(f + g)(2a)$

14. $\left(\dfrac{f}{g}\right)(1)$

15. $(fg)\left(\dfrac{1}{2}\right)$

16. $(f \circ g)(0)$

17. $(g \circ f)(p)$

Which of the functions graphed or defined as follows are one-to-one?

18.

19. $f(x) = \sqrt{3x + 2}$

20. $f(x) = (x - 2)^2$, domain $[2, +\infty)$

For each of the following that defines a one-to-one function, write an equation for the inverse in the form $y = f^{-1}(x)$.

21. $f(x) = \dfrac{1}{x + 1}$

22. $f(x) = 2 + x^2$, domain $(-\infty, 0]$

23. $f(x) = 2x + 8$

24. Are the following two functions inverses of each other? Show why or why not.

$$f(x) = \frac{x}{x + 2} \quad \text{and} \quad g(x) = \frac{2x}{x - 1}$$

25. The force of wind on a sail varies jointly as the area of the sail and the square of the wind velocity. If the force is 8 lb when the velocity is 15 mph and the area is 3 sq ft, find the force when the area is 6 sq ft and the velocity is 22.5 mph.

5

Polynomial and Rational Functions

Many applications of mathematics require *polynomial functions*. Some examples of polynomial functions were discussed earlier; a linear function of the form $P(x) = a_1x + a_0$ is a polynomial function of degree 1, while a quadratic function $P(x) = a_2x^2 + a_1x + a_0$ is a polynomial function of degree 2. In this chapter, we show how to graph polynomial functions of degree 3 or more. We also present several methods for finding the *x*-intercepts, called *zeros,* of these functions. *Rational functions* and their graphs are discussed in the final section of the chapter.

5.1 Graphing Polynomial Functions

Polynomial functions are defined as follows.

Definition of Polynomial Function	A **polynomial function of degree n** is a function defined by $$P(x) = a_nx^n + a_{n-1}x^{n-1} + \ldots + a_1x + a_0,$$ for complex numbers $a_n, a_{n-1}, \ldots, a_1,$ and $a_0,$ where $a_n \neq 0.$

$P(x)$ is used instead of $f(x)$ in this chapter to emphasize that the functions are polynomial functions. The polynomials discussed in Chapter 1 had only real-number coefficients. In this chapter the domain of the coefficients is extended to include all complex numbers.* The number a_n is the **leading coefficient** of $P(x)$. If all the coefficients of a polynomial are 0, the polynomial equals 0 and is called the **zero polynomial.** The zero polynomial has no degree. A polynomial of the form $P(x) = a_0$ for a nonzero complex number a_0 has degree 0, however.

We begin the discussion of graphing polynomial functions with the graphs of polynomial functions defined by equations of the form $P(x) = ax^n$.

Example 1   Graph each function.

(a) $P(x) = x^3$

Choose several values for x, and find the corresponding values of $P(x)$, or y, as shown in the upper table in Figure 5.1. Plot the resulting ordered pairs and connect the points with a smooth curve. The graph of $P(x) = x^3$ is shown in color in Figure 5.1.

(b) $P(x) = x^5$

Work as in part (a) of this example to get the graph shown in black in Figure 5.1. Both the graph of $P(x) = x^3$ and the graph of $P(x) = x^5$ are symmetric with respect to the origin. (See the discussion of odd and even functions at the end of this section.)

(c) $P(x) = x^4$, $P(x) = x^6$

Some typical ordered pairs for the graphs of $P(x) = x^4$ and $P(x) = x^6$ are given in the tables in Figure 5.2. These graphs are symmetric with respect to the y-axis, as is the graph of $P(x) = ax^2$ for a nonzero real number a.   ✢

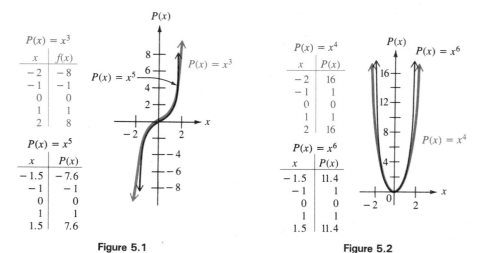

$P(x) = x^3$	
x	$f(x)$
-2	-8
-1	-1
0	0
1	1
2	8

$P(x) = x^5$	
x	$P(x)$
-1.5	-7.6
-1	-1
0	0
1	1
1.5	7.6

Figure 5.1

$P(x) = x^4$	
x	$P(x)$
-2	16
-1	1
0	0
1	1
2	16

$P(x) = x^6$	
x	$P(x)$
-1.5	11.4
-1	1
0	0
1	1
1.5	11.4

Figure 5.2

*Recall that the set of complex numbers includes both real numbers and imaginary numbers such as $5i$ or $6 + i\sqrt{7}$.

As with the graph of $y = ax^2$ in Section 3.3, the value of a in $P(x) = ax^n$ affects the width of the graph. When $|a| > 1$, the graph is narrower than the graph of $P(x) = x^n$; when $0 < |a| < 1$, the graph is broader. The graph of $P(x) = -ax^n$ is reflected about the x-axis as compared to the graph of $P(x) = ax^n$.

Example 2   Graph each of the following functions.

(a) $P(x) = \dfrac{1}{2}x^3$

The graph is broader than that of $P(x) = x^3$ but has the same general shape. It includes the points $(-2, -4)$, $(-1, -1/2)$, $(0, 0)$, $(1, 1/2)$, and $(2, 4)$. See Figure 5.3.

(b) $P(x) = -\dfrac{3}{2}x^4$

The following table gives some ordered pairs for this function.

x	$P(x)$
-2	-24
-1	$-3/2$
0	0
1	$-3/2$
2	-24

The graph is shown in Figure 5.4. This graph is narrower than that of $P(x) = x^4$, since $|-3/2| > 1$, and opens downward instead of upward.   ✛

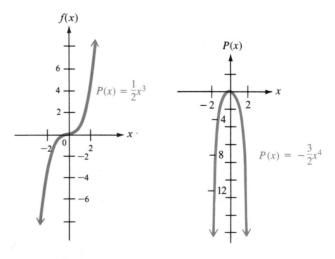

Figure 5.3   Figure 5.4

Compared with the graph of $P(x) = ax^n$, the graph of $P(x) = ax^n + k$ is translated (shifted) k units up if $k > 0$ and $|k|$ units down if $k < 0$. Also, the graph of $P(x) = a(x - h)^n$ is translated h units to the right if $h > 0$ and $|h|$ units to the left if $h < 0$, when compared with the graph of $P(x) = ax^n$.

Example 3   Graph each of the following.

(a) $P(x) = x^5 - 2$

The graph will be the same as that of $P(x) = x^5$, but translated down 2 units. See Figure 5.5.

(b) $P(x) = (x + 1)^6$

This function P has a graph like that of $P(x) = x^6$, but since $x + 1 = x - (-1)$, it is translated one unit to the left as shown in Figure 5.6.

(c) $P(x) = -x^3 + 3$

The negative sign causes the graph to be reflected about the x-axis compared to the graph of $P(x) = x^3$. As shown in Figure 5.7, the graph is also translated up 3 units.   ✚

Figure 5.5   Figure 5.6   Figure 5.7

Generalizing from the graphs in Examples 1–3, the domain of a polynomial function is the set of all real numbers. The range of a polynomial function of odd degree is also the set of all real numbers. Some typical graphs of polynomial functions of odd degree are shown in Figure 5.8.

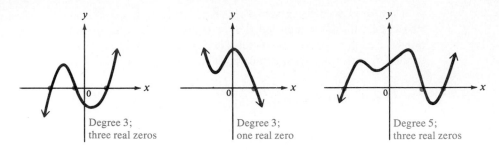

Degree 3;
three real zeros

Degree 3;
one real zero

Degree 5;
three real zeros

Figure 5.8

The values of x that satisfy $P(x) = 0$ (that is, the x-intercepts) are called the **zeros** of $P(x)$. Methods for finding zeros of first- and second-degree polynomials were given in Chapter 2. In later sections of this chapter, several theorems are presented that help to find, or at least approximate, the zeros of polynomials of higher degree. The graphs in Figure 5.8 suggest that every polynomial function P of odd degree has at least one real zero.

A polynomial function of even degree will have a range that takes the form $(-\infty, k]$ or $[k, +\infty)$ for some real number k. Figure 5.9 shows two typical graphs of polynomial functions of even degree.

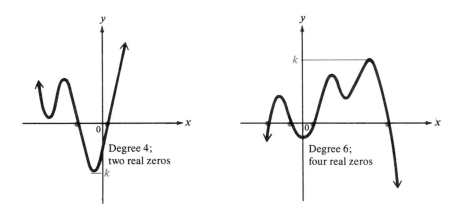

Degree 4;
two real zeros

Degree 6;
four real zeros

Figure 5.9

It is difficult to graph most polynomial functions without using calculus. A large number of points must be plotted to get a reasonably accurate graph. If a polynomial function P can be factored into linear factors, however, its graph can be approximated by plotting only a few points.

Example 4   Graph $P(x) = (2x + 3)(x - 1)(x + 2)$.

Multiplying out the expression on the right would show that P is a third-degree polynomial, also called a *cubic* polynomial. Sketch the graph of P by first setting each of the three factors equal to 0 and solving the resulting equations to find the zeros of the function.

$$2x + 3 = 0 \qquad \text{or} \qquad x - 1 = 0 \qquad \text{or} \qquad x + 2 = 0$$

$$x = -\frac{3}{2} \qquad\qquad\qquad x = 1 \qquad\qquad\qquad x = -2$$

The three zeros, $-3/2$, 1, and -2, divide the x-axis into four intervals:

$$(-\infty, -2), \left(-2, -\frac{3}{2}\right), \left(-\frac{3}{2}, 1\right), (1, +\infty).$$

These intervals are shown in Figure 5.10.

Figure 5.10

Since the three zeros are the only x-intercepts, the values of $P(x)$ in the intervals shown in Figure 5.10 are either always positive or always negative. To find the sign of $P(x)$ in each interval, select a value of x in the interval and determine by substitution whether the function values are positive or negative in that interval. A typical selection of test points and the results of the tests are shown below.

Interval	Test point	Value of $P(x)$	Sign of $P(x)$
$x < -2$	-3	-12	Negative
$-2 < x < -3/2$	$-7/4$	$11/32$	Positive
$-3/2 < x < 1$	0	-6	Negative
$1 < x$	2	28	Positive

When the values of $P(x)$ are negative, the graph is below the x-axis, and when $P(x)$ takes on positive values, the graph is above the x-axis. These results suggest that the graph looks like the sketch in Figure 5.11 on the following page. The sketch could be improved by plotting additional points in each region.   ✦

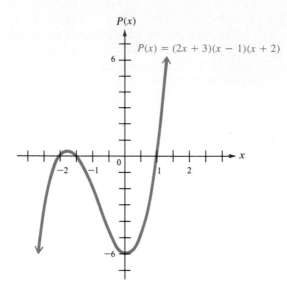

Figure 5.11

Example 5   Sketch the graph of $P(x) = 3x^4 + x^3 - 2x^2$.

The polynomial can be factored as follows:

$$3x^4 + x^3 - 2x^2 = x^2(3x^2 + x - 2)$$
$$= x^2(3x - 2)(x + 1).$$

The zeros, 0, 2/3, and -1, divide the x-axis into four intervals:

$$x < -1, \qquad -1 < x < 0, \qquad 0 < x < 2/3, \qquad \text{and} \qquad 2/3 < x.$$

Determine the sign of $P(x)$ in each region by substituting a value for x from each region to get the following information.

Interval	*Test point*	*Value of* $P(x)$	*Location relative to x-axis*
$x < -1$	-2	32	Above
$-1 < x < 0$	$-1/2$	$-7/16$	Below
$0 < x < 2/3$	$1/3$	$-4/27$	Below
$2/3 < x$	1	2	Above

With the values of x used for the test points and the corresponding values of y, sketch the graph as shown in Figure 5.12.   ✤

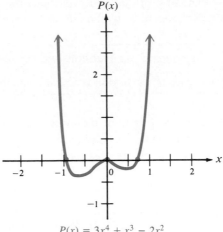

$$P(x) = 3x^4 + x^3 - 2x^2$$

Figure 5.12

Some polynomial functions are examples of *even* or *odd* functions, defined as follows.

Definitions of Even and Odd Functions	A function $y = f(x)$ is
	an **even function** if $f(-x) = f(x)$,
	or an **odd function** if $f(-x) = -f(x)$,
	for every x in the domain of f.

Example 6   Decide whether the given polynomial functions are odd, even, or neither.

(a) $P(x) = 6x^4 - 8x^2 + 1$

Replacing x with $-x$ gives

$$P(-x) = 6(-x)^4 - 8(-x)^2 + 1 = 6x^4 - 8x^2 + 1 = P(x).$$

This function is even.

(b) $P(x) = -25x^5 + 4x^3 + 3x$

Here

$$\begin{aligned}
P(-x) &= -25(-x)^5 + 4(-x)^3 + 3(-x) \\
&= 25x^5 - 4x^3 - 3x \\
&= -P(x).
\end{aligned}$$

The function is odd.

(c) $P(x) = 10x - 3$

Since $P(-x) = -10x - 3$, which equals neither $P(x)$ nor $-P(x)$, this function is neither even nor odd.   ✤

The idea of odd and even functions is useful in graphing because of the following properties, which are based on the tests for symmetry in Chapter 3.

> The graph of an even function is symmetric with respect to the y-axis.
>
> The graph of an odd function is symmetric with respect to the origin.

Look at the graphs of the even functions $P(x) = x^4$ and $P(x) = x^6$ and the odd functions $P(x) = x^3$ and $P(x) = x^5$ at the beginning of this section to see that these statements are reasonable.

5.1   Exercises

Each of the polynomial functions defined below is symmetric with respect to a line or a point. For each function, (a) sketch the graph, and (b) give the line or point of symmetry. See Examples 1–3.

1. $P(x) = \dfrac{1}{4}x^6$

2. $P(x) = -\dfrac{2}{3}x^5$

3. $P(x) = -\dfrac{5}{4}x^5$

4. $P(x) = 2x^4$

5. $P(x) = \dfrac{1}{2}x^3 + 1$

6. $P(x) = -x^4 + 2$

7. $P(x) = -(x + 1)^3$

8. $P(x) = \dfrac{1}{3}(x + 3)^4$

9. $P(x) = (x - 1)^4 + 2$ **10.** $P(x) = (x + 2)^3 - 1$

Graph each of the following. See Examples 4 and 5.

11. $P(x) = 2x(x - 3)(x + 2)$

12. $P(x) = x^2(x + 1)(x - 1)$

13. $P(x) = x^2(x - 2)(x + 3)^2$

14. $P(x) = x^2(x - 5)(x + 3)(x - 1)$

15. $P(x) = x^3 - x^2 - 2x$

16. $P(x) = -x^3 - 4x^2 - 3x$

17. $P(x) = 3x^4 + 5x^3 - 2x^2$

18. $P(x) = 4x^3 + 2x^2 - 12x$

19. $P(x) = 2x^3(x^2 - 4)(x - 1)$

20. $P(x) = 5x^2(x^3 - 1)(x + 2)$

21. $P(x) = x^4 - 4x^2$

22. $P(x) = -x^4 + 16x^2$

23. $P(x) = x^2(x - 3)^3(x + 1)$

24. $P(x) = x(x - 4)^2(x + 2)^2$

25. The polynomial function

$$A(x) = -.015x^3 + 1.058x$$

gives the approximate alcohol concentration (in tenths of a percent) in an average person's bloodstream x hr after drinking about 8 oz of 100-proof whiskey. The function is approximately valid for x in the interval [0, 8].

(a) Graph $A(x)$.

(b) Using the graph you drew for part (a), estimate the time of maximum alcohol concentration.

(c) In one state, a person is legally drunk if the blood alcohol concentration exceeds .15%. Use the graph from part (a) to estimate the period in which the average person is legally drunk.

26. A technique for measuring cardiac output depends on the concentration of a dye in the bloodstream after a known amount is injected into a vein near the heart. For a normal heart, the concentration of dye in the bloodstream at ume x (in seconds) is given by the function

$$g(x) = -.006x^4 + .140x^3 - .053x^2 + 1.79x.$$

Graph $g(x)$.

27. The pressure of the oil in a reservoir tends to drop with time. By taking sample pressure readings for a particular oil reservoir, petroleum engineers have found that the change in pressure is given by

$$P(t) = t^3 - 25t^2 + 200t,$$

where t is time in years from the date of the first reading.

(a) Graph $P(t)$.
(b) For what time period is the change in pressure (drop) increasing? decreasing?

28. During the early part of the twentieth century, the deer population of the Kaibab Plateau in Arizona experienced a rapid increase, because hunters had reduced the number of natural predators. The increase in population depleted the food resources and eventually caused the population to decline. For the period from 1905 to 1930, the deer population was approximated by

$$D(x) = -.125x^5 + 3.125x^4 + 4000,$$

where x is time in years from 1905.

(a) Use a calculator to find enough points to graph $D(x)$.
(b) From the graph, over what period of time (from 1905 to 1930) was the population increasing? relatively stable? decreasing?

Decide whether the polynomial functions defined as follows are odd, even, or neither. See Example 6.

29. $P(x) = 2x^3$ 30. $P(x) = -4x^5$ 31. $P(x) = .2x^4$

32. $P(x) = -x^6$ 33. $P(x) = -x^5$ 34. $P(x) = (x - 1)^3$

35. $P(x) = 2x^3 + 3$ 36. $P(x) = 4x^3 - x$ 37. $P(x) = x^4 + 3x^2 + 5$

38. $P(x) = 11x + 2$ 39. $P(x) = (x + 5)^2$ 40. $P(x) = (3x - 5)^2$

Approximate maximum or minimum values of polynomial functions can be found for given intervals by first evaluating the function at the left endpoint of the interval, then adding 0.1 to the value of x and reevaluating the polynomial. Keep doing this until the right endpoint of the interval is reached. Then identify the approximate maximum and minimum value for the polynomial on the interval.

41. $P(x) = x^3 + 4x^2 - 8x - 8$, $[-3.8, -3]$ 42. $P(x) = x^3 + 4x^2 - 8x - 8$, $[.3, 1]$

43. $P(x) = 2x^3 - 5x^2 - x + 1$, $[-1, 0]$ 44. $P(x) = 2x^3 - 5x^2 - x + 1$, $[1.4, 2]$

45. $P(x) = x^4 - 7x^3 + 13x^2 + 6x - 28$, $[-2, -1]$ 46. $P(x) = x^4 - 7x^3 + 13x^2 + 6x - 28$, $[2, 3]$

47. The graphs of $P(x) = x$, $P(x) = x^3$, and $P(x) = x^5$ all pass through the points $(-1, -1)$, $(0, 0)$, and $(1, 1)$. Draw a graph showing these three functions for the domain $[-1, 1]$. Use your result to decide how the graph of $P(x) = x^{11}$ would look for the same interval.

48. The graphs of $P(x) = x^2$, $P(x) = x^4$, and $P(x) = x^6$ all pass through the points $(-1, 1)$, $(0, 0)$, and $(1, 1)$. Draw a graph showing these three functions for the domain $[-1, 1]$. Use your result to decide how the graph of $P(x) = x^{12}$ would look for the same interval.

5.2   Polynomial Division

In the previous section we graphed polynomial functions that were factored. Factoring depends on division, so in this section we give a shortcut method for polynomial division.

In many polynomial division problems the divisor is a first-degree binomial of the form $x - k$, where the coefficient of x is 1. A shortcut for this type of division problem is developed by omitting all variables and writing only coefficients. Zero is used to represent the coefficient of any missing powers of the variables. Since the coefficient of x in the divisor is always 1 in problems of this type, the 1 can be left out also. Using these omissions, the division problem on the left below is written in the shortened form shown on the right.

$$
\begin{array}{r}
3x^2 + 10x + 40 \\
x - 4\overline{)3x^3 - 2x^2 + 0x - 150} \\
\underline{3x^3 - 12x^2} \\
10x^2 \\
\underline{10x^2 - 40x} \\
40x \\
\underline{40x - 160} \\
10
\end{array}
\qquad
\begin{array}{r}
3 \quad 10 \quad\quad 40 \\
-4\overline{)3 - 2 \quad\quad 0 - 150} \\
\underline{3 - 12} \\
10 \\
\underline{\mathbf{10} - 40} \\
40 \\
\underline{\mathbf{40} - 160} \\
10
\end{array}
$$

The boldface numbers are repetitions of the numbers directly above them and can be omitted also.

$$
\begin{array}{r}
3 \quad 10 \quad\quad 40 \\
-4\overline{)3 - 2 \quad\quad 0 - 150} \\
\underline{- 12} \\
10 \\
\underline{- 40} \\
40 \\
\underline{- 160} \\
10
\end{array}
$$

The entire problem can now be condensed, with the top row of numbers omitted, since it duplicates the bottom row.

$$
\begin{array}{r}
-4\overline{)3 - 2 \quad\quad 0 - 150} \\
\underline{- 12 - 40 - 160} \\
3 \quad 10 \quad 40 \quad 10
\end{array}
$$

The bottom row was obtained by subtracting -12, -40, and -160 from the corresponding terms above them. For reasons that will become clear later in this section, change the -4 at the left to 4, which also changes the sign of the numbers in the second row, and then *add*.

$$
\begin{array}{r}
4\overline{)3 - 2 \quad\quad 0 - 150} \\
\underline{12 \quad 40 \quad 160} \\
3 \quad 10 \quad 40 \quad 10
\end{array}
$$

Read the quotient and remainder from the bottom row. The quotient will have degree one less than the polynomial being divided.

$$\frac{3x^3 - 2x^2 - 150}{x - 4} = 3x^2 + 10x + 40 + \frac{10}{x - 4}$$

This shortcut process is called **synthetic division.**

In summary, to use synthetic division to divide a polynomial by a binomial of the form $x - k$, first write the coefficients of the polynomial in decreasing powers of the variable, using 0 as the coefficient of any missing powers. Write the number k to the left in the same row. (In the example above, $x - k$ is $x - 4$, so k is 4.) Next bring down the leading coefficient of the polynomial (3 in the example above) as the first number in the last row. Multiply by k to get the first number in the second row ($3 \cdot 4 = 12$). Add this result to the next number in the first row to get the second number in the third row ($12 + (-2) = 10$). Multiply by k to get the second number in the second row. Add this to the next number in the first row to get the third number in the third row, and so on. This process of multiplying each result in the third row by k and adding the product to the number in the next column is repeated until there is a number in the last row for each coefficient in the first row. To avoid incorrect results, it is very important to use a 0 for any missing terms (including the constant term) when setting up the division.

Example 1   Use synthetic division to divide $5m^3 - 6m^2 - 28m - 2$ by $m + 2$.
     Begin by writing

$$-2\overline{)5 \quad -6 \quad -28 \quad -2.}$$

The 2 is changed to -2 since k is found by writing $m + 2$ as $m - (-2)$. Next, bring down the 5.

$$
\begin{array}{r}
-2\overline{)\,5 \quad -6 \quad -28 \quad -2} \\
\hline
5
\end{array}
$$

Now, multiply -2 by 5 to get -10, and add it to -6 from the first row. The result is -16.

$$
\begin{array}{r}
-2\overline{)\,5 \quad\; -6 \quad -28 \quad -2} \\
-10 \quad\qquad\qquad \\
\hline
5 \quad -16
\end{array}
$$

Next, multiply -2 by -16 to get 32. Add this to -28 from the first row.

$$
\begin{array}{r}
-2\overline{)\,5 \quad\;\; -6 \quad -28 \quad -2} \\
-10 \quad\; 32 \qquad\quad \\
\hline
5 \quad -16 \quad\;\, 4
\end{array}
$$

Finally, $(-2)(4) = -8$. Add this result to -2 to get -10.

$$
\begin{array}{r}
-2\overline{)\,5 \quad\;\; -6 \quad -28 \quad\;\; -2} \\
-10 \quad\; 32 \quad -8 \\
\hline
5 \quad -16 \quad\;\, 4 \quad -10
\end{array}
$$

The coefficients of the quotient polynomial and the remainder are read directly from the bottom row. Since the degree of the quotient will always be one less than the degree of the polynomial to be divided,

$$\frac{5m^3 - 6m^2 - 28m - 2}{m + 2} = 5m^2 - 16m + 4 + \frac{-10}{m + 2}. \quad \clubsuit$$

The result of the division in Example 1 can be written as

$$5m^3 - 6m^2 - 28m - 2 = (m + 2)(5m^2 - 16m + 4) + (-10)$$

by multiplying both sides by the denominator $m + 2$. The following theorem is a generalization of the division process illustrated above.

Division Algorithm	For any polynomial $P(x)$ and any complex number k, there exists a unique polynomial $Q(x)$ and number r such that $$P(x) = (x - k)Q(x) + r.$$

For example, in the synthetic division above,

$$5m^3 - 6m^2 - 28m - 2 = (m + 2)(5m^2 - 16m + 4) + (-10)$$
$$P(m) \qquad\qquad = (m - k) \cdot \qquad Q(m) \qquad + \quad r.$$

By the division algorithm, $P(x) = (x - k)Q(x) + r$. This equality is true for all complex values of x, so it is true for $x = k$. Thus

$$P(k) = (k - k)Q(k) + r$$
$$P(k) = r.$$

This proves the following theorem.

Remainder Theorem	If the polynomial $P(x)$ is divided by $x - k$, then the remainder is $P(k)$.

As an example of this theorem, the polynomial $P(m) = 5m^3 - 6m^2 - 28m - 2$ was written above as

$$P(m) = (m + 2)(5m^2 - 16m + 4) + (-10).$$

To find $P(-2)$, substitute -2 for m.

$$P(-2) = (-2 + 2)[5(-2)^2 - 16(-2) + 4] + (-10)$$
$$= 0[5(-2)^2 - 16(-2) + 4] + (-10)$$
$$= -10$$

By the remainder theorem, instead of replacing m with -2 to find $P(-2)$, divide $P(m)$ by $m + 2$ using synthetic division. Then $P(-2)$ is the remainder, -10.

Example 2   Let $P(x) = -x^4 + 3x^2 - 4x - 5$. Find $P(-2)$.
Use the remainder theorem and synthetic division.

$$
\begin{array}{r|rrrrr}
-2) & -1 & 0 & 3 & -4 & -5 \\
    &   & 2 & -4 & 2 & 4 \\
\hline
    & -1 & 2 & -1 & -2 & -1
\end{array}
$$

The remainder when $P(x)$ is divided by $x - (-2) = x + 2$ is -1, so $P(-2) = -1$.   ✜

Example 3   Find $P(-2 + i)$ if $P(x) = x^3 - 4x^2 + 2x - 29i$.
Use synthetic division and addition and multiplication of complex numbers. (Recall that the product of two complex numbers can be found in the same way as the product of two binomials.)

$$
\begin{array}{r|rrrr}
-2 + i) & 1 & -4 & 2 & -29i \\
        &   & -2 + i & 11 - 8i & -18 + 29i \\
\hline
        & 1 & -6 + i & 13 - 8i & -18
\end{array}
$$

Since the remainder is -18, $P(-2 + i) = -18$.   ✜

The remainder theorem gives a quick way to decide if a number k is a zero of a polynomial $P(x)$. Use synthetic division to find $P(k)$; if the remainder is zero, then $P(k) = 0$ and k is a zero of $P(x)$. When $P(k) = 0$, the number k is also called a **root** or **solution** of the equation $P(x) = 0$.

Example 4   Decide whether the given number is a zero of the given polynomial.
(a) 2; $P(x) = x^3 - 4x^2 + 9x - 10$
Use synthetic division.

$$
\begin{array}{r|rrrr}
2) & 1 & -4 & 9 & -10 \\
   &   & 2 & -4 & 10 \\
\hline
   & 1 & -2 & 5 & 0
\end{array}
$$

Since the remainder is 0, $P(2) = 0$, and 2 is a zero of the polynomial $P(x) = x^3 - 4x^2 + 9x - 10$.

(b) -2; $P(x) = 3x^3 - 2x^2 + 4x$
Remember to use a coefficient of 0 for the missing constant term in the synthetic division.

$$
\begin{array}{r|rrrr}
-2) & 3 & -2 & 4 & 0 \\
    &   & -6 & 16 & -40 \\
\hline
    & 3 & -8 & 20 & -40
\end{array}
$$

The remainder is not zero, so -2 is not a zero of $P(x) = 3x^3 - 2x^2 + 4x$. In fact, $P(-2) = -40$.   ✜

5.2 Exercises

Use synthetic division to perform each of the following divisions. See Example 1.

1. $\dfrac{x^3 + 2x^2 - 17x - 10}{x + 5}$

2. $\dfrac{a^4 + 4a^3 + 2a^2 + 9a + 4}{a + 4}$

3. $\dfrac{m^4 - 3m^3 - 4m^2 + 12m}{m - 2}$

4. $\dfrac{p^4 - 3p^3 - 5p^2 + 2p - 16}{p + 2}$

5. $\dfrac{3x^3 - 11x^2 - 20x + 3}{x - 5}$

6. $\dfrac{4p^3 + 8p^2 - 16p - 9}{p + 3}$

7. $\dfrac{4m^3 - 3m - 2}{m + 1}$

8. $\dfrac{3q^3 - 4q + 2}{q - 1}$

9. $\dfrac{x^5 + 3x^4 + 2x^3 + 2x^2 + 3x + 1}{x + 2}$

10. $\dfrac{m^6 - 3m^4 + 2m^3 - 6m^2 - 5m + 3}{m + 2}$

11. $\dfrac{\frac{1}{3}x^3 - \frac{2}{9}x^2 + \frac{1}{27}x + 1}{x - \frac{1}{3}}$

12. $\dfrac{x^3 + x^2 + \frac{1}{2}x + \frac{1}{8}}{x + \frac{1}{2}}$

13. $\dfrac{9z^3 - 8z^2 + 7z - 2}{z - 2}$

14. $\dfrac{-11p^4 + 2p^3 - 8p^2 - 4}{p + 1}$

15. $\dfrac{y^3 - 1}{y - 1}$

16. $\dfrac{r^5 - 1}{r - 1}$

17. $\dfrac{x^4 - 1}{x - 1}$

18. $\dfrac{x^7 + 1}{x + 1}$

Express each polynomial in the form $P(x) = (x - k)Q(x) + r$ for the given value of k.

19. $P(x) = x^3 + x^2 + x - 8; \quad k = 1$

20. $P(x) = 2x^3 + 3x^2 + 4x - 10; \quad k = -1$

21. $P(x) = -x^3 + 2x^2 + 4; \quad k = -2$

22. $P(x) = -4x^3 + 2x^2 - 3x - 10; \quad k = 2$

23. $P(x) = x^4 - 3x^3 + 2x^2 - x + 5; \quad k = 3$

24. $P(x) = 2x^4 + x^3 - 15x^2 + 3x; \quad k = -3$

25. $P(x) = 3x^4 + 4x^3 - 10x^2 + 15; \quad k = -1$

26. $P(x) = -5x^4 + x^3 + 2x^2 + 3x + 1; \quad k = 1$

For each of the following polynomials, use the remainder theorem to find $P(k)$.
See Examples 2 and 3.

27. $k = 5; \quad P(x) = -x^2 + 2x + 7$

28. $k = -3; \quad P(x) = 3x^2 + 8x + 5$

29. $k = 3; \quad P(x) = x^2 - 4x + 5$

30. $k = -2; \quad P(x) = x^2 + 5x + 6$

31. $k = -2; \quad P(x) = 5x^3 + 2x^2 - x$

32. $k = 2; \quad P(x) = 2x^3 - 3x^2 - 5x + 4$

33. $k = 2 + i; \quad P(x) = x^2 - 5x + 1$

34. $k = 3 - 2i; \quad P(x) = x^2 - x + 3$

35. $k = 1 - i; \quad P(x) = x^3 + x^2 - x + 1$

36. $k = 2 - 3i; \quad P(x) = x^3 + 2x^2 + x - 5$

Use synthetic division to decide whether the given number is a zero of the given polynomial. See Example 4.

37. $2; \quad P(x) = x^2 + 2x - 8$

38. $-1; \quad P(m) = m^2 + 4m - 5$

39. $2; \quad P(g) = g^3 - 3g^2 + 4g - 4$

40. $-3; \quad P(m) = m^3 + 2m^2 - m + 6$

41. $4; \quad P(r) = 2r^3 - 6r^2 - 9r + 4$

42. $-4; \quad P(y) = 9y^3 + 39y^2 + 12y$

43. $1 - i$; $P(y) = y^2 - 2y + 2$

44. $3 - 2i$; $P(r) = r^2 - 6r + 13$

45. $2 + i$; $P(k) = k^2 + 3k + 4$

46. $1 - 2i$; $P(z) = z^2 - 3z + 5$

47. i; $P(x) = x^3 + 2ix^2 + 2x + i$

48. $-i$; $P(p) = p^3 - ip^2 + 3p + 5i$

49. $1 + i$; $P(p) = p^3 + 3p^2 - p + 1$

50. $2 - i$; $P(r) = 2r^3 - r^2 + 3r - 5$

5.3 Zeros of Polynomials

By the remainder theorem, if $P(k) = 0$, then the remainder when $P(x)$ is divided by $x - k$ is zero. This means that $x - k$ is a factor of $P(x)$. Conversely, if $x - k$ is a factor of $P(x)$, then $P(k)$ must equal 0. This is summarized in the following theorem.

Factor Theorem

> The polynomial $x - k$ is a factor of the polynomial $P(x)$ if and only if $P(k) = 0$.

Example 1 Is $x - 1$ a factor of $P(x) = 2x^4 + 3x^2 - 5x + 7$?

By the factor theorem, $x - 1$ will be a factor of $P(x)$ only if $P(1) = 0$. Use synthetic division and the remainder theorem to decide.

$$
\begin{array}{r|rrrrr}
1 & 2 & 0 & 3 & -5 & 7 \\
  &   & 2 & 2 & 5 & 0 \\
\hline
  & 2 & 2 & 5 & 0 & 7
\end{array}
$$

Since the remainder is 7, $P(1) = 7$, not 0, so $x - 1$ is not a factor of $P(x)$. ✦

Example 2 Is $x - i$ a factor of $P(x) = 3x^3 + (-4 - 3i)x^2 + (5 + 4i)x - 5i$?

The only way $x - i$ can be a factor of $P(x)$ is for $P(i)$ to be 0. Decide if this is the case by using synthetic division.

$$
\begin{array}{r|rrrr}
i & 3 & -4 - 3i & 5 + 4i & -5i \\
  &   & 3i & -4i & 5i \\
\hline
  & 3 & -4 & 5 & 0
\end{array}
$$

Since the remainder is 0, $P(i) = 0$, and $x - i$ is a factor of $P(x)$. The other factor is the quotient $3x^2 - 4x + 5$, so $P(x)$ can be factored as

$$P(x) = (x - i)(3x^2 - 4x + 5). \quad ✦$$

The next theorem says that every polynomial of degree 1 or more has a zero, so that every such polynomial can be factored. This theorem was first proved by the mathematician Carl F. Gauss in his doctoral thesis in 1799 when he was 22 years old. Although many proofs of this result have been given, all of them involve mathematics beyond the algebra in this book, so no proof is included here.

Fundamental Theorem of Algebra	Every polynomial of degree 1 or more has at least one complex zero.

From the fundamental theorem, if $P(x)$ is of degree 1 or more then there is some number k such that $P(k) = 0$. By the factor theorem, then

$$P(x) = (x - k) \cdot Q(x)$$

for some polynomial $Q(x)$. The fundamental theorem and the factor theorem can be used to factor $Q(x)$ in the same way. Assuming that $P(x)$ has degree n and repeating this process n times gives

$$P(x) = a(x - k_1)(x - k_2) \ldots (x - k_n),$$

where a is the leading coefficient of $P(x)$. Each of these factors leads to a zero of $P(x)$, so $P(x)$ has the n zeros $k_1, k_2, k_3, \ldots, k_n$. This result can be used to prove the next theorem. The proof is left for the exercises. See Exercise 61.

A polynomial of degree n has at most n distinct zeros.

The theorem says that there exist *at most n* distinct zeros. For example, the polynomial function $P(x) = x^3 + 3x^2 + 3x + 1 = (x + 1)^3$ is of degree 3 but has only one zero, -1. Actually, the zero -1 occurs three times, since there are three factors of $x + 1$; this zero is called a **zero of multiplicity 3.**

Example 3   Find a polynomial $P(x)$ of degree 3 that satisfies the following conditions.

(a) Zeros of -1, 2, and 4; $P(1) = 3$

These three zeros give $x - (-1) = x + 1$, $x - 2$, and $x - 4$ as factors of $P(x)$. Since $P(x)$ is to be of degree 3, these are the only possible factors by the theorem just above. Therefore, $P(x)$ has the form

$$P(x) = a(x + 1)(x - 2)(x - 4)$$

for some nonzero real number a. To find a, use the fact that $P(1) = 3$.

$$P(1) = a(1 + 1)(1 - 2)(1 - 4) = 3$$
$$a(2)(-1)(-3) = 3$$
$$6a = 3$$
$$a = \frac{1}{2}$$

Thus,
$$P(x) = \frac{1}{2}(x + 1)(x - 2)(x - 4),$$

or
$$P(x) = \frac{1}{2}x^3 - \frac{5}{2}x^2 + x + 4.$$

(b) -2 is a zero of multiplicity 3; $P(-1) = 4$

The polynomial $P(x)$ has the form

$$P(x) = a(x + 2)(x + 2)(x + 2)$$
$$= a(x + 2)^3.$$

Since $P(-1) = 4$,

$$P(-1) = a(-1 + 2)^3 = 4,$$
$$a(1)^3 = 4$$

or $a = 4,$

and $P(x) = 4(x + 2)^3 = 4x^3 + 24x^2 + 48x + 32$.   ✛

The remainder theorem can be used to show that both $2 + i$ and $2 - i$ are zeros of $P(x) = x^3 - x^2 - 7x + 15$. It is not a coincidence that both $2 + i$ and its conjugate $2 - i$ are zeros of this polynomial. If $a + bi$ is a zero of a polynomial function with *real* coefficients, then so is $a - bi$. This is given as the next theorem. The proof is left for the exercises. See Exercise 62.

Conjugate Zeros Theorem	If $P(x)$ is a polynomial having only real coefficients and if $a + bi$ is a zero of $P(x)$, then the conjugate $a - bi$ is also a zero of $P(x)$.

The requirement that the polynomial have only real coefficients is very important. For example, $P(x) = x - (1 + i)$ has $1 + i$ as a zero, but the conjugate $1 - i$ is not a zero.

Example 4   Find a polynomial of lowest degree having real coefficients and zeros 3 and $2 + i$.

The complex number $2 - i$ also must be a zero, so there are at least three zeros, 3, $2 + i$, and $2 - i$. For the polynomial to be of lowest degree these must be the only zeros. Then by the factor theorem there must be three factors, $x - 3$, $x - (2 + i)$, and $x - (2 - i)$. A polynomial of lowest degree is

$$P(x) = (x - 3)[x - (2 + i)][x - (2 - i)]$$
$$= (x - 3)(x - 2 - i)(x - 2 + i)$$
$$= x^3 - 7x^2 + 17x - 15.$$

Other polynomials, such as $2(x^3 - 7x^2 + 17x - 15)$ or $\sqrt{5}(x^3 - 7x^2 + 17x - 15)$, for example, also satisfy the given conditions on zeros. The information on zeros given in the problem is not enough to give a specific value for the leading coefficient.   ✛

The conjugate zeros theorem helps predict the number of real zeros of polynomials with real coefficients. A polynomial of odd degree n, where $n \geq 1$, with real coefficients must have at least one real zero (since zeros of the form $a + bi$, where $b \neq 0$, occur in conjugate pairs.) On the other hand, a polynomial of even degree n with real coefficients need have no real zeros but may have up to n real zeros.

Example 5   Find all zeros of $P(x) = x^4 - 7x^3 + 18x^2 - 22x + 12$, given that $1 - i$ is a zero.

Since $1 - i$ is a zero, by the conjugate zeros theorem $1 + i$ is also a zero. The remaining zeros are found by first dividing the original polynomial by $x - (1 - i)$.

$$
\begin{array}{r|rrrr}
1 - i) & 1 & -7 & 18 & -22 & 12 \\
& & 1 - i & -7 + 5i & 16 - 6i & -12 \\
\hline
& 1 & -6 - i & 11 + 5i & -6 - 6i & 0
\end{array}
$$

Rather than go back to the original equation, divide the quotient from the first division by $x - (1 + i)$ as follows.

$$
\begin{array}{r|rrrr}
1 + i) & 1 & -6 - i & 11 + 5i & -6 - 6i \\
& & 1 + i & -5 - 5i & 6 + 6i \\
\hline
& 1 & -5 & 6 & 0
\end{array}
$$

Find the zeros of the quadratic polynomial $x^2 - 5x + 6$ by solving the equation $x^2 - 5x + 6 = 0$. Factoring the polynomial shows that the zeros are 2 and 3, so the four zeros of $P(x)$ are $1 - i$, $1 + i$, 2, and 3.   ❖

The theorems given in this section can be used to factor a polynomial into linear factors (factors of the form $ax - b$) when one or more of its zeros are known.

Example 6   Factor $P(x)$ into linear factors, given that k is a zero of $P(x)$.

(a) $P(x) = 6x^3 + 19x^2 + 2x - 3$; $k = -3$

Since $k = -3$ is a zero of $P(x)$, $x - (-3) = x + 3$ is a factor. Use synthetic division to divide $P(x)$ by $x + 3$.

$$
\begin{array}{r|rrrr}
-3) & 6 & 19 & 2 & -3 \\
& & -18 & -3 & 3 \\
\hline
& 6 & 1 & -1 & 0
\end{array}
$$

The quotient is $6x^2 + x - 1$, so

$$P(x) = (x + 3)(6x^2 + x - 1).$$

Factor $6x^2 + x - 1$ as $(2x + 1)(3x - 1)$ to get

$$P(x) = (x + 3)(2x + 1)(3x - 1),$$

where all factors are linear.

(b) $P(x) = 3x^3 + (-1 + 3i)x^2 + (-12 + 5i)x + 4 - 2i; \ k = 2 - i$

One factor is $x - (2 - i)$ or $x - 2 + i$. Divide $P(x)$ by $x - (2 - i)$.

$$
\begin{array}{r}
2 - i \overline{)\, 3 \quad -1 + 3i \quad -12 + 5i \quad \ 4 - 2i} \\
\underline{6 - 3i \qquad 10 - 5i \quad -4 + 2i} \\
3 \qquad 5 \qquad\qquad -2 \qquad\qquad 0
\end{array}
$$

By the division algorithm,

$$P(x) = (x - 2 + i)(3x^2 + 5x - 2).$$

Factor $3x^2 + 5x - 2$ as $(3x - 1)(x + 2)$; then a linear factored form of $P(x)$ is

$$P(x) = (x - 2 + i)(3x - 1)(x + 2). \quad \clubsuit$$

5.3   Exercises

Use the factor theorem to decide whether the second polynomial is a factor of the first. See Examples 1 and 2.

1. $4x^2 + 2x + 42; \ x - 3$

2. $-3x^2 - 4x + 2; \ x + 2$

3. $x^3 + 2x^2 - 3; \ x - 1$

4. $2x^3 + x + 2; \ x + 1$

5. $3x^3 - 12x^2 - 11x - 20; \ x - 5$

6. $4x^3 + 6x^2 - 5x - 2; \ x + 2$

7. $2x^4 + 5x^3 - 2x^2 + 5x + 3; \ x + 3$

8. $5x^4 + 16x^3 - 15x^2 + 8x + 16; \ x + 4$

For each of the following, find a polynomial of degree 3 with real coefficients that satisfies the given conditions. See Example 3.

9. Zeros of -3, -1, and 4; $P(2) = 5$

10. Zeros of 1, -1, and 0; $P(2) = -3$

11. Zeros of -2, 1, and 0; $P(-1) = -1$

12. Zeros of 2, 5, and -3; $P(1) = -4$

13. Zeros of 3, i, and $-i$; $P(2) = 50$

14. Zeros of -2, i, and $-i$, $P(-3) = 30$

For each of the following, find a polynomial of lowest degree with real coefficients having the given zeros. See Example 4.

15. $5 + i$ and $5 - i$

16. $3 - 2i$ and $3 + 2i$

17. 2, $1 - i$ and $1 + i$

18. -3, $2 - i$, and $2 + i$

19. $1 + \sqrt{2}$, $1 - \sqrt{2}$, and 1

20. $1 - \sqrt{3}$, $1 + \sqrt{3}$, and -2

21. $2 + i$, $2 - i$, 3, and -1

22. $3 + 2i$, $3 - 2i$, -1, and 2

23. 2 and $3 + i$

24. -1 and $4 - 2i$

25. $1 - \sqrt{2}$, $1 + \sqrt{2}$, and $1 - i$

26. $2 + \sqrt{3}$, $2 - \sqrt{3}$, and $2 + 3i$

27. $2 - i$, $3 + 2i$

28. $5 + i$, $4 - i$

29. 4, $1 - 2i$, $3 + 4i$

30. -1, $1 + \sqrt{2}$, $1 - \sqrt{2}$, $1 + 4i$

31. $1 + 2i$, 2 (multiplicity 2)

32. $2 + i$, -3 (multiplicity 2)

For each of the following polynomials, one zero is given. Find all others. See Example 5.

33. $P(x) = x^3 - x^2 - 4x - 6; \quad 3$

34. $P(x) = x^3 - 5x^2 + 17x - 13; \quad 1$

35. $P(x) = x^3 + x^2 - 4x - 24; \quad -2 + 2i$

36. $P(x) = x^3 + x^2 - 20x - 50; \quad -3 + i$

37. $P(x) = 2x^3 - 2x^2 - x - 6;$ 2

38. $P(x) = 2x^3 - 5x^2 + 6x - 2;$ $1 + i$

39. $P(x) = x^4 + 5x^2 + 4;$ $-i$

40. $P(x) = x^4 + 10x^3 + 27x^2 + 10x + 26;$ i

41. $P(x) = x^4 - 3x^3 + 6x^2 + 2x - 60;$ $1 + 3i$

42. $P(x) = x^4 - 6x^3 - x^2 + 86x + 170;$ $5 + 3i$

43. $P(x) = x^4 - 6x^3 + 15x^2 - 18x + 10;$ $2 + i$

44. $P(x) = x^4 + 8x^3 + 26x^2 + 72x + 153;$ $-3i$

Factor $P(x)$ into linear factors given that k is a zero of $P(x)$. See Example 6.

45. $P(x) = 2x^3 - 3x^2 - 17x + 30;$ $k = 2$

46. $P(x) = 2x^3 - 3x^2 - 5x + 6;$ $k = 1$

47. $P(x) = 6x^3 + 25x^2 + 3x - 4;$ $k = -4$

48. $P(x) = 8x^3 + 50x^2 + 47x - 15;$ $k = -5$

49. $P(x) = x^3 + (7 - 3i)x^2 + (12 - 21i)x - 36i;$ $k = 3i$

50. $P(x) = 2x^3 + (3 + 2i)x^2 + (1 + 3i)x + i;$ $k = -i$

51. $P(x) = 2x^3 + (3 - 2i)x^2 + (-8 - 5i)x + 3 + 3i;$ $k = 1 + i$

52. $P(x) = 6x^3 + (19 - 6i)x^2 + (16 - 7i)x + 4 - 2i;$ $k = -2 + i$

53. Show that -2 is a zero of multiplicity 2 of $P(x) = x^4 + 2x^3 - 7x^2 - 20x - 12$ and find all other complex zeros. Then write $P(x)$ in factored form.

54. Show that -1 is a zero of multiplicity 3 of $P(x) = x^5 + 9x^4 + 33x^3 + 55x^2 + 42x + 12$, and find all other complex zeros. Then write $P(x)$ in factored form.

55. The displacement at time t of a particle moving along a straight line is given by

$$s(t) = t^3 - 2t^2 - 5t + 6,$$

where t is in seconds and s is measured in centimeters. The displacement is 0 after 1 sec has elapsed. At what other times is the displacement 0?

56. For headphone radios, the cost function (in thousands of dollars) is given by

$$C(x) = 2x^3 - 9x^2 + 17x - 4,$$

and the revenue function (in thousands of dollars) is given by $R(x) = 5x$, where x is the number of items (in hundred thousands) produced. Cost equals revenue if 200,000 items are produced ($x = 2$). Find all other break-even points.

If c and d are complex numbers, prove each of the following statements. These statements are used in Exercise 63. (Hint: Let $c = a + bi$ and $d = m + ni$ and form the conjugates, the sums, and the products. The notation $\bar{c}$ represents the conjugate of c.)

57. $\overline{c + d} = \bar{c} + \bar{d}$

58. $\overline{cd} = \bar{c} \cdot \bar{d}$

59. $\bar{a} = a$ for any real number a

60. $\overline{c^n} = (\bar{c})^n$, n is a positive integer

Prove each of the following statements.

61. A polynomial of degree n has at most n distinct zeros. (*Hint:* Use an indirect proof, where you assume the opposite of the statement is true and show that it leads to a contradiction.)

62. If $a + bi$ is a zero of a polynomial $P(x)$ having only real coefficients, then $a - bi$ is also a zero of $P(x)$.

63. Complete the proof of the conjugate zeros theorem, outlined below. Assume that

$$P(x) = a_n x^n + a_{n-1} x^{n-1} + \ldots + a_1 x + a_0,$$

where all coefficients are real numbers.

(a) If the complex number z is a zero of $P(x)$, find $P(z)$.

(b) Take the conjugate of both sides of the result from part (a).

(c) Use generalizations of the properties given in Exercises 57–60 on the result of part (b) to show that $a_n(\bar{z})^n + a_{n-1}(\bar{z})^{n-1} + \ldots + a_1(\bar{z}) + a_0 = 0$.

(d) Why does the result in part (c) mean that $\bar{z}$ is a zero of $P(x)$?

64. Show that the zeros of $P(x) = x^3 + ix^2 - (7 - i)x + (6 - 6i)$ are $1 - i$, 2, and -3. Does the conjugate zeros theorem apply? Why or why not?

5.4 Rational Zeros of Polynomials

By the fundamental theorem of algebra, every polynomial of degree 1 or more has a zero. However, the fundamental theorem merely says that such a zero exists. It gives no help at all in identifying zeros. Other theorems can be used to find any rational zeros of polynomials with rational coefficients or to find decimal approximations of any irrational zeros.

The next theorem gives a useful method for finding a set of possible zeros of a polynomial with integer coefficients.

Rational Zeros Theorem

Let $P(x) = a_nx^n + a_{n-1}x^{n-1} + \ldots + a_1x + a_0$, where $a_n \neq 0$, be a polynomial with integer coefficients. If p/q is a rational number written in lowest terms and if p/q is a zero of $P(x)$, then p is a factor of the constant term a_0 and q is a factor of the leading coefficient a_n.

This theorem can be proved as follows.

$P(p/q) = 0$ since p/q is a zero of $P(x)$, so

$$a_n(p/q)^n + a_{n-1}(p/q)^{n-1} + \ldots + a_1(p/q) + a_0 = 0.$$

This also can be written as

$$a_n(p^n/q^n) + a_{n-1}(p^{n-1}/q^{n-1}) + \ldots + a_1(p/q) + a_0 = 0.$$

Multiply both sides of this last result by q^n and add $-a_0q^n$ to both sides.

$$a_np^n + a_{n-1}p^{n-1}q + \ldots + a_1pq^{n-1} = -a_0q^n$$

Factoring out p gives

$$p(a_np^{n-1} + a_{n-1}p^{n-2}q + \ldots + a_1q^{n-1}) = -a_0q^n.$$

This result shows that $-a_0q^n$ equals the product of the two factors, p and $(a_np^{n-1} + \ldots + a_1q^{n-1})$. For this reason, p must be a factor of $-a_0q^n$. Since it was assumed that p/q is written in lowest terms, p and q have no common factor other than 1, so p is not a factor of q^n. Thus p must be a factor of a_0. In a similar way it can be shown that q is a factor of a_n.

In the next example we use the rational zeros theorem to find all rational zeros of a polynomial with integer coefficients.

Example 1   Find all rational zeros of $P(x) = 2x^4 - 11x^3 + 14x^2 - 11x + 12$.

If p/q is to be a rational zero of $P(x)$, by the rational zeros theorem p must be a factor of $a_0 = 12$ and q must be a factor of $a_4 = 2$. The possible values of p are ± 1, ± 2, ± 3, ± 4, ± 6, or ± 12, while q must be ± 1 or ± 2. The possible rational zeros are found by forming all possible quotients of the form p/q; any rational zero of $P(x)$ will come from the list

$$\pm 1, \ \pm 1/2, \ \pm 2, \ \pm 3, \ \pm 3/2, \ \pm 4, \ \pm 6, \ \text{or} \ \pm 12.$$

Though none of these numbers may be zeros, if $P(x)$ has any rational zeros, they will be in the list above. These proposed zeros can be checked by synthetic division. Doing so shows that 4 is a zero.

$$
\begin{array}{r|rrrrr}
4 & 2 & -11 & 14 & -11 & 12 \\
  &   & 8 & -12 & 8 & -12 \\
\hline
  & 2 & -3 & 2 & -3 & 0 \\
\end{array}
$$

As a fringe benefit of this calculation, the simpler polynomial $Q(x) = 2x^3 - 3x^2 + 2x - 3$ can be used to find the remaining zeros. Any rational zero of $Q(x)$ will have a numerator of ± 3 or ± 1 and a denominator of ± 1 or ± 2 and so will come from the list

$$\pm 3, \ \pm 3/2, \ \pm 1, \ \pm 1/2.$$

Again use synthetic division and trial and error to find that 3/2 is a zero.

$$
\begin{array}{r|rrrr}
3/2 & 2 & -3 & 2 & -3 \\
    &   & 3 & 0 & 3 \\
\hline
    & 2 & 0 & 2 & 0 \\
\end{array}
$$

The quotient is $2x^2 + 2$, which, by the quadratic formula, has i and $-i$ as zeros. They are imaginary zeros, however. The rational zeros of the polynomial function

$$P(x) = 2x^4 - 11x^3 + 14x^2 - 11x + 12$$

are 4 and 3/2.   ✛

Example 2   Find all rational zeros of $P(x) = 6x^4 + 7x^3 - 12x^2 - 3x + 2$, and factor the polynomial.

For a rational number p/q to be a zero of $P(x)$, p must be a factor of $a_0 = 2$ and q must be a factor of $a_4 = 6$. Thus, p can be ± 1 or ± 2 and q can be ± 1, ± 2, ± 3, or ± 6. The rational zeros, p/q, must come from the following list.

$$\pm 1, \ \pm 2, \ \pm 1/2, \ \pm 1/3, \ \pm 1/6, \ \pm 2/3$$

Check 1 first because it is easy.

$$
\begin{array}{r|rrrrr}
1 & 6 & 7 & -12 & -3 & 2 \\
  &   & 6 & 13 & 1 & -2 \\
\hline
  & 6 & 13 & 1 & -2 & 0 \\
\end{array}
$$

The 0 remainder shows that 1 is a zero. Now use the quotient polynomial $6x^3 + 13x^2 + x - 2$ and synthetic division to find that -2 is also a zero.

$$
\begin{array}{r|rrrr}
-2) & 6 & 13 & 1 & -2 \\
 & & -12 & -2 & 2 \\
\hline
 & 6 & 1 & -1 & 0
\end{array}
$$

The new quotient polynomial is $6x^2 + x - 1$. Use the quadratic formula or factor to solve the equation $6x^2 + x - 1 = 0$. The remaining two zeros are $1/3$ and $-1/2$.

Factor the polynomial $P(x)$ in the following way. Since the four zeros of $P(x) = 6x^4 + 7x^3 - 12x^2 - 3x + 2$ are 1, -2, $1/3$, and $-1/2$, the corresponding factors are $x - 1$, $x + 2$, $x - 1/3$, and $x + 1/2$, and

$$P(x) = a(x - 1)(x + 2)\left(x - \frac{1}{3}\right)\left(x + \frac{1}{2}\right).$$

Since $a_4 = 6 = 2 \cdot 3$, fractions can be cleared in the last two factors by writing the product as

$$
\begin{aligned}
P(x) &= 6(x - 1)(x + 2)(x - 1/3)(x + 1/2) \\
&= (x - 1)(x + 2)(3)(x - 1/3)(2)(x + 1/2) \\
&= (x - 1)(x + 2)(3x - 1)(2x + 1). \quad \clubsuit
\end{aligned}
$$

Rational zeros of a polynomial with rational coefficients can be found by first multiplying the polynomial by a number that will clear it of all fractions, then using the rational zeros theorem.

Example 3   Find all rational zeros of

$$P(x) = x^4 - \frac{1}{6}x^3 + \frac{2}{3}x^2 - \frac{1}{6}x - \frac{1}{3}.$$

Find the values of x that make $P(x) = 0$, or

$$x^4 - \frac{1}{6}x^3 + \frac{2}{3}x^2 - \frac{1}{6}x - \frac{1}{3} = 0.$$

Multiply both sides by 6 to eliminate all fractions.

$$6x^4 - x^3 + 4x^2 - x - 2 = 0$$

This polynomial will have the same zeros as $P(x)$. The possible rational zeros are of the form p/q where p is ± 1, or ± 2, and q is ± 1, ± 2, ± 3, or ± 6. Then p/q may be

$$\pm 1, \ \pm 2, \ \pm 1/2, \ \pm 1/3, \ \pm 1/6, \ \text{or} \ \pm 2/3.$$

Use synthetic division to find that $-1/2$ and $2/3$ are zeros.

$$-\frac{1}{2}\overline{)\begin{array}{ccccc} 6 & -1 & 4 & -1 & -2 \\ & -3 & 2 & -3 & 2 \\ \hline 6 & -4 & 6 & -4 & 0 \end{array}}$$

$$\frac{2}{3}\overline{)\begin{array}{cccc} 6 & -4 & 6 & -4 \\ & 4 & 0 & 4 \\ \hline 6 & 0 & 6 & 0 \end{array}}$$

The final quotient is $Q(x) = 6x^2 + 6 = 6(x^2 + 1)$. This polynomial can be solved by the quadratic formula; its zeros are i and $-i$. Since these zeros are imaginary numbers, there are just two rational zeros: $-1/2$ and $2/3$. ✦

5.4 Exercises

Give all possible rational zeros for the following polynomials.

1. $P(x) = 6x^3 + 17x^2 - 31x - 1$

2. $P(x) = 15x^3 + 61x^2 + 2x - 1$

3. $P(x) = 12x^3 + 20x^2 - x - 2$

4. $P(x) = 12x^3 + 40x^2 + 41x + 3$

5. $P(x) = 2x^3 + 7x^2 + 12x - 8$

6. $P(x) = 2x^3 + 20x^2 + 68x - 40$

7. $P(x) = x^4 + 4x^3 + 3x^2 - 10x + 50$

8. $P(x) = x^4 - 2x^3 + x^2 + 18$

Find all rational zeros of the following polynomials. See Examples 1 and 2.

9. $P(x) = x^3 - 2x^2 - 13x - 10$

10. $P(x) = x^3 + 5x^2 + 2x - 8$

11. $P(x) = x^3 + 6x^2 - x - 30$

12. $P(x) = x^3 - x^2 - 10x - 8$

13. $P(x) = x^3 + 9x^2 - 14x - 24$

14. $P(x) = x^3 + 3x^2 - 4x - 12$

15. $P(x) = x^4 + 9x^3 + 21x^2 - x - 30$

16. $P(x) = x^4 + 4x^3 - 7x^2 - 34x - 24$

Find the rational zeros of the following polynomials; then write each in factored form. See Example 2.

17. $P(x) = 6x^3 + 17x^2 - 31x - 12$

18. $P(x) = 15x^3 + 61x^2 + 2x - 8$

19. $P(x) = 12x^3 + 20x^2 - x - 6$

20. $P(x) = 12x^3 + 40x^2 + 41x + 12$

21. $P(x) = 2x^3 + 7x^2 + 12x - 8$

22. $P(x) = 2x^3 + 20x^2 + 68x - 40$

23. $P(x) = 2x^4 + 3x^3 - 4x^2 - 3x + 2$

24. $P(x) = x^4 - 2x^3 + x^2 + 18$

25. $P(x) = 3x^4 + 5x^3 - 10x^2 - 20x - 8$

26. $P(x) = 6x^4 + x^3 - 7x^2 - x + 1$

27. $P(x) = 3x^4 + 4x^3 - x^2 + 4x - 4$

28. $P(x) = x^4 + 8x^3 + 16x^2 - 8x - 17$

29. $P(x) = 2x^5 - 3x^4 - 11x^3 + 13x^2 + 9x - 10$

30. $P(x) = 4x^5 + 4x^4 - 37x^3 - 37x^2 + 9x + 9$

Find all rational zeros of the following polynomials. See Example 3.

31. $P(x) = x^3 - \frac{4}{3}x^2 - \frac{13}{3}x - 2$

32. $P(x) = x^3 + x^2 - \frac{16}{9}x + \frac{4}{9}$

33. $P(x) = x^4 + \frac{1}{4}x^3 + \frac{11}{4}x^2 + x - 5$

34. $P(x) = \frac{10}{7}x^4 - x^3 - 7x^2 + 5x - \frac{5}{7}$

35. $P(x) = \dfrac{1}{3}x^5 + x^4 - \dfrac{5}{3}x^3 - \dfrac{11}{3}x^2 + 4$ **36.** $P(x) = x^5 + x^4 - \dfrac{37}{4}x^2 + \dfrac{9}{4}x + \dfrac{9}{4}$

37. Show that $P(x) = x^2 - 2$ has no rational zeros, so $\sqrt{2}$ must be irrational.

38. Show that $P(x) = x^2 - 5$ has no rational zeros, so $\sqrt{5}$ must be irrational.

39. Show that $P(x) = x^4 + 5x^2 + 4$ has no rational zeros.

40. Show that $P(x) = x^5 - 3x^3 + 5$ has no rational zeros.

41. Show that any integer zeros of a polynomial function must be factors of the constant term a_0.

5.5 Real Zeros of Polynomials

Every polynomial of degree 1 or more has a zero, but the fundamental theorem of algebra does not tell whether a polynomial has *real* zeros. Even if it does have real zeros, often there is no way to find their exact values. In this section we discuss methods of determining how many real zeros a polynomial function may have and then give a method for approximating any real zeros. This method works well using a computer or a calculator.

Descartes' rule of signs, stated below, gives a useful, practical test for finding the number of positive or negative real zeros of a given polynomial.

Descartes' Rule of Signs	Let $P(x)$ be a polynomial with real coefficients and terms in descending powers of x.

(a) The number of positive real zeros of $P(x)$ either equals the number of variations in sign occurring in the coefficients of $P(x)$, or is less than the number of variations by a positive even integer.

(b) The number of negative real zeros of $P(x)$ either equals the number of variations in sign of $P(-x)$, or is less than the number of variations by a positive even integer.

In the theorem, variation in sign is a change from positive to negative or negative to positive in successive terms of the polynomial. Missing terms (those with 0 coefficients) are counted as no change in sign and can be ignored.

For the purposes of this theorem, zeros of multiplicity k count as k zeros. For example,

$$P(x) = (x - 1)^4 = + x^4 - 4x^3 + 6x^2 - 4x + 1$$

has 4 changes of sign. By Descartes' rule of signs, $P(x)$ has either 4, 2, or 0 positive real zeros. In this case there are 4, and each of the 4 positive real zeros is 1.

Example 1   Find the number of positive and negative real zeros possible for the polynomial

$$P(x) = x^4 - 6x^3 + 8x^2 + 2x - 1.$$

$P(x)$ has 3 variations in sign:

$$+ \ x^4 \ \underset{1}{\overset{\frown}{-}} \ 6x^3 \ \underset{2}{\overset{\frown}{+}} \ 8x^2 \ + \ 2x \ \underset{3}{\overset{\frown}{-}} \ 1.$$

Thus, by Descartes' rule of signs, $P(x)$ has either 3 or $3 - 2 = 1$ positive real zeros. Since

$$P(-x) = (-x)^4 - 6(-x)^3 + 8(-x)^2 + 2(-x) - 1$$
$$= x^4 + 6x^3 + 8x^2 - 2x - 1$$

has only one variation in sign, $P(x)$ has only one negative real zero.   ✣

Example 2   Find the number of positive and negative real zeros possible for

$$Q(x) = x^5 + 5x^4 + 3x^2 + 2x + 1.$$

The polynomial $Q(x)$ has no variations in sign and so has no positive real zeros. Here

$$Q(-x) = -x^5 + 5x^4 + 3x^2 - 2x + 1,$$

with three variations in sign, so $Q(x)$ has either 3 or 1 negative real zeros. The other zeros are imaginary numbers.

Much of our work in locating real zeros uses the following result, which is related to the fact that graphs of polynomial functions are unbroken curves, with no gaps or sudden jumps. The proof requires advanced methods, so it is not given here.

Intermediate Value Theorem for Polynomials	If $P(x)$ defines a polynomial function with *only real coefficients,* and if for real numbers a and b, $P(a)$ and $P(b)$ are opposite in sign, then there exists at least one real zero between a and b.

This theorem helps to identify intervals where zeros of polynomials are located. For example, in Figure 5.13, $P(a)$ and $P(b)$ are opposite in sign, so 0 is between $P(a)$ and $P(b)$. Then, by the intermediate value theorem, there must be a number c between a and b such that $P(c) = 0$. If $P(a)$ and $P(b)$ are *not* opposite in sign, however, it does not mean there is no zero between a and b. There might be an even number of zeros in that interval.

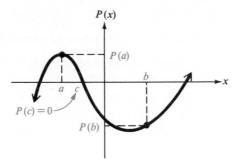

Figure 5.13

Example 3   Does $P(x) = x^3 - 2x^2 - x + 1$ have any real zeros between 2 and 3? Use synthetic division to find $P(2)$ and $P(3)$.

$$
\begin{array}{r|rrrr}
2) & 1 & -2 & -1 & 1 \\
   &   & 2 & 0 & -2 \\
\hline
   & 1 & 0 & -1 & -1
\end{array}
\qquad
\begin{array}{r|rrrr}
3) & 1 & -2 & -1 & 1 \\
   &   & 3 & 3 & 6 \\
\hline
   & 1 & 1 & 2 & 7
\end{array}
$$

The results show that $P(2) = -1$ and $P(3) = 7$. Since $P(2)$ is negative but $P(3)$ is positive, there must be a real zero between 2 and 3.   ❖

The intermediate value theorem for polynomials is helpful in limiting the search for real zeros to a smaller and smaller interval. In Example 3 the theorem was used to show that the polynomial $P(x) = x^3 - 2x^2 - x + 1$ has a real zero between 2 and 3. The theorem then could be used repeatedly to express the zero more accurately.

As suggested by the graphs of Figure 5.14, if the values of $|x|$ in a polynomial get larger and larger, then so will the values of $|y|$. This is used in the next theorem, the boundedness theorem, which shows how the bottom row of a synthetic division can be used to place upper and lower bounds on the possible real zeros of a polynomial.

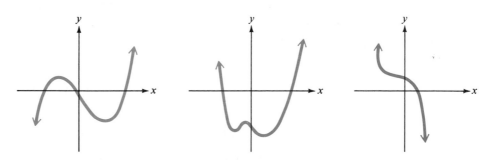

Figure 5.14

Boundedness Theorem

Let $P(x)$ be a polynomial with *real* coefficients and with a *positive* leading coefficient. If $P(x)$ is divided synthetically by $x - c$, and

(a) if $c > 0$ and all numbers in the bottom row of the synthetic division are nonnegative, then $P(x)$ has no zero greater than c;

(b) if $c < 0$ and the numbers in the bottom row of the synthetic division alternate in sign (with 0 considered positive or negative, as needed), then $P(x)$ has no zero less than c.

Example 4 Approximate the real zeros of $P(x) = x^4 - 6x^3 + 8x^2 + 2x - 1$.

Use Descartes' rule of signs first. From Example 1, $P(x)$ has either three or one positive real zeros and one negative real zero. Next, check for rational zeros. The only possible rational zeros are ± 1.

$$
\begin{array}{r|rrrrr}
1) & 1 & -6 & 8 & 2 & -1 \\
  &   & 1 & -5 & 3 & 5 \\
\hline
  & 1 & -5 & 3 & 5 & 4 \\
\end{array}
$$

$$
\begin{array}{r|rrrrr}
-1) & 1 & -6 & 8 & 2 & -1 \\
   &   & -1 & 7 & -15 & 13 \\
\hline
   & 1 & -7 & 15 & -13 & 12 \\
\end{array}
$$

Neither 1 nor -1 is a zero.

Use the two theorems in this section to search in some consistent way for the location of irrational real zeros. The leading coefficient of $P(x)$ is positive and the numbers in the last row of the second synthetic division above alternate in sign. Since $-1 < 0$, by the boundedness theorem -1 is less than or equal to any zero of $P(x)$. By the synthetic division above, -1 is not a zero of $P(x)$. Also, $P(-1) = 12 > 0$. By substitution, or synthetic division, $P(0) = -1 < 0$. Thus the one negative real zero is between -1 and 0.

Try $c = -.5$. Divide $P(x)$ by $x + .5$.

$$
\begin{array}{r|rrrrr}
-.5) & 1 & -6 & 8 & 2 & -1 \\
    &   & -.5 & 3.25 & -5.625 & 1.8125 \\
\hline
    & 1 & -6.5 & 11.25 & -3.625 & .8125 \\
\end{array}
$$

Since $P(-.5) = .8125 > 0$ and $P(0) = -1 < 0$, there is a real zero between $-.5$ and 0.

Now try $c = -.4$.

$$
\begin{array}{r|rrrrr}
-.4) & 1 & -6 & 8 & 2 & -1 \\
    &   & -.4 & 2.56 & -4.224 & .8896 \\
\hline
    & 1 & -6.4 & 10.56 & -2.224 & -.1104 \\
\end{array}
$$

Since $P(-.5)$ is positive, but $P(-.4)$ is negative, there is a zero between $-.5$ and $-.4$. The value of $P(-.4)$ is closer to zero than $P(-.5)$, so it is probably safe to say that, to one decimal place of accuracy, $-.4$ is a real zero of $P(x)$. A more accurate result can be found, if desired, by continuing this process.

Find the remaining real zeros of $P(x)$ by using synthetic division to find $P(1)$, $P(2)$, $P(3)$, and so on, until a change in sign is noted. It is helpful to use the shortened form of synthetic division shown below. Only the last row of the synthetic division is shown for each division. The first row of the chart is used for each division and the work in the second row of the division is done mentally.

x					$P(x)$
	1	-6	8	2	-1
-1	1	-7	15	-13	12
0	1	-6	8	2	-1
1	1	-5	3	5	4
2	1	-4	0	2	3
3	1	-3	-1	-1	-4
4	1	-2	0	2	7

Zero between -1 and 0

Zero between 0 and 1

Zero between 2 and 3

Zero between 3 and 4

Since the polynomial is degree 4, there are no more than four zeros. Expand the table to find the real zeros to the nearest tenth. For example, for the zero between 0 and 1, work as follows. Start halfway between 0 and 1 with $x = .5$. Since $P(.5) > 0$ and $P(0) < 0$, try $x = .4$ next, and so on.

x					$P(x)$
	1	-6	8	2	-1
.5	1	-5.5	5.25	4.63	1.31
.4	1	-5.6	5.76	4.30	.72
.3	1	-5.7	6.29	3.89	.17
.2	1	-5.8	6.84	3.37	$-.33$

Zero between .3 and .2

The value $P(.3) = .17$ is closer to 0 than $P(.2) = -.33$, so to the nearest tenth, the zero is .3. Use synthetic division to verify that the remaining two zeros are approximately 2.4 and 3.7. ✦

If a polynomial of degree 3 or more cannot be factored, the only way to graph it is to find many points to plot. The theorems studied in this chapter are helpful in deciding which points to plot and in finding the ordered pairs.

Example 5 Graph the function defined by $P(x) = 8x^3 - 12x^2 + 2x + 1$.

By Descartes' rule of signs, $P(x)$ has two or zero positive real zeros and one negative real zero. Locate the real zeros using synthetic division. Start with $x =$

0 and then find $P(1)$, $P(2)$, $P(3)$, and so on until a row of the synthetic division is all positive, indicating there are no zeros greater than the x-value for that row. Then do the same thing in the negative direction; find $P(-1)$, $P(-2)$, and so on until alternating signs in the last row of the synthetic division indicate the number that is less than or equal to all the real zeros. As shown in the shortened form of synthetic division below, a sign change in the value of $P(x)$ indicates a zero.

x				$P(x)$	Ordered pair	
	8	-12	2	1		
2	8	4	10	21	$(2, 21)$	
1	8	-4	-2	-1	$(1, -1)$	$\leftarrow$ Zero
0	8	-12	2	1	$(0, 1)$	$\leftarrow$ Zero
-1	8	-20	22	-21	$(-1, -21)$	$\leftarrow$ Zero

All three real zeros have been located. As expected, two are positive and one is negative. Since the numbers in the row for $x = 2$ are all positive and $2 > 0$, 2 is greater than any real zero of $P(x)$. Also, -1 is less than any real zero of $P(x)$.

By the intermediate value theorem, there is a zero between 0 and 1 and between -1 and 0, as well as between 1 and 2. The polynomial function is graphed by plotting the points from the chart and then drawing a continuous curve through them, as shown in Figure 5.15. ✣

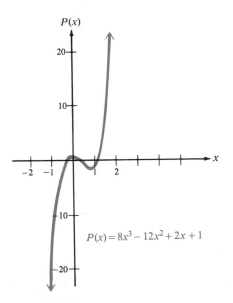

$$P(x) = 8x^3 - 12x^2 + 2x + 1$$

Figure 5.15

Example 6   Graph $P(x) = 3x^4 - 14x^3 + 54x - 3$.

Use Descartes' rule of signs to see that there are three positive real zeros or one positive real zero and one negative real zero. The points to plot are found using synthetic division to make a table like the one shown below. Start with $x = 0$ and work up through the positive integers until a row with all nonnegative numbers is found. Then work down through the negative integers until a row with alternating signs is found.

x					$P(x)$	*Ordered pair*
	3	-14	0	54	-3	
5	3	1	5	79	392	(5, 392)
4	3	-2	-8	22	85	(4, 85)
3	3	-5	-15	9	24	(3, 24)
2	3	-8	-16	22	41	(2, 41)
1	3	-11	-11	43	40	(1, 40)
0	3	-14	0	54	-3	(0, -3)
-1	3	-17	17	37	-40	(-1, -40)
-2	3	-20	40	-26	49	(-2, 49)

Since the row in the chart for $x = 5$ contains all positive numbers, the polynomial has no zero greater than 5. Also, since the row for $x = -2$ has numbers that alternate in sign, there is no zero less than -2. By the changes in sign of $P(x)$, the polynomial has zeros between 0 and 1 and between -2 and -1. Plotting the points found above and drawing a continuous curve through them gives the graph in Figure 5.16.   ❖

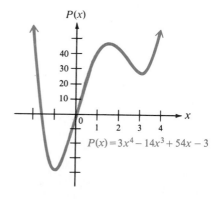

$P(x) = 3x^4 - 14x^3 + 54x - 3$

Figure 5.16

5.5 Exercises

A calculator will be helpful in working many of the exercises in this section. Such exercises are identified with colored numbers.

Use Descartes' rule of signs to find the number of positive and negative real zeros possible for each of the following polynomials. See Examples 1 and 2.

1. $P(x) = 2x^3 - 4x^2 + 2x + 7$

2. $P(x) = x^3 + 2x^2 + x - 10$

3. $P(x) = 5x^4 + 3x^2 + 2x - 9$

4. $P(x) = 3x^4 + 2x^3 - 8x^2 - 10x - 1$

5. $P(x) = x^5 + 3x^4 - x^3 + 2x + 3$

6. $P(x) = 2x^5 - x^4 + x^3 - x^2 + x + 5$

Use the intermediate value theorem for polynomials to show that each of the following polynomials has a real zero between the numbers given. See Example 3.

7. $P(x) = 3x^2 - 2x - 6$;   1 and 2

8. $P(x) = x^3 + x^2 - 5x - 5$;   2 and 3

9. $P(x) = 2x^3 - 8x^2 + x + 16$;   2 and 2.5

10. $P(x) = 3x^3 + 7x^2 - 4$;   1/2 and 1

11. $P(x) = 2x^4 - 4x^2 + 3x - 6$;   2 and 1.5

12. $P(x) = x^4 - 4x^3 - x + 1$;   1 and .3

13. $P(x) = -3x^4 - x^3 + 2x^2 + 4$;   1 and 1.2

14. $P(x) = x^5 + 2x^4 + x^3 + 3$; -1.3 and -2

Show that the real zeros of each of the following polynomial functions satisfy the given conditions.

15. $P(x) = x^4 - x^3 + 3x^2 - 8x + 8$;   no real zero greater than 2

16. $P(x) = 2x^5 - x^4 + 2x^3 - 2x^2 + 4x - 4$;   no real zero greater than 1

17. $P(x) = x^4 + x^3 - x^2 + 3$;   no real zero less than -2

18. $P(x) = x^5 + 2x^3 - 2x^2 + 5x + 5$;   no real zero less than -1

19. $P(x) = 3x^4 + 2x^3 - 4x^2 + x - 1$;   no real zero greater than 1

20. $P(x) = 3x^4 + 2x^3 - 4x^2 + x - 1$;   no real zero less than -2

21. $P(x) = x^5 - 3x^3 + x + 2$;   no real zero greater than 2

22. $P(x) = x^5 - 3x^3 + x + 2$;   no real zero less than -3

For each of the following polynomials, (a) Find the number of positive and negative real zeros that are possible. See Examples 1 and 2. (b) Approximate each zero as a decimal to the nearest tenth. See Example 4.

23. $P(x) = x^3 + 3x^2 - 2x - 6$

24. $P(x) = x^3 + x^2 - 5x - 5$

25. $P(x) = x^3 - 4x^2 - 5x + 14$

26. $P(x) = x^3 + 9x^2 + 34x + 13$

27. $P(x) = x^3 + 6x - 13$

28. $P(x) = 4x^3 - 3x^2 + 4x - 5$

29. $P(x) = 4x^4 - 8x^3 + 17x^2 - 2x - 14$

30. $P(x) = 3x^4 - 4x^3 - x^2 + 8x - 2$

31. $P(x) = -x^4 + 2x^3 + 3x^2 + 6$

32. $P(x) = -2x^4 - x^2 + x - 5$

Graph each of the following polynomials. See Examples 5 and 6.

33. $P(x) = x^3 - 7x - 6$

34. $P(x) = x^3 + x^2 - 4x - 4$

35. $P(x) = x^4 - 5x^2 + 6$

36. $P(x) = x^3 - 3x^2 - x + 3$

37. $P(x) = 6x^3 + 11x^2 - x - 6$

38. $P(x) = x^4 - 2x^2 - 8$

39. $P(x) = -x^3 + 6x^2 - x - 14$

40. $P(x) = 6x^4 - x^3 - 23x^2 - 4x + 12$

The following polynomials have zeros in the given intervals. Approximate these zeros to the nearest hundredth.

41. $P(x) = x^4 + x^3 - 6x^2 - 20x - 16;$ [3.2, 3.3] and [-1.4, -1.1]

42. $P(x) = x^4 - 2x^3 - 2x^2 - 18x + 5;$ [.2, .4] and [3.7, 3.8]

43. $P(x) = x^4 - 4x^3 - 20x^2 + 32x + 12;$ [-4, -3], [-1, 0], [1, 2], and [6, 7]

44. $P(x) = x^4 - 4x^3 - 44x^2 + 160x - 80;$ [-7, -6], [0, 1], [2, 3], and [7, 8]

5.6   Rational Functions

A function of the form

$$f(x) = \frac{p(x)}{q(x)},$$

where $p(x)$ and $q(x)$ are polynomial functions, is called a **rational function.** Since any values of x such that $q(x) = 0$ are excluded from the domain, a rational function usually has a graph that has one or more breaks in it.

The simplest rational function with a variable denominator is

$$f(x) = \frac{1}{x}.$$

The domain of this function is the set of all real numbers except 0. The number 0 cannot be used as a value of x, but for graphing it is helpful to find the values of $f(x)$ for some values of x close to 0. The following table shows what happens to $f(x)$ as x gets closer and closer to 0 from either side.

x approaches 0

x	-1	$-.1$	$-.01$	$-.001$	.001	.01	.1	1
$f(x)$	-1	-10	-100	-1000	1000	100	10	1

$|f(x)|$ gets larger and larger

The table suggests that $|f(x)|$ gets larger and larger as x gets closer and closer to 0, which is written in symbols as

$$|f(x)| \to \infty \text{ as } x \to 0.$$

(The symbol $x \to 0$ means that x approaches as close as desired to 0, without necessarily ever being equal to 0.) Since x cannot equal 0, the graph of $f(x) = 1/x$ will never intersect the vertical line $x = 0$. This line is called a *vertical asymptote*.

On the other hand, as $|x|$ gets larger and larger, the values of $f(x) = 1/x$ get closer and closer to 0, as shown in the table below.

x	$-10{,}000$	-1000	-100	-10	10	100	1000	$10{,}000$
$f(x)$	$-.0001$	$-.001$	$-.01$	$-.1$	$.1$	$.01$	$.001$	$.0001$

Letting $|x|$ get larger and larger without bound (written $|x| \to \infty$) causes the graph of $y = f(x) = 1/x$ to move closer and closer to the horizontal line $y = 0$. This line is called a *horizontal asymptote*.

Since $f(-x) = 1/(-x) = -(1/x) = -f(x)$, f is an odd function, symmetric with respect to the origin. Choosing some positive values of x and finding the corresponding values of $f(x)$ gives the first-quadrant part of the graph shown in Figure 5.17. The other part of the graph (in the third quadrant) can be found by symmetry.

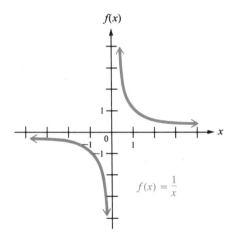

Figure 5.17

Example 1   Graph $f(x) = -\dfrac{2}{x}$.

The expression on the right side of the equation can be rewritten so that

$$f(x) = -2 \cdot \frac{1}{x}.$$

Compared to $f(x) = 1/x$, the graph will be reflected about the x-axis (because of the negative sign), and each point will be twice as far from the x-axis. See Figure 5.18.   ❖

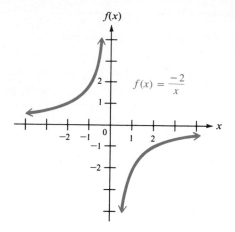

Figure 5.18

Example 2   **Graph** $f(x) = \dfrac{2}{1 + x}$.

The domain of this function is the set of all real numbers except -1. As shown in Figure 5.19, the graph is that of $f(x) = 1/x$, translated 1 unit to the left, with each y-value doubled.

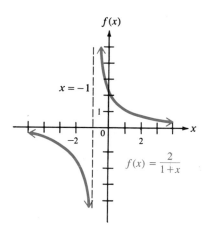

Figure 5.19

The examples above suggest the following definition of vertical and horizontal asymptotes.

Definitions of Vertical and Horizontal Asymptotes

For the rational function with $f(x) = \dfrac{p(x)}{q(x)}$, written in lowest terms,

if $|f(x)| \to \infty$ as $x \to a$, then the line $x = a$ is a **vertical asymptote**;

and if $f(x) \to a$ as $|x| \to \infty$, then the line $y = a$ is a **horizontal asymptote**.

Vertical asymptotes are found by setting the denominator of a rational function equal to zero and then solving. Horizontal asymptotes are found by considering what happens to $f(x)$ as $|x| \to \infty$.

Example 3 Graph $f(x) = \dfrac{1}{(x - 1)(x + 4)}$.

This graph is not symmetric with respect to the x-axis, the y-axis, or the origin. Since the denominator is 0 when x is replaced by either 1 or -4, the lines $x = 1$ and $x = -4$ are vertical asymptotes.

As $|x|$ gets larger and larger, $|(x - 1)(x + 4)|$ also gets larger and larger, bringing y closer and closer to 0. This means that the x-axis is a horizontal asymptote. The vertical asymptotes divide the x-axis into three intervals. It is convenient to consider each interval separately. Find the y-intercept, $-1/4$, by letting $x = 0$, and plot a few points in each interval on the x-axis to get the graph shown in Figure 5.20. ✛

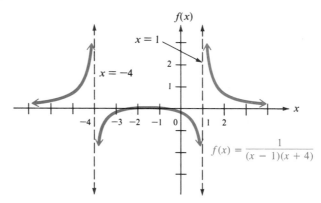

Figure 5.20

Example 4 Graph $f(x) = \dfrac{x + 1}{(2x - 1)(x + 3)}$.

The graph has vertical asymptotes at $x = 1/2$ and $x = -3$. Replacing x with 0 gives $-1/3$ as the y-intercept. For the x-intercept, the value of y must be 0. The only way that y can equal 0 is if the numerator, $x + 1$, is 0. This happens when $x = -1$, making -1 the x-intercept.

The horizontal asymptote is found by multiplying the factors in the denominator to get

$$f(x) = \frac{x+1}{(2x-1)(x+3)} = \frac{x+1}{2x^2 + 5x - 3}.$$

Now divide each term in the numerator and denominator by x^2, since 2 is the largest exponent on x. This gives

$$f(x) = \frac{\dfrac{x}{x^2} + \dfrac{1}{x^2}}{\dfrac{2x^2}{x^2} + \dfrac{5x}{x^2} - \dfrac{3}{x^2}} = \frac{\dfrac{1}{x} + \dfrac{1}{x^2}}{2 + \dfrac{5}{x} - \dfrac{3}{x^2}}.$$

As $|x|$ gets larger and larger, the quotients $1/x$, $1/x^2$, $5/x$, and $3/x^2$ all approach 0, and the value of $f(x)$ approaches

$$\frac{0+0}{2+0-0} = \frac{0}{2} = 0.$$

The line $y = 0$ is therefore a horizontal asymptote. Using these asymptotes and intercepts (the graph has no symmetry with respect to either axis or the origin) and plotting a few points gives the graph in Figure 5.21. Plot at least one point in each of the intervals $(-\infty, -3)$, $(-3, -1)$, $(-1, 1/2)$, and $(1/2, +\infty)$. ✦

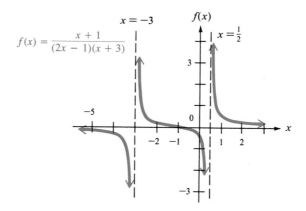

$$f(x) = \frac{x+1}{(2x-1)(x+3)}$$

Figure 5.21

The horizontal asymptote is not always the line $y = 0$. In the next examples we graph rational functions with horizontal asymptotes that are not the line $y = 0$.

Example 5 Graph $f(x) = \dfrac{2x+1}{x-3}$.

Work as in the previous example to find any horizontal asymptote. Since the highest exponent on x is 1, divide each term by x.

$$f(x) = \frac{2x + 1}{x - 3} = \frac{\dfrac{2x}{x} + \dfrac{1}{x}}{\dfrac{x}{x} - \dfrac{3}{x}} = \frac{2 + \dfrac{1}{x}}{1 - \dfrac{3}{x}}$$

As $|x|$ gets larger and larger, both $1/x$ and $3/x$ approach 0, and $f(x)$ approaches

$$\frac{2 + 0}{1 - 0} = \frac{2}{1} = 2,$$

so the line $y = 2$ is the horizontal asymptote. Setting $x - 3 = 0$ shows that the vertical asymptote is the line $x = 3$.

Letting $x = 0$ gives the y-intercept, $-1/3$. The value of y is 0 when the numerator, $2x + 1$, is 0, so the x-intercept is $-1/2$. Use this information to get the graph in Figure 5.22.

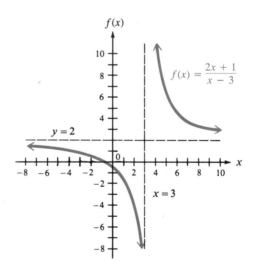

Figure 5.22

Example 6   Graph $f(x) = \dfrac{3(x + 1)(x - 2)}{(x + 4)^2}$.

The only vertical asymptote is the line $x = -4$. To find any horizontal asymptotes, multiply the factors in the numerator and denominator.

$$f(x) = \frac{3x^2 - 3x - 6}{x^2 + 8x + 16}$$

Now divide by x^2.

$$f(x) = \frac{\dfrac{3x^2}{x^2} - \dfrac{3x}{x^2} - \dfrac{6}{x^2}}{\dfrac{x^2}{x^2} + \dfrac{8x}{x^2} + \dfrac{16}{x^2}} = \frac{3 - \dfrac{3}{x} - \dfrac{6}{x^2}}{1 + \dfrac{8}{x} + \dfrac{16}{x^2}}$$

Letting $|x|$ get larger and larger shows that $y = 3/1$, or $y = 3$, is the horizontal asymptote. The y-intercept is $-3/8$, and the x-intercepts are -1 and 2. See the final graph in Figure 5.23. ✤

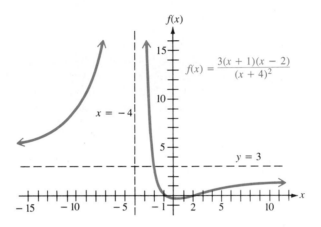

$$f(x) = \frac{3(x + 1)(x - 2)}{(x + 4)^2}$$

Figure 5.23

The graph of the function in Example 6 actually crosses the horizontal asymptote. Remember, the horizontal asymptote indicates what happens to $f(x)$ as $|x|$ gets larger and larger. It does not predict what happens to $f(x)$ for small values of x. While it is possible for the graph of a rational function to cross a horizontal asymptote, the graph never crosses a vertical asymptote.

The next example shows a rational function that has a numerator with higher degree than the denominator. Polynomial division must be used to find the asymptotes for such functions.

Example 7 Graph $f(x) = \dfrac{x^2 + 1}{x - 2}$.

Divide by x^2 as in the examples above to see that the graph has no horizontal asymptote because

$$f(x) = \frac{\dfrac{x^2}{x^2} + \dfrac{1}{x^2}}{\dfrac{x}{x^2} - \dfrac{2}{x^2}} = \frac{1 + \dfrac{1}{x^2}}{\dfrac{1}{x} + \dfrac{2}{x^2}}$$

does not approach any real number as $|x| \to \infty$. Since the numerator has higher degree than the denominator, divide $x^2 + 1$ by $x - 2$ to rewrite the expression in another form. Using synthetic division gives

$$
\begin{array}{r}
2\overline{)1 \quad 0 \quad 1} \\
\underline{2 \quad 4} \\
1 \quad 2 \quad 5.
\end{array}
$$

The function can now be written as

$$
f(x) = \frac{x^2 + 1}{x - 2} = x + 2 + \frac{5}{x - 2}.
$$

For very large values of $|x|$, $5/(x - 2)$ is close to 0, and the graph approaches the line $y = x + 2$. This line is an *oblique asymptote* (neither vertical nor horizontal) for the function. Setting $x - 2 = 0$ gives the vertical asymptote, the line $x = 2$. The y-intercept is $-1/2$. There is no x-intercept. Using the asymptotes, the y-intercept, and additional points as needed leads to the graph in Figure 5.24. ✤

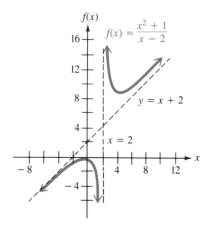

Figure 5.24

In general, if the degree of its numerator is exactly one more than the degree of its denominator, a rational function may have an oblique asymptote. The equation of this asymptote is found by dividing the numerator by the denominator and dropping the remainder.

The following list shows the steps involved in graphing rational functions.

Graphing Rational Functions

Use the following procedure when graphing a rational function *written in lowest terms*.

1. Find any vertical asymptotes by setting the denominator equal to 0 and solving for x. If a is a zero of the denominator, then the line $x = a$ is a vertical asymptote.

2. Determine any other asymptotes. There are three possibilities:
   (a) If the numerator has lower degree than the denominator, there is a horizontal asymptote, $y = 0$.
   (b) If the numerator and denominator have the same degree, and the function is of the form

   $$f(x) = \frac{a_n x^n + \ldots + a_0}{b_n x^n + \ldots + b_0}, \quad \text{where } b_n \neq 0,$$

   dividing by x^n in the numerator and denominator produces the horizontal asymptote

   $$y = \frac{a_n}{b_n}.$$

   (c) If the numerator is of degree exactly one more than the denominator, there may be an oblique asymptote. To find it, divide the numerator by the denominator and drop any remainder. Set the rest of the quotient equal to y to get the equation of the asymptote.

3. Find any intercepts.

4. Check for symmetry.

5. Plot a few selected points—at least one in each interval of the domain as determined by the vertical asymptotes and x-intercepts.

6. Complete the sketch.

As mentioned earlier, a rational function must be in lowest terms before we can use the methods discussed in this section to determine the graph. The next example gives a typical rational function that is not in lowest terms.

Example 8   Graph $f(x) = \dfrac{x^2 - 4}{x - 2}$.

The domain of this function is $x \neq 2$. Dividing the numerator by the denominator gives

$$\begin{array}{r} 2\overline{)\,1 \quad 0 \quad -4} \\ \underline{2 \quad \ 4} \\ 1 \quad 2 \qquad 0. \end{array}$$

The remainder is 0, which indicates that the quotient was not written in lowest terms, so there is no oblique asymptote. Since the domain still has the restriction $x \neq 2$, the function can be written as $f(x) = x + 2$, $x \neq 2$. The graph is the same as that of $f(x) = x + 2$, except for the one point with x-value 2, which is missing. See Figure 5.25. ❖

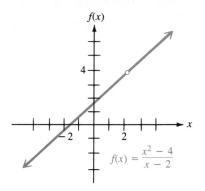

Figure 5.25

5.6 Exercises

Find any vertical, horizontal, or oblique asymptotes for Exercises 1–16.
See Examples 1–7.

1. $f(x) = \dfrac{2}{x - 5}$

2. $f(x) = \dfrac{-1}{x + 2}$

3. $f(x) = \dfrac{-8}{3x - 7}$

4. $f(x) = \dfrac{5}{4x - 9}$

5. $f(x) = \dfrac{2 - x}{x + 2}$

6. $f(x) = \dfrac{x - 4}{5 - x}$

7. $f(x) = \dfrac{3x - 5}{2x + 9}$

8. $f(x) = \dfrac{4x + 3}{3x - 7}$

9. $f(x) = \dfrac{2}{x^2 - 4x + 3}$

10. $f(x) = \dfrac{-5}{x^2 - 3x - 10}$

11. $f(x) = \dfrac{x^2 - 1}{x + 3}$

12. $f(x) = \dfrac{x^2 + 4}{x - 1}$

13. $f(x) = \dfrac{(x - 3)(x + 1)}{(x + 2)(2x - 5)}$

14. $f(x) = \dfrac{3(x + 2)(x - 4)}{(5x - 1)(x - 5)}$

15. $f(x) = \dfrac{2x^2 - 32}{x^2 + 1}$

16. $f(x) = \dfrac{-5x^2 + 20}{x^2 + 9}$

17. Sketch the following graphs and compare them with the graph of $f(x) = 1/x^2$.

(a) $f(x) = \dfrac{1}{(x - 3)^2}$

(b) $f(x) = \dfrac{-2}{x^2}$

(c) $f(x) = \dfrac{-2}{(x - 3)^2}$

Graph each of the following. See Examples 1–8.

18. $f(x) = \dfrac{4}{5 + 3x}$

19. $f(x) = \dfrac{1}{(x - 2)(x + 4)}$

20. $f(x) = \dfrac{3}{(x + 4)^2}$

21. $f(x) = \dfrac{3x}{(x + 1)(x - 2)}$

22. $f(x) = \dfrac{2x + 1}{(x + 2)(x + 4)}$

23. $f(x) = \dfrac{5x}{x^2 - 1}$

24. $f(x) = \dfrac{-x}{x^2 - 4}$

25. $f(x) = \dfrac{3x}{x - 1}$

26. $f(x) = \dfrac{4x}{1 - 3x}$

27. $f(x) = \dfrac{x + 1}{x - 4}$

28. $f(x) = \dfrac{x - 5}{x + 3}$

29. $f(x) = \dfrac{x}{x^2 - 9}$

30. $f(x) = \dfrac{3x}{x^2 - 16}$

31. $f(x) = \dfrac{x^2 - 3x + 2}{x - 3}$

32. $f(x) = \dfrac{x^2 - 5}{x + 2}$

33. $f(x) = \dfrac{2x^2 + 3}{x - 4}$

34. $f(x) = \dfrac{x^2 + 1}{x + 3}$

35. $f(x) = \dfrac{x^2 + 2x}{2x - 1}$

36. $f(x) = \dfrac{x^2 - x}{x + 2}$

37. $f(x) = \dfrac{(x - 3)(x + 1)}{(x - 1)^2}$

38. $f(x) = \dfrac{x(x - 2)}{(x + 3)^2}$

39. $f(x) = \dfrac{(x - 5)(x - 2)}{x^2 + 9}$

40. $f(x) = \dfrac{1}{x^2 + 1}$

41. $f(x) = \dfrac{-9}{x^2 + 9}$

42. $f(x) = \dfrac{x^2 - 16}{x + 4}$

43. $f(x) = \dfrac{x^2 + 2x - 3}{x - 1}$

44. $f(x) = \dfrac{2x^2 - 3x - 2}{2x + 1}$

Work the following problems.

45. Antique-car owners often enter their cars in a *concours d'elegance* in which a maximum of 100 points can be awarded to a particular car. Points are awarded for the general attractiveness of the car. The function

$$C(x) = \frac{10x}{49(101 - x)}$$

expresses the cost, in thousands of dollars, of restoring a car so that it will win x points. Graph the function.

46. Suppose the average cost per unit, $C(x)$, to produce x units of margarine is given by

$$C(x) = \frac{500}{x + 30}.$$

(a) Find $C(10)$, $C(20)$, $C(50)$, $C(75)$, and $C(100)$.
(b) Would a more reasonable domain for C be $(0, +\infty)$ or $[0, +\infty)$? Why?
(c) Graph $C(x)$.

47. In situations involving environmental pollution, a cost-benefit model expresses cost as a function of the percentage of pollutant removed from the environment. Suppose a cost-benefit model is expressed as

$$y = \frac{6.7x}{100 - x},$$

where y is the cost in thousands of dollars of removing x percent of a certain pollutant.
(a) Graph the function.
(b) Is it possible, according to this function, to remove all of the pollutant?

48. In a recent year, the cost per ton, y, to build an oil tanker of x thousand deadweight tons was approximated by

$$y = \frac{110,000}{x + 225}.$$

(a) Find y for $x = 25$, $x = 50$, $x = 100$, $x = 200$, $x = 300$, and $x = 400$.
(b) Graph the function.

*In recent years the economist Arthur Laffer has been a center of controversy because of his **Laffer curve**, an idealized version of which is shown here.*

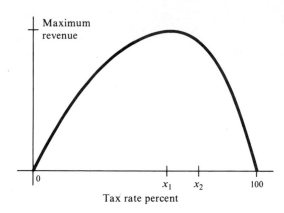

According to this curve, increasing a tax rate, say from x_1 percent to x_2 percent on the graph above, can actually lead to a decrease in government revenue. All economists agree on the endpoints, 0 revenue at tax rates of both 0% and 100%, but there is much disagreement on the location of the rate x_1 that produces maximum revenue.

49. Suppose an economist studying the Laffer curve produces the rational function defined by

$$y = \frac{80x - 8000}{x - 110},$$

with y giving government revenue in tens of millions of dollars for a tax rate of x percent, with the function valid for $55 \le x \le 100$. Find the revenue for the following tax rates.
(a) 55%   **(b)** 60%   **(c)** 70%
(d) 90%   **(e)** 100%   **(f)** Graph the function

50. Suppose an economist studies a different tax, this time producing

$$y = \frac{60x - 6000}{x - 120}.$$

where y is government revenue in millions of dollars from a tax rate of x percent, with the function valid for $50 \le x \le 100$. Find the revenue from the following tax rates.
(a) 50%
(b) 60%
(c) 80%
(d) 100%
(e) Graph the function.

51. Let $f(x) = p(x)/q(x)$ define a rational function. If the degree of $p(x)$ is m, and the degree of $q(x)$ is n, and if $m > n$, tell how to find the horizontal or vertical asymptotes of f.

Chapter 5   *Summary*

Key Words

5.1 polynomial function
leading coefficient
zero polynomial
zeros of a polynomial
even function
odd function

5.2 synthetic division
root of an equation
5.3 multiplicity
5.6 rational function
vertical asymptote
horizontal asymptote
oblique asymptote

Key Ideas

5.2 Division Algorithm

For any polynomial $P(x)$ and any complex number k, there exists a unique polynomial $Q(x)$ and number r such that

$$P(x) = (x - k)Q(x) + r.$$

5.2 Remainder Theorem

If the polynomial $P(x)$ is divided by $x - k$, then the remainder is $P(k)$.

5.3 Factor Theorem

The polynomial $x - k$ is a factor of the polynomial $P(x)$ if and only if $P(k) = 0$.

5.3 Fundamental Theorem of Algebra

Every polynomial of degree 1 or more has at least one complex zero.

A polynomial of degree n has at most n distinct zeros.

5.3 Conjugate Zeros Theorem

If $P(x)$ is a polynomial having only real coefficients and if $a + bi$ is a zero of $P(x)$, then the conjugate $a - bi$ is also a zero of $P(x)$.

5.4 Rational Zeros Theorem

Let $P(x) = a_n x^n + a_{n-1} x^{n-1} + \ldots + a_1 x + a_0$, where $a_n \neq 0$, be a polynomial with integer coefficients. If p/q is a rational number written in lowest terms and if p/q is a zero of $P(x)$, then p is a factor of the constant term a_0 and q is a factor of the leading coefficient a_n.

5.5 Descartes' Rule of Signs	Let $P(x)$ be a polynomial with real coefficients and terms in descending powers of x. **(a)** The number of positive real zeros of $P(x)$ either equals the number of variations in sign occurring in the coefficients of $P(x)$, or is less than the number of variations by a positive even integer. **(b)** The number of negative real zeros of $P(x)$ either equals the number of variations in sign of $P(-x)$, or is less than the number of variations by a positive even integer.
5.5 Intermediate Value Theorem for Polynomials	If $P(x)$ defines a polynomial function with *only real coefficients,* and if for real numbers a and b, $P(a)$ and $P(b)$ are opposite in sign, then there exists at least one real zero between a and b.
5.5 Boundedness Theorem	Let $P(x)$ be a polynomial with *real* coefficients and with a *positive* leading coefficient. If $P(x)$ is divided synthetically by $x - c$, and **(a)** if $c > 0$ and all numbers in the bottom row of the synthetic division are nonnegative, then $P(x)$ has no zero greater than c; **(b)** if $c < 0$ and the numbers in the bottom row of the synthetic division alternate in sign (with 0 considered positive or negative, as needed), then $P(x)$ has no zero less than c.

Chapter 5   *Review Exercises*

Graph P for each of the following.

1. $P(x) = x^3 + 5$

2. $P(x) = 1 - x^4$

3. $P(x) = x^2(2x + 1)(x - 2)$

4. $P(x) = (4x - 3)(3x + 2)(x - 1)$

5. $P(x) = 2x^3 + 13x^2 + 15x$

6. $P(x) = x(x - 1)(x + 2)(x - 3)$

Decide whether the functions defined as follows are odd, even, or neither.

7. $g(x) = \dfrac{x}{|x|}$,   where $x \neq 0$

8. $h(x) = 9x^6 + 5|x|$

9. $f(x) = x$

10. $f(x) = -x^4 + 9x^2 - 3x^6$

Use synthetic division to perform each division.

11. $(2x^3 - x^2 - x - 6) \div (x + 1)$

12. $(3t^3 + 10t^2 - 9t - 4) \div (t + 4)$

13. $(4m^3 + m^2 - 12) \div (m - 2)$

14. $(5y^3 + 24y^2 + 25y - 6) \div (y + 3)$

Express each polynomial in the form $P(x) = (x - k)Q(x) + r$ for the given value of k.

15. $P(z) = 2z^3 + 15z^2 + 30z + 20; \quad k = -2$

16. $P(r) = r^3 + 6r^2 - r - 35; \quad k = -5$

17. $P(m) = 2m^3 - 13m^2 + 17m + 15; \quad k = 4$

18. $P(x) = 3x^3 + 2x^2 - 19x + 16; \quad k = 2$

Use synthetic division to find $P(2)$ for each of the following.

19. $P(x) = x^3 + 3x^2 - 5x + 1$

20. $P(x) = 2x^3 - 4x^2 + 3x - 10$

21. $P(x) = 5x^4 - 12x^2 + 2x - 8$

22. $P(x) = x^5 - 3x^2 + 2x - 4$

In Exercises 23–26, find a polynomial of lowest degree having the following zeros.

23. $-1, 4, 7$

24. $8, 2, 3$

25. $-\sqrt{7}, \sqrt{7}, 2, -1$

26. $1 + \sqrt{5}, 1 - \sqrt{5}, -4, 1$

27. Is -1 a zero of $P(x) = 2x^4 + x^3 - 4x^2 + 3x + 1$?

28. Is -2 a zero of $P(x) = 2x^4 + x^3 - 4x^2 + 3x + 1$?

29. Is $x + 1$ a factor of $P(x) = x^3 + 2x^2 + 3x - 1$?

30. Is $x + 1$ a factor of $P(x) = 2x^3 - x^2 + x + 4$?

31. Find a polynomial of degree 3 with -2, 1, and 4 as zeros, and $P(2) = 16$.

32. Find a polynomial of degree 4 with real coefficients having 1, -1, and $3i$ as zeros, and $P(2) = 39$.

33. Find a lowest-degree polynomial with real coefficients having zeros 2, -2, and $-i$.

34. Find a lowest-degree polynomial with real coefficients having zeros 2, -3, and $5i$.

35. Find a polynomial of lowest degree with real coefficients having -3 and $1 - i$ as zeros.

36. Find all zeros of $P(x) = x^4 - 3x^3 - 8x^2 + 22x - 24$, given that $1 - i$ is a zero, and factor $P(x)$.

37. Find all zeros of $P(x) = x^4 - 6x^3 + 14x^2 - 24x + 40$, given that $3 + i$ is a zero, and factor $P(x)$.

38. Find all zeros of $P(x) = x^4 + x^3 - x^2 + x - 2$, given that 1 is a zero, and factor $P(x)$.

Find all rational zeros of each polynomial in Exercises 39–44.

39. $P(x) = 2x^3 - 9x^2 - 6x + 5$

40. $P(x) = 3x^3 - 10x^2 - 27x + 10$

41. $P(x) = x^3 - \frac{17}{6}x^2 - \frac{13}{3}x - \frac{4}{3}$

42. $P(x) = 8x^4 - 14x^3 - 29x^2 - 4x + 3$

43. $P(x) = -x^4 + 2x^3 - 3x^2 + 11x - 7$

44. $P(x) = x^5 - 3x^4 - 5x^3 + 15x^2 + 4x - 12$

45. Use a polynomial to show that $\sqrt{11}$ is irrational.

46. Show that $P(x) = x^3 - 9x^2 + 2x - 5$ has no rational zeros.

Show that the following polynomials have real zeros satisfying the given conditions.

47. $P(x) = 3x^3 - 8x^2 + x + 2$, zero in $[-1, 0]$ and $[2, 3]$

48. $P(x) = 4x^3 - 37x^2 + 50x + 60$, zero in $[2, 3]$ and $[7, 8]$

49. $P(x) = x^3 + 2x^2 - 22x - 8$, zero in $[-1, 0]$ and $[-6, -5]$

50. $P(x) = 2x^4 - x^3 - 21x^2 + 51x - 36$ has no real zero greater than 4

51. $P(x) = 6x^4 + 13x^3 - 11x^2 - 3x + 5$ has no zero greater than 1 or less than -3

In Exercises 52 and 53 approximate the real zeros of each polynomial as decimals to the nearest tenth.

52. $P(x) = 2x^3 - 11x^2 - 2x + 2$

53. $P(x) = x^4 - 4x^3 - 5x^2 + 14x - 15$

54. **(a)** Find the number of positive and negative zeros of $P(x) = x^3 + 3x^2 - 4x - 2$.
(b) Show that $P(x)$ has a zero between -4 and -3. Approximate this zero to the nearest tenth.

Graph each of the following polynomials.

55. $P(x) = 2x^3 - 11x^2 - 2x + 2$ (See Exercise 52.)

56. $P(x) = x^4 - 4x^3 - 5x^2 + 14x - 15$ (See Exercise 53.)

57. $P(x) = x^3 + 3x^2 - 4x - 2$ (See Exercise 54.)

58. $P(x) = 2x^4 - 3x^3 + 4x^2 + 5x - 1$

Graph f for each of the following.

59. $f(x) = \dfrac{8}{x}$

60. $f(x) = \dfrac{2}{3x - 1}$

61. $f(x) = \dfrac{4x - 2}{3x + 1}$

62. $f(x) = \dfrac{6x}{(x - 1)(x + 2)}$

63. $f(x) = \dfrac{2x}{x^2 - 1}$

64. $f(x) = \dfrac{x^2 + 4}{x + 2}$

65. $f(x) = \dfrac{x^2 - 1}{x}$

66. $f(x) = \dfrac{x^2 + 6x + 5}{x - 3}$

67. $f(x) = \dfrac{4x^2 - 9}{2x + 3}$

Chapter 5   *Test*

Graph P for each of the following.

1. $P(x) = x^3 - 3$

2. $P(x) = (x + 2)^2(x - 1)(x - 2)$

3. Decide whether $g(x) = x + x^{-1}$ is odd, even, or neither.

Use synthetic division to perform each division.

4. $(3x^3 + 4x^2 - 9x + 6) \div (x + 2)$

5. $(2x^3 - 11x^2 + 28) \div (x - 5)$

6. Use synthetic division to find $P(3)$ for $P(x) = 2x^3 - 9x^2 + 4x + 8$.

7. Find a polynomial of lowest degree having zeros of 0, 4, and -7.

8. Is 3 a zero of $P(x) = 6x^4 - 11x^3 - 35x^2 + 34x + 24$? Why or why not?

9. Is $x + 2$ a factor of $P(x) = 6x^4 - 11x^3 - 35x^2 + 34x + 24$? Why or why not?

10. Find a polynomial of degree 3 with real coefficients having -1 and $3i$ as zeros, with $P(0) = 45$.

11. Find a lowest-degree polynomial with real coefficients having zeros 3 and $2i$.

12. Find all zeros of $P(x) = x^4 + x^3 + 10x^2 + 9x + 9$, given that $3i$ is a zero.

Find all rational zeros of each polynomial.

13. $P(x) = 2x^3 - x^2 - 8x - 5$

14. $P(x) = x^3 + 7x - 6$

15. Approximate the two real zeros of $P(x)$ as decimals to the nearest tenth if $P(x) = x^4 + 4x^2 - 6x - 5$.

16. Graph $P(x) = x^3 - 3x + 4$.

17. Show that $P(x) = x^4 + 4x^2 - 6x - 5$ has no zero less than -1.

18. Show that $P(x) = 2x^3 + 9x^2 + 13x + 5$ has a zero in $[-1, 0]$.

Graph f for each of the following.

19. $f(x) = \dfrac{2x^2}{(x - 1)^2}$

20. $f(x) = \dfrac{1}{x^2 - 3x}$

6

Exponential and Logarithmic Functions

In this chapter we introduce two kinds of functions that are quite different from the functions discussed earlier. Until now, we have worked with *algebraic functions* that involve only the basic operations of addition, subtraction, multiplication, division, and taking roots. The *exponential* and *logarithmic* functions in this chapter go beyond (transcend) these basic operations, so they are called **transcendental functions.*** Many applications of mathematics, particularly those involving growth and decay of quantities, require the closely interrelated exponential and logarithmic functions introduced in this chapter.

6.1 Exponential Functions

Recall from Chapter 1 the definition of a^r, where r is a rational number: if $r = m/n$, then for appropriate values of m and n,

$$a^{m/n} = (\sqrt[n]{a})^m.$$

For example,

$$16^{3/4} = (\sqrt[4]{16})^3 = 2^3 = 8,$$

---

*The trigonometric functions studied in trigonometry are also transcendental functions.

$$27^{-1/3} = \frac{1}{27^{1/3}} = \frac{1}{\sqrt[3]{27}} = \frac{1}{3},$$

and
$$64^{-1/2} = \frac{1}{64^{1/2}} = \frac{1}{\sqrt{64}} = \frac{1}{8}.$$

In this section the definition of a^r is extended to include all real (not just rational) values of the exponent r. For example, the new symbol $2^{\sqrt{3}}$ might be evaluated by approximating the exponent $\sqrt{3}$ by the numbers 1.7, 1.73, 1.732, and so on. Since these decimals approach the value of $\sqrt{3}$ more and more closely, it seems reasonable that $2^{\sqrt{3}}$ should be approximated more and more closely by the numbers $2^{1.7}$, $2^{1.73}$, $2^{1.732}$, and so on. (Recall, for example, that $2^{1.7} = 2^{17/10} = \sqrt[10]{2^{17}}$.) In fact, this is exactly how $2^{\sqrt{3}}$ is defined (in a more advanced course).

With this interpretation of real exponents, all rules and theorems for exponents are valid for real-number exponents as well as rational ones. In addition to the rules for exponents presented earlier, several new properties are used in this chapter. For example, if $y = 2^x$, then each real value of x leads to exactly one value of y, and therefore, $y = 2^x$ defines a function. Furthermore,

$$\text{if } 3^x = 3^4, \text{ then } x = 4,$$

and for $p > 0$,

$$\text{if } p^2 = 3^2, \text{ then } p = 3.$$

Also,
$$4^2 < 4^3 \quad \text{but} \quad \left(\frac{1}{2}\right)^2 > \left(\frac{1}{2}\right)^3,$$

so that when $a > 1$, increasing the exponent on a leads to a *larger* number, but if $0 < a < 1$, increasing the exponent on a leads to a *smaller* number.

These properties are generalized below. Proofs of the properties are not given here, as they require more advanced mathematics.

Additional Properties of Exponents

(a) If $a > 0$ and $a \neq 1$, then a^x is a unique real number for all real numbers x.

(b) If $a > 0$ and $a \neq 1$, then $a^b = a^c$ if and only if $b = c$.

(c) If $a > 1$ and $m < n$, then $a^m < a^n$.

(d) If $0 < a < 1$ and $m < n$, then $a^m > a^n$.

Parts (a) and (b) of the theorem require $a > 0$ so that a^x is always defined. For example, $(-6)^x$ is not a real number if $x = 1/2$. This means that a^x will always be positive, since a is positive. In part (a), $a \neq 1$ because $1^x = 1$ for every real-number value of x, so that each value of x does not lead to a unique real number. For part (b) to hold, a must not equal 1 since, for example, $1^4 = 1^5$, even though $4 \neq 5$.

The properties given above are useful in solving equations, as shown by the next examples.

Example 1   Solve $\left(\dfrac{1}{3}\right)^x = 81$.

First, write $1/3$ as 3^{-1}, so that $(1/3)^x = (3^{-1})^x = 3^{-x}$. Since $81 = 3^4$,

$$\left(\frac{1}{3}\right)^x = 81$$

becomes $$3^{-x} = 3^4$$

By the second property above,

$$-x = 4, \quad \text{or} \quad x = -4.$$

The solution set of the given equation is $\{-4\}$. (In Section 6.4 we describe a method for solving exponential equations when this approach is not possible.)   ✛

Example 2   If $81 = b^{4/3}$, find b.

Raise both sides of the equation to the $3/4$ power, since $(b^{4/3})^{3/4} = b^1 = b$.

$$81 = b^{4/3}$$
$$\pm 81^{3/4} = (b^{4/3})^{3/4}$$
$$(\pm \sqrt[4]{81})^3 = b^1$$
$$(\pm 3)^3 = b$$
$$\pm 27 = b$$

Remember that the process of raising both sides of an equation to the same power may result in false "solutions." For this reason, it is necessary to check all proposed solutions in the original equation.

$$27^{4/3} = (\sqrt[3]{27})^4 \qquad\qquad (-27)^{4/3} = (\sqrt[3]{-27})^4$$
$$= 3^4 \qquad\qquad\qquad\qquad = (-3)^4$$
$$= 81 \qquad\qquad\qquad\qquad = 81$$

Both potential solutions check, so the solution set is $\{-27, 27\}$.   ✛

As mentioned above, the expression a^x satisfies all the properties of exponents from Chapter 1. We can now define a function $f(x) = a^x$ whose domain is the set of all real numbers (and not just the rationals).

| Definition of Exponential Function | If $a > 0$ and $a \neq 1$, then $$f(x) = a^x$$ defines the **exponential function** with base a. |

(If $a = 1$, the function is the constant function $f(x) = 1$.)

Example 3 If $f(x) = 2^x$, find each of the following.

(a) $f(-1)$

   Replace x with -1.

$$f(-1) = 2^{-1} = \frac{1}{2}$$

(b) $f(3) = 2^3 = 8$

(c) $f(5/2) = 2^{5/2} = (2^5)^{1/2} = 32^{1/2} = \sqrt{32} = 4\sqrt{2}$ ♦

Figure 6.1(a) shows the graph of $f(x) = 2^x$; the base of this exponential function is 2. This graph was found by obtaining a number of ordered pairs belonging to the function and then drawing a smooth curve through them. The domain of the function is the set of all real numbers, and the range is the set of all positive numbers. The x-axis is a horizontal asymptote. The graph of $f(x) = 2^x$ is typical of graphs of $f(x) = a^x$ where $a > 1$. For larger values of a, the graphs rise more steeply, but the general shape is similar to the graph in Figure 6.1(a), with domain $(-\infty, +\infty)$ and range $(0, +\infty)$.

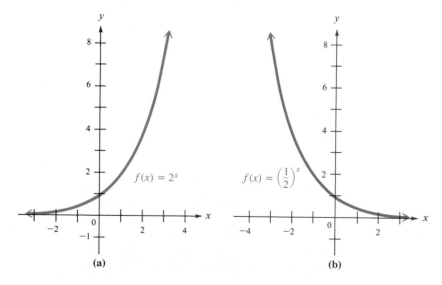

(a)

(b)

Figure 6.1

In Figure 6.1(b), the graph of $f(x) = (1/2)^x$ was obtained in a similar way. The graphs of $f(x) = 2^x$ and $f(x) = (1/2)^x$ are mirror images of each other with respect to the y-axis, because $(1/2)^x = (2^{-1})^x = 2^{-x}$. As shown by the horizontal line test, exponential functions are one-to-one.

Example 4   Graph $f(x) = 2^{-x+2}$.
  A table of values is given below.

x	-1	0	1	2	3
$f(x)$	8	4	2	1	1/2

Plotting these points and then connecting them with a smooth curve gives the graph in Figure 6.2. Notice that this graph goes down as x increases, while the graph of $f(x) = 2^x$ goes up. This happens because the negative sign in front of the x makes $f(x)$ get smaller as x gets larger. The graph is also translated 2 units to the right when compared to the graph of $f(x) = 2^{-x}$.   ✤

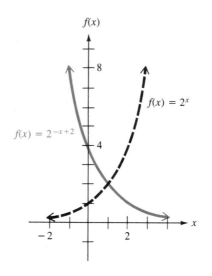

Figure 6.2

Example 5   Graph $f(x) = 2^{-x^2}$.
  Write $f(x) = 2^{-x^2}$ as $f(x) = 1/(2^{x^2})$ to find ordered pairs that belong to the function. Some ordered pairs are shown in the chart below.

x	-2	-1	0	1	2
y	1/16	1/2	1	1/2	1/16

As the chart suggests, $0 \leq y \leq 1$ for all values of x. Plotting these points and drawing a smooth curve through them gives the graph in Figure 6.3. This graph is symmetric with respect to the y-axis and has the x-axis as a horizontal asymptote.   ✛

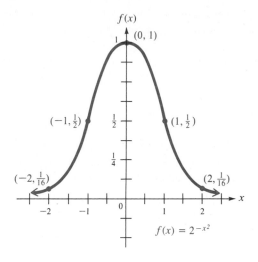

Figure 6.3

Perhaps the single most useful exponential function is the function defined by $f(x) = e^x$, where e is an irrational number that occurs often in practical applications. The number e comes up in a natural way when using the formula for *compound interest* (interest paid on both principal and interest).

Compound Interest

If P dollars is deposited in an account paying a rate of interest i compounded (paid) m times per year, then after n years the account will contain A dollars, where

$$A = P\left(1 + \frac{i}{m}\right)^{nm}.$$

For example, let $1000 be deposited in an account paying 8% per year compounded quarterly, or four times per year. After 10 years the account will contain

$$P\left(1 + \frac{i}{m}\right)^{nm} = 1000\left(1 + \frac{.08}{4}\right)^{10(4)}$$
$$= 1000(1 + .02)^{40}$$
$$= 1000(1.02)^{40}$$

dollars. The number $(1.02)^{40}$ can be found in financial tables or by using a calculator with a y^x key. To five decimal places, $(1.02)^{40} = 2.20804$. The amount on deposit after 10 years is

$$1000(1.02)^{40} = 1000(2.20804) = 2208.04,$$

or $2208.04.

In the formula for compound interest, A is sometimes called the **future value** and P the **present value.**

Example 6   An accountant wants to buy a new computer in three years that will cost $20,000.

(a) How much should be deposited now, at 6% interest compounded annually, to give the required $20,000 in three years?

Since the money deposited should amount to $20,000 in three years, $20,000 is the future value of the money. To find the present value P of $20,000 (the amount to deposit now), use the compound interest formula with $A = 20,000$, $i = .06$, $m = 1$, and $n = 3$.

$$A = P\left(1 + \frac{i}{m}\right)^{mn}$$

$$20,000 = P\left(1 + \frac{.06}{1}\right)^{1(3)} = P(1.06)^3$$

$$\frac{20,000}{(1.06)^3} = P$$

$$P = 16,792.39$$

The accountant must deposit $16,792.39.

(b) If only $15,000 is available to deposit now, what annual interest rate is required for it to increase to $20,000 in three years?

Here $P = 15,000$, $A = 20,000$, $m = 1$, $n = 3$, and i is unknown. Substitute the known values into the compound interest formula and solve for i.

$$A = P\left(1 + \frac{i}{m}\right)^{n}$$

$$20,000 = 15,000\left(1 + \frac{i}{1}\right)^{3}$$

$$\frac{4}{3} = (1 + i)^3 \qquad \text{Divide both sides by 15,000}$$

$$\left(\frac{4}{3}\right)^{1/3} = 1 + i \qquad \text{Take the cube root on both sides}$$

$$\left(\frac{4}{3}\right)^{1/3} - 1 = i \qquad \text{Subtract 1 on both sides}$$

$$i \approx .10 \qquad \text{Use a calculator}$$

An interest rate of 10% will produce enough interest to increase the $15,000 deposit to the $20,000 needed at the end of three years.   ✢

Suppose now that a lucky investment produces annual interest of 100%, so that $i = 1.00$, or $i = 1$. Suppose also that only $1 can be deposited at this rate, and for only one year. Then $P = 1$ and $n = 1$. Substitute into the formula for compound interest:

$$P\left(1 + \frac{i}{m}\right)^{nm} = 1\left(1 + \frac{1}{m}\right)^{1(m)} = \left(1 + \frac{1}{m}\right)^{m}$$

As interest is compounded more and more often, the value of this expression will increase. If interest is compounded annually, making $m = 1$, the total amount on deposit is

$$\left(1 + \frac{1}{m}\right)^{m} = \left(1 + \frac{1}{1}\right)^{1} = 2^{1} = 2,$$

so an investment of $1 becomes $2 in one year.

A calculator with a y^x key gives the results in the following table. These results have been rounded to five decimal places.

m	$\left(1 + \dfrac{1}{m}\right)^{m}$
1	2
2	2.25
5	2.48832
10	2.59374
25	2.66584
50	2.69159
100	2.70481
500	2.71557
1000	2.71692
10,000	2.71815
1,000,000	2.71828

The table suggests that as m increases, the value of $(1 + 1/m)^{m}$ gets closer and closer to some fixed number. It turns out that this is indeed the case. This fixed number is called e.

Value of e

To nine decimal places,

$$e \approx 2.718281828.*$$

---

*The symbol $\approx$ means "is approximately equal to."

Table 2 in Appendix B gives various powers of e. Also, some calculators will give values of e^x. In Figure 6.4 the functions $y = 2^x$, $y = e^x$, and $y = 3^x$ are graphed for comparison.

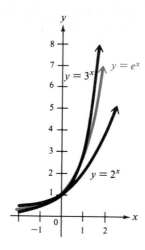

Figure 6.4

As suggested by the table above,

$$\left(1 + \frac{1}{m}\right)^m \to e \quad \text{as} \quad m \to +\infty$$

Because of this, if interest is calculated more and more often ($m \to +\infty$), a process called *continuous compounding,* the compound interest formula leads to a new formula that involves the number e.

Continuous Compounding

> If P dollars is deposited in an account paying a rate of interest i compounded continuously for n years, the final amount on deposit is A dollars, where
> $$A = Pe^{ni}.$$

Example 7   Suppose \$5000 is deposited in an account paying 8% compounded continuously. Find the total amount on deposit after five years.

Let $P = 5000$, $n = 5$, and $i = .08$. Then

$$A = 5000e^{5(.08)} = 5000e^{.4}$$

From Table 4 or a calculator, $e^{.4} \approx 1.49182$, and

$$A = 5000(1.49182) = 7459.10,$$

or $7459.10.* Check that daily compounding would have produced a compound amount about 30¢ less. ✚

Example 8   Suppose $24,000 is deposited at 16% for 9 years. Find the interest earned by **(a)** daily, **(b)** hourly, and **(c)** continuous compounding.
(a) In 9 years there are $9 \times 365 = 3285$ days. The total amount on deposit with daily compounding is

$$24,000\left(1 + \frac{.16}{365}\right)^{3285} = 101,264.74,$$

or $101,264.74. The interest earned is

$$\$101,264.74 - \$24,000 = \$77,264.74.$$

(b) In one year, there are $365 \times 24 = 8760$ hours, while in 9 years there are $9 \times 8760 = 78,840$ hours. The total amount on deposit is

$$24,000\left(1 + \frac{.16}{8760}\right)^{78,840} = 101,295.37,$$

or $101,295.37. This amount includes interest of

$$\$101,295.37 - \$24,000 = \$77,295.37,$$

only $30.63 more than when interest is compounded daily.

(c) The total amount on deposit with continuous compounding is

$$24,000e^{9(.16)} = 24,000e^{1.44}$$
$$= 24,000(4.2206958) \qquad \text{Use a calculator}$$
$$= 101,296.70,$$

or $101,296.70. With continuous compounding, the amount of interest earned, $101,296.70 - $24,000 = $77,296.70, is only $1.33 more than when interest is compounded hourly. ✚

As mentioned above, the number e is important as the base of an exponential function because many practical applications require an exponential function with base e. For example, it can be shown that in situations involving growth or decay of a population, the amount or number present at time t often can be closely approximated by a function of the form

$$y = y_0 e^{kt},$$

where y_0 is the amount or number present at time $t = 0$ and k is a constant. (In other words, once the numbers y_0 and k have been determined, population is a function of time.) The next example illustrates exponential growth.

---

*With some calculators the INV and ln x keys are used to get e^x.

Example 9   Suppose the population of a southwestern city is

$$P(t) = 10,000e^{.04t},$$

where t represents time measured in years. The population at time $t = 0$ is

$$P(0) = 10,000e^{(.04)0}$$
$$= 10,000e^0$$
$$= 10,000(1)$$
$$= 10,000.$$

The population of the city is 10,000 at time $t = 0$, written $P_0 = 10,000$. In year $t = 5$ the population is

$$P(5) = 10,000e^{(.04)5}$$
$$= 10,000e^{.2}.$$

The number $e^{.2}$ can be found in Table 2 or by using a suitable calculator. By either of these methods, $e^{.2} = 1.22140$ (to five decimal places), so

$$P(5) = 10,000(1.22140) = 12,214.$$

Thus, in five years the population of the city will be about 12,200. A graph of $P(t) = 10,000e^{.04t}$ is shown in Figure 6.5.   ✥

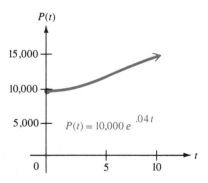

Figure 6.5

6.1   *Exercises*

Evaluate each of the following expressions.

1. 2^3 **2.** 3^{-2} **3.** $25^{1/2}$ **4.** $\left(\dfrac{1}{27}\right)^{1/3}$

Let $f(x) = (2/3)^x$. Find each of the following values. See Example 3.

5. $f(2)$ **6.** $f(-1)$ **7.** $f\left(\dfrac{1}{2}\right)$ **8.** $f(0)$

9. Graph f for each of the following. Compare the graphs to that of $f(x) = 2^x$.
   (a) $f(x) = 2^x + 1$ **(b)** $f(x) = 2^x - 4$ **(c)** $f(x) = 2^{x+1}$ **(d)** $f(x) = 2^{x-4}$

10. Graph f for each of the following. See Example 4.
   (a) $f(x) = 3^{-x} - 2$ **(b)** $f(x) = 3^{-x} + 4$ **(c)** $f(x) = 3^{-x-2}$ **(d)** $f(x) = 3^{-x+4}$

Graph f for each of the following. See Examples 4 and 5.

11. $f(x) = 3^x$ **12.** $f(x) = 4^x$ **13.** $f(x) = (3/2)^x$

14. $f(x) = 3^{-x}$ **15.** $f(x) = e^{-x}$ **16.** $f(x) = e^{3x}$

17. $f(x) = e^{x+1}$ **18.** $f(x) = e^{1-x}$ **19.** $f(x) = 2^{|x|}$

20. $f(x) = 2^{-|x|}$ **21.** $f(x) = 2^x + 2^{-x}$ **22.** $f(x) = (1/2)^x + (1/2)^{-x}$

Solve each of the following equations. See Examples 1 and 2.

23. $4^x = 2$ **24.** $125^r = 5$ **25.** $\left(\dfrac{1}{2}\right)^k = 4$ **26.** $\left(\dfrac{2}{3}\right)^x = \dfrac{9}{4}$

27. $2^{3-y} = 8$ **28.** $5^{2p+1} = 25$ **29.** $\dfrac{1}{27} = b^{-3}$ **30.** $\dfrac{1}{81} = k^{-4}$

31. $4 = r^{2/3}$ **32.** $z^{5/2} = 32$ **33.** $27^{4z} = 9^{z+1}$ **34.** $32^t = 16^{1-t}$

35. $125^{-x} = 25^{3x}$ **36.** $216^{3-a} = 36^a$ **37.** $\left(\dfrac{1}{8}\right)^{-2p} = 2^{p+3}$ **38.** $3^{-h} = \left(\dfrac{1}{27}\right)^{1-2h}$

39. $\left(\dfrac{1}{2}\right)^{-x} = \left(\dfrac{1}{4}\right)^{x+1}$ **40.** $\left(\dfrac{2}{3}\right)^{k-1} = \left(\dfrac{81}{16}\right)^{k+1}$

41. For $a > 1$, how does the graph of $f(x) = a^x$ change as a increases? What if $0 < a < 1$?

Use a calculator to help graph f in Exercises 42–47.

42. $f(x) = \dfrac{e^x - e^{-x}}{2}$ **43.** $f(x) = \dfrac{e^x + e^{-x}}{2}$ **44.** $f(x) = x \cdot 2^x$

45. $f(x) = x^2 \cdot 2^{-x}$ **46.** $f(x) = (1 - x)e^x$ **47.** $f(x) = xe^x$

48. Suppose that the population of a city is given by $P(t)$, where

$$P(t) = 1{,}000{,}000e^{.02t},$$

and t represents time measured in years from some initial year. Find each of the following.
   (a) P_0 **(b)** $P(2)$ **(c)** $P(4)$ **(d)** $P(10)$
   (e) Graph $y = P(t)$.

49. Suppose the quantity in grams of a radioactive substance present at time t is

$$Q(t) = 500e^{-.05t}.$$

Let t be time measured in days from some initial day. Find the quantity present at each of the following times.
   (a) $t = 0$ **(b)** $t = 4$ **(c)** $t = 8$ **(d)** $t = 20$
   (e) Graph $y = Q(t)$.

50. Experiments have shown that the sales of a product, under relatively stable market conditions, but in the absence of promotional activities such as advertising, tend to decline at a constant yearly rate. This rate of sales decline varies considerably from product to product, but seems to remain the same for any particular product. The sales decline can be expressed by a function of the form

$$S(t) = S_0 e^{-at},$$

where $S(t)$ is the rate of sales at time t measured in years, S_0 is the rate of sales at time $t = 0$, and a is the sales decay constant.

(a) Suppose the sales decay constant for a particular product is $a = .10$. Let $S_0 = 50,000$ and find $S(1)$ and $S(3)$.

(b) Find $S(2)$ and $S(10)$ if $S_0 = 80,000$ and $a = .05$.

51. Suppose $10,000 is deposited at 12% interest for 3 years. Find the interest earned if the interest is compounded
(a) annually;   (b) quarterly;   (c) daily (365 days).

Use the formula for compound interest and a calculator to find the future value of each amount.

52. $4292 at 6% compounded annually for 10 years

53. $965.43 at 9% compounded annually for 15 years

54. $10,765 at 11% compounded semiannually for 7 years

55. $8906.54 at 5% compounded semiannually for 9 years

56. $3256 at 8% compounded quarterly for 13 quarters

57. $56,780 at 5.3% compounded quarterly for 23 quarters

58. $45,788 at 6% compounded daily (ignoring leap years) for 11 years of 365 days

Find the present value for the following future values. See Example 6(a).

59. $10,000, if interest is 12% compounded semiannually for 5 years

60. $25,000, if interest is 6% compounded quarterly for 11 quarters

61. $45,678.93, if interest is 9.6% compounded monthly for 11 months

62. $123,788, if interest is 8.7% compounded daily for 195 days

Find the required annual interest rate to the nearest tenth in each of the following. See Example 6(b).

63. $25,000 compounded annually for 2 years to yield $31,360

64. $65,000 compounded monthly for 6 months to yield $65,325

65. $1200 compounded quarterly for 5 years to yield $1780

66. $15,000 compounded semiannually for 4 semiannual periods to yield $19,000

Find the total amount on deposit if the following amounts are deposited at the given rates of interest for the given time periods. Assume continuous compounding. See Examples 7 and 8.

67. $16,500 at 9.2% for 9 years   68. $23,889 at 7.1% for 13 years

69. $67,312 at 7.9% for 18 months   70. $43,789.12 at 5.67% for 30 months

71. Suppose $56,890 is deposited at 5.1% for 3 years. Find the interest earned by
(a) daily,   (b) hourly,   and   (c) continuous compounding.

72. If $34,678.79 is deposited at 5.6% for 5 years, find the interest earned by
(a) daily,   (b) hourly,   and   (c) continuous compounding.

Find the present value of each of the future values in Exercises 73–76. Assume continuous compounding.

73. $18,000 at 10% for 5 years

74. $25,000 at 7% for 10 years

75. $1650 at 6% for 8 years

76. $110,000 at 8% for 5 years

77. How much money must be deposited at 5% compounded continuously to amount to $8000 in 4 years?

78. What amount will grow to $12,000 in 5 years if interest is compounded continuously at 6%?

Solve each of the following problems. See Example 9.

79. *Escherichia coli* is a strain of bacteria that occurs naturally in many different organisms. Under certain conditions, the number of bacteria present in a colony is

$$E(t) = E_0 \cdot 2^{t/30},$$

where $E(t)$ is the number of bacteria present t min after the beginning of an experiment, and E_0 is the number present when $t = 0$. Let $E_0 = 2,400,000$ and find the number of bacteria at the following times.
(a) $t = 5$ (b) $t = 10$ (c) $t = 60$ (d) $t = 120$

80. The higher a student's grade-point average, the fewer applications that the student need send to medical schools (other things being equal). Using information given in a guide-book for prospective medical students, we constructed the function $y = 540e^{-1.3x}$ for the number of applications a student should send out. Here y is the number of applications for a student whose grade-point average is x. The domain of x is the interval $[2.0, 4.0]$. Find the number of applications that should be sent out by students having the following grade-point averages.
(a) 2.0   (b) 3.0   (c) 3.5   (d) 3.9

In calculus, it is shown that

$$e^x = 1 + x + \frac{x^2}{2 \cdot 1} + \frac{x^3}{3 \cdot 2 \cdot 1} + \frac{x^4}{4 \cdot 3 \cdot 2 \cdot 1} + \frac{x^5}{5 \cdot 4 \cdot 3 \cdot 2 \cdot 1} + \cdots .$$

81. Use the terms shown here and replace x with 1 to approximate $e^1 = e$ to three decimal places. Then check your results in Table 2 or with a calculator.

82. Use the terms shown here and replace x with $-.05$ to approximate $e^{-.05}$ to four decimal places. Check your results in Table 2 or with a calculator.

6.2 Logarithmic Functions

The previous section dealt with exponential functions of the form $y = a^x$ for all positive values of a, where $a \neq 1$. As mentioned there, the horizontal line test shows that exponential functions are one-to-one, and thus have inverse functions. In this section we discuss the inverses of exponential functions. The equation defining the inverse of a function is found by exchanging x and y in the equation that defines the function. Doing so with $y = a^x$ gives

$$x = a^y$$

as the equation of the inverse function of the exponential function defined by $y = a^x$. This equation can be solved for y by using the following definition.

Definition of Logarithm	For all real numbers y, and all positive numbers a and x, where $a \neq 1$: $$y = \log_a x \quad \text{if and only if} \quad x = a^y$$

The "log" in the definition above is an abbreviation for *logarithm*. Read $\log_a x$ as "the logarithm of x to the base a." Intuitively, the logarithm to the base a of x is the power to which a must be raised to yield x.

Example 1   The chart below shows several pairs of equivalent statements. The same statement is written in both exponential and logarithmic forms.

Exponential form	*Logarithmic form*
$2^3 = 8$	$\log_2 8 = 3$
$(1/2)^{-4} = 16$	$\log_{1/2} 16 = -4$
$10^5 = 100{,}000$	$\log_{10} 100{,}000 = 5$
$3^{-4} = 1/81$	$\log_3 (1/81) = -4$
$5^1 = 5$	$\log_5 5 = 1$
$(3/4)^0 = 1$	$\log_{3/4} 1 = 0$ ✤

The *logarithmic function with base a* is defined as follows.

Definition of Logarithmic Function	If $a > 0$, $a \neq 1$, and $x > 0$, then $$f(x) = \log_a x$$ defines the **logarithmic function with base a.**

Exponential and logarithmic functions are inverses of each other. Since the domain of an exponential function is the set of all real numbers, the range of a logarithmic function also will be the set of all real numbers. In the same way, both the range of an exponential function and the domain of a logarithmic function are the set of all positive real numbers, so logarithms can be found for positive numbers only.

The graph of $y = 2^x$ is shown in Figure 6.6(a). The graph of its inverse is found by reflecting the graph of $y = 2^x$ about the line $y = x$. The graph of the inverse function, defined by $y = \log_2 x$, shown in color, has the y-axis as a vertical asymptote.

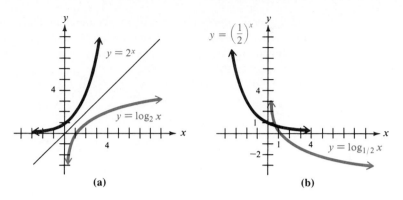

Figure 6.6

The graph of $y = (1/2)^x$ is shown in Figure 6.6(b). The graph of its inverse, defined by $y = \log_{1/2} x$, in color, is found by reflecting the graph of $y = (1/2)^x$ about the line $y = x$. As Figure 6.6(b) suggests, the graph of $y = \log_{1/2} x$ also has the y-axis for a vertical asymptote.

Example 2   Graph $f(x) = \log_2 (x - 1)$.

The domain of this function is $(1, +\infty)$, since logarithms can be found only for positive numbers. The graph of $f(x) = \log_2 (x - 1)$ is the same as the graph of $f(x) = \log_2 x$, translated one unit to the right. See Figure 6.7.   ✦

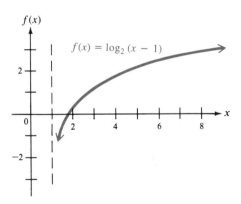

Figure 6.7

Example 3   Graph $y = \log_3 |x|$.

Write $y = \log_3 |x|$ in exponential form as $3^y = |x|$ to help identify some ordered pairs that satisfy the equation. Here, it is easier to choose y-values and find the corresponding x-values. Doing so gives the following ordered pairs.

x	-3	-1	$-1/3$	$1/3$	1	3
y	1	0	-1	-1	0	1

Plotting these points (be careful to get them in the right order) and connecting them with a smooth curve gives the graph in Figure 6.8. Notice that this is an even function, since $y = \log_3 |-x| = \log_3 |x|$. Therefore, the graph is symmetric with respect to the y-axis. ✚

Figure 6.8

The next example shows how to solve equations involving logarithms.

Example 4   Solve each of the following equations.

(a) $\log_x \dfrac{8}{27} = 3$

First, write the expression in exponential form.

$$x^3 = \frac{8}{27}$$

$$x^3 = \left(\frac{2}{3}\right)^3$$

$$x = \frac{2}{3}$$

The solution set is $\{2/3\}$.

(b) $\log_4 x = 5/2$

In exponential form, the given statement becomes

$$4^{5/2} = x$$
$$(4^{1/2})^5 = x$$
$$2^5 = x$$
$$32 = x.$$

The solution set is $\{32\}$.   ✚

Logarithms originally were important as an aid for numerical calculations, but the availability of inexpensive calculators has greatly reduced the need for this application of logarithms. Yet the principles behind the use of logarithms for calculation are important; these principles are based on the properties listed below.

Properties of **Logarithms**	For any positive real numbers x and y, real number r, and any positive real number a, $a \neq 1$:

(a) $\log_a xy = \log_a x + \log_a y$ (b) $\log_a \dfrac{x}{y} = \log_a x - \log_a y$

(c) $\log_a x^r = r\log_a x$ (d) $\log_a a = 1$

(e) $\log_a 1 = 0$

To prove part (a) of the properties of logarithms, let

$$m = \log_a x \quad \text{and} \quad n = \log_a y.$$

By the definition of logarithm,

$$a^m = x \quad \text{and} \quad a^n = y.$$

Multiplication gives

$$a^m \cdot a^n = xy.$$

By a property of exponents,

$$a^{m+n} = xy.$$

Now use the definition of logarithm to write

$$\log_a xy = m + n.$$

Since $m = \log_a x$ and $n = \log_a y$,

$$\log_a xy = \log_a x + \log_a y.$$

Properties (b) and (c) are proven in a similar way. (See Exercises 78 and 79.) Properties (d) and (e) follow directly from the definition of logarithm since $a^1 = a$ and $a^0 = 1$.

The properties of logarithms are useful for rewriting expressions with logarithms in different forms, as shown in the next examples.

Example 5 Assuming that all variables represent positive real numbers, use the properties of logarithms to rewrite each of the following expressions.

(a) $\log_6 7 \cdot 9$

$\log_6 7 \cdot 9 = \log_6 7 + \log_6 9$

(b) $\log_9 \dfrac{15}{7}$

$\log_9 \dfrac{15}{7} = \log_9 15 - \log_9 7$

(c) $\log_5 \sqrt{8}$

$\log_5 \sqrt{8} = \log_5 8^{1/2} = \dfrac{1}{2}\log_5 8$

(d) $\log_a \dfrac{mnq}{p^2} = \log_a m + \log_a n + \log_a q - 2 \log_a p$

(e) $\log_a \sqrt[3]{m^2} = \dfrac{2}{3} \log_a m$

(f) $\log_b \sqrt[n]{\dfrac{x^3 y^5}{z^m}} = \dfrac{1}{n} \log_b \dfrac{x^3 y^5}{z^m}$

$$= \dfrac{1}{n} (\log_b x^3 + \log_b y^5 - \log_b z^m)$$

$$= \dfrac{1}{n} (3 \log_b x + 5 \log_b y - m \log_b z)$$

$$= \dfrac{3}{n} \log_b x + \dfrac{5}{n} \log_b y - \dfrac{m}{n} \log_b z \quad \clubsuit$$

Example 6 Use the properties of logarithms to write each of the following as a single logarithm with a coefficient of 1. Assume that all variables represent positive real numbers.

(a) $\log_3 (x + 2) + \log_3 x - \log_3 2$

$$\log_3 (x + 2) + \log_3 x - \log_3 2 = \log_3 \dfrac{(x + 2)x}{2}$$

(b) $2 \log_a m - 3 \log_a n = \log_a m^2 - \log_a n^3 = \log_a \dfrac{m^2}{n^3}$

(c) $\dfrac{1}{2} \log_b m + \dfrac{3}{2} \log_b 2n - \log_b m^2 n$

$$= \log_b m^{1/2} + \log_b (2n)^{3/2} - \log_b m^2 n$$

$$= \log_b \dfrac{m^{1/2}(2n)^{3/2}}{m^2 n}$$

$$= \log_b \dfrac{2^{3/2} n^{1/2}}{m^{3/2}}$$

$$= \log_b \left(\dfrac{2^3 n}{m^3} \right)^{1/2}$$

$$= \log_b \sqrt{\dfrac{8n}{m^3}} \quad \clubsuit$$

Example 7 Assume that $\log_{10} 2 = .3010$. Find the base 10 logarithms of 4 and 5.
By the properties of logarithms,

$$\log_{10} 4 = \log_{10} 2^2 = 2 \log_{10} 2 = 2(.3010) = .6020$$

$$\log_{10} 5 = \log_{10} \dfrac{10}{2} = \log_{10} 10 - \log_{10} 2 = 1 - .3010 = .6990. \quad \clubsuit$$

Compositions of the exponential and logarithmic functions can be used to get two more useful properties. If $f(x) = a^x$ and $g(x) = \log_a x$, then

$$f[g(x)] = a^{\log_a x}$$

and

$$g[f(x)] = \log_a a^x.$$

By definition, if functions f and g are inverses of each other, then $f[g(x)] = g[f(x)] = x$. Since exponential and logarithmic functions of the same base are inverses of each other, this definition leads to the following theorem.

Theorem on Inverses

For $a > 0$, $a \neq 1$:

$$a^{\log_a x} = x \quad \text{and} \quad \log_a a^x = x.$$

By the results of the last theorem,

$$\log_5 5^3 = 3, \qquad 7^{\log_7 10} = 10, \qquad \text{and} \qquad \log_r r^{k+1} = k + 1.$$

6.2 Exercises

For each of the following statements, write an equivalent statement in logarithmic form. See Example 1.

1. $3^4 = 81$ **2.** $2^5 = 32$ **3.** $10^4 = 10,000$ **4.** $8^2 = 64$

5. $(1/2)^{-4} = 16$ **6.** $(2/3)^{-3} = 27/8$ **7.** $10^{-4} = .0001$ **8.** $(1/100)^{-2} = 10,000$

For each of the following statements, write an equivalent statement in exponential form. See Example 1.

9. $\log_6 36 = 2$ **10.** $\log_5 5 = 1$ **11.** $\log_{\sqrt{3}} 81 = 8$ **12.** $\log_4 (1/64) = -3$

13. $\log_{10} .0001 = -4$ **14.** $\log_3 \sqrt[3]{9} = 2/3$ **15.** $\log_m k = n$ **16.** $\log_2 r = y$

Find the value of each of the following expressions.

17. $\log_5 25$ **18.** $\log_3 81$ **19.** $\log_8 8$ **20.** $\log_7 1$

21. $\log_{10} .001$ **22.** $\log_6 \dfrac{1}{216}$ **23.** $\log_{25} 5$ **24.** $\log_{16} 2$

25. $\log_4 \dfrac{\sqrt[3]{4}}{2}$ **26.** $\log_9 \dfrac{\sqrt[4]{27}}{3}$ **27.** $\log_{1/3} \dfrac{9^{-4}}{3}$ **28.** $\log_{1/4} \dfrac{16^2}{2^{-3}}$

29. $\log_6 36^4$ **30.** $\log_5 125^2$ **31.** $2^{\log_2 9}$ **32.** $8^{\log_8 11}$

Solve each of the following equations. See Example 4.

33. $x = \log_2 32$ **34.** $x = \log_2 128$ **35.** $\log_x 25 = -2$ **36.** $\log_x (1/16) = -2$

Write each of the following expressions as a sum, difference, or product of logarithms. Simplify the result if possible. Assume that all variables represent positive real numbers. See Example 5.

37. $\log_3 (2/5)$ **38.** $\log_4 (6/7)$ **39.** $\log_2 \dfrac{6x}{y}$ **40.** $\log_3 \dfrac{4p}{q}$

41. $\log_5 \dfrac{5\sqrt{7}}{3}$ **42.** $\log_2 \dfrac{2\sqrt{3}}{5}$ **43.** $\log_4 (2x + 5y)$ **44.** $\log_6 (7m + 3q)$

45. $\log_k \dfrac{pq^2}{m}$ **46.** $\log_z \dfrac{x^5 y^3}{3}$ **47.** $\log_m \sqrt{\dfrac{5r^3}{z^5}}$ **48.** $\log_p \sqrt[3]{\dfrac{m^5 n^4}{t^2}}$

Write each of the following expressions as a single logarithm. Assume that all variables represent positive real numbers. See Example 6.

49. $\log_a x + \log_a y - \log_a m$ **50.** $(\log_b k - \log_b m) - \log_b a$

51. $2 \log_m a - 3 \log_m b^2$ **52.** $\dfrac{1}{2} \log_y p^3 q^4 - \dfrac{2}{3} \log_y p^4 q^3$

53. $-\dfrac{3}{4} \log_x a^6 b^8 + \dfrac{2}{3} \log_x a^9 b^3$ **54.** $\log_a (pq^2) + 2 \log_a (p/q)$

55. $\log_b (x + 2) + \log_b 7x - \log_b 8$ **56.** $\log_h (2m - 3) + \log_h 2m - \log_h 3, \, 2m - 3 > 0$

57. $2 \log_a (z - 1) + \log_a (3z + 2), \, z - 1 > 0$ **58.** $\log_b (2y + 5) - \dfrac{1}{2} \log_b (y + 3)$

59. $-\dfrac{2}{3} \log_5 5m^2 + \dfrac{1}{2} \log_5 25m^2$ **60.** $-\dfrac{3}{4} \log_3 16p^4 - \dfrac{2}{3} \log_3 8p^3$

61. Graph each of the following functions. Compare the graphs to that of $f(x) = \log_2 x$. See Examples 2 and 3.
 (a) $f(x) = (\log_2 x) + 3$ (b) $f(x) = \log_2 (x + 3)$ (c) $f(x) = |\log_2 (x + 3)|$

62. Graph each of the following functions. Compare the graphs to that of $f(x) = \log_{1/2} x$. See Examples 2 and 3.
 (a) $f(x) = (\log_{1/2} x) - 2$ (b) $f(x) = \log_{1/2} (x - 2)$ (c) $f(x) = |\log_{1/2} (x - 2)|$

Graph each of the following functions. See Example 2.

63. $f(x) = \log_3 x$ **64.** $f(x) = \log_{10} x$ **65.** $f(x) = \log_{1/2} (1 - x)$ **66.** $f(x) = \log_{1/3} (3 - x)$

67. $f(x) = \log_2 x^2$ **68.** $f(x) = \log_3 (x - 1)$ **69.** $f(x) = x \log_{10} x$ **70.** $f(x) = x^2 \log_{10} x$

Given $\log_{10} 2 = .3010$ and $\log_{10} 3 = .4771$, find each of the following logarithms without using calculators or tables. See Example 7.

71. $\log_{10} 6$ **72.** $\log_{10} 12$ **73.** $\log_{10} 9$

74. $\log_{10} 20$ **75.** $\log_{10} 30$ **76.** $\log_{10} 36$

77. The population of an animal species that is introduced into a certain area may grow rapidly at first but then grow more slowly as time goes on. A logarithmic function can provide an excellent description of such growth. Suppose that the population of foxes, $F(t)$, in an area t months after the foxes were introduced there is

$$F(t) = 500 \log_{10} (2t + 3).$$

Use a calculator with a log key to find the population of foxes at the following times:
 (a) when they are first released into the area (that is, when $t = 0$);
 (b) after 3 months;
 (c) after 15 months.
 (d) Graph $y = F(t)$.

Prove the following properties of logarithms.

78. $\log_a (x/y) = \log_a x - \log_a y$ **79.** $\log_a x^r = r \log_a x$

6.3 Evaluating Logarithms

Common Logarithms Since our number system uses base 10, logarithms to base 10 are most convenient for numerical calculation, historically the main application of logarithms. Base 10 logarithms are called **common logarithms.** The common logarithm of the number x, or $\log_{10} x$, is often abbreviated as just $\log x$, and we will use that convention from now on.

A calculator with a log key or a table of logarithms is needed to find base 10 logarithms of numbers other than powers of 10. The use of tables of common logarithms is discussed in Appendix A.

Example 1 Use a calculator to find the following logarithms.

(a) log 142

Enter 142 and press the log key. (This may be a second function key on some calculators.) The result should be 2.152 to the nearest thousandth.

(b) log .005832

A calculator gives

$$\log .005832 \approx -2.234. \quad \clubsuit$$

Example 2 Find N for each of the following.

(a) $\log N = 1.4932$

Enter 1.4932 and press the INV key, then the log key. You should get $N \approx 31.13$. To check, find log 31.13.

(b) $\log N = -4.106$

Enter 4.106, then press the $\pm$ key to make the entry negative. Using the INV and log keys gives $7.834296 - 05$. The -05 is the power on 10 in exponential notation, so

$$N \approx 7.8 \times 10^{-05} = .000078. \quad \clubsuit$$

In chemistry, the pH of a solution is defined as

$$pH = -\log [H_3O^+],$$

where $[H_3O^+]$ is the hydronium ion concentration in moles per liter.* The pH value is a measure of the acidity or alkalinity of solutions. Pure water has a pH of 7.0, substances with pH values greater than 7.0 are alkaline, and substances with pH values less than 7.0 are acidic.

---

*A *mole* is the amount of a substance that contains the same number of molecules as the number of atoms in exactly 12 grams of carbon 12.

Example 3   **(a)** Find the pH of a solution with $[H_3O^+] = 2.5 \times 10^{-4}$

$$pH = -\log [H_3O^+]$$
$$pH = -\log (2.5 \times 10^{-4})$$
$$= -(\log 2.5 + \log 10^{-4})$$
$$= -(.3979 - 4)$$
$$= -.3979 + 4$$
$$= 3.6$$

It is customary to round pH values to the nearest tenth.

(b) Find the hydronium ion concentration of a solution with pH $= 7.1$.

$$pH = -\log [H_3O^+]$$
$$7.1 = -\log [H_3O^+]$$
$$-7.1 = \log [H_3O^+]$$

Now find the number with a logarithm of -7.1. Enter 7.1, $\pm$, and use the INV and log keys to get

$$[H_3O^+] = 7.9 \times 10^{-8} \quad \clubsuit$$

Natural Logarithms   In most practical applications of logarithms, the number $e \approx 2.718281828$ is used as base. Logarithms to base e are called **natural logarithms**, since they occur in many natural-world applications, such as those involving growth and decay. The abbreviation ln x is used for the natural logarithm of x, so that $\log_e x = \ln x$. (Read ln x as "el-en x" or "the natural logarithm of x.") A graph of the natural logarithm function, defined by $f(x) = \ln x$, is given in Figure 6.9.

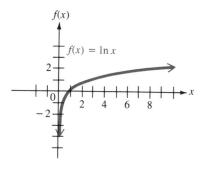

Figure 6.9

Natural logarithms can be found with a calculator that has an ln key or with a table of natural logarithms. A table of natural logarithms is given in Table 2.

Reading directly from this table,

$$\ln 55 = 4.0073,$$
$$\ln 1.9 = 0.6419,$$

and
$$\ln 0.4 = -.9163.$$

Example 4   Use a calculator or Table 2 to find the following logarithms.

(a) ln 85

If you are using a calculator, enter 85, press the ln key, and read the result, 4.4427. Table 2 does not give ln 85. However, the value of ln 85 can be found from the table by first using the properties of logarithms.

$$\ln 85 = \ln (8.5 \times 10)$$
$$= \ln 8.5 + \ln 10$$
$$\approx 2.1401 + 2.3026$$
$$= 4.4427$$

A result found in this way is sometimes slightly different from the answer found using a calculator. Such a difference is due to rounding error.

(b) ln 36

A calculator gives ln 36 = 3.5835. To use the table, first use properties of logarithms, since 36 is not listed in Table 2.

$$\ln 36 = \ln 6^2$$
$$= 2 \ln 6$$
$$\approx 2(1.7918)$$
$$= 3.5836$$

Alternatively, find ln 36 as follows:

$$\ln 36 = \ln 9 \cdot 4 = \ln 9 + \ln 4 = 2.1972 + 1.3863 = 3.5835. \quad \clubsuit$$

Logarithms to Other Bases   A calculator or a table can be used to find the values of either natural logarithms (base *e*) or common logarithms (base 10). However, sometimes it is convenient to use logarithms to other bases. The following theorem can be used to convert logarithms from one base to another.

Change of Base Theorem

For any positive real numbers *x*, *a*, and *b*, where $a \neq 1$ and $b \neq 1$:

$$\log_a x = \frac{\log_b x}{\log_b a}$$

This result is proved by using the definition of logarithm to write $y = \log_a x$ as $x = a^y$ or $x = a^{\log_a x}$ (for positive x and positive a, $a \neq 1$). Now take base b logarithms of both sides of this last equation:

$$x = a^{\log_a x}$$

$$\log_b x = \log_b a^{\log_a x} \qquad \text{Take logs on both sides}$$

$$\log_b x = (\log_a x)(\log_b a), \qquad \text{Power rule}$$

from which

$$\log_a x = \frac{\log_b x}{\log_b a}. \qquad \begin{array}{l}\text{Divide both sides} \\ \text{by } \log_b a\end{array}$$

Example 5 Use natural logarithms to find each of the following. Round to the nearest hundredth.

(a) $\log_5 17$

Let $x = 17$, $a = 5$, and $b = e$. Substituting into the change of base theorem gives

$$\log_5 17 = \frac{\log_e 17}{\log_e 5}$$

$$= \frac{\ln 17}{\ln 5}.$$

Now use a calculator or Table 2.

$$\log_5 17 \approx \frac{2.8332}{1.6094}$$

$$\approx 1.76$$

To check, use a calculator with a y^x key, along with the definition of logarithm, to verify that $5^{1.76} \approx 17$.

(b) $\log_2 .1$

$$\log_2 .1 = \frac{\ln .1}{\ln 2} \approx \frac{-2.3026}{.6931} = -3.32 \quad \clubsuit$$

Example 6 One measure of the diversity of the species in an ecological community is given by the formula

$$H = -[P_1 \log_2 P_1 + P_2 \log_2 P_2 + \cdots + P_n \log_2 P_n],$$

where $P_1, P_2 \ldots, P_n$ are the proportions of a sample belonging to each of n species found in the sample. For example, in a community with two species, where there are 90 of one species and 10 of the other, $P_1 = 90/100 = .9$ and $P_2 = 10/100 = .1$. Thus,

$$H = -[.9 \log_2 .9 + .1 \log_2 .1].$$

In Example 5(b), $\log_2 .1$ was found to be -3.32. Now find $\log_2 .9$.

$$\log_2 .9 = \frac{\ln .9}{\ln 2}$$

$$\approx \frac{-.1054}{.6931}$$

$$\approx -.152$$

Therefore,

$$H \approx -[(.9)(-.152) + (.1)(-3.32)] \approx .469. \quad \clubsuit$$

Recall from Section 6.1 that in the formula $A = Pe^{ni}$, P is the present value of the future value A. The formula can be solved for A, P, n, or i. The following example shows how to find the present value when the other values are given.

Example 7   Find the present value of \$10,000 compounded continuously at 5% for 8 years.

Let $A = 10,000$, $n = 8$, and $i = .05$ in the formula for continuous compounding, and solve for P.

$$A = Pe^{ni}$$
$$10,000 = Pe^{8(.05)}$$
$$10,000 = Pe^{.4}$$
$$10,000 = P(1.49182) \qquad \text{Use a calculator or Table 2}$$
$$P = 6703.22 \qquad \text{Divide by 1.49182}$$

The present value is \$6703.22. $\clubsuit$

6.3   Exercises

Find each of the following logarithms to four decimal places. Use a calculator or Table 2. See Examples 1 and 4.

1. $\log 43$ **2.** $\log 98$ **3.** $\log 1247$ **4.** $\log 783$

5. $\log .014$ **6.** $\log .0069$ **7.** $\log 60,000$ **8.** $\log 640$

9. $\ln 580$ **10.** $\ln 420$ **11.** $\ln .08$ **12.** $\ln .7$

13. $\ln 81,000$ **14.** $\ln 121,000$ **15.** $\ln 350$ **16.** $\ln 900$

Find N for each of the following logarithms. See Example 2.

17. $\log N = 3.421$ **18.** $\log N = 1.007$

19. $\log N = -2.413$ **20.** $\log N = -.7514$

Find each of the following logarithms to the nearest hundredth. See Example 5.

21. $\log_5 10$ **22.** $\log_9 12$ **23.** $\log_{15} 5$ **24.** $\log_{1/2} 3$

25. $\log_{100} 83$ **26.** $\log_{200} 175$ **27.** $\log_{2.9} 7.5$ **28.** $\log_{5.8} 12.7$

Graph f for each of the following.

29. $f(x) = \ln x$ **30.** $f(x) = |\ln x|$ **31.** $f(x) = x \ln x$ **32.** $f(x) = x^2 \ln x$

Use base 10 logarithms to solve each of the following applied problems.

33. Philadelphia Chemicals, Inc., has to store a new gas it plans to produce. The pressure p (lb per cu ft) and volume v (cu ft) of the gas are related by the formula $pv^{1.6} = 800$. What is p if $v = 7.6$ cu ft?

34. The approximate period T (in seconds) of a simple pendulum of length L (in feet) is

$$T \approx 2\pi \sqrt{\frac{L}{32}}.$$

Find the period of a pendulum of length $L = 26$ ft.

35. The number of years, n, since two independently evolving languages split off from a common ancestral language is approximated by $n \approx -7600 \log r$, where r is the proportion of words from the ancestral language common to both languages.
(a) Find n if $r = .9$.   (b) Find n if $r = .3$.
(c) How many years have elapsed since the split if half of the words of the ancestral language are common to both languages?

The loudness of sounds is measured in a unit called a decibel. *To measure with this unit, we first assign an intensity of I_0 to a very faint sound, called the* threshold sound. *If a particular sound has intensity I, then the decibel rating of this louder sound is*

$$d = 10\log_{10} \frac{I}{I_0}.$$

36. Find the decibel ratings of sounds having the following intensities:
(a) $100I_0$ (b) $1000I_0$ (c) $100{,}000I_0$ (d) $1{,}000{,}000I_0$.

37. Find the decibel ratings of the following sounds, having intensities as given. (You will need a calculator with a log key.) Round each answer to the nearest whole number.
(a) Whisper, $115I_0$
(b) Busy street, $9{,}500{,}000I_0$
(c) Heavy truck, 20 m away, $1{,}200{,}000{,}000I_0$
(d) Rock music, $895{,}000{,}000{,}000I_0$
(e) Jetliner at takeoff, $109{,}000{,}000{,}000{,}000I_0$

38. The intensity of an earthquake, measured on the *Richter scale,* is given by

$$\log_{10} \frac{I}{I_0},$$

where I_0 is the intensity of an earthquake of a certain (small) size. Find the Richter scale ratings of earthquakes having the following intensities:
(a) $1000I_0$ (b) $1{,}000{,}000I_0$ (c) $100{,}000{,}000I_0$.

39. The San Francisco earthquake of 1906 had a Richter scale rating of 8.6. Use a calculator with a y^x key to express the intensity of this earthquake as a multiple of I_0 (see Exercise 38).

40. How much more powerful is an earthquake with a Richter scale rating of 8.6 than one with a rating of 8.2?

For each of the following substances, find the pH from the given hydronium ion concentration. See Example 3(a).

41. Grapefruit, 6.3×10^{-4}

42. Crackers, 3.9×10^{-9}

43. Limes, 1.6×10^{-2}

44. Sodium hydroxide (lye), 3.2×10^{-14}

Find the $[H_3O^+]$ for each of the following substances from the given pH. See Example 3(b).

45. Soda pop, 2.7   **46.** Wine, 3.4   **47.** Beer, 4.8   **48.** Drinking water, 6.5

Work the following problems. For Exercises 49 and 50, see Example 6.

49. Suppose a sample of a small community shows two species with 50 individuals each. Find the index of diversity H.

50. A virgin forest in northwestern Pennsylvania has 4 species of large trees with the following proportions of each: hemlock, .521; beech, .324; birch, .081; maple, .074. Find the index of diversity H.

51. The number of species in a sample is given by

$$S(n) = a \ln\left(1 + \frac{n}{a}\right).$$

Here n is the number of individuals in the sample and a is a constant that indicates the diversity of species in the community. If $a = .36$, find $S(n)$ for the following values of n.

(a) 100   **(b)** 200   **(c)** 150   **(d)** 10

52. In Exercise 51, find $S(n)$ if a changes to .88. Use the following values of n.

(a) 50   **(b)** 100   **(c)** 250

53. If an object is fired vertically upward and is subject only to the force of gravity, g, and to air resistance, then the maximum height, H, attained by the object is

$$H = \frac{1}{K}\left(V_0 - \frac{g}{K} \ln \frac{g + V_0 K}{g}\right),$$

where V_0 is the initial velocity of the object and K is a constant. Find H if $K = 2.5$, $V_0 = 1000$ ft per sec, and $g = 32$ ft per sec per sec.

54. The pull, P, of a tracked vehicle on dry sand under certain conditions is approximated by

$$P = W\left[.2 + .16 \ln \frac{G(bl)^{3/2}}{W}\right],$$

where G is an index of sand strength, W is the load on the vehicle, b is the width of the track, and l is the length of the track.* Find P if $W = 10$, $G = 5$, $b = 30.5$ cm, and $l = 61.0$ cm.

---

*Gerald W. Turnage, *Prediction of Track Pull Performance in a Desert Sand,* unpublished MS thesis, The Florida State University, 1971.

55. The following formula can be used to estimate the population of the United States, where t is time in years measured from 1914 (times before 1914 are negative)*:

$$N = \frac{197,273,000}{1 + e^{-.03134t}}.$$

(a) Complete the following chart.

Year	Observed Population	Predicted Population
1790	3,929,000	3,967,000
1810	7,240,000	
1860	31,443,000	30,669,000
1900	75,995,000	
1970	204,000,000	
1980	224,000,000	

(b) Estimate the number of years after 1914 that it would take for the population to increase to 197,273,000.

In the central Sierra Nevada mountains of California, the percent of moisture that falls as snow rather than rain is approximated reasonably well by

$$p = 86.3 \ln h - 680,$$

where p is the percent of snow at an altitude h in feet. (Assume h $\geq$ 3000.)

56. Find the percent of moisture that falls as snow at the following altitudes.
    (a) 3000 ft   **(b)** 4000 ft   **(c)** 7000 ft

57. Graph p.

58. Many environmental situations place effective limits on the population growth of an organism in an area. Many such limited growth situations are described by the *logistic function*

$$G(t) = \frac{mG_0}{G_0 + (m - G_0)e^{-kmt}},$$

where G_0 is the initial number present, m is the maximum possible size of the population, k is a positive constant, and t is time in appropriate units. Assume $G_0 = 1000$, $m = 25,000$, and $k = .04$.
    (a) Find $G(.005)$.   **(b)** Find $G(.01)$.   **(c)** Find a value of t so that $G(t) = m/2$.

To find the maximum permitted levels of certain pollutants in fresh water, the EPA has established the functions of Exercises 59–60, where M(h) is the maximum permitted level

---

*The formula is given in Alfred J. Lotka, *Elements of Mathematical Biology*, reprinted by Dover Press, 1957, p. 67.

of pollutant for a water hardness of h milligrams per liter. Find M(h) in each case.
(These results give the maximum permitted average concentration in micrograms per liter
for a 24-hour period.)

59. Pollutant: copper, $M(h) = e^r$, where $r = .65 \ln h - 1.94$ and $h = 9.7$

60. Pollutant: lead, $M(h) = e^r$, where $r = 1.51 \ln h - 3.37$ and $h = 8.4$

61. Given $g(x) = e^x$, evaluate the following.
  (a) $g(\ln 3)$ **(b)** $g(\ln 5^2)$ **(c)** $g\left(\ln \dfrac{1}{e}\right)$

62. Given $f(x) = 3^x$, evaluate the following.
  (a) $f(\log_3 7)$ **(b)** $f(\log_3 [\ln 3])$ **(c)** $f(\log_3 [2 \ln 3])$

63. Given $f(x) = \ln x$, evaluate the following.
  (a) $f(e^5)$ **(b)** $f(e^{\ln 3})$ **(c)** $f(e^{2 \ln 3})$

64. Given $f(x) = \log_2 x$, evaluate the following.
  (a) $f(2^3)$ **(b)** $f(2^{\log_2 2})$ **(c)** $f(2^{2 \log_2 2})$

6.4   *Exponential and Logarithmic Equations*

As mentioned at the beginning of this chapter, exponential and logarithmic functions are important in many useful applications of mathematics. Using these functions in applications often requires solving exponential and logarithmic equations. Some simple equations were solved in the first two sections of this chapter. More general methods for solving these equations depend on the properties below. These properties follow from the fact that exponential and logarithmic functions are one-to-one. Property 1 was given and used to solve exponential equations in Section 6.1.

Properties of Logarithmic and Exponential Functions

For $b > 0$ and $b \neq 1$:

1. $x = y$ if and only if $b^x = b^y$

2. if $x > 0$ and $y > 0$,

$x = y$ if and only if $\log_b x = \log_b y$

Exponential Equations   The following examples show how these properties are used to solve equations. The first examples illustrate a general method, using Property 2, for solving exponential equations.

Example 1   Solve the equation $7^x = 12$.

While any appropriate base b can be used to apply Property 2, the best practical base to use is often base e. Taking base e (natural) logarithms of both sides gives

$$\ln 7^x = \ln 12$$
$$x \cdot \ln 7 = \ln 12$$
$$x = \frac{\ln 12}{\ln 7}.$$

A decimal approximation for x can be found using Table 2 or a calculator:

$$x = \frac{\ln 12}{\ln 7} \approx \frac{2.4849}{1.9459}$$

Using a calculator to divide 2.4849 by 1.9459 gives

$$x \approx 1.277.$$

A calculator with a y^x key can be used to check this answer. Evaluate $7^{1.277}$; the result should be approximately 12. This step verifies that, to the nearest thousandth, the solution set is $\{1.277\}$.   ✤

Example 2   Solve $e^{-2 \ln x} = \dfrac{1}{16}$.

Use a property of logarithms to rewrite the left side of the equation.

$$e^{\ln x^{-2}} = \frac{1}{16} \qquad \text{Power rule}$$

$$x^{-2} = \frac{1}{16} \qquad \text{Theorem on inverses}$$

$$x^{-2} = 4^{-2} \qquad \frac{1}{16} = \frac{1}{4^2} = 4^{-2}$$

$$x = 4 \qquad \text{Property 1}$$

Check this answer by substituting in the original equation to see that the solution set is $\{4\}$.   ✤

Logarithmic Equations   The properties of logarithms given in Section 6.2 are useful in solving logarithmic equations, as shown in the next examples.

Example 3   Solve $\log_a (x + 6) - \log_a (x + 2) = \log_a x$.

Using a property of logarithms, rewrite the equation as

$$\log_a \frac{x + 6}{x + 2} = \log_a x. \qquad \text{Quotient rule}$$

Then $$\frac{x + 6}{x + 2} = x \qquad \text{Property 2}$$

$$x + 6 = x(x + 2)$$
$$x + 6 = x^2 + 2x$$
$$x^2 + x - 6 = 0$$
$$(x + 3)(x - 2) = 0$$
$$x = -3 \quad \text{or} \quad x = 2.$$

The negative solution ($x = -3$) cannot be used since it is not in the domain of $\log_a x$. For this reason, the only valid solution is the positive number 2, giving the solution set $\{2\}$.   ✚

Example 4   Solve $\log (3x + 2) + \log (x - 1) = 1$.

Since $\log x$ is an abbreviation for $\log_{10} x$, and $1 = \log_{10} 10$, the properties of logarithms give

$$\log (3x + 2)(x - 1) = \log 10 \qquad \text{Product rule}$$
$$(3x + 2)(x - 1) = 10 \qquad \text{Property 2}$$
$$3x^2 - x - 2 = 10$$
$$3x^2 - x - 12 = 0.$$

Now use the quadratic formula to get

$$x = \frac{1 \pm \sqrt{1 + 144}}{6}.$$

If $x = (1 - \sqrt{145})/6$, then $x - 1 < 0$; therefore, $\log (x - 1)$ does not exist and this proposed solution must be discarded, giving the solution set

$$\{(1 + \sqrt{145})/6\}. \quad ✚$$

The definition of logarithm could have been used in Example 4 by first writing

$$\log (3x + 2) + \log (x - 1) = 1$$
$$\log_{10} (3x + 2)(x - 1) = 1 \qquad \text{Product rule}$$
$$(3x - 2)(x - 1) = 10^1, \qquad \text{Definition of logarithm}$$

then continuing as shown above.

Example 5   Solve $\ln e^{\ln x} - \ln (x - 3) = \ln 2$.

On the left, $\ln e^{\ln x}$ can be written as

$$\ln e^{\ln x} = \ln x$$

using the theorem on inverses at the end of Section 6.2. The equation becomes

$$\ln x - \ln (x - 3) = \ln 2$$

$$\ln \frac{x}{x - 3} = \ln 2 \qquad \text{Quotient rule}$$

$$\frac{x}{x - 3} = 2 \qquad \text{Property 2}$$

$$x = 2x - 6$$

$$6 = x.$$

Verify that the solution set is {6}. ✚

A summary of the methods used for solving equations in this section follows.

Solving Exponential and Logarithmic Equations	An exponential or logarithmic equation may be solved by changing the equation into one of the following forms, where a and b are real numbers, $a > 0$, and $a \neq 1$.

1. $a^{f(x)} = b$

Solve by taking logarithms of each side. (Natural logarithms are often a good choice.)

2. $\log_a f(x) = \log_a g(x)$

From the given equation, $f(x) = g(x)$, which is solved algebraically.

3. $\log_a f(x) = b$

Solve by using the definition of logarithm to write the expression in exponential form as $f(x) = a^b$.

The next examples show applications of exponential and logarithmic equations.

Example 6 Suppose that

$$P(t) = 10,000e^{.4t}$$

gives the population of a city at time t measured in years. In how many years will the population of the city double?

Since the population of the city is 10,000 when $t = 0$, the population of the city will be doubled when it reaches 20,000. Find the value of t for which $P(t) = 20,000$. Replace $P(t)$ with 20,000 to get

$$20,000 = 10,000e^{.4t},$$

or
$$2 = e^{.4t}.$$

Take natural logarithms on both sides, then solve for t.

$$\ln 2 = \ln e^{.4t}$$

Since $\ln e^{.4t} = (.4t)\ln e = (.4t)(1) = .4t$,

$$\ln 2 = .4t,$$

with

$$t = \frac{\ln 2}{.4}.$$

Using a calculator or Table 2 to find $\ln 2$ and then using a calculator to find the quotient gives $t = 1.73$ to the nearest hundredth. The population of the city will double in about 1.73 years.   ✛

Example 7   Suppose the amount, y, in grams, of a certain radioactive substance at a time t is given by

$$y = y_0 e^{-.1t},$$

where y_0 is the amount of the substance present initially (when $t = 0$) and t is measured in days. Find the *half-life* of the substance—that is, the time it takes for half a given amount of the substance to decay.

Find the time t for y to be reduced to a value equal to $y_0/2$. That is, solve the equation $(1/2)y_0 = y_0 e^{-.1t}$

$$\frac{1}{2}y_0 = y_0 e^{-.1t}$$

$$\frac{1}{2} = e^{-.1t} \qquad \text{Divide by } y_0 \text{ on both sides}$$

$$\ln \frac{1}{2} = \ln e^{-.1t} \qquad \text{Take natural logarithms on both sides}$$

$$\ln \frac{1}{2} = -.1t \ln e \qquad \text{Property of logarithms}$$

$$\ln \frac{1}{2} = -.1t \qquad \ln e = 1$$

$$t = \frac{\ln 1/2}{-.1}$$

From Table 2 or a calculator, $\ln 1/2 = \ln .5 = -.6931$. Dividing gives

$$t \approx 6.9 \text{ days.}   ✛$$

Example 8   Carbon 14 is a radioactive isotope of carbon that has a half-life of about 5600 years. The earth's atmosphere contains much carbon, mostly in the form of carbon dioxide, with small traces of carbon 14. Most of this atmospheric carbon is in the form of the nonradioactive isotope carbon 12. The ratio of carbon 14 to carbon 12 is virtually constant in the atmosphere. However, as a plant absorbs carbon dioxide from the air in the process of photosynthesis, the carbon 12 stays in the plant while the carbon 14 decays by conversion to nitrogen. Thus, the ratio of carbon 14 to carbon 12 is smaller in the plant than it is in the atmosphere. Even when the plant is eaten by an animal, this ratio will continue to decrease. Based on these facts, a method of dating objects called *carbon 14 dating* has been developed.

Let R be the (nearly constant) ratio of carbon 14 to carbon 12 found in the atmosphere, and let r be the ratio found in a fossil. It can be shown that the relationship between R and r is given by

$$\frac{R}{r} = e^{(t \ln 2)/5600},$$

where t is the age of the fossil in years.

(a) Verify the formula for $t = 0$.

If $t = 0$,

$$\frac{R}{r} = e^0 = 1.$$

Thus $R = r$, so that the ratio in the fossil is the same as the ratio in the atmosphere. This is true only when $t = 0$.

(b) Verify the formula for $t = 5600$.

Substitute 5600 for t. Then

$$\frac{R}{r} = e^{(5600 \ln 2)/5600}$$

$$\frac{R}{r} = e^{\ln 2} = 2 \qquad \text{(Recall that } a^{\log_a x} = x.\text{)}$$

$$r = \frac{1}{2}R.$$

From this last result, the ratio in the fossil is half the ratio in the atmosphere. Since the half-life of carbon 14 is 5600 years, only half of it should remain at the end of that time. Thus, the formula gives the correct result for $t = 5600$. ✛

6.4 Exercises

Solve the following equations. When necessary, give answers as decimals rounded to the nearest thousandth. See Examples 1–5.

1. $3^x = 6$
2. $4^x = 12$
3. $3^{a+2} = 5$
4. $5^{2-x} = 12$

5. $6^{1-2k} = 8$
6. $3^{2m-5} = 13$
7. $e^{k-1} = 4$
8. $e^{2-y} = 12$

9. $2e^{5a+2} = 8$
10. $10e^{3z-7} = 5$
11. $2^x = -3$
12. $(1/4)^p = -4$

13. $e^{2x} \cdot e^{5x} = e^{14}$
14. $e^{\ln x} = 3$
15. $e^{-\ln x} = 2$
16. $e^{3 \ln x} = \frac{1}{8}$

17. $e^{\ln x + \ln (x-2)} = 8$
18. $e^{2 \ln x - \ln (x+2)} = \frac{8}{3}$

19. $100(1 + .02)^{3+n} = 150$
20. $500(1 + .05)^{p/4} = 200$

21. $\log (t - 1) = 1$

22. $\log q^2 = 1$

23. $\log (x - 3) = 1 - \log x$

24. $\log (z - 6) = 2 - \log (z + 15)$

25. $\ln (y + 2) = \ln (y - 7) + \ln 4$

26. $\ln p - \ln (p + 1) = \ln 5$

27. $\ln (5 + 4y) - \ln (3 + y) = \ln 3$

28. $\ln m + \ln (2m + 5) = \ln 7$

29. $\ln x + 1 = \ln (x - 4)$

30. $\ln (4x - 2) = \ln 4 - \ln (x - 2)$

31. $2 \ln (x - 3) = \ln (x + 5) + \ln 4$

32. $\ln (k + 5) + \ln (k + 2) = \ln 14k$

33. $\log_5 (r + 2) + \log_5 (r - 2) = 1$

34. $\log_4 (z + 3) + \log_4 (z - 3) = 1$

35. $\log_3 (a - 3) = 1 + \log_3 (a + 1)$

36. $\log w + \log (3w - 13) = 1$

37. $\ln e^x - \ln e^3 = \ln e^5$

38. $\ln e^x - 2 \ln e = \ln e^4$

39. $\log_2 \sqrt{2y^2} - 1 = 1/2$

40. $\log_2 (\log_2 x) = 1$

41. $\log z = \sqrt{\log z}$

42. $\log x^2 = (\log x)^2$

Find the solution of each of the following equations. Round to the nearest thousandth.

43. $\log_x 5.87 = 2$ **44.** $\log_x 11.9 = 3$ **45.** $1.8^{p+4} = 9.31$ **46.** $3.7^{5z-1} = 5.88$

47. The amount of a radioactive substance present at time t measured in seconds is $A(t) = 5000e^{-.02t}$, where $A(t)$ is measured in grams. Find the half-life of the substance, that is, the time it takes until only half the substance remains.

48. Find the half-life of a radioactive specimen if the amount y in grams present at time t in days is $y = 1000e^{-.4t}$.

49. Suppose the number of rabbits in a colony is $y = y_0 e^{.4t}$, where t represents time in months and y_0 is the rabbit population when $t = 0$.
   (a) If $y_0 = 100$, find the number of rabbits present at time $t = 4$.
   (b) How long will it take for the number of rabbits to triple?

50. A midwestern city finds its residents moving to the suburbs. Its population is declining according to the relationship $P = P_0 e^{-.04t}$, where t is time measured in years and P_0 is the population at time $t = 0$. Assume that $P_0 = 1,000,000$.
   (a) Find the population at time $t = 1$.
   (b) Estimate the time it will take for the population to be reduced to 750,000.
   (c) How long will it take for the population to be cut in half?

For Exercises 51–56, refer to Example 8 in this section of the text.

51. Suppose an Egyptian mummy is discovered in which the ratio of carbon 14 to carbon 12 is only about half the ratio found in the atmosphere. About how long ago did the Egyptian die?

52. If the ratio of carbon 14 to carbon 12 in an object is 1/4 the atmospheric ratio, how old is the object? How old if the ratio is 1/8?

53. Paint from the Lascaux caves of France contains 15% of the normal amount of carbon 14. Estimate the age of the caves.

54. Solve the formula for t.

55. Suppose a specimen is found in which $r = (2/3)R$. Estimate the age of the specimen.

56. Estimate the age of a specimen having $r = R/12$.

A large cloud of radioactive debris from a nuclear explosion has floated over the Pacific Northwest, contaminating much of the hay supply. Consequently, farmers in the area are concerned that the cows who eat this hay will give contaminated milk. (The tolerance level for radioactive iodine in milk is 0.) The percent of the initial amount of radioactive iodine still present in the hay after t days is approximated by $P(t) = 100e^{-.1t}$.

57. Some scientists feel that the hay is safe after the percent of radioactive iodine has declined to 10% of the original amount. Find the number of days before the hay can be used.

58. Other scientists believe that the hay is not safe until the level of radioactive iodine has declined to only 1% of the original level. Find the number of days this would take.

Solve each of the following equations for the indicated variables. Use logarithms to the appropriate bases.

59. $P = P_0 e^{kt/1000}$ for t

60. $I = \dfrac{E}{R}(1 - e^{-Rt/2})$ for t

61. $T = T_0 + (T_1 - T_0) 10^{-kt}$ for t

62. $A = \dfrac{Pi}{1 - (1 + i)^{-n}}$ for n

Recall (from the exercises for Section 6.3) the formula for the decibel rating of the loudness of a sound is

$$d = 10 \log \frac{I}{I_0}.$$

63. Solve this formula for I.

64. A few years ago, there was a controversy about a proposed government limit on factory noise. One group wanted a maximum of 89 decibels, while another group wanted 86. This difference seemed very small to many people. Find the percent by which the 89-decibel intensity exceeds that for 86 decibels.

The formula for compound interest,

$$A = P\left(1 + \frac{i}{m}\right)^{nm},$$

was given in Section 6.1. Use natural logarithms and solve for n to the nearest tenth in Exercises 65–66.

65. $A = \$10{,}000,\ P = \$7500,\ i = 6\%,\ m = 2$

66. $A = \$8400,\ P = \$5000,\ i = 8\%,\ m = 4$

67. Find the number of years to the nearest hundredth for $27,000 to become $35,140 at 5% interest compounded quarterly.

68. Find n to the nearest hundredth if $1786 becomes $2063.40 at 11.6%, with interest compounded monthly.

69. Find the number of years for $2000 to amount to $2500 at 6% compounded semiannually.

70. Find the number of years for $16,000 to become $25,000 at 8% compounded quarterly.

In Exercises 71–72, use the formula for continuous compounding from Section 6.1,
$A = Pe^{ni}$. *Round answers to the nearest tenth.*

71. Find the interest rate that will cause $5000 to grow to $7250 in 4 years.

72. Find the interest rate at which $3200 will grow to $4500 in 6 years.

73. The growth of bacteria in food products makes it necessary to time-date some products
(such as milk) so that they will be sold and consumed before the bacteria count is too
high. Suppose for a certain product that the number of bacteria present is given by

$$f(t) = 500e^{.1t},$$

under certain storage conditions, where t is time in days after packing of the product
and the value of $f(t)$ is in millions. Find the number of bacteria present at each of the
following times.

(a) 2 days   (b) 1 week   (c) 2 weeks

74. (a) Suppose the product in Exercise 73 cannot be safely eaten after the bacteria count
reaches 3,000,000,000. How long will this take?

(b) If $t = 0$ corresponds to January 1, what date should be placed on the product?

Chapter 6   *Summary*

Key Words

6.1 exponential function
compound interest
future value
present value
continuous compounding

6.2 logarithm
logarithmic function
6.3 common logarithm
natural logarithm

Key Ideas

6.1 Additional Properties of Exponents

(a) If $a > 0$ and $a \neq 1$, then a^x is a unique real number for all real numbers x.

(b) If $a > 0$ and $a \neq 1$, then $a^b = a^c$ if and only if $b = c$.

(c) If $a > 1$ and $m < n$, then $a^m < a^n$.

(d) If $0 < a < 1$ and $m < n$, then $a^m > a^n$.

6.2 Properties of Logarithms

For any positive real numbers x and y, real number r, and positive real number a, $a \neq 1$:

(a) $\log_a xy = \log_a x + \log_a y$

(b) $\log_a \dfrac{x}{y} = \log_a x - \log_a y$

(c) $\log_a x^r = r \cdot \log_a x$

(d) $\log_a a = 1$

(e) $\log_a 1 = 0$

6.3 Change of Base Theorem

For any positive real numbers x, a, and b, where $a \neq 1$ and $b \neq 1$:

$$\log_a x = \frac{\log_b x}{\log_b a}$$

Chapter 6 *Review Exercises*

Graph each of the following functions.

1. $y = 2^x$ **2.** $y = 2^{-x}$ **3.** $y = (1/2)^{x+1}$ **4.** $y = \log_3 x$

Solve each of the following equations.

5. $8^p = 32$ **6.** $9^{2y-1} = 27^y$ **7.** $\dfrac{8}{27} = b^{-3}$ **8.** $\dfrac{1}{2} = \left(\dfrac{b}{4}\right)^{1/4}$

The amount of a certain radioactive material, in grams, present after t days is given by

$$A(t) = 800e^{-.04t}.$$

Use Table 2 or a calculator to find A(t) if t has the following values.

9. $t = 0$ **10.** $t = 5$

Suppose P dollars is compounded continuously at rate of interest i annually; the final amount on deposit would be $A = Pe^{ni}$ dollars after n years. How much would $1200 amount to at 10% compounded continuously for the following numbers of years?

11. 4 yr   **12.** 10 yr

13. Historically, the consumption of electricity has increased at a continuous rate of 6% per year. If it continued to increase at this rate, find the number of years before exactly twice as much electricity would be needed.

14. Suppose a conservation campaign together with higher rates caused demand for electricity to increase at only 2% per year. (See Exercise 13.) Find the number of years before twice as much electricity would be needed.

Write each of the following expressions in logarithmic form.

15. $2^5 = 32$ **16.** $100^{1/2} = 10$ **17.** $(1/16)^{1/4} = 1/2$
18. $(3/4)^{-1} = 4/3$ **19.** $10^{.4771} = 3$ **20.** $e^{2.4849} = 12$

Write each of the following logarithms in exponential form.

21. $\log_{10} .001 = -3$ **22.** $\log_2 \sqrt{32} = 5/2$ **23.** $\log 3.45 = .537819$ **24.** $\ln 45 = 3.806662$

Use properties of logarithms to write each of the following logarithms as a sum, difference, or product of logarithms.

25. $\log_3 \dfrac{mn}{5r}$ **26.** $\log_2 \dfrac{\sqrt{7}}{15}$ **27.** $\log_5 x^2 y^4 \sqrt[5]{m^3 p}$ **28.** $\log_7 (7k + 5r^2)$

Find each of the following logarithms. Round to the nearest thousandth.

29. $\log 45.6$ **30.** $\log 1,230,000$ **31.** $\log .00056$ **32.** $\log .0411$
33. $\ln 35$ **34.** $\ln 470$ **35.** $\ln 144,000$ **36.** $\ln 98,000$
37. $\log_{3.4} 15.8$ **38.** $\log_{1/2} 9.45$ **39.** $\log_3 769$ **40.** $\log_{2/3} 5/8$

Find N, given the following.

41. $\log N = 3.498$ **42.** $\log N = .0043$ **43.** $\log N = -1.8297$ **44.** $\log N = -2.501$

The height in meters of the members of a certain tribe is approximated by

$$h = .5 + \log t,$$

where t is the tribe member's age in years, and $1 \le t \le 20$. *Estimate the heights of tribe members of the following ages.*

45. 2 yr   **46.** 5 yr   **47.** 10 yr   **48.** 20 yr

Use a formula for compound interest to find the total amount on deposit if the following amounts are placed on deposit at the given interest rates for the given numbers of years.

49. $1000 at 8% compounded annually for 9 years

50. $312.45 at 6% compounded semiannually for 16 years

51. $1900 at 8% compounded quarterly for 9 years

52. $57,809.34 at 12% compounded quarterly for 5 years

53. $2500 at 12% compounded monthly for 3 years

54. $11,702.55 at 18% compounded monthly for 4 years

55. $2800 at 6% compounded continuously for 10 years

56. $19,000 at 10% compounded continuously for 7 years

Find the present value of each of the following future values.

57. $4500 at 6% compounded annually for 9 years

58. $11,500 at 4% compounded annually for 12 years

59. $2000 at 4% compounded semiannually for 11 years

60. $2000 at 6% compounded quarterly for 8 years

61. $3456.78 at 5.6% compounded monthly for 30 months

62. How long would it take for $1000 at 5% compounded continuously to double?

63. In Exercise 62, how long would it take at 10%?

64. If the inflation rate were 10%, use the formula for continuous compounding to find the number of years for a $1 item to cost $2.

65. In Exercise 64, find the number of years if the rate of inflation were 13%.

66. What annual interest rate, to the nearest tenth, will produce $8780 if $3500 is left at interest for 10 years?

67. Find the annual interest rate to the nearest tenth that will produce $27,208.60 if $12,700 is left at interest compounded semiannually for 5 years.

68. How many years (to the nearest tenth) will be needed for $7800 to increase to $14,088 at 12% compounded quarterly?

69. Find the number of years (to the nearest tenth) for $48,000 to become $58,344 at 10% interest compounded semiannually.

If R dollars is deposited at the end of each year in an account paying a rate of interest i per year compounded annually, then after n years the account will contain a total of

$$R\left[\frac{(1 + i)^n - 1}{i}\right]$$

dollars. *(This sequence of equal payments is called an* annuity. *Annuities are discussed in Section 9.3.) Find the final amount on deposit for each of the following. Use logarithms or a calculator.*

70. $800, 12%, 10 years   **71.** $1500, 6%, 7 years   **72.** $375, 10%, 12 years

73. Manuel deposits $10,000 at the end of each year for 12 years in an account paying 12% compounded annually. He then puts this total amount on deposit in another account paying 10% compounded semiannually for another 9 years. Find the total amount on deposit after the entire 21-year period.

74. Scott Hardy deposits $12,000 at the end of each year for 8 years in an account paying 5% compounded annually. He then leaves the money alone with no further deposits for an additional 6 years. Find the total amount on deposit after the entire 14-year period.

Solve each of the following equations. Round to the nearest thousandth if necessary.

75. $5^r = 11$

76. $10^{2r-3} = 17$

77. $e^{p+1} = 10$

78. $(1/2)^{3k+1} = 3$

79. $\log_{64} y = 1/3$

80. $\ln 6x - \ln (x + 1) = \ln 4$

81. $\log_{16} \sqrt{x + 1} = 1/4$

82. $\ln x + 3 \ln 2 = \ln \dfrac{2}{x}$

83. $\ln (\ln e^{-x}) = \ln 3$

Chapter 6 *Test*

Graph each of the following functions.

1. $y = (1.5)^{x+2}$

2. $y = \log_{1/2} x$

Solve each of the following equations.

3. $25^{2x-1} = 125^{x+1}$

4. $\left(\dfrac{a}{2}\right)^{-1/3} = 4$

5. The amount of radioactive material, in grams, present after t days is given by $A(t) = 600\, e^{-.05t}$. Find the amount present after 12 days.

6. Suppose a deposit of P dollars is compounded continuously at a rate of interest i annually. The final amount on deposit will be $A = Pe^{ni}$ dollars after n years. How much will $2500 amount to at 8% compounded continuously for 12 years?

7. If a population increases at a continuous rate of 4% per year, how many years will it take for the population to double in size?

Write each of the following in logarithmic form.

8. $a^2 = b$

9. $e^c = 4.82$

Write each of the following in exponential form.

10. $\log_3 \sqrt{27} = \dfrac{3}{2}$

11. $\ln 5 = a$

12. Use properties of logarithms to write the following as a sum, difference, or product of logarithms.

$$\log_7 \frac{x^2 \sqrt[4]{y}}{z^3}$$

Find each of the following logarithms. Round to the nearest thousandth.

13. ln 2300

14. $\log_{2.7} 94.6$

15. Find N to the nearest thousandth if $\log N = 2.3766$.

16. Use the formula for compound interest to find the total amount on deposit if $2400 is deposited at 10% interest compounded quarterly for 9 years.

17. How many years, to the nearest tenth, will be needed for $5000 to increase to $18,000 at 12% compounded monthly?

Solve the following equations. Round answers to the nearest thousandth.

18. $8^{2w-4} = 100$

19. $\log_3 (m + 2) = 2$

20. $\ln x - 4 \ln 3 = \ln \dfrac{5}{x}$

7

Systems of Equations and Inequalities

Many applications of mathematics require the simultaneous solution of a large number of equations or inequalities having many variables. Any set of equations is called a **system of equations.** The solution set of a system of equations is the intersection of the solution sets of the individual equations. It is customary to write a system by listing its equations. For example, the system of equations $2x + y = 4$ and $x - y = 6$ is written as

$$2x + y = 4$$
$$x - y = 6.$$

Systems of inequalities are defined similarly. Methods are presented in this chapter for solving systems of equations or inequalities.

7.1 Linear Systems with Two Variables

A **first-degree equation in n unknowns** is any equation of the form

$$a_1x_1 + a_2x_2 + \cdots + a_nx_n = k,$$

where $a_1, a_2, \ldots, a_n$, and k are constants and $x_1, x_2, \ldots, x_n$ are variables. Such equations are also called *linear equations.* Generally, only systems of linear equations with two or three variables will be discussed in this book, although the methods used can be extended to systems with more variables.

The solution set of a linear equation in two variables is an infinite set of ordered pairs. Since the graph of such an equation is a straight line, there are three possibilities for the solution set of a system of two linear equations in two variables. An example of each possibility is shown in Figure 7.1.

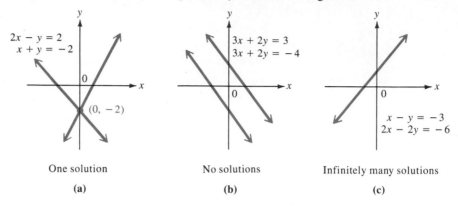

One solution | No solutions | Infinitely many solutions
(a) | (b) | (c)

Figure 7.1

Possible Graphs of a Linear System with Two Equations and Two Variables	**1.** The graphs of the two equations intersect in a single point. The coordinates of this point give the solution of the system. This is the most common case. See Figure 7.1(a).
	2. The graphs are distinct parallel lines. In this case, the system is said to be **inconsistent.** That is, there is no solution common to both equations. The solution set of the linear system is empty. See Figure 7.1(b).
	3. The graphs are the same line. In this case, the equations are said to be **dependent,** and any solution of one equation is also a solution of the other. Thus, there are infinitely many solutions. See Figure 7.1(c).

Although the *number* of solutions of a linear system can often be seen from the graph of the equations of the system, it is usually difficult to determine an exact solution from the graph. The most general algebraic method of finding the solution of a system of two linear equations, called the **substitution method,** is illustrated in the following example and used again in Section 7.4.

Example 1 Solve the system

$$2x + 5y = 21 \tag{1}$$
$$3x - 2y = -16. \tag{2}$$

Begin by solving one equation for one of the variables in terms of the other. For example, solving equation (1) for x gives

$$2x + 5y = 21 \tag{1}$$
$$2x = 21 - 5y$$
$$x = \frac{21 - 5y}{2}. \tag{3}$$

Now substitute this result for x into equation (2).

$$3x - 2y = -16 \tag{2}$$

$$3\left(\frac{21 - 5y}{2}\right) - 2y = -16$$

To eliminate the fraction on the left, multiply both sides of the equation by 2 and then solve for y.

$$2 \cdot 3\left(\frac{21 - 5y}{2}\right) - 2 \cdot 2y = 2(-16)$$

$$3(21 - 5y) - 4y = -32$$

$$63 - 15y - 4y = -32$$

$$-19y = -95$$

$$y = 5$$

Substitute $y = 5$ back into equation (3) to find x.

$$x = \frac{21 - 5y}{2} \tag{3}$$

$$x = \frac{21 - 5 \cdot 5}{2}$$

$$x = -2$$

The solution set for the system is $\{(-2, 5)\}$. Check by substituting -2 for x and 5 for y in each of the equations of the system. ✦

Some systems of two equations can be solved by the **addition method.** With this method, we first multiply the equations on both sides by suitable numbers, so that when they are added, one variable is eliminated. The result is an equation in one variable that can be solved by methods from Chapter 2. The solution is then substituted into one of the original equations, making it possible to solve for the other variable. In this process the given system is replaced by new systems that have the same solution set as the original system. Systems that have the same solution set are called **equivalent systems.** The addition method is illustrated by the following examples.

Example 2 Solve the system

$$3x - 4y = 1 \tag{4}$$

$$2x + 3y = 12. \tag{5}$$

To eliminate x, multiply both sides of equation (4) by -2 and both sides of equation (5) by 3 to get equations (6) and (7).

$$-6x + 8y = -2 \qquad \text{(6)}$$
$$6x + 9y = 36 \qquad \text{(7)}$$

Although this new system is not the same as the given system, it will have the same solution set.

Now add the two equations to eliminate x, and then solve the result for y.

$$-6x + 8y = -2$$
$$\underline{6x + 9y = 36}$$
$$17y = 34$$
$$y = 2$$

Substitute 2 for y in equation (4) or (5). Choosing equation (4) gives

$$3x - 4(2) = 1$$
$$3x = 9$$
$$x = 3.$$

The solution set of the given system is $\{(3, 2)\}$, which can be checked by substituting 3 for x and 2 for y in equation (5). ✛

Since the addition method of solution results in the elimination of one variable from the system, it is also called the **elimination method.**

Example 3 Solve the system

$$3x - 2y = 4$$
$$-6x + 4y = 7.$$

The variable x can be eliminated by multiplying both sides of the first equation by 2 and then adding.

$$6x - 4y = 8$$
$$\underline{-6x + 4y = 7}$$
$$0 = 15$$

Both variables were eliminated here, leaving the false statement $0 = 15$, a signal that these two equations have no solutions in common. The system is inconsistent, with an empty solution set. In slope-intercept form, the equations would show that the graphs are parallel lines because they have the same slope but different y-intercepts. ✛

Example 4 Solve the system

$$-4x + y = 2$$
$$8x - 2y = -4.$$

Eliminate x by multiplying both sides of the first equation by 2 and then adding it to the second equation.

$$-8x + 2y = 4$$
$$\underline{8x - 2y = -4}$$
$$0 = 0$$

This true statement, $0 = 0$, indicates that a solution of one equation is also a solution of the other, so the solution set is an infinite set of ordered pairs. The graphs of the equations are the same line, since the slopes and y-intercepts are equal. The two equations are dependent.

We will write the solution of a system of dependent equations as an ordered pair by expressing x in terms of y as follows. Choose either equation and solve for x. Choosing the first equation gives

$$-4x + y = 2$$
$$x = \frac{2 - y}{-4} = \frac{y - 2}{4}.$$

Write the solution set as

$$\left\{ \left(\frac{y - 2}{4}, y \right) \right\}.$$

By selecting values for y and calculating the corresponding values for x, individual ordered pairs of the solution set can be found. For example, if $y = -2$, $x = (-2 - 2)/4 = -1$ and the ordered pair $(-1, -2)$ is a solution.

Examples 3 and 4 suggest the following summary.

Consider the system of equations

$$a_1x + b_1y = c_1$$
$$a_2x + b_2y = c_2,$$

with $a_2 \neq 0$, $b_2 \neq 0$, and $c_2 \neq 0$. The graphs of the equations are parallel lines and the system is inconsistent if

$$\frac{a_1}{a_2} = \frac{b_1}{b_2} \neq \frac{c_1}{c_2}.$$

(See Figure 7.1(b).) The graphs of the equations are the same line and the equations are dependent if

$$\frac{a_1}{a_2} = \frac{b_1}{b_2} = \frac{c_1}{c_2}.$$

(See Figure 7.1(c).)

Applications of mathematics often require the solution of a system of equations, as the next example shows.

Example 5   The manager of a shoe store purchased 10 pairs of style 1501 shoes and 12 pairs of style 1470 shoes for $551. A later purchase of 4 pairs of style 1501 and 3 pairs of style 1470 cost $170. Find the cost per pair for each style.

A system of linear equations can be written from the information in the problem. To begin, let

$$x = \text{cost of a pair of style 1501 shoes}$$
$$y = \text{cost of a pair of style 1470 shoes.}$$

Since 10 pairs of 1501 and 12 pairs of 1470 shoes cost $551,

$$10x + 12y = 551. \tag{8}$$

Also,
$$4x + 3y = 170. \tag{9}$$

To solve this system by the addition method, multiply both sides of equation (9) by -4 and add the result to equation (8).

$$10x + 12y = 551$$
$$\underline{-16x - 12y = -680}$$
$$-6x \qquad = -129$$
$$x \qquad = 21.5$$

Substitute 21.5 for x in equation (8) or (9). Using (8),

$$10(\mathbf{21.5}) + 12y = 551$$
$$215 + 12y = 551$$
$$12y = 336$$
$$y = \mathbf{28.}$$

Style 1501 shoes cost $21.50 a pair and style 1470 shoes cost $28 a pair.

Check by using the information of the original problem: 10 pairs of style 1501 and 12 pairs of style 1470 cost

$$10(\mathbf{21.50}) + 12(\mathbf{28}) = 215 + 336 = 551,$$

while 4 pairs of style 1501 and 3 pairs of style 1470 cost

$$4(\mathbf{21.50}) + 3(\mathbf{28}) = 86 + 84 = 170,$$

as required.   ✛

Usually, as the price of an item goes up, demand for the item goes down and the supply of the item goes up. Changes in gasoline prices illustrate this situation. The price where supply and demand are equal is called the **equilibrium price,** and the resulting supply or demand is called the **equilibrium supply** or **equilibrium demand.**

Example 6   **(a)** Suppose that the supply of a product is related to its price by the equation

$$p = \frac{2}{3}x,$$

where p is price in dollars and x is supply in appropriate units. Find the price for the supply levels $x = 9$ and $x = 18$.

When $x = 9$,

$$p = \frac{2}{3}x = \frac{2}{3}(9) = 6.$$

When $x = 18$,

$$p = \frac{2}{3}x = \frac{2}{3}(18) = 12.$$

(b) Suppose demand and price for the same product are related by

$$p = -\frac{1}{3}x + 18,$$

where p is price and x is demand. Find the price for the demand levels $x = 6$ and $x = 18$.

When $x = 6$,

$$p = -\frac{1}{3}x + 18 = -\frac{1}{3}(6) + 18 = 16,$$

and when $x = 18$,

$$p = -\frac{1}{3}x + 18 = -\frac{1}{3}(18) + 18 = 12.$$

(c) Graph both functions on the same axes.

Use the ordered pairs found in parts (a) and (b) and the y-intercepts to get the graphs shown in Figure 7.2.

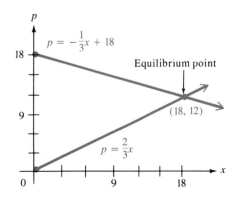

Figure 7.2

(d) Find the equilibrium price, supply, and demand. Solve the system

$$p = \frac{2}{3}x$$

$$p = -\frac{1}{3}x + 18$$

using substitution.

$$\frac{2}{3}x = -\frac{1}{3}x + 18$$

$$x = 18$$

This gives 18 units as the equilibrium supply or demand. Find the equilibrium price by substituting 18 for x in either equation. Using $p = (2/3)x$ gives

$$p = (2/3)(18) = 12$$

or \$12, the equilibrium price. The point (18, 12) that gives the equilibrium values is shown in Figure 7.2. ✚

7.1 Exercises

Use the substitution method to solve each system. Check your answers. See Example 1.

1. $y = 2x + 3$
$3x + 4y = 78$

2. $y = 4x - 6$
$2x + 5y = -8$

3. $3x - 2y = 12$
$5x = 4 - 2y$

4. $8x + 3y = 2$
$5x = 17 + 6y$

5. $2x - 3y + 1 = 0$
$3y - 25 = -4x$

6. $3x - 5y = 0$
$2x = 3y + 3$

7. $\frac{x}{2} = \frac{7}{6} - \frac{y}{3}$ *LCD*
$\frac{2x}{3} = \frac{3y}{2} - \frac{7}{3}$

8. $\frac{3x}{4} = 4 + \frac{y}{2}$
$\frac{x}{3} + \frac{5y}{4} = -\frac{7}{6}$

9. $4.12x - 3.73y = 21.6$
$6.54x + 1.71y = 26.76$

10. $2.01x + 5.27y = 20.32$
$7.93x = 4.68y - 47.19$

(elimination)

Use the addition method to solve each system. Check your answers. (Hint for Exercises 27–30: Let $1/x = t$, $1/y = u$.) See Examples 2–4.

11. $x + y = 9$
$2x - y = 0$

12. $4x + y = 9$
$3x - y = 5$

13. $5x + 3y = 7$
$7x - 3y = -19$

14. $2x + 7y = -8$
$-2x + 3y = -12$

15. $3x + 2y = 5$
$6x + 4y = 8$

16. $9x - 5y = 1$
$-18x + 10y = 1$

17. $2x - 3y = -7$
$5x + 4y = 17$

18. $4x + 3y = -1$
$2x + 5y = 3$

19. $5x + 7y = 6$
$10x - 3y = 46$

20. $12x - 5y = 9$
$3x - 8y = -18$

21. $4x - y = 9$
$-8x + 2y = -18$

22. $3x + 5y + 2 = 0$
$9x + 15y + 6 = 0$

23. $\frac{x}{2} + \frac{y}{3} = 8$
$\frac{2x}{3} + \frac{3y}{2} = 17$

24. $\frac{x}{5} + 3y = 31$
$2x - \frac{y}{5} = 8$

25. $\frac{3x}{2} - \frac{y}{3} = 5$
$\frac{5x}{2} + \frac{2y}{3} = 12$

26. $\frac{4x}{5} + \frac{y}{4} = -2$
$\frac{x}{5} + \frac{y}{8} = 0$

L.C.D. get rid of fractions

27. $\dfrac{2}{x} + \dfrac{1}{y} = \dfrac{3}{2}$

$\dfrac{3}{x} - \dfrac{1}{y} = 1$

28. $\dfrac{1}{x} + \dfrac{3}{y} = \dfrac{16}{5}$

$\dfrac{5}{x} + \dfrac{4}{y} = 5$

29. $\dfrac{2}{x} + \dfrac{1}{y} = 11$

$\dfrac{3}{x} - \dfrac{5}{y} = 10$

30. $\dfrac{2}{x} + \dfrac{3}{y} = 18$

$\dfrac{4}{x} - \dfrac{5}{y} = -8$

31. $.05x - .02y = -.18$
$.04x + .06y = -.22$

32. $.08x + .03y = -.24$
$.04x - .01y = -.32$

33. $.6x + .3y = .087$
$.5x - .4y = .378$

34. $.7x - .5y = -.884$
$.1x + .3y = .13$

Write each problem as a system of equations and then solve. See Example 5.

35. Find values of a and b so that the line with equation $ax + by = 5$ passes through the points $(-2, 1)$ and $(-1, -2)$.

36. Find m and b so that the line $y = mx + b$ passes through $(4, 6)$ and $(-5, -3)$.

37. Find m and b so that the line $y = mx + b$ passes through $(2, 5)$ and $(-1, 4)$.

38. Find values of a and b so that the line with equation $ax + by = 23$ passes through $(-3, 2)$ and $(-4, -5)$.

39. A company borrowed money from two sources to develop a new product. One loan was at 12% interest, and the other was for half as much money at 11% interest. Find the amount of each loan if the total annual interest was $2800.

40. Ed invests $18,000, the profit from a real estate sale, in two ways. Part of the money is invested in certificates of deposit at 7.5% annual interest. With the rest of the money, he buys bonds that pay 9% interest per year. The two investments earn annual interest of $1507.50. How much was invested in each way?

41. A plant food contains two chemicals, A and B. Seven hundred kilograms of these chemicals will be used to make a batch of the plant food. The two chemicals must be in the ratio of 4 to 3 (by weight). How much of each chemical should be used?

42. A television manufacturer must ship 100 televisions to its two Michigan warehouses. It costs $42 to ship a television to warehouse A and $30 to ship a television to Warehouse B. The manufacturer has allotted $3408 for shipping costs. How many televisions should be shipped to each warehouse?

43. Octane ratings show the percent of isooctane in gasoline. An octane rating of 98, for example, indicates a gasoline that is 98% isooctane. How many gallons of 98-octane gasoline should be mixed with 92-octane gasoline to produce 40 gallons of 94-octane gasoline?

44. A pharmacist needs 5 liters of a 24% alcohol solution. She has on hand 30% alcohol solution and 18% alcohol solution. How many liters of each should be mixed to get the required solution?

45. At the Mustang Ranch, 6 goats and 5 sheep sell for $305, while 2 goats and 9 sheep cost $285. Find the cost of a goat and a sheep.

46. Linda Ramirez is a building contractor. If she hires 7 day laborers and 2 concrete finishers, her payroll for the day is $692, while 1 day laborer and 5 concrete finishers cost $476. Find the daily wage charge of each type of worker.

47. During summer vacation Hector and Ann earned a total of $2176. Hector worked 8 days less than Ann and earned $4 per day less. Find the number of days he worked and the daily wage made, if the total number of days worked by both was 72.

48. A bank teller has ten-dollar bills and twenty-dollar bills. He has 25 more twenties than tens. The value of the bills is $2900. How many of each kind does he have?

49. Mike Karelius plans to invest $30,000 he won in a lottery. With part of the money he buys a mutual fund, paying 9% a year. The rest he invests in utility bonds paying 10% per year. The first year his investments bring a return of $2820. How much is invested at each rate?

50. How much lowfat milk that is 3% butterfat should be mixed with milk that is 18% butterfat to get 25 gal of a lowfat milk that is 4.8% butterfat?

The break-even point for a company is the point where its costs equal its revenues. If both cost and revenue are expressed as linear equations, the break-even point is the solution of a linear system. In each of the following exercises, C represents the cost to produce x items, and R represents the revenue from the sale of x items. Use the substitution method to find the break-even point in each case, that is, the point where C = R.

51. $C = 1.5x + 252$
  $R = 5.5x$

52. $C = 2.5x + 90$
  $R = 3x$

53. $C = 20x + 10,000$
  $R = 30x - 11,000$

54. $C = 4x + 125$
  $R = 9x - 200$

55. $C = 18x + 240$
  $R = 20x - 500$

56. $C = 4x + 48$
  $R = 6x - 80$

In each of the following exercises p is the price of an item, while x represents the supply in one equation and the demand in the other. Find the equilibrium price and the equilibrium supply/demand. See Example 6.

57. $p = 80 - \dfrac{3}{5}x$

  $p = \dfrac{2}{5}x$

58. $p = 630 - \dfrac{3}{4}x$

  $p = \dfrac{3}{4}x$

59. $3p \quad\ = 84 - 2x$
  $3p - x = 0$

60. $4p + x = 80$
  $3p - 2x = 5$

61. $5p + 3x = 75$
  $3x - 10p = 0$

62. $3p + 2x = 315$
  $5x - 6p = 0$

63. Suppose that the demand and price for a certain model of electric can opener are related by

$$p = 16 - \frac{5}{4}x,$$

where p is price and x is demand, in appropriate units. Find the price when the demand is at the following levels.

(a) 0 units   **(b)** 4 units   **(c)** 8 units

Find the demand for the electric can opener at the following prices.

(d) $6   **(e)** $11   **(f)** $16

(g) Graph $p = 16 - (5/4)x$.

Suppose that the price and supply of the item above are related by

$$p = \frac{3}{4}x,$$

where x represents the supply and p the price. Find the supply at the following prices.

(h) $0   **(i)** $10   **(j)** $20

(k) Graph $p = (3/4)x$ on the same axes used for part (g).

(l) Find the equilibrium supply.

(m) Find the equilibrium price.

64. Let the supply and demand equations for strawberry-flavored licorice be

$$\text{supply: } p = \frac{3}{2}x \quad \text{and} \quad \text{demand: } p = 81 - \frac{3}{4}x.$$

(a) Graph these on the same axes.

(b) Find the equilibrium demand.

(c) Find the equilibrium price.

65. Let the supply and demand equations for butter pecan ice cream be given by

$$\text{supply: } p = \frac{2}{5}x \quad \text{and} \quad \text{demand: } p = 100 - \frac{2}{5}x.$$

(a) Graph these on the same axes.

(b) Find the equilibrium demand.

(c) Find the equilibrium price.

66. Let the supply and demand equations for sugar be given by

$$\text{supply: } p = 1.4x - .6 \quad \text{and} \quad \text{demand: } p = -2x + 3.2.$$

(a) Graph these on the same axes.

(b) Find the equilibrium demand.

(c) Find the equilibrium price.

7.2 Linear Systems with Three Variables

A solution of a linear equation $ax + by + cz = k$ with three variables is an **ordered triple** (x, y, z). For example, $(1, 2, -4)$ is a solution of $2x + 5y - 3z = 24$. The solution set of such an equation is an infinite set of ordered triples. It is shown in geometry that the graph of a linear equation in three variables is a plane in three-dimensional space. Considering the possible intersections of the planes representing three equations in three unknowns shows that the solution set of such a system may be either a single ordered triple (x, y, z), an infinite set of ordered triples (dependent equations), or the empty set (an inconsistent system). See Figure 7.3 on the following page.

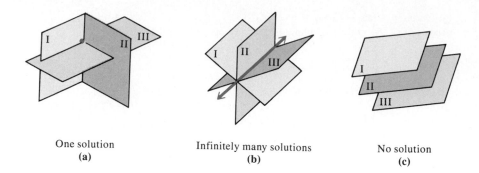

One solution
(a)

Infinitely many solutions
(b)

No solution
(c)

Figure 7.3

A system of equations with three or more variables can be solved with repeated use of the addition method, as shown in the next example.

Example 1   Solve the system

$$2x + y - z = 2 \tag{1}$$
$$x + 3y + 2z = 1 \tag{2}$$
$$x + y + z = 2. \tag{3}$$

Several steps are needed to obtain an equation in one variable. First, eliminate the same variable from each of two pairs of equations. Then eliminate another variable from the two resulting equations. For example, to eliminate x, multiply both sides of equation (2) by -2 and add to equation (1).

$$
\begin{array}{r}
2x + y - z = 2 \\
-2x - 6y - 4z = -2 \\
\hline
-5y - 5z = 0
\end{array}
\tag{4}
$$

The variable x must be eliminated again from a different pair of equations, say (2) and (3). Multiply both sides of (2) by -1, then add the result to equation (3).

$$
\begin{array}{r}
-x - 3y - 2z = -1 \\
x + y + z = 2 \\
\hline
-2y - z = 1
\end{array}
\tag{5}
$$

Now solve the system formed by the two equations (4) and (5). To eliminate z, multiply both sides of equation (4) by -1 and both sides of equation (5) by 5 and add.

$$
\begin{array}{r}
5y + 5z = 0 \\
-10y - 5z = 5 \\
\hline
-5y \quad\quad = 5 \\
y = -1
\end{array}
$$

Substitute y into equation (5) to find z. (Equation (4) could have been used.)

$$-2(-1) - z = 1$$
$$2 - z = 1$$
$$-z = -1$$
$$z = 1$$

Now use equation (3) and the values for y and z to find x. (Either equation (1) or (2) could have been used.)

$$x + y + z = 2$$
$$x - 1 + 1 = 2$$
$$x = 2$$

The solution set of the system is $\{(2, -1, 1)\}$. ✤

Example 2 Solve the system

$$x + 2y + z = 4 \qquad \textbf{(6)}$$
$$3x - y - 4z = -9. \qquad \textbf{(7)}$$

Geometrically, the solution is the intersection of the two planes given by equations (6) and (7). The intersection of two different nonparallel planes is a line. Thus there will be an infinite number of ordered triples in the solution set, representing the points on the line of intersection. To describe these ordered triples, proceed as follows.

To eliminate x, multiply both sides of equation (6) by -3 and add to equation (7). (Either y or z could have been eliminated instead.)

$$-3x - 6y - 3z = -12$$
$$\underline{3x - y - 4z = -9}$$
$$-7y - 7z = -21 \qquad \textbf{(8)}$$

Now solve equation (8) for z.

$$-7y - 7z = -21$$
$$-7z = 7y - 21$$
$$z = -y + 3$$

This gives z in terms of y. Now, express x also in terms of y by solving equation (6) for x and substituting $-y + 3$ for z in the result.

$$x + 2y + z = 4$$
$$x = -2y - z + 4$$
$$x = -2y - (-y + 3) + 4$$
$$x = -y + 1$$

The system has an infinite number of solutions. For any value of y, the value of z is given by $-y + 3$ and x equals $-y + 1$. For example, if $y = 1$, then $x = -1 + 1 = 0$ and $z = -1 + 3 = 2$, giving the solution $(0, 1, 2)$. Verify that another solution is $(-1, 2, 1)$. ✣

Had equation (8) in Example 2 been solved for y instead of z, the solution would have had a different form but would have led to the same set of solutions. The form of the solution set found above, $\{(-y + 1, y, -y + 3)\}$, is said to have y **arbitrary.**

A system like the one in Example 2 occurs when two of the equations in a system of three equations with three variables are dependent—that is, when one is a multiple of the other. In such a case, there are really only two equations in three variables, and Example 2 illustrates the method of solution. On the other hand, an inconsistent system is indicated by a false statement at some point in the solution, as in Example 3 of Section 7.1.

Applications with three unknowns often require solving systems of three linear equations. As an example, the equation of the parabola $y = ax^2 + bx + c$ that passes through three given points can be found by solving a system of three equations with three unknowns.

Example 3   Find the equation of the parabola $y = ax^2 + bx + c$ that passes through $(2, 4)$, $(-1, 1)$, and $(-2, 5)$.

Since the three points lie on the graph of the equation $y = ax^2 + bx + c$, they must satisfy the equation. Substituting each ordered pair into the equation gives three equations with three variables.

$$4 = a(2)^2 + b(2) + c \qquad \text{or} \qquad 4 = 4a + 2b + c \qquad \textbf{(9)}$$
$$1 = a(-1)^2 + b(-1) + c \qquad \text{or} \qquad 1 = a - b + c \qquad \textbf{(10)}$$
$$5 = a(-2)^2 + b(-2) + c \qquad \text{or} \qquad 5 = 4a - 2b + c \qquad \textbf{(11)}$$

This system can be solved by the addition method. First eliminate c using equations (9) and (10). Then eliminate c using equations (10) and (11).

$$
\begin{array}{ll}
4 = 4a + 2b + c & 1 = a - b + c \\
\underline{-1 = -a + b - c} & \underline{-5 = -4a + 2b - c} \\
3 = 3a + 3b & -4 = -3a + b
\end{array}
$$

Now solve the system of two equations in two variables by eliminating a.

$$
\begin{array}{l}
3 = 3a + 3b \\
\underline{-4 = -3a + b} \\
-1 = \ 4b \\
-\dfrac{1}{4} = b
\end{array}
$$

Find a by substituting $-1/4$ for b in $3 = 3a + 3b$, which is equivalent to $1 = a + b$.

$$1 = a + b$$

$$1 = a - \frac{1}{4}$$

$$\frac{5}{4} = a$$

Finally, find c by substituting $a = 5/4$ and $b = -1/4$ in equation (10).

$$1 = a - b + c$$

$$1 = \frac{5}{4} - \left(-\frac{1}{4}\right) + c$$

$$1 = \frac{6}{4} + c = \frac{3}{2} + c$$

$$-\frac{1}{2} = c$$

The equation of the parabola is

$$y = \frac{5}{4}x^2 - \frac{1}{4}x - \frac{1}{2} \quad \text{or} \quad 4y = 5x^2 - x - 2. \quad \clubsuit$$

7.2 Exercises

Use the addition method to solve each of the following systems of three equations in three unknowns. Check your answers. See Example 1.

1. $x + y + z = 2$
$2x + y - z = 5$
$x - y + z = -2$

2. $2x + y + z = 9$
$-x - y + z = 1$
$3x - y + z = 9$

3. $x + 3y + 4z = 14$
$2x - 3y + 2z = 10$
$3x - y + z = 9$

4. $4x - y + 3z = -2$
$3x + 5y - z = 15$
$-2x + y + 4z = 14$

5. $x + 2y + 3z = 8$
$3x - y + 2z = 5$
$-2x - 4y - 6z = 5$

6. $3x - 2y - 8z = 1$
$9x - 6y - 24z = -2$
$x - y + z = 1$

7. $x + 4y - z = 6$
$2x - y + z = 3$
$3x + 2y + 3z = 16$

8. $4x - 3y + z = 9$
$3x + 2y - 2z = 4$
$x - y + 3z = 5$

9. $5x + y - 3z = -6$
$2x + 3y + z = 5$
$-3x - 2y + 4z = 3$

10. $2x - 5y + 4z = -35$
$5x + 3y - z = 1$
$x + y + z = 1$

11. $x - 3y - 2z = -3$
$3x + 2y - z = 12$
$-x - y + 4z = 3$

12. $x + y + z = 3$
$3x - 3y - 4z = -1$
$x + y + 3z = 11$

13. $2x + 6y - z = 6$
$4x - 3y + 5z = -5$
$6x + 9y - 2z = 11$

14. $8x - 3y + 6z = -2$
$4x + 9y + 4z = 18$
$12x - 3y + 8z = -2$

15. $\dfrac{1}{x} + \dfrac{1}{y} - \dfrac{1}{z} = \dfrac{1}{4}$
$\dfrac{2}{x} - \dfrac{1}{y} + \dfrac{3}{z} = \dfrac{9}{4}$
$-\dfrac{1}{x} - \dfrac{2}{y} + \dfrac{4}{z} = 1$

16. $\dfrac{3}{x} + \dfrac{2}{y} - \dfrac{1}{z} = \dfrac{11}{6}$

$\dfrac{1}{x} - \dfrac{1}{y} + \dfrac{3}{z} = -\dfrac{11}{12}$

$\dfrac{2}{x} + \dfrac{1}{y} + \dfrac{1}{z} = \dfrac{7}{12}$

17. $\dfrac{2}{x} - \dfrac{2}{y} + \dfrac{1}{z} = -1$

$\dfrac{4}{x} + \dfrac{1}{y} - \dfrac{2}{z} = -9$

$\dfrac{1}{x} + \dfrac{1}{y} - \dfrac{3}{z} = -9$

18. $\dfrac{5}{x} - \dfrac{1}{y} - \dfrac{2}{z} = -6$

$-\dfrac{1}{x} + \dfrac{3}{y} - \dfrac{3}{z} = -12$

$\dfrac{2}{x} - \dfrac{1}{y} - \dfrac{1}{z} = 6$

19. $1.2x + 3y - 4.8z = -5.28$
$-3x + 5.8y + 2.4z = -4.16$
$.7x - .5y + .1z = .75$

20. $-.6x + 2.5y - 3.2z = -1.25$
$1.1x + .4y + .8z = 1.22$
$-.3x - .7y - .4z = -.91$

21. $5.3x - 4.7y + 5.9z = 1.14$
$-2.5x + 3.2y - 1.4z = 7.22$
$2.25x - 2.88y + 1.26z = 4.88$

22. $7.77x - 8.61y + 4.2z = 15.96$
$11.6x - 4.9y + .8z = 12.4$
$3.7x - 4.1y + 2z = 7.6$

Solve each of the following systems in terms of the arbitrary variable x. (Hint: Begin by eliminating either y or z.) See Example 2.

23. $x - 2y + 3z = 6$
$2x - y + 2z = 5$

24. $3x + 4y - z = 13$
$x + y + 2z = 15$

25. $5x - 4y + z = 9$
$x + y = 15$

26. $x - y + z = -6$
$4x + y + z = 7$

27. $3x - 5y - 4z = -7$
$y - z = -13$

28. $3x - 2y + z = 15$
$x + 4y - z = 11$

Solve the following word problems. For Exercises 29 and 30, see Example 3.

29. Find a, b, and c so that the graph of the equation $y = ax^2 + bx + c$ passes through the points $(2, 3)$, $(-1, 0)$, and $(-2, 2)$.

30. Find a, b, and c so that $(2, 14)$, $(0, 0)$, and $(-1, -1)$ lie on the graph of $y = ax^2 + bx + c$.

31. Find the equation of the circle $x^2 + y^2 + ax + by + c = 0$ that passes through the points $(2, 1)$, $(-1, 0)$, $(3, 3)$.

32. Find the equation of the circle $x^2 + y^2 + ax + by + c = 0$ that passes through the points $(4, 2)$, $(-5, -2)$, and $(0, 3)$.

33. A cashier has a total of 30 bills, made up of ones, fives, and twenties. The number of twenties is 9 more than the number of ones. The total value of the money is $351. How many of each type of bill are there?

34. A wine merchant wishes to get 300 liters of wine that sells for $3 per liter. She wishes to mix wines selling for $4.50, $1.50, and $2.25, respectively. She must use twice as much of the $2.25 wine as the $1.50 wine. How many liters of each should she use?

35. A glue company needs to make some glue that it can sell for $120 per barrel. It wants to use 150 barrels of glue worth $100 per barrel, along with some glue worth $150 per barrel, and glue worth $190 per barrel. It must use the same number of barrels of $150 and $190 glue. How much of the $150 and $190 glue will be needed? How many barrels of $120 glue will be produced?

36. The Rolling Rocks sell three kinds of concert tickets, "up close," "middle," and "farther back." "Up close" tickets cost $2 more than "middle" tickets, while "middle" tickets cost $1 more than "farther back" tickets. Twice the cost of an "up close" ticket is $1 more than 3 times the cost of a "farther back" seat. Find the price of each kind of ticket.

37. The perimeter of a triangle is 33 cm. The longest side is 3 cm longer than the medium side. The medium side is twice the shortest side. Find the length of each side of the triangle.

38. The sum of the angles of a triangle is 180°. The largest angle is 10° more than twice the smallest. The third angle is 50° less than the sum of the other two angles. Find the measures of each of the three angles.

39. Jay Davis wins $100,000 in a lottery. He invests part of the money in real estate with an annual return of 10% and another part in a money market account at 9% interest. He invests the third part, which amounts to $20,000 less than the sum of the other two parts, in certificates of deposit at 7.5%. The total annual interest on the money is $8900. How much was invested in each way?

40. Betsy Mackay invests $10,000, received from her grandmother, in three ways. With one part, she buys mutual funds which offer a return of 8% per year. The second part, which amounts to twice the first, is used to buy government bonds at 9% per year. She puts the rest in the bank at 5% annual interest. The first year her investments bring a return of $830. How much did she invest in each way?

7.3   Solution of Linear Systems by Matrices

The elimination method used to solve linear systems of equations in the last two sections can be streamlined to a systematic method by using *matrices* (singular, *matrix*). In this section we describe one way matrices are used to solve linear systems. Matrix methods are particularly suitable for computer solutions of large systems of equations having many unknowns.

To begin, consider a system of three equations and three unknowns such as

$$a_1x + b_1y + c_1z = d_1$$
$$a_2x + b_2y + c_2z = d_2$$
$$a_3x + b_3y + c_3z = d_3.$$

This system can be written in an abbreviated form as

$$\begin{bmatrix} a_1 & b_1 & c_1 & d_1 \\ a_2 & b_2 & c_2 & d_2 \\ a_3 & b_3 & c_3 & d_3 \end{bmatrix}$$

Such a rectangular array of numbers enclosed by brackets is called a **matrix.** Each number in the array is an **element** or **entry.** The constants in the last column of the matrix can be set apart from the coefficients of the variables by using a vertical line, as shown in the following **augmented matrix.**

$$\text{Rows} \begin{bmatrix} a_1 & b_1 & c_1 & | & d_1 \\ a_2 & b_2 & c_2 & | & d_2 \\ a_3 & b_3 & c_3 & | & d_3 \end{bmatrix}$$

Columns

The rows of this augmented matrix can be treated the same as the equations of a system of equations, since the augmented matrix is actually a short form of the system. Any transformation of the matrix that will result in an equivalent system is permitted. The following **matrix row operations** produce such transformations.

**Matrix Row
Operations**

> For any real number k and any augmented matrix of a system of linear equations, the following operations will produce the matrix of an equivalent system:
>
> 1. **interchanging any two rows of a matrix,**
> 2. **multiplying the elements of any row of a matrix by the same nonzero number k, and**
> 3. **adding a multiple of the elements of one row to the elements of another row.**

Example 1 **(a)** The first row operation is used to change the matrix

$$\begin{bmatrix} 1 & 3 & 5 \\ 0 & 1 & 2 \\ 1 & -1 & -2 \end{bmatrix} \quad \text{to} \quad \begin{bmatrix} 0 & 1 & 2 \\ 1 & 3 & 5 \\ 1 & -1 & -2 \end{bmatrix}$$

by interchanging the first two rows.

(b) Using the second row operation with $k = -2$ changes

$$\begin{bmatrix} 1 & 3 & 5 \\ 0 & 1 & 2 \\ 1 & -1 & -2 \end{bmatrix} \quad \text{to} \quad \begin{bmatrix} -2 & -6 & -10 \\ 0 & 1 & 2 \\ 1 & -1 & -2 \end{bmatrix},$$

where the elements of the first row of the original matrix were multiplied by -2.

(c) The third row operation is used to change

$$\begin{bmatrix} 1 & 3 & 5 \\ 0 & 1 & 2 \\ 1 & -1 & -2 \end{bmatrix} \quad \text{to} \quad \begin{bmatrix} 0 & 4 & 7 \\ 0 & 1 & 2 \\ 1 & -1 & -2 \end{bmatrix},$$

by multiplying each element in the third row of the original matrix by -1 and adding the results to the corresponding elements in the first row of that matrix. That is, the elements in the new first row were found as follows.

$$\begin{bmatrix} 1 + 1(-1) & 3 + (-1)(-1) & 5 + (-2)(-1) \\ 0 & 1 & 2 \\ 1 & -1 & -2 \end{bmatrix} = \begin{bmatrix} 0 & 4 & 7 \\ 0 & 1 & 2 \\ 1 & -1 & -2 \end{bmatrix}$$

Rows two and three were left unchanged.   ✜

If the word "row" is replaced by "equation," it can be seen that the three row operations also apply to a system of equations, so that a system of equations can be solved by transforming its corresponding matrix into the matrix of an equivalent, simpler system.

The **Gauss-Jordan method** is a systematic way of using the matrix row operations to solve a system of equations. The equations should all be in the same form, with variable terms in the same order on the left, and the constant term on the right. The next example illustrates the method.

Example 2   Use the Gauss-Jordan method to solve the linear system

$$3x - 4y = 1$$
$$5x + 2y = 19.$$

The equations are already in the proper form, so begin with the augmented matrix

$$\left[\begin{array}{cc|c} 3 & -4 & 1 \\ 5 & 2 & 19 \end{array}\right]$$

The goal is to transform this augmented matrix into one in which the value of the variables will be easy to see. That is, since each column in the matrix represents the coefficients of one variable, the matrix should be transformed to the form

$$\left[\begin{array}{cc|c} 1 & 0 & k \\ 0 & 1 & j \end{array}\right]$$

for real numbers k and j. Once the augmented matrix is in this form, the matrix can be rewritten as a linear system to get

$$x = k$$
$$y = j.$$

The necessary transformations are performed as follows. It is best to work in columns, beginning in each column with the element that is to become 1. In the augmented matrix,

$$\left[\begin{array}{cc|c} 3 & -4 & 1 \\ 5 & 2 & 19 \end{array}\right],$$

there is a 3 in the first row, first column position. To get a 1 in this position, use operation 2 and multiply each entry in the first row by 1/3. (We abbreviate this as (1/3)R1.)

$$\begin{bmatrix} 1 & -4/3 & \bigm| & 1/3 \\ 5 & 2 & \bigm| & 19 \end{bmatrix} \quad \frac{1}{3}\,R1$$

To get 0 in the second-row, first-column position, multiply each element of the first row by -5 and add the result to the corresponding element in the second row, using operation 3.

$$\begin{bmatrix} 1 & -4/3 & \bigm| & 1/3 \\ 0 & 26/3 & \bigm| & 52/3 \end{bmatrix} \quad -5R1 + R2$$

To get 1 in the second row, second column, multiply each element of the second row by 3/26, using operation 2.

$$\begin{bmatrix} 1 & -4/3 & \bigm| & 1/3 \\ 0 & 1 & \bigm| & 2 \end{bmatrix} \quad \frac{3}{26}\,R2$$

Finally, get 0 in the first row, second column by multiplying each element of the second row by 4/3 and adding the result to the corresponding element in the first row.

$$\begin{bmatrix} 1 & 0 & \bigm| & 3 \\ 0 & 1 & \bigm| & 2 \end{bmatrix} \quad \frac{4}{3}\,R2 + R1$$

This last matrix leads to the system

$$x = 3$$
$$y = 2,$$

which gives the solution set $\{(3, 2)\}$. This solution could have been read directly from the third column of the final matrix. ✛

Example 3   Use the Gauss-Jordan method to solve the system

$$x - y + 5z = -6$$
$$3x + 3y - z = 10$$
$$x + 3y + 2z = 5.$$

Since the system is in proper form, begin by writing the augmented matrix of the linear system.

$$\begin{bmatrix} 1 & -1 & 5 & \bigm| & -6 \\ 3 & 3 & -1 & \bigm| & 10 \\ 1 & 3 & 2 & \bigm| & 5 \end{bmatrix}$$

The final matrix is to be of the form

$$\begin{bmatrix} 1 & 0 & 0 & \bigm| & m \\ 0 & 1 & 0 & \bigm| & n \\ 0 & 0 & 1 & \bigm| & p \end{bmatrix},$$

where m, n, and p are real numbers. This final form of the matrix gives the solution: $x = m$, $y = n$, and $z = p$.

There is already a 1 in the first row, first column. To get a 0 in the second row of the first column, multiply each element in the first row by -3 and add the result to the corresponding element in the second row, using operation 3.

$$\begin{bmatrix} 1 & -1 & 5 & | & -6 \\ 0 & 6 & -16 & | & 28 \\ 1 & 3 & 2 & | & 5 \end{bmatrix} \quad -3R1 + R2$$

Now, change the last element in the first column to 0 by using operation 3 and multiplying each element of the first row by -1, then adding the results to the corresponding elements of the third row.

$$\begin{bmatrix} 1 & -1 & 5 & | & -6 \\ 0 & 6 & -16 & | & 28 \\ 0 & 4 & -3 & | & 11 \end{bmatrix} \quad -1R1 + R3$$

The same procedure is used to transform the second and third columns. For both of these columns, perform the additional step of getting 1 in the appropriate position of each column.

$$\begin{bmatrix} 1 & -1 & 5 & | & -6 \\ 0 & 1 & -8/3 & | & 14/3 \\ 0 & 4 & -3 & | & 11 \end{bmatrix} \quad \frac{1}{6}R2$$

$$\begin{bmatrix} 1 & 0 & 7/3 & | & -4/3 \\ 0 & 1 & -8/3 & | & 14/3 \\ 0 & 4 & -3 & | & 11 \end{bmatrix} \quad R2 + R1$$

$$\begin{bmatrix} 1 & 0 & 7/3 & | & -4/3 \\ 0 & 1 & -8/3 & | & 14/3 \\ 0 & 0 & 23/3 & | & -23/3 \end{bmatrix} \quad -4R2 + R3$$

$$\begin{bmatrix} 1 & 0 & 7/3 & | & -4/3 \\ 0 & 1 & -8/3 & | & 14/3 \\ 0 & 0 & 1 & | & -1 \end{bmatrix} \quad \frac{3}{23}R3$$

$$\begin{bmatrix} 1 & 0 & 0 & | & 1 \\ 0 & 1 & -8/3 & | & 14/3 \\ 0 & 0 & 1 & | & -1 \end{bmatrix} \quad -\frac{7}{3}R3 + R1$$

$$\begin{bmatrix} 1 & 0 & 0 & | & 1 \\ 0 & 1 & 0 & | & 2 \\ 0 & 0 & 1 & | & -1 \end{bmatrix} \quad \frac{8}{3}R3 + R2$$

The linear system associated with this final matrix is

$$x = 1$$
$$y = 2$$
$$z = -1,$$

and the solution set is $\{(1, 2, -1)\}$. ✚

Example 4 Use the Gauss-Jordan method to solve the system

$$x + y = 2$$
$$2x + 2y = 5.$$

Write the augmented matrix,

$$\begin{bmatrix} 1 & 1 & | & 2 \\ 2 & 2 & | & 5 \end{bmatrix}$$

Multiply the elements in the first row by -2 and add the result to the corresponding elements in the second row.

$$\begin{bmatrix} 1 & 1 & | & 2 \\ 0 & 0 & | & 1 \end{bmatrix} \qquad -2R1 + R2$$

The next step would be to get a 1 in the second-row, second-column position. Because of the zeros, it is impossible to go further. Since the second row corresponds to the equation

$$0x + 0y = 1,$$

which has no solution, the system is inconsistent. ✚

Example 5 Use the Gauss-Jordan method to solve

$$3x + 4y = 7$$
$$-6x - 8y = -14.$$

The augmented matrix is

$$\begin{bmatrix} 3 & 4 & | & 7 \\ -6 & -8 & | & -14 \end{bmatrix}$$

Multiply the first row by 1/3 to get 1 in the first column of the first row.

$$\begin{bmatrix} 1 & 4/3 & | & 7/3 \\ -6 & -8 & | & -14 \end{bmatrix} \qquad \frac{1}{3}R1$$

Multiply the first row by 6 and add the results to the second row entries to get 0 in the first-column, second-row position.

$$\begin{bmatrix} 1 & 4/3 & | & 7/3 \\ 0 & 0 & | & 0 \end{bmatrix} \quad 6R1 + R2$$

Since the second row contains only 0s, which indicates a dependent system, the method cannot be continued. The corresponding system of equations

$$x + \frac{4}{3}y = \frac{7}{3}$$

$$0x + 0y = 0$$

reduces to the single equation $x + (4/3)y = 7/3$. Solve this equation for x:

$$x + \frac{4}{3}y = \frac{7}{3}$$

$$x = \frac{7}{3} - \frac{4}{3}y \quad \text{or} \quad x = \frac{7 - 4y}{3}.$$

The solution is written with y arbitrary as

$$\left\{ \left(\frac{7 - 4y}{3}, y \right) \right\}. \quad \clubsuit$$

Example 6   Use the Gauss-Jordan method to solve the system

$$2x - 5y + 3z = 1$$
$$x + 4y - 2z = 8.$$

Recall from Section 7.2 that a system with two equations and three variables has an infinite number of solutions. The Gauss-Jordan method can be used to indicate the solution with an arbitrary variable. Start with the augmented matrix

$$\begin{bmatrix} 2 & -5 & 3 & | & 1 \\ 1 & 4 & -2 & | & 8 \end{bmatrix}.$$

Exchange rows to get a 1 in the first-row, first-column position.

$$\begin{bmatrix} 1 & 4 & -2 & | & 8 \\ 2 & -5 & 3 & | & 1 \end{bmatrix}$$

Now multiply each element in the first row by -2 and add to the corresponding element in the second row.

$$\begin{bmatrix} 1 & 4 & -2 & | & 8 \\ 0 & -13 & 7 & | & -15 \end{bmatrix} \quad -2R1 + R2$$

Multiply each element in the second row by $-1/13$.

$$\begin{bmatrix} 1 & 4 & -2 & | & 8 \\ 0 & 1 & -7/13 & | & 15/13 \end{bmatrix} \quad -\frac{1}{13}R2$$

Multiply each element in the second row by -4 and add to the corresponding element in the first row.

$$\begin{bmatrix} 1 & 0 & 2/13 & | & 44/13 \\ 0 & 1 & -7/13 & | & 15/13 \end{bmatrix} \quad -4R2 + R1$$

It is not possible to continue with the Gauss-Jordan method. The equations that correspond to the final matrix are

$$x + \frac{2}{13}z = \frac{44}{13} \quad \text{and} \quad y - \frac{7}{13}z = \frac{15}{13}.$$

Solve these equations for x and y, respectively.

$$x + \frac{2}{13}z = \frac{44}{13} \qquad\qquad y - \frac{7}{13}z = \frac{15}{13}$$

$$x = \frac{44}{13} - \frac{2}{13}z \qquad\qquad y = \frac{15}{13} + \frac{7}{13}z$$

$$x = \frac{44 - 2z}{13} \qquad\qquad y = \frac{15 + 7z}{13}$$

The solution can now be written with z arbitrary as

$$\left\{ \left(\frac{44 - 2z}{13}, \frac{15 + 7z}{13}, z \right) \right\}. \quad \clubsuit$$

Although only examples with two or three variables have been shown, the Gauss-Jordan method can be extended to solve linear systems with more variables. As the number of variables increases, the process quickly becomes very tedious and the opportunity for error increases. The method is suitable for use by computers, however, and a fairly large system can be solved quickly in that way.

7.3 Exercises

Use the third row operation to change each of the following matrices as indicated. See Example 1.

1. $\begin{bmatrix} 2 & 4 \\ 4 & 7 \end{bmatrix}$; -2 times row 1 added to row 2

2. $\begin{bmatrix} -1 & 4 \\ 7 & 0 \end{bmatrix}$; 7 times row 1 added to row 2

3. $\begin{bmatrix} 1 & 1/5 \\ 5 & 2 \end{bmatrix}$; -5 times row 1 added to row 2

4. $\begin{bmatrix} 5 & 7 \\ 2 & -4 \end{bmatrix}$; 4/7 times row 1 added to row 2

5. $\begin{bmatrix} 1 & 5 & 6 \\ -2 & 3 & -1 \\ 4 & 7 & 0 \end{bmatrix}$; 2 times row 1 added to row 2

6. $\begin{bmatrix} 2 & 5 & 6 \\ 4 & -1 & 2 \\ 3 & 7 & 1 \end{bmatrix}$; -6 times row 3 added to row 1

7. $\begin{bmatrix} -3 & 1 & -4 \\ 2 & 1 & 3 \\ -7 & 5 & 2 \end{bmatrix}$; -5 times row 2 added to row 3

8. $\begin{bmatrix} 4 & 10 & -8 \\ 7 & 4 & 3 \\ -1 & 1 & 0 \end{bmatrix}$; -4 times row 3 added to row 2

Write the augmented matrix for each of the following systems. Do not solve the system.

9. $2x + 3y = 11$
$\quad x + 2y = 8$

10. $3x + 5y = -13$
$\quad\; 2x + 3y = -9$

11. $x + 5y = 6$
$\quad x + 2y = 8$

12. $2x + 7y = 1$
$\qquad\quad 5x = -15$

13. $2x + \;y + \;z = 3$
$\quad 3x - 4y + 2z = -7$
$\quad\; x + \;y + \;z = 2$

14. $4x - 2y + 3z = 4$
$\quad\; 3x + 5y + \;z = 7$
$\quad\; 5x - \;y + 4z = 7$

15. $\;x + y = 2$
$\quad 2y + z = -4$
$\qquad\quad z = 2$

16. $\qquad x = 6$
$\quad\; y + 2z = 2$
$\quad x - 3z = 6$

17. $x = 5$
$\quad y = -2$
$\quad z = 3$

18. $\quad\; x = 8$
$\quad y + z = 6$
$\qquad z = 2$

Write the system of equations associated with each of the following augmented matrices.
Do not try to solve.

19. $\begin{bmatrix} 1 & 0 & | & 2 \\ 0 & 1 & | & 3 \end{bmatrix}$

20. $\begin{bmatrix} 1 & 0 & | & 5 \\ 0 & 1 & | & -3 \end{bmatrix}$

21. $\begin{bmatrix} 2 & 1 & | & 1 \\ 3 & -2 & | & -9 \end{bmatrix}$

22. $\begin{bmatrix} 1 & -5 & | & -18 \\ 6 & 2 & | & 20 \end{bmatrix}$

23. $\begin{bmatrix} 1 & 0 & 0 & | & 2 \\ 0 & 1 & 0 & | & 3 \\ 0 & 0 & 1 & | & -2 \end{bmatrix}$

24. $\begin{bmatrix} 1 & 0 & 1 & | & 4 \\ 0 & 1 & 0 & | & 2 \\ 0 & 0 & 1 & | & 3 \end{bmatrix}$

25. $\begin{bmatrix} 3 & 2 & 1 & | & 1 \\ 0 & 2 & 4 & | & 22 \\ -1 & -2 & 3 & | & 15 \end{bmatrix}$

26. $\begin{bmatrix} 2 & 1 & 3 & | & 12 \\ 4 & -3 & 0 & | & 10 \\ 5 & 0 & -4 & | & -11 \end{bmatrix}$

Use the Gauss-Jordan method to solve each of the following systems of equations. See
Examples 2–5.

27. $x + y = 5$
$\quad x - y = -1$

28. $\;x + 2y = 5$
$\quad 2x + \;y = -2$

29. $\;x + \;y = -3$
$\quad 2x - 5y = -6$

30. $3x - 2y = 4$
$\quad 3x + \;y = -2$

31. $2x - 3y = 10$
$\quad 2x + 2y = 5$

32. $4x + y = 5$
$\quad 2x + y = 3$

33. $2x - 5y = 10$
$\quad 4x - 5y = 15$

34. $\quad 4x - \;y = 3$
$\quad -2x + 3y = 1$

35. $2x - 3y = 2$
$\quad 4x - 6y = 1$

36. $\;x + 2y = 1$
$\quad 2x + 4y = 3$

37. $\quad 6x - 3y = 1$
$\quad -12x + 6y = -2$

38. $\quad x - y = 1$
$\quad -x + y = -1$

39. $x + y \qquad = -1$
$\qquad\; y + z = 4$
$\quad x \qquad + z = 1$

40. $x \qquad - z = -3$
$\qquad y + z = 9$
$\quad x \qquad + z = 7$

41. $\;x + \;y - \;z = 6$
$\quad 2x - \;y + \;z = -9$
$\quad\; x - 2y + 3z = 1$

42. $\;x + 3y - 6z = 7$
$\quad 2x - \;y + 2z = 0$
$\quad\; x + \;y + 2z = -1$

43. $-x + y \qquad = -1$
$\qquad\quad y - z = 6$
$\quad\;\; x \qquad + z = -1$

44. $\;x + y \qquad = 1$
$\quad 2x \qquad - z = 0$
$\qquad\; y + 2z = -2$

45. $2x - \;y + 3z = 0$
$\quad\; x + 2y - \;z = 5$
$\qquad\; 2y + \;z = 1$

46. $\quad 4x + 2y - 3z = 6$
$\quad\;\; x - 4y + \;z = -4$
$\quad -x \qquad + 2z = 2$

47.
$$x - 2y + z = 5$$
$$2x + y - z = 2$$
$$-2x + 4y - 2z = 2$$

48.
$$3x + 5y - z = 4$$
$$4x - y + 2z = 1$$
$$-6x - 10y + 2z = 0$$

Use the Gauss-Jordan method to solve each of the following systems.

49.
$$3x + 2y - w = 0$$
$$2x + z + 2w = 5$$
$$x + 2y - z = -2$$
$$2x - y + z + w = 2$$

50.
$$x + 3y - 2z - w = 9$$
$$4x + y + z + 2w = 2$$
$$-3x - y + z - w = -5$$
$$x - y - 3z - 2w = 2$$

51.
$$3.5x + 2.9y = 12.91$$
$$1.7x - 3.8y = -6.23$$

52.
$$-4.3x + 1.1y = 10.1$$
$$4.8x - 2.7y = -17.31$$

53.
$$9.1x + 2.3y - z = -3.06$$
$$4.7x + y - 3.8z = -21.15$$
$$5.1y - 4.7z = -31.77$$

54.
$$7.2x + 4.8y + 3.1z = -14.96$$
$$3x - y + 11.5z = -15.6$$
$$1.1x + 2.4y = -1.24$$

Solve each of the following word problems by first setting up a system of equations. Use the Gauss-Jordan method.

55. Juanita deposits some money in a bank account paying 8% per year. She uses some additional money, amounting to 1/3 the amount placed in the bank, to buy bonds paying 5% per year. With the balance of her funds she buys a 9% certificate of deposit. The first year her investments bring a return of $830. If the total of the investments is $10,000, how much is invested at each rate?

56. To get the necessary funds for a planned expansion, a small company took out three loans totaling $25,000. The company was able to borrow some of the money at 8%. They borrowed $2000 more than 1/2 the amount of the 8% loan at 10%, and the rest at 9%. The total annual interest was $2220. How much did they borrow at each rate?

57. One day a service station sold 400 gal of unleaded gasoline, 150 gal of regular gasoline, and 130 gal of premium gasoline for a total of $842.90. The next day 380 gal of unleaded, 170 gal of regular, and 150 gal of premium were sold for $866.90. The difference in price per gallon of unleaded and regular is three times the difference in price per gallon of premium and regular. How much does this station charge for each type of gasoline?

58. A chemist has three solutions containing acid: some 5% solution, some 15% solution, and some 25% solution. She needs to mix some of each to get 50 liters of 20% solution. Since she has a lot of the 5% solution, she wants to use twice as much of that as the 15% solution. How much of each solution should she use?

Solve each of the following systems by the Gauss-Jordan method. Let z be the arbitrary variable. See Example 6.

59.
$$x - 3y + 2z = 10$$
$$2x - y - z = 8$$

60.
$$3x + y - z = 12$$
$$x + 2y + z = 10$$

61.
$$x + 2y - z = 0$$
$$3x - y + z = 6$$
$$-2x - 4y + 2z = 0$$

62.
$$3x + 5y - z = 0$$
$$4x - y + 2z = 1$$
$$-6x - 10y + 2z = 0$$

63.
$$x - 2y + z = 5$$
$$-2x + 4y - 2z = 2$$
$$2x + y - z = 2$$

64.
$$3x + 6y - 3z = 12$$
$$-x - 2y + z = 16$$
$$x + y - 2z = 20$$

65. At rush hours, substantial traffic congestion is encountered at the traffic intersections shown in the figure.

The city wishes to improve the signals at these corners so as to speed the flow of traffic. The traffic engineers first gather data. As the figure shows, 700 cars per hour come down M Street to intersection A; 300 cars per hour come to intersection A on 10th Street. A total of x_1 of these cars leave A on M Street, while x_4 cars leave A on 10th Street. The number of cars entering A must equal the number leaving, so that

$$x_1 + x_4 = 700 + 300$$

or
$$x_1 + x_4 = 1000. \qquad (1)$$

For intersection B, x_1 cars enter B on M Street, and x_2 cars enter B on 11th Street. The figure shows that 900 cars leave B on 11th while 200 leave on M. We have

$$x_1 + x_2 = 900 + 200$$

$$x_1 + x_2 = 1100. \qquad (2)$$

At intersection C, 400 cars enter on N Street, 300 on 11th Street, while x_2 leave on 11th Street and x_3 leave on N Street. This gives

$$x_2 + x_3 = 400 + 300$$

$$x_2 + x_3 = 700. \qquad (3)$$

Finally, intersection D has x_3 cars entering on N and x_4 entering on 10th. There are 400 leaving D on 10th and 200 leaving on N.

(a) Set up an equation for intersection D.

(b) Use equations (1), (2), (3), and your answer to part (a) to set up a system of equations. Solve the system by the Gauss-Jordan method.

(c) Since you got a row of zeros, the system of equations is dependent and does not have a unique solution. Solve equation (1) for x_1 and substitute the result into equation (2).

(d) Solve equation (1) for x_4. What is the largest possible value of x_1 so that x_4 is not negative?

(e) Your answer to part (c) should have been $x_2 - x_4 = 100$. Solve the equation for x_4 and then find the smallest possible value of x_2 so that x_4 is not negative.

(f) Your answer to part (a) should have been $x_3 + x_4 = 600$. Solve the equation for x_4 and then find the largest possible values of x_3 and x_4 so that neither is negative.

(g) From your answers to parts (d) $-$ (f), what is the maximum value of x_4 so that all the equations are satisfied and all variables are nonnegative? Of x_3? Of x_2? Of x_1?

7.4 Nonlinear Systems of Equations

A **nonlinear system of equations** is one in which at least one of the equations is not a first-degree equation. Since nonlinear systems vary, depending upon the type of equations in the system, different solution methods are required for different systems. Although the substitution method discussed in Section 7.1 can almost always be used, it may not be the most efficient approach to the solution. This section illustrates some of the methods best suited to solving certain types of nonlinear systems.

Some nonlinear systems can be solved by the addition method, as in the following example.

Example 1   Solve the system

$$x^2 + y^2 = 4 \tag{1}$$
$$2x^2 - y^2 = 8. \tag{2}$$

Add the two equations to eliminate y^2.

$$\begin{array}{r} x^2 + y^2 = 4 \\ \underline{2x^2 - y^2 = 8} \\ 3x^2 \quad\;\; = 12 \\ x^2 = 4 \end{array}$$

$$x = 2 \quad \text{or} \quad x = -2.$$

Substituting into equation (1) gives the corresponding values of y.

$$\begin{array}{ll} 2^2 + y^2 = 4 & \quad \text{or} \quad & (-2)^2 + y^2 = 4 \\ y^2 = 0 & & y^2 = 0 \\ y = 0 & & y = 0 \end{array}$$

The solution set of the system is $\{(2, 0), (-2, 0)\}$. Check these solutions in the original system. The graph shown in Figure 7.4 also verifies the result. ✤

Figure 7.4

The substitution method is particularly useful in solving a nonlinear system that includes a linear equation. This method is illustrated by the next example.

Example 2   Use the substitution method to solve the system

$$3x^2 - 2y = 5 \qquad \textbf{(3)}$$
$$x + 3y = -4. \qquad \textbf{(4)}$$

Although either equation could be solved for either variable, it is best to use the linear equation. Here, it is simpler to solve equation (4) for y, since x is squared in equation (3).

$$x + 3y = -4$$
$$3y = -4 - x$$
$$y = \frac{-4 - x}{3} \qquad \textbf{(5)}$$

Substituting this value of y into equation (3) gives

$$3x^2 - 2\left(\frac{-4 - x}{3}\right) = 5.$$

Multiply both sides by 3 to eliminate the denominator.

$$9x^2 - 2(-4 - x) = 15$$
$$9x^2 + 8 + 2x = 15$$
$$9x^2 + 2x - 7 = 0$$
$$(9x - 7)(x + 1) = 0$$
$$x = \frac{7}{9} \qquad \text{or} \qquad x = -1$$

Substitute both values of x into equation (5) to find the corresponding y-values.

$$y = \frac{-4 - \left(\frac{7}{9}\right)}{3} \quad \text{or} \quad y = \frac{-4 - (-1)}{3}$$

$$y = -\frac{43}{27} \qquad\qquad y = -1$$

Check in the original system that the solution set is $\{(7/9, -43/27), (-1, -1)\}$. Figure 7.5 shows the graphs of the equations in the system. ✛

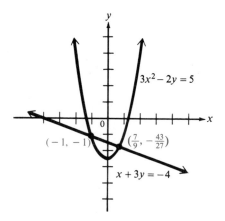

Figure 7.5

Sometimes a combination of the elimination method and the substitution method is effective in solving a system, as illustrated in Example 3.

Example 3 Solve the system

$$x^2 + 3xy + y^2 = 22 \qquad\qquad \textbf{(6)}$$
$$x^2 - xy + y^2 = 6. \qquad\qquad \textbf{(7)}$$

Multiply both sides of equation (7) by -1, and then add the result to equation (6), as follows.

$$x^2 + 3xy + y^2 = 22$$
$$\underline{-x^2 + xy - y^2 = -6}$$
$$4xy \qquad\quad = 16 \qquad\qquad \textbf{(8)}$$

Now solve equation (8) for either x or y and substitute the result into one of the given equations. Solving for y gives

$$y = \frac{4}{x}, \quad \text{if } x \neq 0. \qquad\qquad \textbf{(9)}$$

(The restriction $x \neq 0$ is included since if $x = 0$ there is no value of y that satisfies the system.) Substituting for y in equation (7) (equation (6) could have been used) and simplifying gives

$$x^2 - x\left(\frac{4}{x}\right) + \left(\frac{4}{x}\right)^2 = 6.$$

Simplify the resulting equation as follows.

$$x^2 - x\left(\frac{4}{x}\right) + \left(\frac{4}{x}\right)^2 = 6$$

$$x^2 - 4 + \frac{16}{x^2} = 6$$

$$x^4 - 4x^2 + 16 = 6x^2$$

$$x^4 - 10x^2 + 16 = 0$$

$$(x^2 - 2)(x^2 - 8) = 0$$

$$x^2 = 2 \quad \text{or} \quad x^2 = 8$$

$$x = \sqrt{2} \quad \text{or} \quad x = -\sqrt{2} \quad \text{or} \quad x = 2\sqrt{2} \quad \text{or} \quad x = -2\sqrt{2}$$

Substitute these x values into equation (9) to find corresponding values of y.

$$\text{If } x = \sqrt{2}, y = \frac{4}{\sqrt{2}} = 2\sqrt{2}.$$

$$\text{If } x = -\sqrt{2}, y = \frac{4}{-\sqrt{2}} = -2\sqrt{2}.$$

$$\text{If } x = 2\sqrt{2}, y = \frac{4}{2\sqrt{2}} = \sqrt{2}.$$

$$\text{If } x = -2\sqrt{2}, y = \frac{4}{-2\sqrt{2}} = -\sqrt{2}.$$

The solution set of the system contains four ordered pairs:

$$\{(\sqrt{2}, 2\sqrt{2}), (-\sqrt{2}, -2\sqrt{2}), (2\sqrt{2}, \sqrt{2}), (-2\sqrt{2}, -\sqrt{2})\}.$$

Verify these solutions by substitution in the original system.   ✦

Example 4   Solve the system

$$x^2 + y^2 = 16 \tag{10}$$

$$|x| + y = 4. \tag{11}$$

The substitution method is required here. Equation (11) can be rewritten as $|x| = 4 - y$, then the definition of absolute value can be used to get

$$x = 4 - y \quad \text{or} \quad x = -(4 - y) = y - 4. \tag{12}$$

(Since $|x| \geq 0$ for all real x, $4 - y \geq 0$, or $4 \geq y$.) Substituting from either part of (12) into equation (10) gives the same result.

$$(4 - y)^2 + y^2 = 16 \qquad \text{or} \qquad (y - 4)^2 + y^2 = 16$$

Since $(4 - y)^2 = (y - 4)^2 = 16 - 8y + y^2$, either equation becomes

$$(16 - 8y + y^2) + y^2 = 16$$
$$2y^2 - 8y = 0$$
$$2y(y - 4) = 0$$
$$y = 0 \qquad \text{or} \qquad y = 4.$$

From equation (12),

$$\text{If } y = 0, \text{ then } x = 4 - 0 \quad \text{or} \quad x = 0 - 4.$$
$$x = 4 \qquad\qquad x = -4$$
$$\text{If } y = 4, \text{ then } x = 4 - 4 = 0.$$

The solution set, $\{(4, 0), (-4, 0), (0, 4)\}$, includes the points of intersection shown in Figure 7.6. Be sure to check the solutions in the original system. ✦

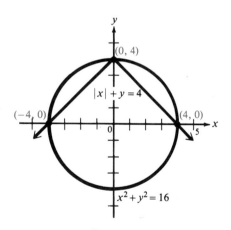

Figure 7.6

Nonlinear systems sometimes lead to solutions that are imaginary numbers.

Example 5 Solve the system

$$x^2 + y^2 = 5$$
$$4x^2 + 3y^2 = 11.$$

Multiplying the first equation on both sides by -3 and adding the result to the second equation gives

$$
\begin{aligned}
-3x^2 - 3y^2 &= -15 \\
\underline{4x^2 + 3y^2 &= 11} \\
x^2 &= -4.
\end{aligned}
$$

By the square root property,

$$x = \pm\sqrt{-4}$$

$$x = 2i \quad \text{or} \quad x = -2i.$$

Find y by substitution. Using $x^2 + y^2 = 5$ gives

$$
\begin{aligned}
-4 + y^2 &= 5 \\
y^2 &= 9 \\
y = 3 \quad &\text{or} \quad y = -3,
\end{aligned}
$$

for either
$$x = 2i \quad \text{or} \quad x = -2i.$$

Checking the solutions in the given system shows that the solution set is $\{(2i, 3), (2i, -3), (-2i, 3), (-2i, -3)\}$. As the graph in Figure 7.7 suggests, imaginary solutions may occur when the graphs of the equations do not intersect.   ✚

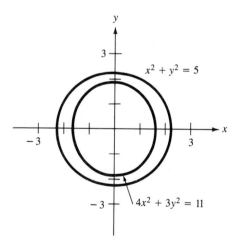

Figure 7.7

A system with imaginary solutions, such as the system in Example 5, is inconsistent if the solutions are restricted to the real numbers. Allowing imaginary-number solutions means most *nonlinear* systems have solutions. It is possible to have a nonlinear system with dependent equations, however.

7.4 Exercises

Give all complex solutions of the following nonlinear systems of equations. See Examples 1–4.

1. $y = x^2$
  $x + y = 2$

2. $y = -x^2 + 2$
  $x - y = 0$

3. $y = (x - 1)^2$
  $x - 3y = -1$

4. $y = (x + 3)^2$
  $x + 2y = -2$

5. $y = x^2 + 4x$
  $2x - y = -8$

6. $y = 6x + x^2$
  $3x - 2y = 10$

7. $x^2 + y^2 = 13$
  $2x + y = 1$

8. $2x^2 - y^2 = 7$
  $y = 3x - 7$

9. $3x^2 + 2y^2 = 5$
  $x - y = -2$

10. $x^2 + y^2 = 5$
  $-3x + 4y = 2$

11. $x^2 + y^2 = 8$
  $x^2 - y^2 = 0$

12. $x^2 + y^2 = 10$
  $2x^2 - y^2 = 17$

13. $5x^2 - y^2 = 0$
  $3x^2 + 4y^2 = 0$

14. $x^2 + y^2 = 4$
  $2x^2 - 3y^2 = -12$

15. $3x^2 + y^2 = 3$
  $4x^2 + 5y^2 = 26$

16. $x^2 + 2y^2 = 9$
  $3x^2 - 4y^2 = 27$

17. $2x^2 + 3y^2 = 5$
  $3x^2 - 4y^2 = -1$

18. $3x^2 + 5y^2 = 17$
  $2x^2 - 3y^2 = 5$

19. $5x^2 - y^2 = 9$
  $x^2 + y^2 = 45$

20. $x^2 + 3y^2 = 91$
  $2x^2 - y^2 = 119$

21. $2x^2 + 2y^2 = 20$
  $3x^2 + 3y^2 = 30$

22. $x^2 + y^2 = 4$
  $5x^2 + 5y^2 = 28$

23. $9x^2 + 4y^2 = 1$
  $x^2 + y^2 = 1$

24. $2x^2 - 3y^2 = 8$
  $6x^2 + 5y^2 = 24$

25. $xy = 6$
  $x + y = 5$

26. $xy = -4$
  $2x + y = -7$

27. $xy = -15$
  $4x + 3y = 3$

28. $xy = 8$
  $3x + 2y = -16$

29. $2xy + 1 = 0$
  $x + 16y = 2$

30. $-5xy + 2 = 0$
  $x - 15y = 5$

31. $x^2 + 4y^2 = 25$
  $xy = 6$

32. $5x^2 - 2y^2 = 6$
  $xy = 2$

33. $x^2 + 2xy - y^2 = 14$
  $x^2 - y^2 = -16$

34. $3x^2 + xy + 3y^2 = 7$
  $x^2 + y^2 = 2$

35. $x^2 - xy + y^2 = 5$
  $2x^2 + xy - y^2 = 10$

36. $3x^2 + 2xy - y^2 = 9$
  $x^2 - xy + y^2 = 9$

37. $x = |y|$
  $x^2 + y^2 = 18$

38. $2x + |y| = 4$
  $x^2 + y^2 = 5$

39. $y = |x - 1|$
  $y = x^2 - 4$

40. $2x^2 - y^2 = 4$
  $|x| = |y|$

41. The sum of two numbers is 6. Their product is -16. Find the numbers.

42. The sum of the squares of two numbers is 100. One number is 2 more than the other. Find the numbers.

43. Does the straight line $3x - 2y = 9$ intersect the circle $x^2 + y^2 = 25$? (To find out, solve the system made up of these two equations.)

44. Do the parabola $y = x^2 + 4$ and the ellipse $2x^2 + y^2 - 4x - 4y = 0$ have any points in common?

45. The area of a right triangle is 216 sq cm. It has a hypotenuse of 30 cm. Find the lengths of the two legs.

46. A rectangle has an area of 240 sq m. The diagonal is 26 m. Find the length and width.

47. In skeet shooting, a person fires at a clay disc that has been thrown into the air. Suppose the disc follows a parabolic path with the height in meters at time t in seconds given by $y = -t^2 + 4t$, while the bullet follows the parabolic path $y = -2t^2 + 12t - 13$. Further, suppose the two paths are in the same plane with common axes. Find the number of seconds for the bullet to meet the disc.

48. In electronics, circuit gain is given by

$$G = \frac{Bt}{R + R_t},$$

where R is the value of a resistor, t is temperature, and B is a constant. The sensitivity of the circuit to temperature is given by

$$S = \frac{BR}{(R + R_t)^2}.$$

If $B = 3.7$ and t is $90°K$ (Kelvin), find the values of R and R_t that will make $G = .4$ and $S = .001$.

7.5   *Systems of Inequalities and Linear Programming*

As shown in Chapter 3, the solution set of an inequality in two variables is usually an infinite set whose graph is one or more regions of the coordinate plane. The solution set of a **system of inequalities,** such as

$$x + y < 3$$
$$x^2 < 2y,$$

is the intersection of the solution sets of its members. The solution is best visualized by its graph. To graph the solution set of the system, graph both inequalities on the same axes and identify the solution by shading heavily the region common to both graphs.

Example 1   Graph the solution set of the system

$$x + y < 3$$
$$x^2 < 2y.$$

Figure 7.8 shows the graphs of both $x + y < 3$ and $x^2 < 2y$. The solution set of the system is the heavily shaded area. The points on the boundaries of $x + y < 3$ and $x^2 < 2y$ do not belong to the graph of the solution. For this reason, the boundaries are dashed lines.   ✛

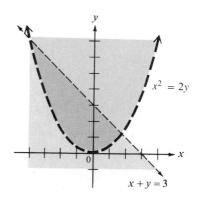

Figure 7.8

Example 2   Graph the solution set of the system

$$y \geq 2^x$$
$$9x^2 + 4y^2 \leq 36$$
$$2x + y < 1.$$

Graph the three inequalities on the same axes and shade the region common to all three, as shown in Figure 7.9. Two boundary lines are solid and one is dashed. The points where the dashed line intersects the solid lines are shown as open circles, since they do not belong to the solution set.   ✛

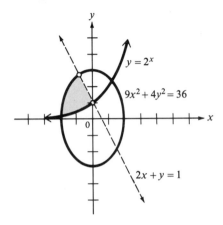

Linear Programming   An important application of mathematics to business and social science is called linear programming. **Linear programming** is used to find such things as minimum cost and maximum profit. The basic ideas of this technique are shown in the following example.

The Smith Company makes two products, tape decks and amplifiers. Each tape deck gives a profit of $3, while each amplifier produces $7 profit. The company must manufacture at least one tape deck per day to satisfy one of its customers, but no more than five because of production problems. Also, the number of amplifiers produced cannot exceed six per day. As a further requirement, the number of tape decks cannot exceed the number of amplifiers. How many of each should the company manufacture in order to obtain the maximum profit?

To begin, translate the statements of the problem into symbols by assuming

x = number of tape decks to be produced daily
y = number of amplifiers to be produced daily.

According to the statement of the problem given above, the company must produce at least one tape deck (one or more), so

$$x \geq 1.$$

Since no more than 5 tape decks may be produced,

$$x \leq 5.$$

Since no more than 6 amplifiers may be made in one day,

$$y \leq 6.$$

The requirement that the number of tape decks may not exceed the number of amplifiers translates as

$$x \leq y.$$

The number of tape decks and of amplifiers cannot be negative, so

$$x \geq 0 \quad \text{and} \quad y \geq 0.$$

These restrictions, or **constraints,** that are placed on production form the system of inequalities

$$x \geq 1, \quad x \leq 5, \quad y \leq 6, \quad x \leq y, \quad x \geq 0, \quad y \geq 0.$$

The maximum possible profit that the company can make, subject to these constraints, is found by sketching the graph of the solution of the system. See Figure 7.10. The only feasible values of x and y are those that satisfy all constraints. These values correspond to points that lie on the boundary or in the shaded region, called the **region of feasible solutions.**

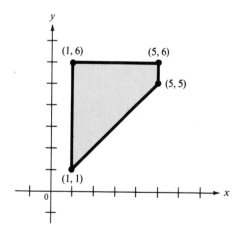

Figure 7.10

Since each tape deck gives a profit of $3, the daily profit from the production of x tape decks is $3x$ dollars. Also, the profit from the production of y amplifiers will be $7y$ dollars per day. The total daily profit is thus

$$\text{profit} = 3x + 7y.$$

The problem of the Smith Company may now be stated as follows: find values of x and y in the shaded region of Figure 7.10 that will produce the maximum possible value of $3x + 7y$.

It can be shown that any optimum value (maximum or minimum) will always occur at a **vertex** (or corner) point of the region of feasible solutions. Locate the point (x, y) that gives the maximum profit by checking the coordinates of the vertex points, shown in Figure 7.10 and listed below. Find the profit that corresponds to each coordinate pair and choose the one that gives the maximum profit.

Point	Profit $= 3x + 7y$
$(1, 1)$	$3(1) + 7(1) = 10$
$(1, 6)$	$3(1) + 7(6) = 45$
$(5, 6)$	$3(5) + 7(6) = 57 \leftarrow$ Maximum
$(5, 5)$	$3(5) + 7(5) = 50$

The maximum profit of \$57 is obtained when 5 tape decks and 6 amplifiers are produced each day.

Example 3   Robin, who is ill, takes vitamin pills. Each day, she must have at least 16 units of vitamin A, at least 5 units of vitamin B_1, and at least 20 units of vitamin C. She can choose between red pills, costing 10¢ each, which contain 8 units of A, 1 of B_1, and 2 of C; and blue pills, costing 20¢ each, which contain 2 units of A, 1 of B_1, and 7 of C. How many of each pill should she buy in order to minimize her cost and yet fulfill her daily requirements?

Let x represent the number of red pills to buy, and let y represent the number of blue pills to buy. Then the cost in pennies per day is given by

$$\text{cost} = 10x + 20y.$$

Since Robin buys x of the 10¢ pills and y of the 20¢ pills, she gets vitamin A as follows: 8 units from each red pill and 2 units from each blue pill. Altogether, she gets $8x + 2y$ units of A per day. Since she must get at least 16 units,

$$8x + 2y \geq 16.$$

Each red pill or each blue pill supplies 1 unit of vitamin B_1. Robin needs at least 5 units per day, so

$$x + y \geq 5.$$

For vitamin C, the inequality is

$$2x + 7y \geq 20.$$

Also, $x \geq 0$ and $y \geq 0$, since Robin cannot buy negative numbers of the pills.

The total cost of the pills can be minimized by finding the solution of the system of inequalities formed by the constraints. (See Figure 7.11.) The solution to this minimizing problem will also occur at a vertex point. Check the coordinates of the vertex points in the cost function to find the lowest cost.

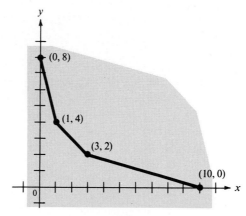

Figure 7.11

Point	Cost $= 10x + 20y$	
$(10, 0)$	$10(10) + 20(0) = 100$	
$(3, 2)$	$10(3) + 20(2) = $ **70**	$\leftarrow$ Minimum
$(1, 4)$	$10(1) + 20(4) = 90$	
$(0, 8)$	$10(0) + 20(8) = 160$	

Robin's best bet is to buy 3 red pills and 2 blue ones, for a total cost of 70¢ per day. She receives minimum amounts of vitamins B_1 and C but an excess of vitamin A. Even with an excess of A, this is still the best buy. ✛

7.5 Exercises

Graph the solution set of each system of inequalities. See Examples 1 and 2.

1. $x + y \leq 4$
$x - 2y \geq 6$

2. $2x + y > 2$
$x - 3y < 6$

3. $4x + 3y < 12$
$y + 4x > -4$

4. $3x + 5y \leq 15$
$x - 3y \geq 9$

5. $x + y \leq 6$
$2x + 2y \geq 12$

6. $3x + 4y < 15$
$6x + 8y > 30$

7. $x + 2y \leq 4$
$y \geq x^2 - 1$

8. $4x - 3y \leq 12$
$y \leq x^2$

9. $y \leq -x^2$
$y \geq x^2 - 6$

10. $x^2 + y^2 \leq 9$
$x \leq -y^2$

11. $x^2 - y^2 < 1$
$-1 < y < 1$

12. $x^2 + y^2 \leq 36$
$-4 \leq x \leq 4$

13. $2x^2 - y^2 > 4$
$2y^2 - x^2 > 4$

14. $y \geq x^2 + 4x + 4$
$y < -x^2$

15. $\dfrac{x^2}{16} + \dfrac{y^2}{9} \leq 1$
$\dfrac{x^2}{4} - \dfrac{y^2}{16} \geq 1$

16. $\dfrac{x^2}{36} - \dfrac{y^2}{9} \geq 1$
$\dfrac{x^2}{81} + y^2 \leq 1$

17. $x + y \leq 4$
$x - y \leq 5$
$4x + y \leq -4$

18. $3x - 2y \geq 6$
$x + y \leq -5$
$y \leq 4$

19. $2y + x \leq -5$
$y \geq 3 + x$
$x \geq 0$
$y \geq 0$

20. $2x + 3y \leq 12$
$2x + 3y > -6$
$3x + y < 4$
$x \geq 0$
$y \geq 0$

21. $\dfrac{x^2}{4} + \dfrac{y^2}{9} > 1$
$x^2 - y^2 \geq 1$
$-4 \leq x \leq 4$

22. $2x - 3y < 6$
$4x^2 + 9y^2 < 36$
$x \geq -1$

23. $y \geq 3^x$
$y \geq 2$

24. $y \leq \left(\dfrac{1}{2}\right)^x$
$y \geq 4$

25. $|x| \geq 2$
$|y| \geq 4$
$y < x^2$

26. $|x| + 2 \geq 4$
$|y| \leq 1$
$\dfrac{x^2}{9} + \dfrac{y^2}{16} \leq 1$

27. $y \leq |x + 2|$
$\dfrac{x^2}{16} - \dfrac{y^2}{9} \leq 1$

28. $y \leq \log x$
$y \geq |x - 2|$

29. $y \geq |4 - x|$
$y \geq |x|$

30. $|x + 2| < y$
$|x| \leq 3$

In Exercises 31–34, use graphical methods to find values of x and y satisfying the given conditions. Find the value of the maximum or minimum.

31. Find $x \geq 0$ and $y \geq 0$ such that
$$2x + 3y \leq 6$$
$$4x + y \leq 6$$
and $5x + 2y$ is maximized.

32. Find $x \geq 0$ and $y \geq 0$ such that
$$x + y \leq 10$$
$$5x + 2y \geq 20$$
$$2y \geq x$$
and $x + 3y$ is minimized.

33. Find $x \geq 2$ and $y \geq 5$ such that
$$3x - y \geq 12$$
$$x + y \leq 15$$
and $2x + y$ is minimized.

34. Find $x \geq 10$ and $y \geq 20$ such that
$$2x + 3y \leq 100$$
$$5x + 4y \leq 200$$
and $x + 3y$ is maximized.

Solve each of the following linear programming problems. See Example 3.

35. Farmer Jones raises only pigs and geese. She wants to raise no more than 16 animals with no more than 12 geese. She spends $5 to raise a pig and $2 to raise a goose. She has $50 available for this purpose. Find the maximum profit she can make if she makes a profit of $8 per goose and $4 per pig.

36. A wholesaler of party goods wishes to display her products at a convention of social secretaries in such a way that she gets the maximum number of inquiries about her whistles and hats. Her booth at the convention has 12 sq m of floor space to be used for display purposes. A display unit for hats requires 2 sq m; and for whistles, 4 sq m. Experience tells the wholesaler that she should never have more than a total of 5 units of whistles and hats on display at one time. If she receives three inquiries for each unit of hats and two inquiries for each unit of whistles on display, how many of each should she display in order to get the maximum number of inquiries?

37. An office manager wants to buy some filing cabinets. She knows that cabinet #1 costs $10 each, requires 6 sq ft of floor space, and holds 8 cu ft of files. Cabinet #2 costs $20 each, requires 8 sq ft of floor space, and holds 12 cu ft. She can spend no more than $140 due to budget limitations, while her office has room for no more than 72 sq ft of cabinets. She wants the maximum storage capacity within the limits imposed by funds and space. How many of each type of cabinet should she buy?

38. In a small town in South Carolina, zoning rules require that the window space (in square feet) in a house be at least one-sixth of the space used up by solid walls. The cost to heat the house is 20¢ for each square foot of solid walls and 80¢ for each square foot of windows. Find the maximum total area (windows plus walls) if $160 is available to pay for heat.

39. The manufacturing process requires that oil refineries manufacture at least 2 gal of gasoline for each gallon of fuel oil. To meet the winter demand for fuel oil, at least 3 million gal a day must be produced. The demand for gasoline is no more than 6.4 million gal per day. If the price of gasoline is $1.90 and the price of fuel oil is $1.50 per gal, how much of each should be produced to maximize revenue?

40. Seall Manufacturing Company makes color television sets. It produces a bargain set that sells for $100 profit and a deluxe set that sells for $150 profit. On the assembly line the bargain set requires 3 hr, while the deluxe set takes 5 hr. The cabinet shop spends 1 hr on the cabinet for the bargain set and 3 hr on the cabinet for the deluxe set. Both sets require 2 hr of time for testing and packing. On a particular production run the Seall Company has available 3900 work hours on the assembly line, 2100 work hours in the cabinet shop, and 2200 work hours in the testing and packing department. How many sets of each type should it produce to make maximum profit? What is the maximum profit?

Chapter 7 *Summary*

Key Words

7.1 system of equations
first-degree equation in n
unknowns
inconsistent system
dependent equations
substitution method
addition method
equivalent systems
elimination method
7.2 ordered triple
y arbitrary

7.3 matrix, matrices
element
entry
augmented matrix
Gauss-Jordan method
7.4 nonlinear system of
equations
7.5 system of inequalities
linear programming
constraints
region of feasible
solutions

Key Ideas

7.1 Possible Graphs of a Linear System with Two Equations and Two Variables

1. The graphs of the two equations intersect in a single point. The coordinates of this point give the solution of the system.
2. The graphs are distinct parallel lines. In this case, the system is said to be inconsistent. That is, there is no solution common to both equations. The solution set of the linear system is empty.
3. The graphs are the same line. In this case, the equations are said to be dependent, and any solution of one equation is also a solution of the other. Thus, there are infinitely many solutions.

7.2 Solution Set of a System with Three Equations and Three Variables

The solution set of a system of three equations in three unknowns may be either a single ordered triple, an infinite set of ordered triples (dependent equations), or the empty set (an inconsistent system).

7.3 Matrix Row Operations

For any real number k and any augmented matrix of a system of linear equations, the following operations will result in the matrix of an equivalent system:

1. interchanging any two rows of a matrix,
2. multiplying the elements of any row of a matrix by the same nonzero number k, and
3. adding a multiple of the elements of one row to the elements of another row.

Chapter 7   *Review Exercises*

Use the substitution method to solve each of the following linear systems. Identify any systems with dependent equations or any inconsistent systems.

1. $4x - 3y = -1$
    $3x + 5y = 50$

2. $10x + 3y = 8$
    $5x - 4y = 26$

3. $7x - 10y = 11$
    $\dfrac{3x}{2} - 5y = 8$

4. $\dfrac{x}{2} + \dfrac{2y}{3} = -8$
    $\dfrac{3x}{4} + \dfrac{y}{3} = 0$

5. $\dfrac{x}{2} - \dfrac{y}{5} = \dfrac{11}{10}$
    $2x - \dfrac{4y}{5} = \dfrac{22}{5}$

6. $4x + 5y - 5 = 0$
    $3x + 7y + 6 = 0$

Use the addition method to solve each of the following linear systems. Identify any systems with dependent equations or any inconsistent systems.

7. $3x - 5y = -18$
    $2x + 7y = 19$

8. $6x + 5y = 53$
    $4x - 3y = 29$

9. $\dfrac{2}{3}x - \dfrac{3}{4}y = 13$
    $\dfrac{1}{2}x + \dfrac{2}{3}y = -5$

10. $3x + 7y = 10$
    $18x + 42y = 50$

11. $\dfrac{-1}{x} + \dfrac{1}{y} = \dfrac{7}{10}$
    $\dfrac{3}{x} - \dfrac{5}{y} = \dfrac{1}{2}$

12. $.9x - .2y = .8$
    $.3x + .7y = 4.1$

Solve each of the following problems by writing a system of equations and then solving it.

13. A student bought some candy bars, paying 25¢ each for some and 50¢ each for others. The student bought a total of 22 bars, paying a total of $8.50. How many of each kind of bar did he buy?

14. Ink worth $25 a bottle is to be mixed with ink worth $18 per bottle to get 12 bottles of ink worth $20 each. How much of each type of ink should be used?

15. Donna Sharp wins $50,000 in a lottery. She invests part of the money at 6%, twice as much at 7%, with $10,000 more than the amount invested at 6% invested at 9%. Total annual interest is $3800. How much is invested at each rate?

16. The sum of three numbers is 23. The second number is 3 more than the first. The sum of the first and twice the third is 4. Find the three numbers.

17. A gold merchant has some 12 carat gold (12/24 pure gold), and some 22 carat gold (22/24 pure). How many grams of each should be mixed to get 25 g of 15 carat gold?

18. A chemist has some 40% acid solution and some 60% solution. How many liters of each should be used to get 40 liters of a 45% solution?

Use the addition method to solve each of the following linear systems.

19. $2x - 3y + z = -5$
     $x + 4y + 2z = 13$
     $5x + 5y + 3z = 14$

20. $x - 3y = 12$
     $2y + 5z = 1$
     $4x + z = 25$

21. $x + y - z = 5$
$2x + y + 3z = 2$
$4x - y + 2z = -1$

22. $5x - 3y + 2z = -5$
$2x + 2y - z = 4$
$4x - y + z = -1$

Solve each of the following problems by writing a system of equations and then solving it.

23. Three kinds of tea worth $4.60, $5.75, and $6.50 per pound are to be mixed to get 20 lb of tea worth $5.25 per pound. The amount of $4.60 tea used is to be equal to the total amount of the other two kinds together. How many pounds of each tea should be used?

24. A 5% solution of a drug is to be mixed with some 15% solution and some 10% solution to get 20 ml of 8% solution. The amount of 5% solution used must be 2 ml more than the sum of the other two solutions. How many milliliters of each solution should be used?

25. The cashier at an amusement park has a total of $2480, made up of fives, tens, and twenties. The total number of bills is 290, and the value of the tens is $60 more than the value of the twenties. How many of each type of bill does the cashier have?

26. Ms. Levy invests $50,000 three ways—at 8%, 8 1/2%, and 11%. In total, she receives $4436.25 per year in interest. The interest from the 11% investment is $80 more than the interest on the 8% investment. Find the amount she has invested at each rate.

Find solutions for the following systems in terms of the arbitrary variable z.

27. $3x - 4y + z = 2$
$2x + y - 4z = 1$

28. $2x + 3y - z = 5$
$-x + 2y + 4z = 8$

Use the Gauss-Jordan method to solve each of the following systems.

29. $2x + 3y = 10$
$-3x + y = 18$

30. $5x + 2y = -10$
$3x - 5y = -6$

31. $3x + y = -7$
$x - y = -5$

32. $x - z = -3$
$y + z = 6$
$2x - 3z = -9$

33. $2x - y + 4z = -1$
$-3x + 5y - z = 5$
$2x + 3y + 2z = 3$

34. $5x - 8y + z = 1$
$3x - 2y + 4z = 3$
$10x - 16y + 2z = 3$

Solve each of the following nonlinear systems of equations.

35. $y = x^2 - 1$
$x + y = 1$

36. $x^2 + y^2 = 2$
$3x + y = 4$

37. $x^2 + 2y^2 = 22$
$2x^2 - y^2 = -1$

38. $x^2 - 4y^2 = 19$
$x^2 + y^2 = 29$

39. $xy = 4$
$x - 6y = 2$

40. $x^2 + 2xy + y^2 = 4$
$x = 3y - 2$

41. Do the circle $x^2 + y^2 = 144$ and the line $x + 2y = 8$ have any points in common? If so, what are they?

42. Find a value of b so that the straight line $3x - y = b$ touches the circle $x^2 + y^2 = 25$ in only one point.

43. The difference of the squares of two real numbers is 64. Their product is 60. Find the numbers.

Graph the solution of each of the following systems of inequalities.

44. $x + y \leq 6$
$2x - y \geq 3$

45. $x - 3y \geq 6$
$y^2 \leq 16 - x^2$

46. $9x^2 + 16y^2 \geq 144$
$x^2 - y^2 \geq 16$

47. A bakery makes both cakes and cookies. Each batch of cakes requires 2 hr in the oven and 3 hr in the decorating room. Each batch of cookies needs 1 1/2 hr in the oven and 2/3 hr in the decorating room. The oven is available no more than 16 hr a day, while the decorating room can be used no more than 12 hr per day. Set up a system of inequalities expressing the given information, and then graph the system.

48. A candy company has 100 kg of chocolate-covered nuts and 125 kg of chocolate-covered raisins to be sold as two different mixtures. One mix will contain 1/2 nuts and 1/2 raisins, while the other mix will contain 1/3 nuts and 2/3 raisins. Set up a system of inequalities expressing the given information, and then graph the system.

Solve the following linear programming problems.

49. Find $x \geq 0$ and $y \geq 0$ such that

$$3x + 2y \leq 12$$
$$5x + y \geq 5$$

and $2x + 4y$ is maximized.

50. Find $x \geq 0$ and $y \geq 0$ such that

$$x + y \leq 50$$
$$2x + y \geq 20$$
$$x + 2y \geq 30$$

and $4x + 2y$ is minimized.

Chapter 7 *Test*

Use the substitution method to solve each of the following linear systems. Identify any systems with dependent equations or any inconsistent systems.

1. $x - 3y = -5$
    $2x + y = 4$

2. $3m - n = 9$
    $m + 2n = 10$

3. $6x + 9y = -21$
    $4x + 6y = -14$

Use the addition method to solve each of the following linear systems. Identify any systems with dependent equations or any inconsistent systems.

4. $4a + 6b = 31$
    $3a + 4b = 22$

5. $\dfrac{1}{4}x - \dfrac{1}{3}y = \dfrac{-5}{12}$
    $\dfrac{1}{10}x + \dfrac{1}{5}y = \dfrac{1}{2}$

6. $x - 2y = 4$
    $-2x + 4y = 6$

Solve each of the following problems by writing a system of equations and then solving it.

7. Cathy Novak has $1.80 in nickels and dimes, with three more dimes than nickels. How many dimes and nickels does she have?

8. Jack Creagh has $1000 more invested at 5% than he has invested at 4%. If his annual income from the two investments together is $698, how much does Jack have invested at each rate?

Use the addition method to solve the following linear systems.

9. $3a - 4b + 2c = 15$
$2a - b + c = 13$
$a + 2b - c = 5$

10. $2x + y + z = 3$
$x + 2y - z = 3$
$3x - y + z = 5$

11. The sum of three numbers is 2. The first number is equal to the sum of the other two, and the third number is the result of subtracting the first from the second. Find the numbers.

Use the Gauss-Jordan method to solve each of the following systems.

12. $3a - 2b = 13$
$4a - b = 19$

13. $2x + 3y - 6z = 1$
$x - y + 2z = 3$
$4x + y - 2z = 7$

14. $x + 2y - z = 1$
$3x + y + z = 3$
$2x - y + 2z = -1$

Solve each of the following nonlinear systems of equations.

15. $x^2 + y^2 - 2x = 9$
$x^2 + y^2 - 4y = 1$

16. $x^2 + y^2 = 25$
$x + y = 7$

17. The sum of the squares of two real numbers is 9, and the sum of those two numbers is 6. Find the numbers.

18. Graph the solution of the following system of inequalities.

$$9x^2 + 4y^2 \geq 36$$

$$y < x^2$$

19. Solve the following linear programming problem.
Find $x \geq 0$ and $y \geq 0$ such that

$$3x + 2y \geq 15$$

$$4x + y \geq 10$$

and $2x + y$ is minimized.

20. Find a solution for the following system in terms of the arbitrary variable z.

$$x - 2y + 3z = 2$$

$$4x + y - z = 1$$

Matrices

Suppose you are the manager of a health food store and you receive the following shipments of vitamins from two suppliers: from Dexter, 2 cartons of vitamin A pills, 7 cartons of vitamin E pills, and 5 cartons of vitamin K pills; from Sullivan, 4 cartons of vitamin A pills, 6 cartons of vitamin E pills, and 9 cartons of vitamin K pills. It might be helpful to rewrite the information in a chart to make it more understandable.

Manufacturer	Cartons of Vitamin		
	A	E	K
Dexter	2	7	5
Sullivan	4	6	9

The information is clearer when presented this way. In fact, as long as you remember what each number represents, you can remove all the labels and write the numbers as a matrix.

$$\begin{bmatrix} 2 & 7 & 5 \\ 4 & 6 & 9 \end{bmatrix}$$

This array of numbers gives all the information needed

Matrices have been of interest to mathematicians for some time. Recently, the use of matrices has gained importance in the life sciences, management, and the social sciences. As mentioned in Chapter 7, matrices are useful in solving systems of equations with computers.

8.1   Basic Properties of Matrices

Matrices are classified by their **dimension** or **order,** that is, by the number of rows and columns that they contain. For example, the matrix

$$\begin{bmatrix} 2 & 7 & 5 \\ 4 & 6 & 9 \end{bmatrix}$$

has two rows and three columns. This matrix has dimension 2 × 3 (read "2 by 3"), or order 2 × 3. By definition, a matrix with m rows and n columns is of **dimension m × n** or **order m × n.** The number of rows is always given first.

Example 1   **(a)** The matrix $\begin{bmatrix} 6 & 5 \\ 3 & 4 \\ 5 & -1 \end{bmatrix}$ is of order 3 × 2.

(b) $\begin{bmatrix} 5 & 8 & 9 \\ 0 & 5 & -3 \\ -4 & 0 & 5 \end{bmatrix}$ is of order 3 × 3.

(c) [1   6   5   -2   5] is of order 1 × 5.

(d) $\begin{bmatrix} 3 \\ -5 \\ 0 \\ 2 \end{bmatrix}$ is of order 4 × 1.   ✚

A matrix having the same number of rows as columns is called a **square matrix.** The matrix given in Example 1(b) above is a square matrix, as are

$$\begin{bmatrix} -5 & 6 \\ 8 & 3 \end{bmatrix} \quad \text{and} \quad \begin{bmatrix} 0 & 0 & 0 & 0 \\ -2 & 4 & 1 & 3 \\ 0 & 0 & 0 & 0 \\ -5 & -4 & 1 & 8 \end{bmatrix}.$$

A matrix containing only one row is called a **row matrix.** The matrix in Example 1(c) is a row matrix, as are

[5   8],   [6   -9   2],   and   [-4   0   0   0].

Finally, a matrix of only one column, as in part (d) of Example 1, is a **column matrix.**

It is customary to use capital letters to name matrices. Subscript notation is used to name the elements of a matrix, as follows.

$$A = \begin{bmatrix} a_{11} & a_{12} & a_{13} & \cdots & a_{1n} \\ a_{21} & a_{22} & a_{23} & \cdots & a_{2n} \\ a_{31} & a_{32} & a_{33} & \cdots & a_{3n} \\ \cdot & \cdot & \cdot & & \cdot \\ \cdot & \cdot & \cdot & & \cdot \\ \cdot & \cdot & \cdot & & \cdot \\ a_{m1} & a_{m2} & a_{m3} & \cdots & a_{mn} \end{bmatrix}$$

Using this notation, the first row, first column element is denoted a_{11}, the second row, third column element is denoted a_{23}, and the ith row, jth column element is denoted a_{ij}.

Two matrices are **equal** if they are of the same order and if each corresponding element, position by position, is equal. Using this definition, the matrices

$$\begin{bmatrix} 2 & 1 \\ 3 & -5 \end{bmatrix} \quad \text{and} \quad \begin{bmatrix} 1 & 2 \\ -5 & 3 \end{bmatrix}$$

are *not* equal (even though they contain the same elements and are of the same order), since the corresponding elements differ.

Example 2   From the definition of equality given above, the only way that the statement

$$\begin{bmatrix} 2 & 1 \\ p & q \end{bmatrix} = \begin{bmatrix} x & y \\ -1 & 0 \end{bmatrix}$$

can be true is if $2 = x$, $1 = y$, $p = -1$, and $q = 0$.   ❖

Example 3   The statement

$$\begin{bmatrix} x \\ y \end{bmatrix} = \begin{bmatrix} 1 \\ 4 \\ 0 \end{bmatrix}$$

can never be true, since the two matrices are of different order. (One is 2×1 and the other is 3×1.)   ❖

At the beginning of this chapter we used the matrix

$$\begin{bmatrix} 2 & 7 & 5 \\ 4 & 6 & 9 \end{bmatrix},$$

where the columns represent the numbers of cartons of three different types of vitamins (A, E, and K, respectively), and the rows represent two different manufacturers (Dexter and Sullivan, respectively). For example, the element 7 represents 7 cartons of vitamin E pills from Dexter, and so on. Suppose another shipment from these two suppliers is described by the following matrix.

$$\begin{bmatrix} 3 & 12 & 10 \\ 15 & 11 & 8 \end{bmatrix}$$

Here, for example, 8 cartons of vitamin K pills arrived from Sullivan. The number of cartons of each kind of pill that were received from these two shipments can be found from these two matrices.

In the first shipment, 2 cartons of vitamin A pills were received from Dexter, and in the second shipment, 3 cartons of vitamin A pills were received. Altogether, $2 + 3$, or 5, cartons of these pills were received. Corresponding elements can be added to find the total number of cartons of each type of pill received.

$$\begin{bmatrix} 2 & 7 & 5 \\ 4 & 6 & 9 \end{bmatrix} + \begin{bmatrix} 3 & 12 & 10 \\ 15 & 11 & 8 \end{bmatrix} = \begin{bmatrix} 2 + 3 & 7 + 12 & 5 + 10 \\ 4 + 15 & 6 + 11 & 9 + 8 \end{bmatrix}$$
$$= \begin{bmatrix} 5 & 19 & 15 \\ 19 & 17 & 17 \end{bmatrix}$$

The last matrix gives the total number of cartons of each type of pill that were received. For example, 15 cartons of vitamin K pills were received from Dexter. Generalizing from this example leads to the following definition.

Definition of Matrix Addition

> The sum of two $m \times n$ matrices A and B is the $m \times n$ matrix $A + B$ in which each element is the sum of the corresponding elements of A and B.

Only matrices of the same order can be added.

Example 4   Find each of the following sums.

(a) $\begin{bmatrix} 5 & -6 \\ 8 & 9 \end{bmatrix} + \begin{bmatrix} -4 & 6 \\ 8 & -3 \end{bmatrix} = \begin{bmatrix} 5 + (-4) & -6 + 6 \\ 8 + 8 & 9 + (-3) \end{bmatrix} = \begin{bmatrix} 1 & 0 \\ 16 & 6 \end{bmatrix}$

(b) $\begin{bmatrix} 2 \\ 5 \\ 8 \end{bmatrix} + \begin{bmatrix} -6 \\ 3 \\ 12 \end{bmatrix} = \begin{bmatrix} -4 \\ 8 \\ 20 \end{bmatrix}$

(c) The matrices

$$A = \begin{bmatrix} 5 & 8 \\ 6 & 2 \end{bmatrix} \quad \text{and} \quad B = \begin{bmatrix} 3 & 9 & 1 \\ 4 & 2 & 5 \end{bmatrix}$$

are of different orders. Therefore, the sum $A + B$ does not exist.   ✣

A matrix containing only zero elements is called a **zero matrix.** For example, [0 0 0] is the 1×3 zero matrix, while

$$\begin{bmatrix} 0 & 0 & 0 \\ 0 & 0 & 0 \end{bmatrix}$$

is the 2×3 zero matrix. A zero matrix can be written for any order.

In Chapter 1 the additive inverse of a real number a was defined as the real number $-a$ such that $a + (-a) = 0$ and $-a + a = 0$. Given a matrix A, a matrix $-A$ can be found such that $A + (-A) = O$, where O is the appropriate zero matrix, and $-A + A = O$. For example, if

$$A = \begin{bmatrix} -5 & 2 & -1 \\ 3 & 4 & -6 \end{bmatrix},$$

then the elements of matrix $-A$ are the additive inverses of the corresponding elements of A. (Remember that each element of A is a real number and thus has an additive inverse.)

$$-A = \begin{bmatrix} 5 & -2 & 1 \\ -3 & -4 & 6 \end{bmatrix}$$

To check, first test that $A + (-A)$ equals O, the appropriate zero matrix.

$$A + (-A) = \begin{bmatrix} -5 & 2 & -1 \\ 3 & 4 & -6 \end{bmatrix} + \begin{bmatrix} 5 & -2 & 1 \\ -3 & -4 & 6 \end{bmatrix} = \begin{bmatrix} 0 & 0 & 0 \\ 0 & 0 & 0 \end{bmatrix} = O$$

Then test that $-A + A$ is also O. Matrix $-A$ is the **additive inverse,** or **negative,** of matrix A. Every matrix of any order has an additive inverse.

Subtraction of real numbers was defined in Chapter 1 by saying that $a - b = a + (-b)$. The same definition is used for subtraction of matrices.

Definition of Matrix Subtraction	If A and B are matrices of the same order, then $$A - B = A + (-B).$$

Example 5   Find each of the following differences.

(a) $\begin{bmatrix} -5 & 6 \\ 2 & 4 \end{bmatrix} - \begin{bmatrix} -3 & 2 \\ 5 & -8 \end{bmatrix} = \begin{bmatrix} -5 & 6 \\ 2 & 4 \end{bmatrix} + \begin{bmatrix} 3 & -2 \\ -5 & 8 \end{bmatrix} = \begin{bmatrix} -2 & 4 \\ -3 & 12 \end{bmatrix}$

(b) [8 6 -4] - [3 5 -8] = [5 1 4]

(c) The matrices

$$\begin{bmatrix} -2 & 5 \\ 0 & 1 \end{bmatrix} \quad \text{and} \quad \begin{bmatrix} 3 \\ 5 \end{bmatrix}$$

are of different order and cannot be subtracted.   ❖

In work with matrices, a real number is called a **scalar** to distinguish it from a matrix. Multiply a scalar and a matrix as follows.

Definition of Multiplication of a Matrix by a Scalar	The product of a scalar k and a matrix X is the matrix kX, each of whose elements is k times the corresponding element of X.

Example 6

(a) $5\begin{bmatrix} 2 & -3 \\ 0 & 4 \end{bmatrix} = \begin{bmatrix} 10 & -15 \\ 0 & 20 \end{bmatrix}$

(b) $\dfrac{3}{4}\begin{bmatrix} 20 & 36 \\ 12 & -16 \end{bmatrix} = \begin{bmatrix} 15 & 27 \\ 9 & -12 \end{bmatrix}$ ✛

8.1 Exercises

Find the order of each of the following matrices. Identify any square, column, or row matrices. See Example 1.

1. $\begin{bmatrix} -4 & 8 \\ 2 & 3 \end{bmatrix}$

2. $\begin{bmatrix} -9 & 6 & 2 \\ 4 & 1 & 8 \end{bmatrix}$

3. $\begin{bmatrix} -6 & 8 & 0 & 0 \\ 4 & 1 & 9 & 2 \\ 3 & -5 & 7 & 1 \end{bmatrix}$

4. $[8 \ -2 \ 4 \ 6 \ 3]$

5. $\begin{bmatrix} 2 \\ 4 \end{bmatrix}$

6. $\begin{bmatrix} -9 & 6 & 5 & 1 & 2 \\ -4 & 0 & 8 & 7 & 3 \end{bmatrix}$

7. $\begin{bmatrix} -5 \\ -8 \\ -4 \\ -6 \end{bmatrix}$

8. $[-9]$

9. $\begin{bmatrix} -4 & 2 & 3 \\ -8 & 2 & 1 \\ 4 & 6 & 8 \end{bmatrix}$

10. $\begin{bmatrix} -4 & 2 \\ 3 & 5 \end{bmatrix}$

Find the values of the variables in each of the following matrices. See Examples 2 and 3.

11. $\begin{bmatrix} 2 & 1 \\ 4 & 8 \end{bmatrix} = \begin{bmatrix} x & 1 \\ y & z \end{bmatrix}$

12. $\begin{bmatrix} -5 \\ y \end{bmatrix} = \begin{bmatrix} -5 \\ 8 \end{bmatrix}$

13. $\begin{bmatrix} 2 & 5 & 6 \\ 1 & m & n \end{bmatrix} = \begin{bmatrix} z & y & w \\ 1 & 8 & -2 \end{bmatrix}$

14. $\begin{bmatrix} 0 & 0 \\ a & b \end{bmatrix} = \begin{bmatrix} p & q \\ 5 & -7 \end{bmatrix}$

15. $\begin{bmatrix} x+6 & y+2 \\ 8 & 3 \end{bmatrix} = \begin{bmatrix} -9 & 7 \\ 8 & k \end{bmatrix}$

16. $\begin{bmatrix} 9 & 7 \\ r & 0 \end{bmatrix} = \begin{bmatrix} m-3 & n+5 \\ 8 & 0 \end{bmatrix}$

17. $\begin{bmatrix} 3 & 5 \\ 8 & 9 \end{bmatrix} + \begin{bmatrix} m & 3 \\ 5 & n \end{bmatrix} = \begin{bmatrix} 9 & 8 \\ 13 & 0 \end{bmatrix}$

18. $[8 \quad p+9 \quad q+5] + [9 \quad -3 \quad 12] = [k-2 \quad 12 \quad 2q]$

19. $\begin{bmatrix} -7+z & 4r & 8s \\ 6p & 2 & 5 \end{bmatrix} + \begin{bmatrix} -9 & 8r & 3 \\ 2 & 5 & 4 \end{bmatrix} = \begin{bmatrix} 2 & 36 & 27 \\ 20 & 7 & 12a \end{bmatrix}$

20. $\begin{bmatrix} a+2 & 3z+1 & 5m \\ 4k & 0 & 3 \end{bmatrix} + \begin{bmatrix} 3a & 2z & 5m \\ 2k & 5 & 6 \end{bmatrix} = \begin{bmatrix} 10 & -14 & 80 \\ 10 & 5 & 9 \end{bmatrix}$

Perform each of the following operations, whenever possible. See Examples 4 and 5.

21. $\begin{bmatrix} 6 & -9 & 2 \\ 4 & 1 & 3 \end{bmatrix} - \begin{bmatrix} -8 & 2 & 5 \\ 6 & -3 & 4 \end{bmatrix}$

22. $\begin{bmatrix} 9 & 4 \\ -8 & 2 \end{bmatrix} + \begin{bmatrix} -3 & 2 \\ -4 & 7 \end{bmatrix}$

23. $2\begin{bmatrix} -6 & 8 \\ 0 & 0 \end{bmatrix} - \begin{bmatrix} 0 & 2 \\ -4 & -2 \end{bmatrix}$

24. $3\begin{bmatrix} 6 & -1 & 4 \\ 2 & 8 & -3 \\ -4 & 5 & 6 \end{bmatrix} + 5\begin{bmatrix} -2 & -8 & -6 \\ 4 & 1 & 3 \\ 2 & -1 & 5 \end{bmatrix}$

25. $4\begin{bmatrix} 1 & -4 \\ 2 & -3 \\ -8 & 4 \end{bmatrix} - 3\begin{bmatrix} -6 & 9 \\ -2 & 5 \\ -7 & -12 \end{bmatrix}$

26. $\begin{bmatrix} -8 & -4 & -2 & -3 \\ 2 & -1 & 5 & 9 \end{bmatrix} - 2\begin{bmatrix} 4 & 6 & 8 & -3 \\ -2 & -5 & 1 & 6 \end{bmatrix}$

27. $\begin{bmatrix} -8 & 4 & 0 \\ 2 & 5 & 0 \end{bmatrix} + \begin{bmatrix} 6 & 3 \\ 8 & 9 \end{bmatrix}$

28. $\begin{bmatrix} 2 \\ 3 \end{bmatrix} - \begin{bmatrix} 8 & 1 \\ 9 & 4 \end{bmatrix}$

29. $\begin{bmatrix} 9 & 4 & 1 & -2 \\ 5 & -6 & 3 & 4 \\ 2 & -5 & 1 & 2 \end{bmatrix} - \begin{bmatrix} -2 & 5 & 1 & 3 \\ 0 & 1 & 0 & 2 \\ -8 & 3 & 2 & 1 \end{bmatrix} + \begin{bmatrix} 2 & 4 & 0 & 3 \\ 4 & -5 & 1 & 6 \\ 2 & -3 & 0 & 8 \end{bmatrix}$

30. $\begin{bmatrix} 6 & -2 & 4 \\ -2 & 5 & 8 \\ 1 & 0 & 2 \end{bmatrix} + \begin{bmatrix} 3 & 0 & 8 \\ 1 & -2 & 4 \\ 6 & 9 & -2 \end{bmatrix} - \begin{bmatrix} -4 & 2 & 1 \\ 0 & 3 & -2 \\ 4 & 2 & 0 \end{bmatrix}$

31. $\begin{bmatrix} -4x+2y & -3x+y \\ 6x-3y & 2x-5y \end{bmatrix} + \begin{bmatrix} -8x+6y & 2x \\ 3y-5x & 6x+4y \end{bmatrix}$

32. $\begin{bmatrix} 4k-8y \\ 6z-3x \\ 2k+5a \\ -4m+2n \end{bmatrix} - \begin{bmatrix} 5k+6y \\ 2z+5x \\ 4k+6a \\ 4m-2n \end{bmatrix}$

33. $\begin{bmatrix} 1.003 & 4.728 \\ 2.341 & 1.572 \\ 4.918 & 2.763 \end{bmatrix} + \begin{bmatrix} 9.871 & 12.283 \\ 10.268 & 14.001 \\ 3.974 & 11.754 \end{bmatrix}$

34. $\begin{bmatrix} .0012 & .9164 & .0915 \\ .3571 & .3752 & .1134 \\ .4263 & .8036 & .2042 \end{bmatrix} + \begin{bmatrix} .6813 & .1119 & .2971 \\ .2471 & .0743 & .7642 \\ .0065 & .3685 & .6603 \end{bmatrix}$

Let $A = \begin{bmatrix} -2 & 4 \\ 0 & 3 \end{bmatrix}$ and $B = \begin{bmatrix} -6 & 2 \\ 4 & 0 \end{bmatrix}$

Find each of the following matrices. See Example 6.

35. $2A$ **36.** $-3B$ **37.** $-4A$ **38.** $5B$

39. $2A - B$ **40.** $-4A + 5B$ **41.** $3A - 11B$ **42.** $-2A + 4B$

43. $3A - 2B$ **44.** $-9A + B$

Work each of the following problems.

45. Mary found that her new car averaged 18.2 miles per gallon the first week, 19.0 the second week, 17.6 the third week, and 18.5 the fourth week. Write this information as a row matrix and then as a column matrix.

46. Rick's scores on the first three tests in his math class were 82, 77, and 85. Paul scored 91, 80, and 82 on the same three tests. Steve's scores on the three tests were 90, 82, and 79. Write this information as a 3 × 3 square matrix in two different ways.

47. Richard Macias bought 7 shares of Sears stock, 9 shares of IBM stock, and 8 shares of Chrysler stock. The following month, he bought 2 shares of Sears stock, no IBM, and 6 shares of Chrysler. Write this information first as a 3 × 2 matrix and then as a 2 × 3 matrix.

48. Margie Bezzone works in a computer store. The first week she sold 5 computers, 3 printers, 4 disc drives, and 6 monitors. The next week she sold 4 computers, 2 printers, 6 disc drives, and 5 monitors. Write this information first as a 2 × 4 matrix and then as a 4 × 2 matrix.

Addition of real numbers satisfies certain properties, such as the commutative, associative, identity, inverse, and closure properties. In the following exercises, you can check to see which of these properties are satisfied by addition of matrices.

Let $A = \begin{bmatrix} a & b \\ c & d \end{bmatrix}$, $B = \begin{bmatrix} e & f \\ g & h \end{bmatrix}$, and $C = \begin{bmatrix} j & m \\ k & n \end{bmatrix}$

Decide which of the following statements are true for these square matrices. Then decide if a similar property holds for any matrices of the same order.

49. $A + B = B + A$ (commutative property)

50. $A + (B + C) = (A + B) + C$ (associative property)

51. $A + B$ is a 2 × 2 matrix.   (closure property)

52. There exists a matrix O such that $A + O = A$ and $O + A = A$.   (identity property)

53. There exists a matrix $-A$ such that $A + (-A) = O$ and $-A + A = O$.   (inverse property)

54. Are these properties valid for matrices that are not square?

8.2   Matrix Products

In the last section we multiplied a matrix by a scalar. Multiplication of two matrices is more complicated, but it is important in practical problems. To show the reasoning behind matrix multiplication, the example about the vitamin pills from Section 8.1 is continued here.

The matrix below, from the previous section, shows the number of cartons of each type of vitamin received from each company.

$$\begin{bmatrix} 2 & 7 & 5 \\ 4 & 6 & 9 \end{bmatrix}$$

The two rows of this matrix represent the number of cartons of each kind of vitamin received from Dexter and Sullivan respectively. Now suppose that each carton of vitamin A pills costs the store $12, each carton of vitamin E pills costs $18, and each carton of vitamin K costs $9.

To find the total cost of the pills from Dexter, multiply as follows.

Vitamin	Number of cartons	Cost per carton	Total cost
A	2	$12	$ 24
E	7	$18	$126
K	5	$ 9	$ 45
			$195 Total from Dexter

The Dexter pills cost a total of $195.

This result is the sum of three products:

$$2(\$12) + 7(\$18) + 5(\$9) = \$195.$$

In the same way, using the second row of the matrix and the three costs, the total cost of the Sullivan pills is

$$4(\$12) + 6(\$18) + 9(\$9) = \$237.$$

The costs, $12, $18, and $9, can be written as a column matrix.

$$\begin{bmatrix} 12 \\ 18 \\ 9 \end{bmatrix}$$

The total costs for each supplier, $195 and $237, also can be written as a column matrix.

$$\begin{bmatrix} 195 \\ 237 \end{bmatrix}$$

The product of the matrices

$$\begin{bmatrix} 2 & 7 & 5 \\ 4 & 6 & 9 \end{bmatrix} \quad \text{and} \quad \begin{bmatrix} 12 \\ 18 \\ 9 \end{bmatrix}$$

can be written as follows.

$$\begin{bmatrix} 2 & 7 & 5 \\ 4 & 6 & 9 \end{bmatrix} \begin{bmatrix} 12 \\ 18 \\ 9 \end{bmatrix} = \begin{bmatrix} 2 \cdot 12 + 7 \cdot 18 + 5 \cdot 9 \\ 4 \cdot 12 + 6 \cdot 18 + 9 \cdot 9 \end{bmatrix} = \begin{bmatrix} 195 \\ 237 \end{bmatrix}$$

Each element of the product was found by multiplying the elements of the *rows* of the matrix on the left and the corresponding elements of the *columns* of the matrix on the right, and then finding the sum of these products. Notice that the product of a 2×3 matrix and a 3×1 matrix is a 2×1 matrix.

Generalizing from this example gives the following definition of matrix multiplication.

Definition of Matrix Multiplication	The product AB of an $m \times n$ matrix A and an $n \times k$ matrix B is found as follows. To get the ith row, jth column element of AB, multiply each element in the ith row of A by the corresponding element in the jth column of B. The sum of these products will give the element of row i, column j of AB.

Example 1 Find the product AB, where

$$A = \begin{bmatrix} 2 & 4 \\ 5 & 6 \end{bmatrix} \quad \text{and} \quad B = \begin{bmatrix} -3 & 5 \\ 4 & -2 \end{bmatrix}$$

Step 1 Multiply the elements of the first row of A and the corresponding elements of the first column of B, and add these products.

$$AB = \begin{bmatrix} \mathbf{2} & \mathbf{4} \\ 5 & 6 \end{bmatrix} \begin{bmatrix} -3 & 5 \\ 4 & -2 \end{bmatrix} \qquad \mathbf{2}(-3) + \mathbf{4}(4) = -6 + 16 = 10$$

The first-row, first-column entry of the product matrix AB is 10.

Step 2 Multiply the elements of the first row of A and the second column of B and then add the products to get the first-row, second-column entry of the product matrix.

$$AB = \begin{bmatrix} \mathbf{2} & \mathbf{4} \\ 5 & 6 \end{bmatrix} \begin{bmatrix} -3 & 5 \\ -4 & -2 \end{bmatrix} \qquad \mathbf{2}(5) + \mathbf{4}(-2) = 10 + (-8) = 2$$

Step 3

$$AB = \begin{bmatrix} 2 & 4 \\ 5 & 6 \end{bmatrix} \begin{bmatrix} -3 & 5 \\ 4 & -2 \end{bmatrix} \qquad 5(-3) + 6(4) = -15 + 24 = 9$$

The second-row, first-column entry of the product matrix is 9.

Step 4

$$AB = \begin{bmatrix} 2 & 4 \\ 5 & 6 \end{bmatrix} \begin{bmatrix} -3 & 5 \\ 4 & -2 \end{bmatrix} \qquad 5(5) + 6(-2) = 25 + (-12) = 13.$$

Finally, 13 is the second-row, second-column entry.

Step 5 Write the product. The four entries in the product matrix come from the four steps above.

$$AB = \begin{bmatrix} 2 & 4 \\ 5 & 6 \end{bmatrix} \begin{bmatrix} -3 & 5 \\ 4 & -2 \end{bmatrix} = \begin{bmatrix} 10 & 2 \\ 9 & 13 \end{bmatrix} \quad \clubsuit$$

Example 2   Find the product

$$\begin{bmatrix} -3 & 4 & 2 \\ 5 & 0 & 4 \end{bmatrix} \begin{bmatrix} -6 & 4 \\ 2 & 3 \\ 3 & -2 \end{bmatrix}$$

Step 1 $\begin{bmatrix} -3 & 4 & 2 \\ 5 & 0 & 4 \end{bmatrix} \begin{bmatrix} -6 & 4 \\ 2 & 3 \\ 3 & -2 \end{bmatrix}$ $(-3)(-6) + 4(2) + 2(3) = 32$

Step 2 $\begin{bmatrix} -3 & 4 & 2 \\ 5 & 0 & 4 \end{bmatrix} \begin{bmatrix} -6 & 4 \\ 2 & 3 \\ 3 & -2 \end{bmatrix}$ $(-3)(4) + 4(3) + 2(-2) = -4$

Step 3 $\begin{bmatrix} -3 & 4 & 2 \\ 5 & 0 & 4 \end{bmatrix} \begin{bmatrix} -6 & 4 \\ 2 & 3 \\ 3 & -2 \end{bmatrix}$ $5(-6) + 0(2) + 4(3) = -18$

Step 4 $\begin{bmatrix} -3 & 4 & 2 \\ 5 & 0 & 4 \end{bmatrix} \begin{bmatrix} -6 & 4 \\ 2 & 3 \\ 3 & -2 \end{bmatrix}$ $5(4) + 0(3) + 4(-2) = 12$

Step 5   Write the product.

$$\begin{bmatrix} -3 & 4 & 2 \\ 5 & 0 & 4 \end{bmatrix} \begin{bmatrix} -6 & 4 \\ 2 & 3 \\ 3 & -2 \end{bmatrix} = \begin{bmatrix} 32 & -4 \\ -18 & 12 \end{bmatrix}$$

As this example shows, the product of a 2×3 matrix and a 3×2 matrix is a 2×2 matrix. ✛

The examples suggest the following restriction on matrix multiplication.

> The product AB of two matrices A and B can be found only if the number of *columns* of A is the same as the number of *rows* of B. The final product will have as many rows as A and as many columns as B.

Example 3   Suppose matrix A is 2×2, while matrix B is 2×4. Can the product AB be calculated? What is the order of the product?

The following diagram helps answer these questions.

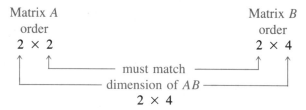

The product AB can be calculated because A has two columns and B has two rows. The order of the product is 2×4. (However, the product BA could not be found.) ✛

Example 4   Find AB and BA if

$$A = \begin{bmatrix} 1 & -3 \\ 7 & 2 \end{bmatrix} \quad \text{and} \quad B = \begin{bmatrix} 1 & 0 & -1 & 2 \\ 3 & 1 & 4 & -1 \end{bmatrix}$$

Use the definition of matrix multiplication to find AB.

$$
\begin{aligned}
AB &= \begin{bmatrix} 1 & -3 \\ 7 & 2 \end{bmatrix}\begin{bmatrix} 1 & 0 & -1 & 2 \\ 3 & 1 & 4 & -1 \end{bmatrix} \\
&= \begin{bmatrix} 1(1) + (-3)(3) & 1(0) + (-3)1 & 1(-1) + (-3)4 & 1(2) + (-3)(-1) \\ 7(1) + 2(3) & 7(0) + 2(1) & 7(-1) + 2(4) & 7(2) + 2(-1) \end{bmatrix} \\
&= \begin{bmatrix} -8 & -3 & -13 & 5 \\ 13 & 2 & 1 & 12 \end{bmatrix}
\end{aligned}
$$

Since B is a 2×4 matrix, and A is a 2×2 matrix, the product BA cannot be found. ✛

Example 5   Find MN and NM, given

$$M = \begin{bmatrix} 1 & 3 \\ -2 & 4 \end{bmatrix} \quad \text{and} \quad N = \begin{bmatrix} 2 & 5 \\ 10 & -3 \end{bmatrix}$$

By the definition of matrix multiplication,

$$MN = \begin{bmatrix} 1 & 3 \\ -2 & 4 \end{bmatrix}\begin{bmatrix} 2 & 5 \\ 10 & -3 \end{bmatrix}$$

$$= \begin{bmatrix} 1(2) + 3(10) & 1(5) + 3(-3) \\ -2(2) + 4(10) & -2(5) + 4(-3) \end{bmatrix}$$

$$= \begin{bmatrix} 32 & -4 \\ 36 & -22 \end{bmatrix}$$

Similarly,

$$NM = \begin{bmatrix} 2 & 5 \\ 10 & -3 \end{bmatrix}\begin{bmatrix} 1 & 3 \\ -2 & 4 \end{bmatrix}$$

$$= \begin{bmatrix} 2(1) + 5(-2) & 2(3) + 5(4) \\ 10(1) + (-3)(-2) & 10(3) + (-3)4 \end{bmatrix}$$

$$= \begin{bmatrix} -8 & 26 \\ 16 & 18 \end{bmatrix} \quad ❖$$

In Example 4 the product AB could be found, but not BA. In Example 5, although both MN and NM could be found, they were not equal, showing that multiplication of matrices is not commutative. This fact distinguishes matrix arithmetic from the arithmetic of real numbers. Other properties of matrix arithmetic are given as Exercises 39–45 at the end of this section.

Example 6   A contractor builds three kinds of houses, models A, B, and C, with a choice of two styles, colonial or ranch. Matrix P below shows the number of each kind of house the contractor is planning to build for a new 100-home subdivision. The amounts for each of the main materials used depend on the style of the house. These amounts are shown in matrix Q below, while matrix R gives the cost in dollars for each kind of material. Concrete is measured here in cubic yards, lumber in 1000 board feet, brick in 1000's, and shingles in 100 square feet.

	Colonial	**Ranch**	
Model A	0	30	
Model B	10	20	$= P$
Model C	20	20	

	Concrete	Lumber	Brick	Shingles	
Colonial	10	2	0	2	
Ranch	50	1	20	2	$= Q$

	Cost per unit	
Concrete	20	
Lumber	180	
Brick	60	$= R$
Shingles	25	

(a) What is the total cost of materials for all houses of each model?

To find the materials cost for each model, first find matrix PQ, which will show the total amount of each material needed for all houses of each model.

$$PQ = \begin{bmatrix} 0 & 30 \\ 10 & 20 \\ 20 & 20 \end{bmatrix} \begin{bmatrix} 10 & 2 & 0 & 2 \\ 50 & 1 & 20 & 2 \end{bmatrix} = \begin{array}{cccc} \text{Concrete} & \text{Lumber} & \text{Brick} & \text{Shingles} \end{array} \begin{bmatrix} 1500 & 30 & 600 & 60 \\ 1100 & 40 & 400 & 60 \\ 1200 & 60 & 400 & 80 \end{bmatrix} \begin{array}{l} \text{Model A} \\ \text{Model B} \\ \text{Model C} \end{array}$$

Multiplying PQ and the cost matrix R gives the total cost of materials for each model.

$$(PQ)R = \begin{bmatrix} 1500 & 30 & 600 & 60 \\ 1100 & 40 & 400 & 60 \\ 1200 & 60 & 400 & 80 \end{bmatrix} \begin{bmatrix} 20 \\ 180 \\ 60 \\ 25 \end{bmatrix} = \begin{array}{c} \text{Cost} \end{array} \begin{bmatrix} 72,900 \\ 54,700 \\ 60,800 \end{bmatrix} \begin{array}{l} \text{Model A} \\ \text{Model B} \\ \text{Model C} \end{array}$$

(b) How much of each of the four kinds of material must be ordered?

The totals of the columns of matrix PQ will give a matrix whose elements represent the total amounts of each material needed for the subdivision. Call this matrix T and write it as a row matrix.

$$T = \begin{bmatrix} 3800 & 130 & 1400 & 200 \end{bmatrix}$$

(c) What is the total cost of the materials?

The total cost of all the materials is given by the product of matrix R, the cost matrix, and matrix T, the total amounts matrix. To multiply these and get a 1×1 matrix, representing the total cost, requires multiplying a 1×4 matrix and a 4×1 matrix. This is why in (b) a row matrix was written rather than a column matrix. The total materials cost is given by TR, so

$$TR = \begin{bmatrix} 3800 & 130 & 1400 & 200 \end{bmatrix} \begin{bmatrix} 20 \\ 180 \\ 60 \\ 25 \end{bmatrix} = \begin{bmatrix} 188,400 \end{bmatrix}.$$

The total cost of the materials is $188,400. ✛

To help keep track of the quantities a matrix represents, let matrix P, from Example 6, represent models/styles, matrix Q represent styles/materials, and matrix R represent materials/cost. In each case the meaning of the rows is written first and that of the columns second. When the product PQ was found in Example 6, the rows of the matrix represented models and the columns represented materials. Therefore, the matrix product PQ represents models/materials. The common quantity, styles, in both P and Q was eliminated in the product PQ. Do you see that the product $(PQ)R$ represents models/cost?

In practical problems this notation helps to identify the order in which two matrices should be multiplied so that the results are meaningful. In Example 6(c), either product RT or product TR could have been found. However, since T represents subdivisions/materials and R represents materials/cost, only TR gave the required matrix representing subdivisions/cost.

8.2 Exercises

In each of the following exercises, the order of two matrices A and B is given. Find the order of the product AB and the product BA, whenever these products exist. See Example 3.

1. A is 2×2, B is 2×2

2. A is 3×3, B is 3×3

3. A is 4×2, B is 2×4

4. A is 3×1, B is 1×3

5. A is 3×5, B is 5×2

6. A is 4×3, B is 3×6

7. A is 4×2, B is 3×4

8. A is 7×3, B is 2×7

9. A is 4×3, B is 2×5

10. A is 1×6, B is 2×4

Find each of the following matrix products, whenever possible. See Examples 1 and 2.

11. $\begin{bmatrix} 1 & 2 \\ 3 & 4 \end{bmatrix} \begin{bmatrix} -1 \\ 7 \end{bmatrix}$

12. $\begin{bmatrix} -1 & 5 \\ 7 & 0 \end{bmatrix} \begin{bmatrix} 6 \\ 2 \end{bmatrix}$

13. $\begin{bmatrix} 3 & -4 & 1 \\ 5 & 0 & 2 \end{bmatrix} \begin{bmatrix} -1 \\ 4 \\ 2 \end{bmatrix}$

14. $\begin{bmatrix} -6 & 3 & 5 \\ 2 & 9 & 1 \end{bmatrix} \begin{bmatrix} -2 \\ 0 \\ 3 \end{bmatrix}$

15. $\begin{bmatrix} 5 & 2 \\ -1 & 4 \end{bmatrix} \begin{bmatrix} 3 & -2 \\ 1 & 0 \end{bmatrix}$

16. $\begin{bmatrix} -4 & 0 \\ 1 & 3 \end{bmatrix} \begin{bmatrix} -2 & 4 \\ 0 & 1 \end{bmatrix}$

17. $\begin{bmatrix} 2 & 2 & -1 \\ 3 & 0 & 1 \end{bmatrix} \begin{bmatrix} 0 & 2 \\ -1 & 4 \\ 0 & 2 \end{bmatrix}$

18. $\begin{bmatrix} -9 & 2 & 1 \\ 3 & 0 & 0 \end{bmatrix} \begin{bmatrix} 2 \\ -1 \\ 4 \end{bmatrix}$

19. $\begin{bmatrix} -1 & 2 & 0 \\ 0 & 3 & 2 \\ 0 & 1 & 4 \end{bmatrix} \begin{bmatrix} 2 & -1 & 2 \\ 0 & 2 & 1 \\ 3 & 0 & -1 \end{bmatrix}$

20. $\begin{bmatrix} -2 & -3 & -4 \\ 2 & -1 & 0 \\ 4 & -2 & 3 \end{bmatrix} \begin{bmatrix} 0 & 1 & 4 \\ 1 & 2 & -1 \\ 3 & 2 & -2 \end{bmatrix}$

21. $[-2 \quad 4 \quad 1] \begin{bmatrix} 3 & -2 & 4 \\ 2 & 1 & 0 \\ 0 & -1 & 4 \end{bmatrix}$

22. $[0 \quad 3 \quad -4] \begin{bmatrix} -2 & 6 & 3 \\ 0 & 4 & 2 \\ -1 & 1 & 4 \end{bmatrix}$

23. $\begin{bmatrix} -2 & 1 & 4 \\ 0 & 1 & 2 \end{bmatrix} \begin{bmatrix} -2 & 1 & 0 \\ 0 & -2 & 0 \\ 4 & 1 & 2 \end{bmatrix}$

24. $\begin{bmatrix} -1 & 0 & 0 \\ 2 & 1 & 4 \end{bmatrix} \begin{bmatrix} 4 & -2 & 5 \\ 0 & 1 & 4 \\ 2 & -9 & 0 \end{bmatrix}$

25. $\begin{bmatrix} -3 & 0 & 2 & 1 \\ 4 & 0 & 2 & 6 \end{bmatrix} \begin{bmatrix} -4 & 2 \\ 0 & 1 \end{bmatrix}$

26. $\begin{bmatrix} -1 & 2 & 4 & 1 \\ 0 & 2 & -3 & 5 \end{bmatrix} \begin{bmatrix} 1 & 2 & 4 \\ -2 & 5 & 1 \end{bmatrix}$

27. $[-2 \quad 4 \quad 6] \begin{bmatrix} 3 \\ -2 \\ 1 \end{bmatrix}$

28. $[4 \quad 0 \quad 2] \begin{bmatrix} -5 \\ 1 \\ 6 \end{bmatrix}$

29. $\begin{bmatrix} 3 \\ -2 \\ 1 \end{bmatrix} [-2 \quad 4 \quad 6]$

30. $\begin{bmatrix} -5 \\ 1 \\ 6 \end{bmatrix} [4 \quad 0 \quad 2]$

31. $\begin{bmatrix} 1.25 & 2.83 \\ 6.94 & 5.07 \end{bmatrix} \begin{bmatrix} 1.04 & 2.93 \\ 3.68 & 6.55 \end{bmatrix}$

32. $\begin{bmatrix} .041 & .002 \\ .610 & .345 \\ .992 & .613 \end{bmatrix} \begin{bmatrix} .116 & .581 \\ .423 & .704 \end{bmatrix}$

Let $A = \begin{bmatrix} -2 & 4 \\ 1 & 3 \end{bmatrix}$, $B = \begin{bmatrix} -2 & 1 \\ 3 & 6 \end{bmatrix}$, and $C = \begin{bmatrix} 5 & -2 & 1 \\ 0 & 3 & 7 \end{bmatrix}$

Find each of the following products. See Examples 4 and 5.

33. AB **34.** BA **35.** AC **36.** CA

37. Did you get the same answer in Exercises 33 and 34? Do you think that matrix multiplication is commutative?

38. In general, for matrices A and B such that AB and BA both exist, does AB always equal BA?

For the following exercises, let

$$A = \begin{bmatrix} a & b \\ c & d \end{bmatrix} \quad B = \begin{bmatrix} e & f \\ g & h \end{bmatrix} \quad and \quad C = \begin{bmatrix} j & m \\ k & n \end{bmatrix}$$

Decide which of the following statements are true for these three matrices. Then decide if a similar property holds for any square matrices of the same order.

39. $(AB)C = A(BC)$ (associative property)

40. $A(B + C) = AB + AC$ (distributive property)

41. AB is a 2×2 matrix   (closure property)

42. $k(A + B) = kA + kB$ for any real number k

43. $(k + p)A = kA + pA$ for any real numbers k and p

44. $(A + B)(A - B) = A^2 - B^2$ $(A^2 = AA)$

45. $(A + B)^2 = A^2 + 2AB + B^2$

46. The Bread Box, a small neighborhood bakery, sells four main items: sweet rolls, bread, cakes, and pies. The amount of each ingredient (in cups, except for eggs) required for these items is given by matrix A.

	Eggs	Flour	Sugar	Shortening	Milk
Rolls (doz)	1	4	1/4	1/4	1
Bread (loaves)	0	3	0	1/4	0
Cakes	4	3	2	1	1
Pies (crust)	0	1	0	1/3	0

$= A$

The cost (in cents) for each ingredient when purchased in large lots or small lots is given in matrix B.

Cost

	Large lot	Small lot
Eggs	5	5
Flour	8	10
Sugar	10	12
Shortening	12	15
Milk	5	6

$= B$

(a) Use matrix multiplication to find a matrix giving the comparative cost per item for the two purchase options.

Suppose a day's orders consist of 20 dozen sweet rolls, 200 loaves of bread, 50 cakes and 60 pies.

(b) Writing the orders as a 1×4 matrix and using matrix multiplication, write as a matrix the amount of each ingredient needed to fill the day's orders.

(c) Use matrix multiplication to find a matrix giving the costs under the two purchase options to fill the day's orders.

47. For an article in the *Daily News,* reporter Mary Starr compared prices at the three leading supermarkets in town on four items: a quart of milk, a head of lettuce, a loaf of bread, and a pound of hamburger. Her results are summarized in the following matrix.

Market

	I	II	III
Milk	.45	.46	.45
Lettuce	.59	.63	.58
Bread	.95	.89	.92
Hamburger	1.65	1.68	1.60

For her survey, Starr bought 3 quarts of milk, 1 head of lettuce, 2 loaves of bread and 4 pounds of hamburger at each market.

(a) Write the amount of each item purchased as a 1×4 row matrix.

(b) Multiply the two matrices to find Starr's total bill at each market.

8.3   Solution of Linear Systems by Inverses

In this section multiplicative identity elements and multiplicative inverses are introduced and used to solve matrix equations. This leads to another method for solving systems of equations.

The identity property for real numbers says that $a \cdot 1 = a$ and $1 \cdot a = a$ for any real number a. If there is to be a multiplicative identity matrix I, such that

$$AI = A \quad \text{and} \quad IA = A,$$

for any matrix A, then A and I must be square matrices of the same order. Otherwise it would not be possible to find both products. For example, let A be the 2×2 matrix

$$A = \begin{bmatrix} a_{11} & a_{12} \\ a_{21} & a_{22} \end{bmatrix},$$

and let

$$I = \begin{bmatrix} x_{11} & x_{12} \\ x_{21} & x_{22} \end{bmatrix}$$

represent the 2×2 identity matrix. To find I, use the fact that $IA = A$, or

$$\begin{bmatrix} x_{11} & x_{12} \\ x_{21} & x_{22} \end{bmatrix} \begin{bmatrix} a_{11} & a_{12} \\ a_{21} & a_{22} \end{bmatrix} = \begin{bmatrix} a_{11} & a_{12} \\ a_{21} & a_{22} \end{bmatrix}.$$

Multiplying the two matrices on the left side of this equation and setting the elements of the product matrix equal to the corresponding elements of A gives the following system of equations with variables x_{11}, x_{12}, x_{21}, and x_{22}.

$$x_{11}a_{11} + x_{12}a_{21} = a_{11}$$
$$x_{11}a_{12} + x_{12}a_{22} = a_{12}$$
$$x_{21}a_{11} + x_{22}a_{21} = a_{21}$$
$$x_{21}a_{12} + x_{22}a_{22} = a_{22}$$

Notice that this is really two systems of equations in two variables. Use the methods of Section 7.1 to find the solution of this system: $x_{11} = 1$, $x_{12} = x_{21} = 0$, and $x_{22} = 1$. From the solution of the system, the 2×2 identity matrix is

$$I = \begin{bmatrix} 1 & 0 \\ 0 & 1 \end{bmatrix}.$$

Check that with this definition of I, both $AI = A$ and $IA = A$.

Example 1   Let $M = \begin{bmatrix} -2 & 6 \\ 3 & 5 \end{bmatrix}$. Verify that $MI = M$ and $IM = M$.

$$MI = \begin{bmatrix} -2 & 6 \\ 3 & 5 \end{bmatrix} \begin{bmatrix} 1 & 0 \\ 0 & 1 \end{bmatrix} = \begin{bmatrix} -2(1) + 6(0) & -2(0) + 6(1) \\ 3(1) + 5(0) & 3(0) + 5(1) \end{bmatrix}$$

$$= \begin{bmatrix} -2 & 6 \\ 3 & 5 \end{bmatrix} = M$$

$$IM = \begin{bmatrix} 1 & 0 \\ 0 & 1 \end{bmatrix} \begin{bmatrix} -2 & 6 \\ 3 & 5 \end{bmatrix} = \begin{bmatrix} 1(-2) + 0(3) & 1(6) + 0(5) \\ 0(-2) + 1(3) & 0(6) + 1(5) \end{bmatrix}$$

$$= \begin{bmatrix} -2 & 6 \\ 3 & 5 \end{bmatrix} = M \quad \clubsuit$$

The 2×2 identity matrix found above suggests the following generalization.

$n \times n$ Identity Matrix

For any value of n there is an $n \times n$ identity matrix having 1's down the diagonal and 0's elsewhere. The **$n \times n$ identity matrix** is given by I, where

$$I = \begin{bmatrix} 1 & 0 & \cdots & 0 \\ 0 & 1 & \cdots & 0 \\ & & & \\ & & a_{ij} & \\ & & & \\ 0 & 0 & \cdots & 1 \end{bmatrix}.$$

Here $a_{ij} = 1$ when $i = j$ (the diagonal elements) and $a_{ij} = 0$ otherwise.

Example 2   Let $K = \begin{bmatrix} -2 & 4 & 0 \\ 3 & 5 & 9 \\ 0 & 8 & -6 \end{bmatrix}$. Give the 3×3 identity matrix I and show that

$KI = K$.

The 3×3 identity matrix is

$$I = \begin{bmatrix} 1 & 0 & 0 \\ 0 & 1 & 0 \\ 0 & 0 & 1 \end{bmatrix}.$$

By the definition of matrix multiplication,

$$KI = \begin{bmatrix} -2 & 4 & 0 \\ 3 & 5 & 9 \\ 0 & 8 & -6 \end{bmatrix} \begin{bmatrix} 1 & 0 & 0 \\ 0 & 1 & 0 \\ 0 & 0 & 1 \end{bmatrix} = \begin{bmatrix} -2 & 4 & 0 \\ 3 & 5 & 9 \\ 0 & 8 & -6 \end{bmatrix} = K. \quad \clubsuit$$

For every nonzero real number a, there is a multiplicative inverse $1/a$ such that

$$a \cdot \frac{1}{a} = 1 \quad \text{and} \quad \frac{1}{a} \cdot a = 1.$$

(Recall that $1/a$ can also be written a^{-1}.) In the rest of this section, a method is developed for finding a multiplicative inverse for square matrices. The **multiplicative inverse** of a matrix A is written A^{-1}. This matrix must satisfy the statements

$$AA^{-1} = I \quad \text{and} \quad A^{-1}A = I.$$

The multiplicative inverse of a matrix can be found using the matrix row operations given in Section 7.3 and repeated here for convenience.

The **matrix row operations** are:

1. interchanging any two rows of a matrix.

2. multiplying the elements of any row of a matrix by the same nonzero scalar k, and

3. adding a multiple of the elements of one row to the elements of another row.

As an example, let us find the inverse of

$$A = \begin{bmatrix} 2 & 4 \\ 1 & -1 \end{bmatrix}.$$

Let the unknown inverse matrix be

$$A^{-1} = \begin{bmatrix} x & y \\ z & w \end{bmatrix}.$$

By the definition of matrix inverse, $AA^{-1} = I$, or

$$AA^{-1} = \begin{bmatrix} 2 & 4 \\ 1 & -1 \end{bmatrix}\begin{bmatrix} x & y \\ z & w \end{bmatrix} = \begin{bmatrix} 1 & 0 \\ 0 & 1 \end{bmatrix}.$$

By matrix multiplication,

$$\begin{bmatrix} 2x + 4z & 2y + 4w \\ x - z & y - w \end{bmatrix} = \begin{bmatrix} 1 & 0 \\ 0 & 1 \end{bmatrix}.$$

Setting corresponding elements equal gives the system of equations

$$2x + 4z = 1 \qquad \textbf{(1)}$$
$$2y + 4w = 0 \qquad \textbf{(2)}$$

$$x - z = 0 \tag{3}$$
$$y - w = 1. \tag{4}$$

Since equations (1) and (3) involve only x and z, while equations (2) and (4) involve only y and w, these four equations lead to two systems of equations,

$$
\begin{aligned}
2x + 4z &= 1 \\
x - z &= 0
\end{aligned}
\quad \text{and} \quad
\begin{aligned}
2y + 4w &= 0 \\
y - w &= 1.
\end{aligned}
$$

Writing the two systems as augmented matrices gives

$$
\left[\begin{array}{rr|r} 2 & 4 & 1 \\ 1 & -1 & 0 \end{array}\right]
\quad \text{and} \quad
\left[\begin{array}{rr|r} 2 & 4 & 0 \\ 1 & -1 & 1 \end{array}\right].
$$

Each of these systems can be solved by the Gauss-Jordan method. However, since the elements to the left of the vertical bar are identical, the two systems can be combined into one matrix,

$$
\left[\begin{array}{rr|rr} 2 & 4 & 1 & 0 \\ 1 & -1 & 0 & 1 \end{array}\right],
$$

and solved simultaneously as follows. Exchange the two rows to get a 1 in the upper left corner.

$$
\left[\begin{array}{rr|rr} 1 & -1 & 0 & 1 \\ 2 & 4 & 1 & 0 \end{array}\right]
$$

Multiply the first row by -2 and add the results to the second row to get

$$
\left[\begin{array}{rr|rr} 1 & -1 & 0 & 1 \\ 0 & 6 & 1 & -2 \end{array}\right]. \qquad -2\,R_1 + R_2
$$

Now, to get a 1 in the second-row, second-column position, multiply the second row by 1/6.

$$
\left[\begin{array}{rr|rr} 1 & -1 & 0 & 1 \\ 0 & 1 & \frac{1}{6} & -\frac{1}{3} \end{array}\right] \qquad \frac{1}{6}\,R_2
$$

Finally, add the second row to the first row to get a 0 in the second column above the 1.

$$
\left[\begin{array}{rr|rr} 1 & 0 & \frac{1}{6} & \frac{2}{3} \\ 0 & 1 & \frac{1}{6} & -\frac{1}{3} \end{array}\right] \qquad R_2 + R_1
$$

The numbers in the first column to the right of the vertical bar give the values of x and z. The second column gives the values of y and w. That is,

$$
\left[\begin{array}{rr|rr} 1 & 0 & x & y \\ 0 & 1 & z & w \end{array}\right]
=
\left[\begin{array}{rr|rr} 1 & 0 & \frac{1}{6} & \frac{2}{3} \\ 0 & 1 & \frac{1}{6} & -\frac{1}{3} \end{array}\right]
$$

so that

$$A^{-1} = \begin{bmatrix} x & y \\ z & w \end{bmatrix} = \begin{bmatrix} \frac{1}{6} & \frac{2}{3} \\ \frac{1}{6} & -\frac{1}{3} \end{bmatrix}.$$

To check, multiply A by A^{-1}. The result should be I.

$$AA^{-1} = \begin{bmatrix} 2 & 4 \\ 1 & -1 \end{bmatrix}\begin{bmatrix} \frac{1}{6} & \frac{2}{3} \\ \frac{1}{6} & -\frac{1}{3} \end{bmatrix} = \begin{bmatrix} \frac{1}{3} + \frac{2}{3} & \frac{4}{3} - \frac{4}{3} \\ \frac{1}{6} - \frac{1}{6} & \frac{2}{3} + \frac{1}{3} \end{bmatrix} = \begin{bmatrix} 1 & 0 \\ 0 & 1 \end{bmatrix} = I.$$

Verify that $A^{-1}A = I$, also. Finally,

$$A^{-1} = \begin{bmatrix} \frac{1}{6} & \frac{2}{3} \\ \frac{1}{6} & -\frac{1}{3} \end{bmatrix}.$$

The process for finding the multiplicative inverse A^{-1} for any $n \times n$ matrix A that has an inverse is summarized below.

Finding an Inverse Matrix	To obtain A^{-1} for any $n \times n$ matrix A for which A^{-1} exists, follow these steps. **1.** Form the augmented matrix $[A\|I]$, where I is the $n \times n$ identity matrix. **2.** Perform row operations on $[A\|I]$ to get a matrix of the form $[I\|B]$. **3.** Matrix B is A^{-1}. **4.** Verify by showing that $BA = AB = I$.

Example 3   Given

$$A = \begin{bmatrix} 2 & 3 \\ 1 & -1 \end{bmatrix},$$

find A^{-1}.

Find A^{-1} by going through the following steps.

Step 1   Form the augmented matrix $[A\|I]$.

$$[A\|I] = \begin{bmatrix} 2 & 3 & 1 & 0 \\ 1 & -1 & 0 & 1 \end{bmatrix}$$

Perform row operations on $[A\|I]$ until a new matrix of the form $[I\|B]$ is obtained, as follows.

Step 2   Interchanging the first and second rows gives a 1 in the upper left-hand corner.

$$\begin{bmatrix} 1 & -1 & 0 & 1 \\ 2 & 3 & 1 & 0 \end{bmatrix}$$

Step 3   Multiplying the elements of the first row by -2 and adding the results to the second row gives a 0 in the lower left-hand corner.

$$\left[\begin{array}{rr|rr} 1 & -1 & 0 & 1 \\ 0 & 5 & 1 & -2 \end{array}\right] \quad -2\,R_1 + R_2$$

Step 4   Multiplying the elements of the second row by 1/5 gives a 1 for the second element in that row.

$$\left[\begin{array}{rr|rr} 1 & -1 & 0 & 1 \\ 0 & 1 & \frac{1}{5} & -\frac{2}{5} \end{array}\right] \quad \frac{1}{5}\,R_2$$

Step 5   Replacing the first row by the sum of the first row and the second row gives a 0 as the second element in the first row.

$$\left[\begin{array}{rr|rr} 1 & 0 & \frac{1}{5} & \frac{3}{5} \\ 0 & 1 & \frac{1}{5} & -\frac{2}{5} \end{array}\right] \quad R_2 + R_1$$

This last matrix is in the form $[I|B]$, where

$$B = \left[\begin{array}{rr} \frac{1}{5} & \frac{3}{5} \\ \frac{1}{5} & -\frac{2}{5} \end{array}\right],$$

which should equal A^{-1}. To check, multiply B and A. The result should be I.

$$BA = \left[\begin{array}{rr} \frac{1}{5} & \frac{3}{5} \\ \frac{1}{5} & -\frac{2}{5} \end{array}\right]\left[\begin{array}{rr} 2 & 3 \\ 1 & -1 \end{array}\right] = \left[\begin{array}{rr} \frac{2}{5} + \frac{3}{5} & \frac{3}{5} - \frac{3}{5} \\ \frac{2}{5} - \frac{2}{5} & \frac{3}{5} + \frac{2}{5} \end{array}\right] = \left[\begin{array}{rr} 1 & 0 \\ 0 & 1 \end{array}\right] = I$$

Verify that AB also equals I, so that

$$A^{-1} = \left[\begin{array}{rr} \frac{1}{5} & \frac{3}{5} \\ \frac{1}{5} & -\frac{2}{5} \end{array}\right]. \quad \clubsuit$$

Example 4   Find A^{-1} if $A = \left[\begin{array}{rrr} 1 & 0 & 1 \\ 2 & -2 & -1 \\ 3 & 0 & 0 \end{array}\right].$

Use row transformations as follows.

Step 1   Write the augmented matrix $[A|I]$.

$$\left[\begin{array}{rrr|rrr} 1 & 0 & 1 & 1 & 0 & 0 \\ 2 & -2 & -1 & 0 & 1 & 0 \\ 3 & 0 & 0 & 0 & 0 & 1 \end{array}\right]$$

Step 2   Since 1 is already in the upper left-hand corner as desired, begin by using the row operation that will result in a 0 for the first element in the second row. Multiply the elements of the first row by -2, and add the results to the second row.

$$\begin{bmatrix} 1 & 0 & 1 & | & 1 & 0 & 0 \\ 0 & -2 & -3 & | & -2 & 1 & 0 \\ 3 & 0 & 0 & | & 0 & 0 & 1 \end{bmatrix} \quad -2\,R_1 + R_2$$

Step 3 To get 0 for the first element in the third row, multiply the elements of the first row by -3 and add to the third row.

$$\begin{bmatrix} 1 & 0 & 1 & | & 1 & 0 & 0 \\ 0 & -2 & -3 & | & -2 & 1 & 0 \\ 0 & 0 & -3 & | & -3 & 0 & 1 \end{bmatrix} \quad -3\,R_1 + R_3$$

Step 4 To get 1 for the second element in the second row, multiply the elements of the second row by $-1/2$.

$$\begin{bmatrix} 1 & 0 & 1 & | & 1 & 0 & 0 \\ 0 & 1 & \frac{3}{2} & | & 1 & -\frac{1}{2} & 0 \\ 0 & 0 & -3 & | & -3 & 0 & 1 \end{bmatrix} \quad -\frac{1}{2}\,R_2$$

Step 5 To get 1 for the third element in the third row, multiply the elements of the third row by $-1/3$.

$$\begin{bmatrix} 1 & 0 & 1 & | & 1 & 0 & 0 \\ 0 & 1 & \frac{3}{2} & | & 1 & -\frac{1}{2} & 0 \\ 0 & 0 & 1 & | & 1 & 0 & -\frac{1}{3} \end{bmatrix} \quad -\frac{1}{3}\,R_3$$

Step 6 To get 0 for the third element in the first row, multiply the elements of the third row by -1 and add to the first row.

$$\begin{bmatrix} 1 & 0 & 0 & | & 0 & 0 & \frac{1}{3} \\ 0 & 1 & \frac{3}{2} & | & 1 & -\frac{1}{2} & 0 \\ 0 & 0 & 1 & | & 1 & 0 & -\frac{1}{3} \end{bmatrix} \quad -1\,R_3 + R_1$$

Step 7 To get 0 for the third element in the second row, multiply the elements of the third row by $-3/2$ and add to the second row.

$$\begin{bmatrix} 1 & 0 & 0 & | & 0 & 0 & \frac{1}{3} \\ 0 & 1 & 0 & | & -\frac{1}{2} & -\frac{1}{2} & \frac{1}{2} \\ 0 & 0 & 1 & | & 1 & 0 & -\frac{1}{3} \end{bmatrix} \quad -\frac{3}{2}\,R_3 + R_2$$

The last transformation shows that the inverse is

$$A^{-1} = \begin{bmatrix} 0 & 0 & \frac{1}{3} \\ -\frac{1}{2} & -\frac{1}{2} & \frac{1}{2} \\ 1 & 0 & -\frac{1}{3} \end{bmatrix}.$$

Confirm this by forming the products $A^{-1}A$ and AA^{-1}, each of which should equal the matrix I. ✚

As illustrated by the examples, the most efficient order of steps is to make the changes column by column from left to right, so that for each column the required 1 is the result of the first change. Next, perform the steps that obtain the zeros in that column. Then proceed to another column. Since it is tedious to find an inverse with paper and pencil, these same steps can be adapted for a computer program. A computer can produce the inverse of a large matrix, even a 40×40 matrix, in a few seconds.

Example 5   Find A^{-1} given $A = \begin{bmatrix} 2 & -4 \\ 1 & -2 \end{bmatrix}$.

Using row operations to transform the first column of the augmented matrix

$$\begin{bmatrix} 2 & -4 & 1 & 0 \\ 1 & -2 & 0 & 1 \end{bmatrix}$$

results in the following matrices:

$$\begin{bmatrix} 1 & -2 & \frac{1}{2} & 0 \\ 1 & -2 & 0 & 1 \end{bmatrix} \quad \text{and} \quad \begin{bmatrix} 1 & -2 & \frac{1}{2} & 0 \\ 0 & 0 & -\frac{1}{2} & 1 \end{bmatrix}.$$

At this point, the matrix should be changed so that the second-row, second-column element will be 1. Since that element is now 0, there is no way to complete the desired transformation, so A^{-1} does not exist for this matrix A. What is wrong? Just as there is no multiplicative inverse for the real number 0, not every matrix has a multiplicative inverse. Matrix A is an example of such a matrix.   ✦

Matrix equations are solved in much the same way as real-number equations. Properties of matrices are used, together with a multiplication property of equality for matrices that allows both sides of a matrix equation to be multiplied by any appropriate matrix. The next example shows how to solve a matrix equation.

Example 6   Given

$$A = \begin{bmatrix} 2 & 2 \\ -1 & -2 \end{bmatrix} \quad \text{and} \quad B = \begin{bmatrix} 2 \\ 3 \end{bmatrix},$$

find a matrix X so that $AX = B$.

Since A is 2×2 and B is 2×1, matrix X will be a 2×1 matrix, like B. The matrix equation $AX = B$ can be solved as a linear equation is solved, with the properties of matrices (given in 8.2 Exercises) and the facts that if A^{-1} exists, $A^{-1}A = I$ and $IX = X$, as follows.

$AX = B$	Given
$A^{-1}(AX) = A^{-1}B$	Multiplication property of equality
$(A^{-1}A)X = A^{-1}B$	Associative property of multiplication
$IX = A^{-1}B$	Multiplication inverse property
$X = A^{-1}B$	Multiplication identity property

In using the multiplication property of equality for matrices, be careful to multiply on the left on both sides of the equation. Unlike multiplication of real numbers, multiplication of matrices is not commutative.

To find X, first find A^{-1}. Verify that

$$A^{-1} = \begin{bmatrix} 1 & 1 \\ -\frac{1}{2} & -1 \end{bmatrix}.$$

As shown above,

$$X = A^{-1}B = \begin{bmatrix} 1 & 1 \\ -\frac{1}{2} & -1 \end{bmatrix}\begin{bmatrix} 2 \\ 3 \end{bmatrix} = \begin{bmatrix} 5 \\ -4 \end{bmatrix}.$$

Check: $$AX = \begin{bmatrix} 2 & 2 \\ -1 & -2 \end{bmatrix}\begin{bmatrix} 5 \\ -4 \end{bmatrix} = \begin{bmatrix} 2 \\ 3 \end{bmatrix} = B. \quad \clubsuit$$

The result from Example 6 can be used to solve linear systems by first writing the system as a matrix equation $AX = B$, where X is a matrix of the variables of the system. Then the solution is $X = A^{-1}B$.

Example 7   Use the inverse of the coefficient matrix to solve the system

$$2x - 3y = 4$$
$$x + 5y = 2.$$

To represent the system as a matrix equation, use one matrix for the coefficients, one for the variables, and one for the constants, as follows.

$$A = \begin{bmatrix} 2 & -3 \\ 1 & 5 \end{bmatrix}, \quad X = \begin{bmatrix} x \\ y \end{bmatrix}, \quad \text{and} \quad B = \begin{bmatrix} 4 \\ 2 \end{bmatrix}$$

The system can then be written in matrix form as the equation $AX = B$, since

$$AX = \begin{bmatrix} 2 & -3 \\ 1 & 5 \end{bmatrix}\begin{bmatrix} x \\ y \end{bmatrix} = \begin{bmatrix} 2x - 3y \\ x + 5y \end{bmatrix} = \begin{bmatrix} 4 \\ 2 \end{bmatrix} = B.$$

To solve the system, first find A^{-1}.

$$A^{-1} = \begin{bmatrix} \frac{5}{13} & \frac{3}{13} \\ -\frac{1}{13} & \frac{2}{13} \end{bmatrix}$$

Next, find the product $A^{-1}B$.

$$A^{-1}B = \begin{bmatrix} \frac{5}{13} & \frac{3}{13} \\ -\frac{1}{13} & \frac{2}{13} \end{bmatrix}\begin{bmatrix} 4 \\ 2 \end{bmatrix} = \begin{bmatrix} 2 \\ 0 \end{bmatrix}$$

Since $X = A^{-1}B$,

$$X = \begin{bmatrix} x \\ y \end{bmatrix} = \begin{bmatrix} 2 \\ 0 \end{bmatrix},$$

and the solution set of the system is $\{(2, 0)\}$. $\clubsuit$

Example 8   Use the inverse of the coefficient matrix to solve the system

$$2x - 5y = 20$$
$$3x + 4y = 7.$$

The required matrices are

$$A = \begin{bmatrix} 2 & -5 \\ 3 & 4 \end{bmatrix}, \quad X = \begin{bmatrix} x \\ y \end{bmatrix}, \quad \text{and} \quad B = \begin{bmatrix} 20 \\ 7 \end{bmatrix}.$$

Begin by finding A^{-1}.

$$A^{-1} = \begin{bmatrix} \frac{4}{23} & \frac{5}{23} \\ -\frac{3}{23} & \frac{2}{23} \end{bmatrix}$$

Now find $A^{-1}B$.

$$A^{-1}B = \begin{bmatrix} \frac{4}{23} & \frac{5}{23} \\ -\frac{3}{23} & \frac{2}{23} \end{bmatrix} \begin{bmatrix} 20 \\ 7 \end{bmatrix} = \begin{bmatrix} 5 \\ -2 \end{bmatrix}$$

Then, since $X = A^{-1}B$,

$$X = \begin{bmatrix} x \\ y \end{bmatrix} = \begin{bmatrix} 5 \\ -2 \end{bmatrix},$$

from which $x = 5$, $y = -2$, with solution set $\{(5, -2)\}$.   ✦

8.3   Exercises

Decide whether the given matrices are inverses of each other.

1. $\begin{bmatrix} 2 & 3 \\ 1 & 1 \end{bmatrix}, \begin{bmatrix} -1 & 3 \\ 1 & -2 \end{bmatrix}$

2. $\begin{bmatrix} 5 & 7 \\ 2 & 3 \end{bmatrix}, \begin{bmatrix} 3 & -7 \\ -2 & 5 \end{bmatrix}$

3. $\begin{bmatrix} 2 & 1 \\ 3 & 2 \end{bmatrix}, \begin{bmatrix} 2 & 1 \\ -3 & 2 \end{bmatrix}$

4. $\begin{bmatrix} -1 & 2 \\ 3 & -5 \end{bmatrix}, \begin{bmatrix} -5 & -2 \\ -3 & -1 \end{bmatrix}$

5. $\begin{bmatrix} 1 & -2 & -3 \\ 2 & -2 & -5 \\ -1 & 1 & 4 \end{bmatrix}, \begin{bmatrix} -1 & \frac{5}{3} & \frac{4}{3} \\ -1 & \frac{1}{3} & -\frac{1}{3} \\ 0 & \frac{1}{3} & \frac{2}{3} \end{bmatrix}$

6. $\begin{bmatrix} 1 & 2 & -1 \\ 2 & -1 & 3 \\ 3 & -2 & 3 \end{bmatrix}, \begin{bmatrix} \frac{3}{10} & -\frac{2}{5} & \frac{1}{2} \\ \frac{3}{10} & \frac{3}{5} & -\frac{1}{2} \\ -\frac{1}{10} & \frac{4}{5} & -\frac{1}{2} \end{bmatrix}$

7. $\begin{bmatrix} 1 & 2 & -1 \\ 0 & 1 & 3 \\ 2 & 1 & -2 \end{bmatrix}, \begin{bmatrix} 1 & 1 & 2 \\ 1 & 1 & 1 \\ 2 & 3 & 4 \end{bmatrix}$

8. $\begin{bmatrix} 2 & -1 & 4 \\ 0 & 5 & 0 \\ 3 & 2 & -1 \end{bmatrix}, \begin{bmatrix} 1 & 0 & 1 \\ 6 & 4 & 2 \\ 1 & 1 & 0 \end{bmatrix}$

9. $\begin{bmatrix} 4 & 3 & 3 \\ -1 & 0 & -1 \\ -4 & -4 & -3 \end{bmatrix}, \begin{bmatrix} 4 & 3 & 3 \\ -1 & 0 & -1 \\ -4 & -4 & -3 \end{bmatrix}$

10. $\begin{bmatrix} 1 & 0 & 2 \\ -1 & 0 & -2 \\ 1 & 1 & 1 \end{bmatrix}, \begin{bmatrix} 1 & 0 & -2 \\ -1 & 0 & 2 \\ 1 & 1 & 1 \end{bmatrix}$

Use row operations to find any inverses that exist for the following matrices. See Examples 3–5.

11. $\begin{bmatrix} 1 & -1 \\ 2 & 0 \end{bmatrix}$

12. $\begin{bmatrix} 3 & -1 \\ -5 & 2 \end{bmatrix}$

13. $\begin{bmatrix} -6 & 4 \\ -3 & 2 \end{bmatrix}$

14. $\begin{bmatrix} -1 & 2 \\ -2 & -1 \end{bmatrix}$

15. $\begin{bmatrix} -1 & -2 \\ 3 & 4 \end{bmatrix}$

16. $\begin{bmatrix} 5 & 10 \\ -3 & -6 \end{bmatrix}$

17. $\begin{bmatrix} .6 & .2 \\ .5 & .1 \end{bmatrix}$

18. $\begin{bmatrix} .8 & -.3 \\ .5 & -.2 \end{bmatrix}$

19. $\begin{bmatrix} .4 & .3 \\ .2 & .1 \end{bmatrix}$

20. $\begin{bmatrix} -.5 & .7 \\ -.1 & .2 \end{bmatrix}$

21. $\begin{bmatrix} 1 & 0 & 0 \\ 0 & -1 & 0 \\ 1 & 0 & 1 \end{bmatrix}$

22. $\begin{bmatrix} 1 & 3 & 3 \\ -1 & 0 & 0 \\ -4 & -4 & -3 \end{bmatrix}$

23. $\begin{bmatrix} 3 & 6 & 3 \\ 6 & 4 & -2 \\ 0 & 1 & -1 \end{bmatrix}$

24. $\begin{bmatrix} 2 & 6 & 0 \\ 1 & 5 & 3 \\ 0 & 0 & 1 \end{bmatrix}$

25. $\begin{bmatrix} -1 & -1 & -1 \\ 4 & 5 & 0 \\ 0 & 1 & -3 \end{bmatrix}$

26. $\begin{bmatrix} 2 & 0 & 4 \\ 3 & 1 & 5 \\ -1 & 1 & -2 \end{bmatrix}$

27. $\begin{bmatrix} -.4 & .1 & .2 \\ 0 & .6 & .8 \\ .3 & 0 & -.2 \end{bmatrix}$

28. $\begin{bmatrix} .8 & .2 & .1 \\ -.2 & 0 & .3 \\ 0 & 0 & .5 \end{bmatrix}$

Write the matrix of coefficients, A, the matrix of variables, X, and the matrix of constants, B, for each of the following systems. Do not solve.

29. $x + y = 8$
$2x - y = 4$

30. $2x + y = 9$
$x + 3y = 17$

31. $4x + 5y = 7$
$2x + 3y = 5$

32. $2x + 3y = -2$
$5x + 4y = -12$

33. $x + y + z = 9$
$2x + y + 3z = 17$
$5x - 2y + 2z = 16$

34. $3x + y - 5z = 8$
$2x + 6y - 3z = 6$
$x + 5y - 9z = -12$

Solve each of the following systems by using the inverse of the coefficient matrix. See Examples 6–8.

35. $-x + y = 1$
$2x - y = 1$

36. $x + y = 5$
$x - y = -1$

37. $2x - y = -8$
$3x + y = -2$

38. $x + 3y = -12$
$2x - y = 11$

39. $2x + 3y = -10$
$3x + 4y = -12$

40. $2x - 3y = 10$
$2x + 2y = 5$

41. $2x - 5y = 10$
$4x - 5y = 15$

42. $2x - 3y = 2$
$4x - 6y = 1$

Use the inverse of the coefficient matrix to solve each of the following systems of equations. The inverses of the coefficient matrices for the first four problems are found in the text and exercises of this section.

43. $x + z = 3$
$2x - 2y - z = -2$
$3x = 3$

44. $x + 3y + 3z = 11$
$-x = -2$
$-4x - 4y - 3z = -10$

45. $3x + 6y + 3z = 12$
$6x + 4y - 2z = -4$
$y - z = -3$

46. $2x + 4z = 14$
$3x + y + 5z = 19$
$-x + y - 2z = -7$

47. $x + 3y + z = 2$
$x - 2y + 3z = -3$
$2x - 3y - z = 34$

48. $x + y - z = 6$
$2x - y + z = -9$
$x - 2y + 3z = 1$

In Exercises 49 and 50, write a system of equations and solve it by using the inverse of the coefficient matrix.

49. Midtown Manufacturing Company makes two products, plastic plates and plastic cups. Both require time on two machines: plates require 1 hr on machine A and 2 hr on machine $B;$ cups require 3 hr on machine A and 1 hr on machine B. Both machines operate 15 hr a day. How many of each product can be produced in a day under these conditions?

50. A company produces two models of bicycles, model 201 and model 301. Model 201 requires 2 hr of assembly time and model 301 requires 3 hr of assembly time. The parts for model 201 cost $25 per bike and the parts for model 301 cost $30 per bike. If the company has a total of 34 hr of assembly time and $365 available per day for these two models, how many of each can be made in a day?

Let $A = \begin{bmatrix} a & b \\ c & d \end{bmatrix}$. *Show that each of the following statements is true.*

51. $IA = A$ **52.** $AI = A$ **53.** $A \cdot 0 = 0$

54. Find A^{-1}. (Assume $ad - bc \neq 0$.) Show that $AA^{-1} = I$.

55. Show that $A^{-1}A = I$.

8.4   Determinants

Every square matrix A is associated with a real number called the determinant of A, written $\delta(A)$.

Definition of Determinant of a 2 × 2 Matrix	The **determinant of a 2 × 2 matrix** A, $$A = \begin{bmatrix} a_{11} & a_{12} \\ a_{21} & a_{22} \end{bmatrix},$$ is defined as $$\delta(A) = |A| = \begin{vmatrix} a_{11} & a_{12} \\ a_{21} & a_{22} \end{vmatrix} = a_{11}a_{22} - a_{21}a_{12}.$$

Example 1   If $P = \begin{bmatrix} -3 & 4 \\ 6 & 8 \end{bmatrix}$, then

$$\delta(P) = \begin{vmatrix} -3 & 4 \\ 6 & 8 \end{vmatrix} = -3(8) - 6(4) = -48. \quad \clubsuit$$

The definition of a determinant can be extended to a 3 × 3 matrix as follows.

Definition of Determinant of a 3 × 3 Matrix	The **determinant of a 3 × 3 matrix** A, $$A = \begin{bmatrix} a_{11} & a_{12} & a_{13} \\ a_{21} & a_{22} & a_{23} \\ a_{31} & a_{32} & a_{33} \end{bmatrix},$$ is defined as $$\delta(A) = \begin{vmatrix} a_{11} & a_{12} & a_{13} \\ a_{21} & a_{22} & a_{23} \\ a_{31} & a_{32} & a_{33} \end{vmatrix} = (a_{11}a_{22}a_{33} + a_{12}a_{23}a_{31} + a_{13}a_{21}a_{32}) \\ - (a_{31}a_{22}a_{13} + a_{32}a_{23}a_{11} + a_{33}a_{21}a_{12}).$$

An easy method for calculating 3×3 determinants is found by rearranging and factoring the terms given above to get

$$\begin{vmatrix} a_{11} & a_{12} & a_{13} \\ a_{21} & a_{22} & a_{23} \\ a_{31} & a_{32} & a_{33} \end{vmatrix} = a_{11}(a_{22}a_{33} - a_{32}a_{23}) - a_{21}(a_{12}a_{33} - a_{32}a_{13}) \\ + a_{31}(a_{12}a_{23} - a_{22}a_{13}). \tag{*}$$

Each of the quantities in parentheses represents the determinant of a 2×2 matrix that is the part of the 3×3 matrix remaining when the row and column of the multiplier are eliminated, as shown below.

$$a_{11}(a_{22}a_{33} - a_{32}a_{23}) \qquad \begin{bmatrix} a_{11} & a_{12} & a_{13} \\ a_{21} & a_{22} & a_{23} \\ a_{31} & a_{32} & a_{33} \end{bmatrix}$$

$$a_{21}(a_{12}a_{33} - a_{32}a_{13}) \qquad \begin{bmatrix} a_{11} & a_{12} & a_{13} \\ a_{21} & a_{22} & a_{23} \\ a_{31} & a_{32} & a_{33} \end{bmatrix}$$

$$a_{31}(a_{12}a_{23} - a_{22}a_{13}) \qquad \begin{bmatrix} a_{11} & a_{12} & a_{13} \\ a_{21} & a_{22} & a_{23} \\ a_{31} & a_{32} & a_{33} \end{bmatrix}$$

The determinant of each 2×2 matrix is called a **minor** of the associated element in the 3×3 matrix. Thus, the minors of a_{11}, a_{21}, and a_{31} are as shown below.

Element	a_{11}	a_{21}	a_{31}
Minor	$\begin{vmatrix} a_{22} & a_{23} \\ a_{32} & a_{33} \end{vmatrix}$	$\begin{vmatrix} a_{12} & a_{13} \\ a_{32} & a_{33} \end{vmatrix}$	$\begin{vmatrix} a_{12} & a_{13} \\ a_{22} & a_{23} \end{vmatrix}$

Example 2   For the matrix

$$\begin{bmatrix} 6 & 2 & 4 \\ 8 & 9 & 3 \\ 1 & 2 & 0 \end{bmatrix},$$

find the minor of each of the following elements.

(a) 6

Since 6 is in the first row and first column of the matrix, mentally remove that row and column. The minor is the determinant of the remaining 2×2 matrix.

$$\text{Minor of } 6 = \begin{vmatrix} 9 & 3 \\ 2 & 0 \end{vmatrix} = 9 \cdot 0 - 2 \cdot 3 = -6$$

(b) 9

Eliminate the second row and second column, which include the element 9. The determinant of the remaining 2×2 matrix is the minor of 9.

$$\text{Minor of } 9 = \begin{vmatrix} 6 & 4 \\ 1 & 0 \end{vmatrix} = 6 \cdot 0 - 1 \cdot 4 = -4$$

(c) 1

The minor of this element is

$$\text{Minor of } 1 = \begin{vmatrix} 2 & 4 \\ 9 & 3 \end{vmatrix} = 2 \cdot 3 - 9 \cdot 4 = -30. \quad \clubsuit$$

The determinant of the 3×3 matrix can be evaluated by multiplying each element in the first column by its minor and combining the products as indicated in equation (*) given above. This is called the **expansion of the determinant by minors about the first column.**

Example 3   Evaluate the determinant of

$$\begin{bmatrix} 1 & 3 & -2 \\ -1 & -2 & -3 \\ 1 & 1 & 2 \end{bmatrix}$$

by expanding about the first column.

Using the procedure above,

$$\begin{vmatrix} 1 & 3 & -2 \\ -1 & -2 & -3 \\ 1 & 1 & 2 \end{vmatrix} = 1 \begin{vmatrix} -2 & -3 \\ 1 & 2 \end{vmatrix} - (-1) \begin{vmatrix} 3 & -2 \\ 1 & 2 \end{vmatrix} + 1 \begin{vmatrix} 3 & -2 \\ -2 & -3 \end{vmatrix}$$

$$= 1[(-2)(2) - (1)(-3)] + 1[(3)(2) - (1)(-2)]$$
$$+ 1[(3)(-3) - (-2)(-2)]$$
$$= 1(-1) + 1(8) + 1(-13)$$
$$= -1 + 8 - 13 = -6. \quad \clubsuit$$

To get equation (*) the terms could have been rearranged and factored differently by factoring out the three elements of the second or third columns or of any of the three rows of the array. Therefore, expanding by minors about any row or any column results in the same value for the determinant. The following array of signs is helpful in determining the correct signs for the terms of other expansions.

Array of Signs

$$
\begin{array}{ccc}
+ & - & + \\
- & + & - \\
+ & - & +
\end{array}
$$

The signs alternate in each row and each column, beginning with + in the first-row, first-column position, so the array of signs can be reproduced when needed. If the determinant is expanded about the second column, for example, the first term would have a minus sign, the second a plus sign, and the third a minus sign. These signs are independent of the sign of the corresponding element of the matrix. The sign array can be extended for larger matrices.

Example 4 Evaluate the determinant of Example 3 using expansion by minors about the second column.

Using the methods described above,

$$
\begin{vmatrix} 1 & 3 & -2 \\ -1 & -2 & -3 \\ 1 & 1 & 2 \end{vmatrix} = -3 \begin{vmatrix} -1 & -3 \\ 1 & 2 \end{vmatrix} + (-2) \begin{vmatrix} 1 & -2 \\ 1 & 2 \end{vmatrix} - 1 \begin{vmatrix} 1 & -2 \\ -1 & -3 \end{vmatrix}
$$

$$
= -3(1) - 2(4) - 1(-5)
$$

$$
= -3 - 8 + 5
$$

$$
= -6. \quad \clubsuit
$$

The determinant of Examples 3 and 4 was evaluated by expanding by minors, first about the first column and then about the second column. In evaluating the determinant by expansion about the second column, the minors were evaluated using the following information.

Element of second column	*Minor*	*Sign associated with minor*
3	$\begin{vmatrix} -1 & -3 \\ 1 & 2 \end{vmatrix} = 1$	−
−2	$\begin{vmatrix} 1 & -2 \\ 1 & 2 \end{vmatrix} = 4$	+
1	$\begin{vmatrix} 1 & -2 \\ -1 & -3 \end{vmatrix} = -5$	−

The minor of an element, together with the sign associated with the element, is called the **cofactor** of the element. The cofactors for the expansion above are as follows.

Element	Minor	Sign	Cofactor
3	$\begin{vmatrix} -1 & -3 \\ 1 & 2 \end{vmatrix} = 1$	$-$	-1
-2	$\begin{vmatrix} 1 & -2 \\ 1 & 2 \end{vmatrix} = 4$	$+$	4
1	$\begin{vmatrix} 1 & -2 \\ -1 & -3 \end{vmatrix} = -5$	$-$	5

The determinant is then calculated by multiplying each element in the column times its cofactor and adding these products. Doing this with the results above gives

$$3(-1) + (-2)(4) + 1(5) = -3 - 8 + 5 = -6.$$

Compare this calculation with the sum of the products found in Example 4.

The method of expansion by minors or cofactors can be extended to find determinants of any $n \times n$ matrix. For a larger matrix, the sign array is extended. For example, the signs for a 4×4 matrix are arranged as follows.

$$\begin{matrix} + & - & + & - \\ - & + & - & + \\ + & - & + & - \\ - & + & - & + \end{matrix}$$

Example 5 Evaluate

$$\begin{vmatrix} -1 & -2 & 3 & 2 \\ 0 & 1 & 4 & -2 \\ 3 & -1 & 4 & 0 \\ 2 & 1 & 0 & 3 \end{vmatrix}$$

Expanding by minors about the fourth row gives

$$-2\begin{vmatrix} -2 & 3 & 2 \\ 1 & 4 & -2 \\ -1 & 4 & 0 \end{vmatrix} + 1\begin{vmatrix} -1 & 3 & 2 \\ 0 & 4 & -2 \\ 3 & 4 & 0 \end{vmatrix} - 0\begin{vmatrix} -1 & -2 & 2 \\ 0 & 1 & -2 \\ 3 & -1 & 0 \end{vmatrix} + 3\begin{vmatrix} -1 & -2 & 3 \\ 0 & 1 & 4 \\ 3 & -1 & 4 \end{vmatrix}$$

$$= -2(6) + 1(-50) - 0 + 3(-41)$$
$$= -185. \quad \clubsuit$$

Each of the four determinants in Example 5 must be evaluated by expansion of three minors, with much work required to get the final value. Determinants of large matrices can be evaluated quickly and easily, however, with the aid of a computer.

8.4 Exercises

Evaluate each of the following determinants. See Example 1.

1. $\begin{vmatrix} 5 & 8 \\ 2 & -4 \end{vmatrix}$

2. $\begin{vmatrix} -3 & 0 \\ 0 & 9 \end{vmatrix}$

3. $\begin{vmatrix} -1 & -2 \\ 5 & 3 \end{vmatrix}$

4. $\begin{vmatrix} 6 & -4 \\ 0 & -1 \end{vmatrix}$

5. $\begin{vmatrix} 9 & 3 \\ -3 & -1 \end{vmatrix}$

6. $\begin{vmatrix} 0 & 2 \\ 1 & 5 \end{vmatrix}$

7. $\begin{vmatrix} 3 & 4 \\ 5 & -2 \end{vmatrix}$

8. $\begin{vmatrix} -9 & 7 \\ 2 & 6 \end{vmatrix}$

9. $\begin{vmatrix} 0 & 4 \\ 4 & 0 \end{vmatrix}$

10. $\begin{vmatrix} 1 & 0 \\ 0 & 2 \end{vmatrix}$

11. $\begin{vmatrix} 8 & 3 \\ 8 & 3 \end{vmatrix}$

12. $\begin{vmatrix} 9 & -4 \\ -4 & 9 \end{vmatrix}$

13. $\begin{vmatrix} x & 4 \\ 8 & 2 \end{vmatrix}$

14. $\begin{vmatrix} k & 3 \\ 0 & 4 \end{vmatrix}$

15. $\begin{vmatrix} y & 2 \\ 8 & y \end{vmatrix}$

16. $\begin{vmatrix} 3 & 8 \\ m & n \end{vmatrix}$

17. $\begin{vmatrix} x & y \\ y & x \end{vmatrix}$

18. $\begin{vmatrix} 2m & 8n \\ 8n & 2m \end{vmatrix}$

19. $\begin{vmatrix} 1.4 & 2.5 \\ 3.7 & 6.2 \end{vmatrix}$

20. $\begin{vmatrix} .123 & .054 \\ .691 & .302 \end{vmatrix}$

Evaluate each of the following determinants. See Examples 2–4.

21. $\begin{vmatrix} 1 & 0 & 0 \\ 0 & -1 & 0 \\ 1 & 0 & 1 \end{vmatrix}$

22. $\begin{vmatrix} -2 & 0 & 1 \\ 0 & 1 & 0 \\ 0 & 0 & -1 \end{vmatrix}$

23. $\begin{vmatrix} -2 & 0 & 0 \\ 4 & 0 & 1 \\ 3 & 4 & 2 \end{vmatrix}$

24. $\begin{vmatrix} 3 & -2 & 0 \\ 0 & -1 & 1 \\ 4 & 0 & 2 \end{vmatrix}$

25. $\begin{vmatrix} 1 & 2 & 0 \\ -1 & 2 & -1 \\ 0 & 1 & 4 \end{vmatrix}$

26. $\begin{vmatrix} 2 & 1 & -1 \\ 4 & 7 & -2 \\ 2 & 4 & 0 \end{vmatrix}$

27. $\begin{vmatrix} 10 & 2 & 1 \\ -1 & 4 & 3 \\ -3 & 8 & 10 \end{vmatrix}$

28. $\begin{vmatrix} 7 & -1 & 1 \\ 1 & -7 & 2 \\ -2 & 1 & 1 \end{vmatrix}$

29. $\begin{vmatrix} 1 & -2 & 3 \\ 0 & 0 & 0 \\ 1 & 10 & -12 \end{vmatrix}$

30. $\begin{vmatrix} 2 & 3 & 0 \\ 1 & 9 & 0 \\ -1 & -2 & 0 \end{vmatrix}$

31. $\begin{vmatrix} 3 & 3 & -1 \\ 2 & 6 & 0 \\ -6 & -6 & 2 \end{vmatrix}$

32. $\begin{vmatrix} 5 & -3 & 2 \\ -5 & 3 & -2 \\ 1 & 0 & 1 \end{vmatrix}$

33. $\begin{vmatrix} 3 & 2 & 0 \\ 0 & 1 & x \\ 2 & 0 & 0 \end{vmatrix}$

34. $\begin{vmatrix} 0 & 3 & y \\ 0 & 4 & 2 \\ 1 & 0 & 1 \end{vmatrix}$

35. $\begin{vmatrix} .12 & .43 & .18 \\ .65 & .09 & .71 \\ .42 & .55 & .68 \end{vmatrix}$

36. $\begin{vmatrix} 1.04 & 2.87 & 3.69 \\ 5.42 & 3.66 & 7.81 \\ 10.3 & 6.41 & 1.17 \end{vmatrix}$

Solve each of the following equations for x.

37. $\begin{vmatrix} -2 & 0 & 1 \\ -1 & 3 & x \\ 5 & -2 & 0 \end{vmatrix} = 3$

38. $\begin{vmatrix} 4 & 3 & 0 \\ 2 & 0 & 1 \\ -3 & x & -1 \end{vmatrix} = 5$

39. $\begin{vmatrix} 5 & 3x & -3 \\ 0 & 2 & -1 \\ 4 & -1 & x \end{vmatrix} = -7$

40. $\begin{vmatrix} 2x & 1 & -1 \\ 0 & 4 & x \\ 3 & 0 & 2 \end{vmatrix} = x$

For the following determinants, find the cofactor of each element in the second row of the corresponding matrix.

41. $\begin{vmatrix} -2 & 0 & 1 \\ 3 & 2 & -1 \\ 1 & 0 & 2 \end{vmatrix}$

42. $\begin{vmatrix} 0 & -1 & 2 \\ 1 & 0 & 2 \\ 0 & -3 & 1 \end{vmatrix}$

43. $\begin{vmatrix} 1 & 2 & -1 \\ 2 & 3 & -2 \\ -1 & 4 & 1 \end{vmatrix}$

44. $\begin{vmatrix} 2 & -1 & 4 \\ 3 & 0 & 1 \\ -2 & 1 & 4 \end{vmatrix}$

Find the value of each of the following determinants. See Example 5.

45. $\begin{vmatrix} 4 & 0 & 0 & 2 \\ -1 & 0 & 3 & 0 \\ 2 & 4 & 0 & 1 \\ 0 & 0 & 1 & 2 \end{vmatrix}$

46. $\begin{vmatrix} -2 & 0 & 4 & 2 \\ 3 & 6 & 0 & 4 \\ 0 & 0 & 0 & 3 \\ 9 & 0 & 2 & -1 \end{vmatrix}$

47. $\begin{vmatrix} 1 & 1 & 0 & 1 \\ 2 & 1 & 0 & 2 \\ 0 & 1 & -1 & 1 \\ 1 & -1 & 1 & 1 \end{vmatrix}$

48. $\begin{vmatrix} 2 & 7 & 0 & -1 \\ 1 & 0 & 1 & 3 \\ 2 & 4 & -1 & -1 \\ -1 & 1 & 0 & 8 \end{vmatrix}$

Determinants can be used to find the area of a triangle given the coordinates of its vertices. Given a triangle PQR with vertices (x_1, y_1), (x_3, y_3), and (x_2, y_2), as shown in the following figure, we can introduce segments PM, RN, and QS perpendicular to the x-axis, forming trapezoids PMNR, RNSQ, and PMSQ. Recall that the area of a trapezoid is given by half the sum of the parallel bases times the altitude. For example, the area of trapezoid PMSQ equals $(1/2)(y_1 + y_3)(x_3 - x_1)$. The area of triangle PQR can be found by subtracting the area of PMSQ from the sum of the areas of PMNR and RNSQ. Thus, for the area A of triangle PQR,

$$A = (1/2)(x_3y_1 - x_1y_3 + x_2y_3 - x_3y_2 + x_1y_2 - x_2y_1).$$

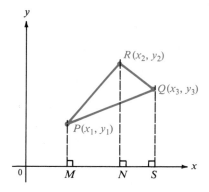

By evaluating the determinant

$$\begin{vmatrix} x_1 & y_1 & 1 \\ x_2 & y_2 & 1 \\ x_3 & y_3 & 1 \end{vmatrix},$$

it can be shown that the area of the triangle is given by the absolute value *of A, where*

$$A = \frac{1}{2} \begin{vmatrix} x_1 & y_1 & 1 \\ x_2 & y_2 & 1 \\ x_3 & y_3 & 1 \end{vmatrix}.$$

Use the given formula to find the area of the following triangles.

49. $P(0, 1)$, $Q(2, 0)$, $R(1, 3)$ **50.** $P(2, 5)$, $Q(-1, 3)$, $R(4, 0)$

51. $P(2, -2)$, $Q(0, 0)$, $R(-3, -4)$ **52.** $P(4, 7)$, $Q(5, -2)$, $R(1, 1)$

53. $P(3, 8)$, $Q(-1, 4)$, $R(0, 1)$ **54.** $P(-3, -1)$, $Q(4, 2)$, $R(3, -3)$

Determinants can be used to find the equation of a line passing through two given points. The next two exercises show how to do this.

55. Expand the determinant below and show that the result is the equation of the line through $(2, 3)$, and $(-1, 4)$.

$$\begin{vmatrix} x & y & 1 \\ 2 & 3 & 1 \\ -1 & 4 & 1 \end{vmatrix} = 0$$

56. (a) Write the equation of the line through the points (x_1, y_1) and (x_2, y_2) using the point-slope formula.

    (b) Expanding the determinant in the equation below, show that the equation is equivalent to the equation in part (a).

$$\begin{vmatrix} x & y & 1 \\ x_1 & y_1 & 1 \\ x_2 & y_2 & 1 \end{vmatrix} = 0$$

57. The inverse of matrix $A = \begin{bmatrix} a & b \\ c & d \end{bmatrix}$ is

$$A^{-1} = \begin{bmatrix} \dfrac{d}{ad - bc} & \dfrac{-b}{ad - bc} \\ \dfrac{-c}{ad - bc} & \dfrac{a}{ad - bc} \end{bmatrix}.$$

Show that A^{-1} does not exist if $\delta(A) = 0$.

58. Show that if $A = \begin{bmatrix} a & b \\ c & d \end{bmatrix}$ and $\delta(A) \neq 0$, then

$$A^{-1} = \begin{bmatrix} \dfrac{d}{\delta(A)} & \dfrac{-b}{\delta(A)} \\ \dfrac{-c}{\delta(A)} & \dfrac{a}{\delta(A)} \end{bmatrix}.$$

This is an alternative method for finding the inverse of a 2×2 matrix A, where $\delta(A) \neq 0$. Unfortunately, the method does not generalize to any $n \times n$ matrix.

8.5   Properties of Determinants

Some determinants can be evaluated only after much tedious calculation. The more calculation is involved, the greater the chance for error. To evaluate the determinant of a 3×3 matrix, it is necessary to evaluate three different determinants and then combine them correctly. For a determinant of a 4×4 matrix, twelve determinants must be found. This section gives several properties of determinants that make them easier to calculate. These properties of determinants are summarized at the end of this section.

The value of

$$\begin{vmatrix} 3 & 0 \\ 5 & 0 \end{vmatrix}$$

is $3 \cdot 0 - 5 \cdot 0 = 0$. Also, expanded about the second row,

$$\begin{vmatrix} 1 & 2 & 3 \\ 0 & 0 & 0 \\ 4 & 6 & 8 \end{vmatrix} = -0 \begin{vmatrix} 2 & 3 \\ 6 & 8 \end{vmatrix} + 0 \begin{vmatrix} 1 & 3 \\ 4 & 8 \end{vmatrix} - 0 \begin{vmatrix} 1 & 2 \\ 4 & 6 \end{vmatrix} = 0.$$

The fact that the matrix for each of these determinants has a row or column of zeros leads to a determinant of 0. These examples suggest the following result.

Determinant Property 1	If every element in a row or column of a matrix is 0, then the determinant equals 0.

To prove Property 1, expand the determinant about the 0 row or column. Each term of this expansion has a 0 factor, so each term is 0. For example,

$$\begin{vmatrix} a & b & c \\ 0 & 0 & 0 \\ d & e & f \end{vmatrix} = -0 \begin{vmatrix} b & c \\ e & f \end{vmatrix} + 0 \begin{vmatrix} a & c \\ d & f \end{vmatrix} - 0 \begin{vmatrix} a & b \\ d & e \end{vmatrix} = 0.$$

Example 1 $\begin{vmatrix} 2 & 4 \\ 0 & 0 \end{vmatrix} = 0$ and $\begin{vmatrix} -3 & 7 & 0 \\ 4 & 9 & 0 \\ -6 & 8 & 0 \end{vmatrix} = 0$ ❖

Suppose the matrix

$$\begin{bmatrix} 1 & 5 \\ 3 & 7 \end{bmatrix}$$

is rewritten as

$$\begin{bmatrix} 1 & 3 \\ 5 & 7 \end{bmatrix},$$

so the rows of the first matrix are the columns of the second matrix. Evaluating the determinant of each matrix gives

$$\begin{vmatrix} 1 & 5 \\ 3 & 7 \end{vmatrix} = 1 \cdot 7 - 3 \cdot 5 = -8$$

$$\begin{vmatrix} 1 & 3 \\ 5 & 7 \end{vmatrix} = 1 \cdot 7 - 5 \cdot 3 = -8.$$

The two determinants are the same, suggesting the next property of determinants.

Determinant Property 2

> If corresponding rows and columns of a matrix are interchanged, the determinant is not changed.

A proof of Property 2 is requested in Exercise 35.

Example 2   Given

$$A = \begin{bmatrix} 2 & 1 & 6 \\ 3 & 0 & 5 \\ -4 & 6 & 9 \end{bmatrix} \quad \text{and} \quad B = \begin{bmatrix} 2 & 3 & -4 \\ 1 & 0 & 6 \\ 6 & 5 & 9 \end{bmatrix},$$

by Property 2, since the rows of the first matrix are the columns of the second,

$$\begin{vmatrix} 2 & 1 & 6 \\ 3 & 0 & 5 \\ -4 & 6 & 9 \end{vmatrix} = \begin{vmatrix} 2 & 3 & -4 \\ 1 & 0 & 6 \\ 6 & 5 & 9 \end{vmatrix}. \quad ❖$$

Exchanging the rows of the matrix

$$\begin{bmatrix} 1 & 6 \\ 5 & 3 \end{bmatrix} \quad \text{gives} \quad \begin{bmatrix} 5 & 3 \\ 1 & 6 \end{bmatrix}.$$

While the first determinant is

$$\begin{vmatrix} 1 & 6 \\ 5 & 3 \end{vmatrix} = 1 \cdot 3 - 5 \cdot 6 = -27,$$

the second determinant is

$$\begin{vmatrix} 5 & 3 \\ 1 & 6 \end{vmatrix} = 5 \cdot 6 - 1 \cdot 3 = 27.$$

The second determinant is the negative of the first. This suggests the third property of determinants.

Determinant Property 3	Interchanging two rows (or columns) of a matrix reverses the sign of the determinant.

Exercise 36 asks for a proof of Property 3.

Example 3   **(a)** Interchange the two columns of

$$\begin{bmatrix} 2 & 5 \\ 3 & 4 \end{bmatrix} \quad \text{to get} \quad \begin{bmatrix} 5 & 2 \\ 4 & 3 \end{bmatrix}.$$

By Property 3, since

$$\begin{vmatrix} 2 & 5 \\ 3 & 4 \end{vmatrix} = -7,$$

then

$$\begin{vmatrix} 5 & 2 \\ 4 & 3 \end{vmatrix} = 7.$$

(b)
$$\begin{vmatrix} 2 & 1 & 6 \\ 3 & 0 & 5 \\ -4 & 6 & 9 \end{vmatrix} = - \begin{vmatrix} -4 & 6 & 9 \\ 3 & 0 & 5 \\ 2 & 1 & 6 \end{vmatrix} \quad \clubsuit$$

Multiplying each element of the second row of the matrix

$$\begin{bmatrix} 2 & -3 \\ 4 & 1 \end{bmatrix}$$

by -5 gives the new matrix

$$\begin{bmatrix} 2 & -3 \\ 4(-5) & 1(-5) \end{bmatrix} = \begin{bmatrix} 2 & -3 \\ -20 & -5 \end{bmatrix}.$$

The determinants of these two matrices are

$$\begin{vmatrix} 2 & -3 \\ 4 & 1 \end{vmatrix} = 2 \cdot 1 - 4(-3) = 14$$

and

$$\begin{vmatrix} 2 & -3 \\ -20 & -5 \end{vmatrix} = 2(-5) - (-20)(-3) = -70.$$

Since $-70 = -5(14)$, the new determinant is -5 times the original determinant. The next property generalizes this idea. A proof is requested in Exercise 37.

Determinant Property 4	If every element of a row (or column) of a matrix is multiplied by the real number k, then the determinant of the new matrix is k times the determinant of the original matrix.

Example 4 By Property 4, if

$$A = \begin{bmatrix} -2 & 3 & 7 \\ 0 & 5 & 2 \\ -16 & 6 & 4 \end{bmatrix} \quad \text{and} \quad B = \begin{bmatrix} 1 & 3 & 7 \\ 0 & 5 & 2 \\ 8 & 6 & 4 \end{bmatrix},$$

then $\delta(A) = -2 \cdot \delta(B)$, since the first column of matrix A is -2 times the first column of matrix B. ✛

The matrix

$$\begin{bmatrix} 2 & 1 & 2 \\ 5 & 10 & 5 \\ 3 & 6 & 3 \end{bmatrix}$$

has first and third columns that are identical. Expanding the determinant about the second column gives

$$\begin{vmatrix} 2 & 1 & 2 \\ 5 & 10 & 5 \\ 3 & 6 & 3 \end{vmatrix} = -1 \begin{vmatrix} 5 & 5 \\ 3 & 3 \end{vmatrix} + 10 \begin{vmatrix} 2 & 2 \\ 3 & 3 \end{vmatrix} - 6 \begin{vmatrix} 2 & 2 \\ 5 & 5 \end{vmatrix}$$

$$= -1(15 - 15) + 10(6 - 6) - 6(10 - 10)$$

$$= 0.$$

This result suggests the next property.

Determinant Property 5	The determinant of a matrix with two identical rows (or columns) equals 0.

Example 5 Since two rows are identical, the determinant of

$$\begin{bmatrix} -4 & 2 & 3 \\ 0 & 1 & 6 \\ -4 & 2 & 3 \end{bmatrix}$$

equals 0. ✛

Multiply each element of the second row of the matrix

$$\begin{bmatrix} -3 & 5 \\ 1 & 2 \end{bmatrix}$$

by 3 and add the result to the first row.

$$\begin{bmatrix} -3 + 1(3) & 5 + 2(3) \\ 1 & 2 \end{bmatrix} = \begin{bmatrix} -3 + 3 & 5 + 6 \\ 1 & 2 \end{bmatrix} = \begin{bmatrix} 0 & 11 \\ 1 & 2 \end{bmatrix}$$

Verify that the determinant of the new matrix is the same as the determinant of the original matrix, -11. This idea, which is generalized below, is perhaps the most useful property of determinants presented in this section.

Determinant Property 6	The determinant of a matrix is unchanged if a multiple of a row (or column) of the matrix is added to the corresponding elements of another row (or column).

Example 6 Multiply each element of the first column of the matrix

$$\begin{bmatrix} -2 & 4 & 1 \\ 2 & 1 & 5 \\ 3 & 0 & 2 \end{bmatrix}$$

by 3, and add the results to the third column to get the new matrix

$$\begin{bmatrix} -2 & 4 & 1 + 3(-2) \\ 2 & 1 & 5 + 3(2) \\ 3 & 0 & 2 + 3(3) \end{bmatrix} = \begin{bmatrix} -2 & 4 & -5 \\ 2 & 1 & 11 \\ 3 & 0 & 11 \end{bmatrix}.$$

By Property 6, the determinants of these two matrices are both 37. ✤

The following examples show how the properties of determinants can be used to simplify the calculation of determinants.

Example 7 Without expanding, show that the determinant of the following matrix is zero.

$$\begin{bmatrix} 2 & 5 & -1 \\ 1 & -15 & 3 \\ -2 & 10 & -2 \end{bmatrix}$$

Each element in the second column equals -5 times the corresponding element in the third column. Multiply the elements of the third column by 5 and then add the results to the corresponding elements in the second column to get a matrix with an equivalent determinant,

$$\begin{bmatrix} 2 & 0 & -1 \\ 1 & 0 & 3 \\ -2 & 0 & -2 \end{bmatrix}.$$

By Property 1, the determinant of this matrix is zero. ✤

Example 8   Evaluate

$$\begin{vmatrix} 4 & 2 & 1 & 0 \\ -2 & 4 & -1 & 7 \\ -5 & 2 & 3 & 1 \\ 6 & 4 & -3 & 2 \end{vmatrix}.$$

Use Property 6 to change the first row (any row or column could be used) of the matrix to a row in which every element but one is zero. Begin by multiplying the elements of the second column by -2 and adding the results to the first column, replacing the first column with this new column. The result is a 0 for the first element in the first row.

$$\begin{bmatrix} 0 & 2 & 1 & 0 \\ -10 & 4 & -1 & 7 \\ -9 & 2 & 3 & 1 \\ -2 & 4 & -3 & 2 \end{bmatrix}$$ The first column replaced by the sum of -2 times the second column and the first column

To get a zero for the second element in the first row, multiply the elements of the third column by -2 and add to the second column.

$$\begin{bmatrix} 0 & 0 & 1 & 0 \\ -10 & 6 & -1 & 7 \\ -9 & -4 & 3 & 1 \\ -2 & 10 & -3 & 2 \end{bmatrix}$$ The second column replaced by the sum of -2 times the third column and the second column

Since the first row now has only one nonzero number, expand the determinant about the first row to get

$$1\begin{vmatrix} -10 & 6 & 7 \\ -9 & -4 & 1 \\ -2 & 10 & 2 \end{vmatrix} = \begin{vmatrix} -10 & 6 & 7 \\ -9 & -4 & 1 \\ -2 & 10 & 2 \end{vmatrix}.$$

Repeat the process with the 3×3 matrix

$$\begin{bmatrix} -10 & 6 & 7 \\ -9 & -4 & 1 \\ -2 & 10 & 2 \end{bmatrix}.$$

Change the third column to a column with two zeros, as shown below.

$$\begin{bmatrix} 53 & 34 & 0 \\ -9 & -4 & 1 \\ -2 & 10 & 2 \end{bmatrix}$$ The first row replaced by the sum of -7 times the second row and the first row

$$\begin{bmatrix} 53 & 34 & 0 \\ -9 & -4 & 1 \\ 16 & 18 & 0 \end{bmatrix}$$

The third row replaced by the sum of -2 times the second row and the third row

Expand the determinant about the third column to evaluate.

$$-1\begin{vmatrix} 53 & 34 \\ 16 & 18 \end{vmatrix} = -1(954 - 544) = -410$$

When applying Property 6, work with sums of *rows* to get a *column* with only one nonzero number or with sums of *columns* to get a *row* with one nonzero number. ✦

The following summary lists the properties of determinants presented in this section.

Properties of Determinants

1. If every element in a row or column of a matrix is 0, then the determinant equals 0.

2. If corresponding rows and columns of a matrix are interchanged, the determinant is not changed.

3. Interchanging two rows (or columns) of a matrix reverses the sign of the determinant.

4. If every element of a row (or column) of a matrix is multiplied by the real number k, then the determinant of the new matrix is k times the determinant of the original matrix.

5. The determinant of a matrix with two identical rows (or columns) equals 0.

6. The determinant of a matrix is unchanged if a multiple of a row (or column) of the matrix is added to the corresponding elements of another row (or column).

8.5 Exercises

Tell why each of the following determinants equals 0.

1. $\begin{vmatrix} 2 & 3 \\ 2 & 3 \end{vmatrix}$ **2.** $\begin{vmatrix} -5 & -5 \\ 6 & 6 \end{vmatrix}$ **3.** $\begin{vmatrix} 2 & 0 \\ 3 & 0 \end{vmatrix}$ **4.** $\begin{vmatrix} -8 & 0 \\ -6 & 0 \end{vmatrix}$

5. $\begin{vmatrix} -1 & 2 & 4 \\ 4 & -8 & -16 \\ 3 & 0 & 5 \end{vmatrix}$ **6.** $\begin{vmatrix} 2 & -8 & 3 \\ 0 & 2 & -1 \\ -6 & 24 & -9 \end{vmatrix}$ **7.** $\begin{vmatrix} 3 & 6 & 6 \\ 2 & 0 & 4 \\ 1 & 4 & 2 \end{vmatrix}$ **8.** $\begin{vmatrix} 1 & 0 & 0 \\ 1 & 0 & 1 \\ 3 & 0 & 0 \end{vmatrix}$

9. $\begin{vmatrix} m & 2 & 2m \\ 3n & 1 & 6n \\ 5p & 6 & 10p \end{vmatrix}$ **10.** $\begin{vmatrix} 7z & 8x & 2y \\ z & x & y \\ 7z & 7x & 7y \end{vmatrix}$

Use the appropriate properties from this section to tell why each of the following is true.
Do not evaluate the determinants.

11. $\begin{vmatrix} 2 & 1 & 6 \\ 3 & 0 & 2 \\ 4 & 1 & 8 \end{vmatrix} = \begin{vmatrix} 2 & 3 & 4 \\ 1 & 0 & 1 \\ 6 & 2 & 8 \end{vmatrix}$

12. $\begin{vmatrix} 4 & -2 \\ 3 & 8 \end{vmatrix} = \begin{vmatrix} 4 & 3 \\ -2 & 8 \end{vmatrix}$

13. $\begin{vmatrix} 2 & 6 \\ 3 & 5 \end{vmatrix} = -\begin{vmatrix} 3 & 5 \\ 2 & 6 \end{vmatrix}$

14. $\begin{vmatrix} -1 & 8 & 9 \\ 0 & 2 & 1 \\ 3 & 2 & 0 \end{vmatrix} = -\begin{vmatrix} 8 & -1 & 9 \\ 2 & 0 & 1 \\ 2 & 3 & 0 \end{vmatrix}$

15. $3\begin{vmatrix} 6 & 0 & 2 \\ 4 & 1 & 3 \\ 2 & 8 & 6 \end{vmatrix} = \begin{vmatrix} 6 & 0 & 2 \\ 4 & 3 & 3 \\ 2 & 24 & 6 \end{vmatrix}$

16. $-\dfrac{1}{2}\begin{vmatrix} 5 & -8 & 2 \\ 3 & -6 & 9 \\ 2 & 4 & 4 \end{vmatrix} = \begin{vmatrix} 5 & 4 & 2 \\ 3 & 3 & 9 \\ 2 & -2 & 4 \end{vmatrix}$

17. $\begin{vmatrix} 3 & -4 \\ 2 & 5 \end{vmatrix} = \begin{vmatrix} 3 & -4 \\ 5 & 1 \end{vmatrix}$

18. $\begin{vmatrix} -1 & 6 \\ 3 & -5 \end{vmatrix} = \begin{vmatrix} -1 & 6 \\ 2 & 1 \end{vmatrix}$

19. $\begin{vmatrix} 5 & 8 \\ 2 & -1 \end{vmatrix} = \begin{vmatrix} 5 & -2 \\ 2 & -5 \end{vmatrix}$

20. $\begin{vmatrix} 13 & 5 \\ 6 & 1 \end{vmatrix} = \begin{vmatrix} -2 & 5 \\ 3 & 1 \end{vmatrix}$

21. $\begin{vmatrix} 2 & 5 & 8 \\ 1 & 0 & 2 \\ 4 & 3 & 5 \end{vmatrix} = \begin{vmatrix} 2 & 5 & 8 \\ 1 & 0 & 2 \\ 7 & 3 & 11 \end{vmatrix}$

22. $2\begin{vmatrix} 4 & 2 & -1 \\ m & 2n & 3p \\ 5 & 1 & 0 \end{vmatrix} = \begin{vmatrix} 4 & 2 & -1 \\ 2m & 4n & 6p \\ 5 & 1 & 0 \end{vmatrix}$

23. $\begin{vmatrix} 3 & 5 & 0 \\ 2 & 1 & 3 \\ -5 & 1 & 6 \end{vmatrix} = \begin{vmatrix} 3 & 5 & 0 \\ 2+3k & 1+5k & 3+0k \\ -5 & 1 & 6 \end{vmatrix}$

24. $\begin{vmatrix} -4 & 2 & 1 \\ 3 & 0 & 5 \\ -1 & 4 & -2 \end{vmatrix} = \begin{vmatrix} -4 & 2 & 1+(-4)k \\ 3 & 0 & 5+3k \\ -1 & 4 & -2+(-1)k \end{vmatrix}$

Use Property 6 to find each of the following determinants. See Examples 7 and 8.

25. $\begin{vmatrix} 2 & 4 \\ 3 & 6 \end{vmatrix}$

26. $\begin{vmatrix} -5 & 10 \\ 6 & -12 \end{vmatrix}$

27. $\begin{vmatrix} 4 & 8 & 0 \\ -1 & -2 & 1 \\ 2 & 4 & 3 \end{vmatrix}$

28. $\begin{vmatrix} 6 & 8 & -12 \\ -1 & 16 & 2 \\ 4 & 0 & -8 \end{vmatrix}$

29. $\begin{vmatrix} 3 & 1 & 2 \\ 2 & 3 & 1 \\ 1 & 0 & -2 \end{vmatrix}$

30. $\begin{vmatrix} -2 & 2 & 3 \\ 0 & 2 & 1 \\ -1 & 4 & 0 \end{vmatrix}$

31. $\begin{vmatrix} -4 & 2 & 3 \\ 2 & 0 & 1 \\ 0 & 4 & 2 \end{vmatrix}$

32. $\begin{vmatrix} 6 & 3 & 2 \\ 1 & 0 & 2 \\ -1 & 4 & 1 \end{vmatrix}$

33. $\begin{vmatrix} 1 & 0 & 2 & 2 \\ 2 & 4 & 1 & -1 \\ 1 & -3 & 1 & 0 \\ 1 & 1 & 0 & 1 \end{vmatrix}$
34. $\begin{vmatrix} 2 & -1 & 1 & 0 \\ 1 & 1 & 0 & 1 \\ 0 & -1 & 1 & 1 \\ 1 & 2 & 1 & 2 \end{vmatrix}$

Use the determinant

$$\begin{vmatrix} a & b & c \\ d & e & f \\ g & h & j \end{vmatrix}$$

to prove the following properties of determinants by finding the values of the original determinant and the new determinant and comparing them.

35. Property 2   **36.** Property 3   **37.** Property 4

8.6 Solution of Linear Systems by Cramer's Rule

The elimination method is used in this section to solve the general system of two equations with two variables,

$$a_1x + b_1y = c_1 \tag{1}$$
$$a_2x + b_2y = c_2. \tag{2}$$

The result is a formula that can be used for any system of two equations with two unknowns. To get this general solution, eliminate y and solve for x by first multiplying both sides of equation (1) by b_2 and both sides of equation (2) by $-b_1$. Then add these results and solve for x.

$$a_1b_2x + b_1b_2y = c_1b_2$$
$$\underline{-a_2b_1x - b_1b_2y = -c_2b_1}$$
$$(a_1b_2 - a_2b_1)x \qquad\qquad = c_1b_2 - c_2b_1$$
$$x = \frac{c_1b_2 - c_2b_1}{a_1b_2 - a_2b_1}$$

Solve for y by multiplying both sides of equation (1) by $-a_2$ and equation (2) by a_1 and then adding the two equations.

$$-a_1a_2x - a_2b_1y = -a_2c_1$$
$$\underline{a_1a_2x + a_1b_2y = a_1c_2}$$
$$(a_1b_2 - a_2b_1)y = a_1c_2 - a_2c_1$$
$$y = \frac{a_1c_2 - a_2c_1}{a_1b_2 - a_2b_1}$$

Both numerators and the common denominator of these values for x and y can be written as determinants, since

$$c_1 b_2 - c_2 b_1 = \begin{vmatrix} c_1 & b_1 \\ c_2 & b_2 \end{vmatrix}, \qquad a_1 c_2 - a_2 c_1 = \begin{vmatrix} a_1 & c_1 \\ a_2 & c_2 \end{vmatrix},$$

$$\text{and} \qquad a_1 b_2 - a_2 b_1 = \begin{vmatrix} a_1 & b_1 \\ a_2 & b_2 \end{vmatrix}.$$

Using these determinants, the solutions for x and y become

$$x = \frac{\begin{vmatrix} c_1 & b_1 \\ c_2 & b_2 \end{vmatrix}}{\begin{vmatrix} a_1 & b_1 \\ a_2 & b_2 \end{vmatrix}} \qquad \text{and} \qquad y = \frac{\begin{vmatrix} a_1 & c_1 \\ a_2 & c_2 \end{vmatrix}}{\begin{vmatrix} a_1 & b_1 \\ a_2 & b_2 \end{vmatrix}}, \qquad \text{if} \begin{vmatrix} a_1 & b_1 \\ a_2 & b_2 \end{vmatrix} \neq 0.$$

For convenience, denote the three determinants in the solution as

$$\begin{vmatrix} a_1 & b_1 \\ a_2 & b_2 \end{vmatrix} = D, \qquad \begin{vmatrix} c_1 & b_1 \\ c_2 & b_2 \end{vmatrix} = D_x, \qquad \text{and} \qquad \begin{vmatrix} a_1 & c_1 \\ a_2 & c_2 \end{vmatrix} = D_y.$$

The elements of D are the four coefficients of the variables in the given system, the elements of D_x are obtained by replacing the coefficients of x in D by the respective constants, and the elements of D_y are obtained by replacing the coefficients of y in D by the respective constants.

These results are summarized as **Cramer's rule.**

Cramer's Rule for 2 × 2 Systems

Given the system

$$a_1 x + b_1 y = c_1$$
$$a_2 x + b_2 y = c_2,$$

with $a_1 b_2 - a_2 b_1 \neq 0$,

then
$$x = \frac{D_x}{D} \qquad \text{and} \qquad y = \frac{D_y}{D},$$

where

$$D_x = \begin{vmatrix} c_1 & b_1 \\ c_2 & b_2 \end{vmatrix}, \qquad D_y = \begin{vmatrix} a_1 & c_1 \\ a_2 & c_2 \end{vmatrix}, \qquad \text{and} \qquad D = \begin{vmatrix} a_1 & b_1 \\ a_2 & b_2 \end{vmatrix} \neq 0.$$

Cramer's rule is used to solve a system of linear equations by evaluating the three determinants D, D_x, and D_y and then writing the appropriate quotients for x and y.

Example 1   Use Cramer's rule to solve the system

$$5x + 7y = -1$$
$$6x + 8y = 1.$$

By Cramer's rule, $x = D_x/D$ and $y = D_y/D$. It is a good idea to find D first, since if $D = 0$, Cramer's rule does not apply. If $D \neq 0$, then find D_x and D_y.

$$D = \begin{vmatrix} 5 & 7 \\ 6 & 8 \end{vmatrix} = 5(8) - 6(7) = -2$$

$$D_x = \begin{vmatrix} -1 & 7 \\ 1 & 8 \end{vmatrix} = (-1)(8) - (1)(7) = -15$$

$$D_y = \begin{vmatrix} 5 & -1 \\ 6 & 1 \end{vmatrix} = 5(1) - (6)(-1) = 11$$

From Cramer's rule,

$$x = \frac{D_x}{D} = \frac{-15}{-2} = \frac{15}{2}$$

and

$$y = \frac{D_y}{D} = \frac{11}{-2} = -\frac{11}{2}.$$

The solution set is $\{(15/2, -11/2)\}$, as can be verified by substituting in the given system.   ✚

Cramer's rule can be generalized to systems of three equations in three variables (or n equations in n variables).

Cramer's Rule for 3 × 3 Systems

Given the system

$$a_1x + b_1y + c_1z = d_1$$
$$a_2x + b_2y + c_2z = d_2$$
$$a_3x + b_3y + c_3z = d_3,$$

then

$$x = \frac{D_x}{D}, \qquad y = \frac{D_y}{D}, \qquad \text{and} \qquad z = \frac{D_z}{D},$$

where

$$D_x = \begin{vmatrix} d_1 & b_1 & c_1 \\ d_2 & b_2 & c_2 \\ d_3 & b_3 & c_3 \end{vmatrix}, \qquad D_y = \begin{vmatrix} a_1 & d_1 & c_1 \\ a_2 & d_2 & c_2 \\ a_3 & d_3 & c_3 \end{vmatrix},$$

$$D_z = \begin{vmatrix} a_1 & b_1 & d_1 \\ a_2 & b_2 & d_2 \\ a_3 & b_3 & d_3 \end{vmatrix}, \qquad \text{and} \qquad D = \begin{vmatrix} a_1 & b_1 & c_1 \\ a_2 & b_2 & c_2 \\ a_3 & b_3 & c_3 \end{vmatrix} \neq 0.$$

Example 2   Use Cramer's rule to solve the system

$$x + y - z + 2 = 0$$
$$2x - y + z + 5 = 0$$
$$x - 2y + 3z - 4 = 0.$$

For Cramer's rule, the system must be rewritten in the form

$$x + y - z = -2$$
$$2x - y + z = -5$$
$$x - 2y + 3z = 4.$$

Verify that the required determinants are as follows:

$$D = \begin{vmatrix} 1 & 1 & -1 \\ 2 & -1 & 1 \\ 1 & -2 & 3 \end{vmatrix} = -3, \qquad D_x = \begin{vmatrix} -2 & 1 & -1 \\ -5 & -1 & 1 \\ 4 & -2 & 3 \end{vmatrix} = 7,$$

$$D_y = \begin{vmatrix} 1 & -2 & -1 \\ 2 & -5 & 1 \\ 1 & 4 & 3 \end{vmatrix} = -22, \qquad D_z = \begin{vmatrix} 1 & 1 & -2 \\ 2 & -1 & -5 \\ 1 & -2 & 4 \end{vmatrix} = -21.$$

Thus

$$x = \frac{D_x}{D} = \frac{7}{-3} = -\frac{7}{3}, \qquad y = \frac{D_y}{D} = \frac{-22}{-3} = \frac{22}{3},$$

and

$$z = \frac{D_z}{D} = \frac{-21}{-3} = 7,$$

so the solution set is $\{(-7/3, 22/3, 7)\}$. ✛

Example 3   Use Cramer's rule to solve the system

$$2x - 3y + 4z = 10$$
$$6x - 9y + 12z = 24$$
$$x + 2y - 3z = 5.$$

First find D.

$$D = \begin{vmatrix} 2 & -3 & 4 \\ 6 & -9 & 12 \\ 1 & 2 & -3 \end{vmatrix} = 2 \begin{vmatrix} -9 & 12 \\ 2 & -3 \end{vmatrix} - 6 \begin{vmatrix} -3 & 4 \\ 2 & -3 \end{vmatrix} + 1 \begin{vmatrix} -3 & 4 \\ -9 & 12 \end{vmatrix}$$

$$= 2(3) - 6(1) + 1(0)$$
$$= 0$$

As mentioned above, Cramer's rule does not apply if $D = 0$. When $D = 0$, the system either is inconsistent or contains dependent equations. Use the elimination method or the Gauss-Jordan method to tell which is the case. Verify that this system is inconsistent.   ✜

8.6   Exercises

Use Cramer's rule to solve each of the following systems of equations. If $D = 0$, use another method to determine the solution. See Example 1.

1. $x + y = 4$
   $2x - y = 2$

2. $3x + 2y = -4$
   $2x - y = -5$

3. $4x + 3y = -7$
   $2x + 3y = -11$

4. $4x - y = 0$
   $2x + 3y = 14$

5. $5x + 4y = 10$
   $3x - 7y = 6$

6. $3x + 2y = -4$
   $5x - y = 2$

7. $2x - 3y = -5$
   $x + 5y = 17$

8. $x + 9y = -15$
   $3x + 2y = 5$

9. $3x + 2y = 4$
   $6x + 4y = 8$

10. $1.5x + 3y = 5$
    $2x + 4y = 3$

11. $12x + 8y = 3$
    $15x + 10y = 9$

12. $15x - 10y = 5$
    $9x + 6y = 3$

Use Cramer's rule to solve each of the following systems of equations. If $D = 0$, use another method to complete the solution. See Examples 2 and 3.

13. $4x - y + 3z = -3$
    $3x + y + z = 0$
    $2x - y + 4z = 0$

14. $5x + 2y + z = 15$
    $2x - y + z = 9$
    $4x + 3y + 2z = 13$

15. $2x - y + 4z = -2$
    $3x + 2y - z = -3$
    $x + 4y + 2z = 17$

16. $x + y + z = 4$
    $2x - y + 3z = 4$
    $4x + 2y - z = -15$

17. $4x - 3y + z = -1$
    $5x + 7y + 2z = -2$
    $3x - 5y - z = 1$

18. $2x - 3y + z = 8$
    $-x - 5y + z = -4$
    $3x - 5y + 2z = 12$

19. $x + 2y + 3z = 4$
    $4x + 3y + 2z = 1$
    $-x - 2y - 3z = 0$

20. $2x - y + 3z = 1$
    $-2x + y - 3z = 2$
    $5x - y + z = 2$

21. $-2x - 2y + 3z = 4$
    $5x + 7y - z = 2$
    $2x + 2y - 3z = -4$

22. $-3x + 2y - 2z = 4$
    $4x + y + z = 5$
    $3x - 2y + 2z = 1$

23. $2x + 3y = 13$
    $2y - z = 5$
    $x + 2z = 4$

24. $3x - z = -10$
    $y + 4z = 8$
    $x + 2z = -1$

25. $5x - y = -4$
    $3x + 2z = 4$
    $4y + 3z = 22$

26. $3x + 5y = -7$
    $2x + 7z = 2$
    $4y + 3z = -8$

27. $x + 2y = 10$
    $3x + 4z = 7$
    $-y - z = 1$

28. $5x - 2y = 3$
    $4y + z = 8$
    $x + 2z = 4$

Use Cramer's rule to solve each of the following systems.

29. $x + z = 0$
    $y + 2z + w = 0$
    $2x - w = 0$
    $x + 2y + 3z = -2$

30. $x - y + z + w = 6$
    $2y - w = -7$
    $x - z = -1$
    $y + w = 1$

31. $.4x - .6y = .4$
    $.3x + .2y = -.22$

32. $-.5x + .4y = .43$
    $.1x + .2y = .11$

33. $.4x - .2y + .3z = -.08$
    $-.5x + .3y - .7z = .19$
    $.8x - .7y + .1z = -.21$

34. $-.6x + .2y - .5z = -.16$
    $.8x - .3y - .2z = -.48$
    $.4x + .6y - .8z = -.8$

Solve each system for x and y using Cramer's rule. Assume a and b are nonzero constants.

35. $bx + y = a^2$
      $ax + y = b^2$

36. $ax + by = \dfrac{b}{a}$

      $x + y = \dfrac{1}{b}$

37. $b^2x + a^2y = b^2$
      $ax + by = a$

38. $x + \dfrac{1}{b}y = b$

      $\dfrac{1}{a}x + y = a$

For the following two exercises, use the system of equations

$$a_1x + b_1y = c_1$$

$$a_2x + b_2y = c_2.$$

39. Assume $D_x = 0$ and $D_y = 0$. Show that if $c_1c_2 \neq 0$, then $D = 0$, and the equations are dependent.

40. Assume $D = 0$, $D_x = 0$, and $b_1b_2 \neq 0$. Show that $D_y = 0$.

Chapter 8 *Summary*

Key Words   **8.1** dimension   **8.3** identity matrix
order   inverse of a matrix
square matrix   **8.4** determinant
row matrix   minor
column matrix   cofactor
zero matrix
scalar

Key Ideas

8.1 Addition of   The sum of two $m \times n$ matrices A and B is the $m \times n$ matrix $A + B$ in which
Matrices   each element is the sum of the corresponding elements of A and B.

8.1 Subtraction   If A and B are matrices of the same order, then
of Matrices
$$A - B = A + (-B).$$

8.2 Multiplication   The product AB of an $m \times n$ matrix A and an $n \times k$ matrix B is found as follows.
of Matrices   To get the ith row, jth column element of AB, multiply each element in the ith
row of A by the corresponding element in the jth column of B. The sum of these
products will give the element of row i, column j of AB.

8.4 Determinant   The determinant of a 2×2 matrix A,
of a 2 × 2
Matrix
$$A = \begin{bmatrix} a_{11} & a_{12} \\ a_{21} & a_{22} \end{bmatrix},$$

is defined as

$$\delta(A) = |A| = \begin{vmatrix} a_{11} & a_{12} \\ a_{21} & a_{22} \end{vmatrix} = a_{11}a_{22} - a_{21}a_{12}.$$

8.4 Determinant of a 3 × 3 Matrix

The determinant of a 3 × 3 matrix A,

$$A = \begin{bmatrix} a_{11} & a_{12} & a_{13} \\ a_{21} & a_{22} & a_{23} \\ a_{31} & a_{32} & a_{33} \end{bmatrix},$$

is defined as

$$\delta(A) = \begin{vmatrix} a_{11} & a_{12} & a_{13} \\ a_{21} & a_{22} & a_{23} \\ a_{31} & a_{32} & a_{33} \end{vmatrix} = (a_{11}a_{22}a_{33} + a_{12}a_{23}a_{31} + a_{13}a_{21}a_{32}) \\ - (a_{31}a_{22}a_{13} + a_{32}a_{23}a_{11} + a_{33}a_{21}a_{12}).$$

8.5 Properties of Determinants

1. If every element in a row or column of a matrix is 0, then the determinant equals 0.

2. If corresponding rows and columns of a matrix are interchanged, the determinant is not changed.

3. Interchanging two rows (or columns) of a matrix reverses the sign of the determinant.

4. If every element of a row (or column) of a matrix is multiplied by the real number k, then the determinant of the new matrix is k times the determinant of the original matrix.

5. The determinant of a matrix with two identical rows (or columns) equals 0.

6. The determinant of a matrix is unchanged if a multiple of a row (or column) of the matrix is added to the corresponding elements of another row (or column).

8.6 Cramer's Rule for 2 × 2 Systems

Given the system

$$a_1x + b_1y = c_1$$
$$a_2x + b_2y = c_2,$$

then

$$x = \frac{D_x}{D} \quad \text{and} \quad y = \frac{D_y}{D},$$

where

$$D_x = \begin{vmatrix} c_1 & b_1 \\ c_2 & b_2 \end{vmatrix}, \quad D_y = \begin{vmatrix} a_1 & c_1 \\ a_2 & c_2 \end{vmatrix}, \quad \text{and} \quad D = \begin{vmatrix} a_1 & b_1 \\ a_2 & b_2 \end{vmatrix} \neq 0.$$

8.6 Cramer's Rule for 3 × 3 Systems

Given the system

$$a_1x + b_1y + c_1z = d_1$$
$$a_2x + b_2y + c_2z = d_2$$
$$a_3x + b_3y + c_3z = d_3,$$

then

$$x = \frac{D_x}{D}, \qquad y = \frac{D_y}{D}, \qquad \text{and} \qquad z = \frac{D_z}{D},$$

where

$$D_x = \begin{vmatrix} d_1 & b_1 & c_1 \\ d_2 & b_2 & c_2 \\ d_3 & b_3 & c_3 \end{vmatrix}, \qquad D_y = \begin{vmatrix} a_1 & d_1 & c_1 \\ a_2 & d_2 & c_2 \\ a_3 & d_3 & c_3 \end{vmatrix},$$

$$D_z = \begin{vmatrix} a_1 & b_1 & d_1 \\ a_2 & b_2 & d_2 \\ a_3 & b_3 & d_3 \end{vmatrix}, \quad \text{and} \quad D = \begin{vmatrix} a_1 & b_1 & c_1 \\ a_2 & b_2 & c_2 \\ a_3 & b_3 & c_3 \end{vmatrix} \neq 0.$$

Chapter 8   *Review Exercises*

Find the values of all variables in the following statements.

1. $\begin{bmatrix} 2 & z & 1 \\ m & 9 & -7 \end{bmatrix} = \begin{bmatrix} x & 5 & 1 \\ -8 & y & p \end{bmatrix}$

2. $\begin{bmatrix} 5 & -10 \\ a & 0 \end{bmatrix} = \begin{bmatrix} b-3 & c+2 \\ 4 & 0 \end{bmatrix}$

3. $\begin{bmatrix} 5 & x+2 \\ -6y & z \end{bmatrix} = \begin{bmatrix} a & 3x-1 \\ 5y & 9 \end{bmatrix}$

4. $\begin{bmatrix} 6+k & 2 & a+3 \\ -2+m & 3p & 2r \end{bmatrix} + \begin{bmatrix} 3-2k & 5 & 7 \\ 5 & 8p & 5r \end{bmatrix} = \begin{bmatrix} 5 & y & 6a \\ 2m & 11 & -35 \end{bmatrix}$

Perform each of the following operations whenever possible.

5. $\begin{bmatrix} 3 & -4 & 2 \\ 5 & -1 & 6 \end{bmatrix} + \begin{bmatrix} -3 & 2 & 5 \\ 1 & 0 & 4 \end{bmatrix}$

6. $\begin{bmatrix} 3 \\ 2 \\ 5 \end{bmatrix} - \begin{bmatrix} 8 \\ -4 \\ 6 \end{bmatrix} + \begin{bmatrix} 1 \\ 0 \\ 2 \end{bmatrix}$

7. $\begin{bmatrix} 2 & 5 & 8 \\ 1 & 9 & 2 \end{bmatrix} - \begin{bmatrix} 3 & 4 \\ 7 & 1 \end{bmatrix}$

8. $3\begin{bmatrix} 2 & 4 \\ -1 & 4 \end{bmatrix} - 2\begin{bmatrix} 5 & 8 \\ 2 & -2 \end{bmatrix}$

9. $-1\begin{bmatrix} 3 & -5 & 2 \\ 1 & 7 & -4 \end{bmatrix} + 5\begin{bmatrix} 0 & 2 \\ -1 & 3 \end{bmatrix}$

10. $10\begin{bmatrix} 2x+3y & 4x+y \\ x-5y & 6x+2y \end{bmatrix} + 2\begin{bmatrix} -3x-y & x+6y \\ 4x+2y & 5x-y \end{bmatrix}$

11. The speed limits in Italy, Britain, and the U.S. are 87 mph, 70 mph, and 55 mph, respectively. The corresponding fatalities per 100 million miles driven (in 1977) were 6.4, 4.0, and 3.3. Write this information as a 3 × 2 matrix and then as a 2 × 3 matrix.

12. In the years 1983 through 1987, a company had sales (in billions) of 13.0, 14.0, 14.2, 14.9, and 16.7. The company had income (in millions) of 1010, 1000, 979, 1079, and 1228. They declared dividends per common share (in dollars) of 1.32, 1.46, 1.54, 1.64, and 1.82. Write this information as a matrix in two ways.

Find each of the following matrix products, whenever possible.

13. $\begin{bmatrix} -3 & 4 \\ 2 & 8 \end{bmatrix}\begin{bmatrix} -1 & 0 \\ 2 & 5 \end{bmatrix}$

14. $\begin{bmatrix} 3 & 2 & -1 \\ 4 & 0 & 6 \end{bmatrix}\begin{bmatrix} -2 & 0 \\ 0 & 2 \\ 3 & 1 \end{bmatrix}$

15. $\begin{bmatrix} 1 & -2 & 4 & 2 \\ 0 & 1 & -1 & 8 \end{bmatrix}\begin{bmatrix} -1 \\ 2 \\ 0 \\ 1 \end{bmatrix}$

16. $\begin{bmatrix} 1 & 2 & 5 \\ -3 & 4 & 7 \\ 0 & 2 & -1 \end{bmatrix}\begin{bmatrix} 4 & 2 & 3 \\ 10 & -5 & 6 \end{bmatrix}$

17. $\begin{bmatrix} 4 & 2 & 3 \\ 10 & -5 & 6 \end{bmatrix}\begin{bmatrix} 1 & 2 & 5 \\ -3 & 4 & 7 \\ 0 & 2 & -1 \end{bmatrix}$

18. $\begin{bmatrix} 3 & -1 & 0 \end{bmatrix}\begin{bmatrix} 1 & 3 & 2 \\ 2 & -4 & 0 \\ 5 & 7 & 3 \end{bmatrix}$

Decide whether or not each of the following pairs of matrices are inverses.

19. $\begin{bmatrix} 2 & -3 \\ 1 & -2 \end{bmatrix}, \begin{bmatrix} 2 & -3 \\ 1 & -2 \end{bmatrix}$

20. $\begin{bmatrix} 1 & 0 \\ 2 & -3 \end{bmatrix}, \begin{bmatrix} 1 & 0 \\ \frac{2}{3} & -\frac{1}{3} \end{bmatrix}$

21. $\begin{bmatrix} 2 & 0 & 6 \\ 0 & 1 & 0 \\ 1 & 0 & 1 \end{bmatrix}, \begin{bmatrix} -1 & 0 & \frac{3}{2} \\ 0 & 1 & 0 \\ \frac{1}{4} & 0 & -1 \end{bmatrix}$

22. $\begin{bmatrix} 1 & 0 & 2 \\ 0 & 2 & 4 \\ 0 & 0 & 1 \end{bmatrix}, \begin{bmatrix} 1 & 0 & -2 \\ 0 & \frac{1}{2} & -2 \\ 0 & 0 & 1 \end{bmatrix}$

Find the inverse, if it exists, for each of the following matrices.

23. $\begin{bmatrix} 2 & 1 \\ 5 & 3 \end{bmatrix}$

24. $\begin{bmatrix} -4 & 2 \\ 0 & 3 \end{bmatrix}$

25. $\begin{bmatrix} 2 & 0 \\ -1 & 5 \end{bmatrix}$

26. $\begin{bmatrix} 2 & 0 & 4 \\ 1 & -1 & 0 \\ 0 & 1 & -2 \end{bmatrix}$

27. $\begin{bmatrix} 2 & -1 & 0 \\ 1 & 0 & 1 \\ 1 & -2 & 0 \end{bmatrix}$

28. $\begin{bmatrix} 2 & 3 & 5 \\ -2 & -3 & -5 \\ 1 & 4 & 2 \end{bmatrix}$

Use matrix inverses to solve each of the following.

29. $x + y = 4$
$2x + 3y = 10$

30. $5x - 3y = -2$
$2x + 7y = -9$

31. $2x + y = 5$
$3x - 2y = 4$

32. $x - 2y = 7$
$3x + y = 7$

33. $x + y + z = 1$
$2x - y \quad\quad = -2$
$3y + z = 2$

34. $x \quad\quad\quad = -3$
$y + z = 6$
$2x \quad - 3z = -9$

35. $3x - 2y + 4z = 1$
$4x + y - 5z = 2$
$-6x + 4y - 8z = -2$

36. $x + 2y \quad\quad = -1$
$3y - z = -5$
$x + 2y - z = -3$

Evaluate each of the following determinants. Identify any systems with dependent equations or any inconsistent systems.

37. $\begin{vmatrix} -1 & 8 \\ 2 & 9 \end{vmatrix}$

38. $\begin{vmatrix} -2 & 4 \\ 0 & 3 \end{vmatrix}$

39. $\begin{vmatrix} -2 & 4 & 1 \\ 3 & 0 & 2 \\ -1 & 0 & 3 \end{vmatrix}$

40. $\begin{vmatrix} -1 & 2 & 3 \\ 4 & 0 & 3 \\ 5 & -1 & 2 \end{vmatrix}$

Solve each of the following determinant equations for x.

41. $\begin{vmatrix} -3 & 2 \\ 1 & x \end{vmatrix} = 5$

42. $\begin{vmatrix} 3x & 7 \\ -x & 4 \end{vmatrix} = 8$

43. $\begin{vmatrix} 2 & 5 & 0 \\ 1 & 3x & -1 \\ 0 & 2 & 0 \end{vmatrix} = 4$

44. $\begin{vmatrix} 6x & 2 & 0 \\ 1 & 5 & 3 \\ x & 2 & -1 \end{vmatrix} = 2x$

Explain why each of the following statements is true.

45. $\begin{vmatrix} 8 & 9 & 2 \\ 0 & 0 & 0 \\ 3 & 1 & 4 \end{vmatrix} = 0$

46. $\begin{vmatrix} 4 & 6 \\ 3 & 5 \end{vmatrix} = \begin{vmatrix} 4 & 3 \\ 6 & 5 \end{vmatrix}$

47. $\begin{vmatrix} 8 & 2 \\ 4 & 3 \end{vmatrix} = 2 \begin{vmatrix} 4 & 1 \\ 4 & 3 \end{vmatrix}$

48. $\begin{vmatrix} 4 & 6 & 2 \\ -3 & 8 & -5 \\ 4 & 6 & 2 \end{vmatrix} = 0$

49. $\begin{vmatrix} 5 & -1 & 2 \\ 3 & -2 & 0 \\ -4 & 1 & 2 \end{vmatrix} = \begin{vmatrix} 5 & -1 & 2 \\ 8 & -3 & 2 \\ -4 & 1 & 2 \end{vmatrix}$ **50.** $\begin{vmatrix} 8 & 2 & -5 \\ -3 & 1 & 4 \\ 2 & 0 & 5 \end{vmatrix} = -\begin{vmatrix} 8 & -5 & 2 \\ -3 & 4 & 1 \\ 2 & 5 & 0 \end{vmatrix}$

Solve each of the following systems by Cramer's rule. Identify any systems with dependent equations or any inconsistent systems.

51. $3x + y = -1$
$5x + 4y = 10$

52. $3x + 7y = 2$
$5x - y = -22$

53. $3x + 2y + z = 2$
$4x - y + 3z = -16$
$x + 3y - z = 12$

54. $5x - 2y - z = 8$
$-5x + 2y + z = -8$
$x - 4y - 2z = 0$

Chapter 8 *Test*

1. Find the values of all variables in the following equation.

$$\begin{bmatrix} 5 & x + 6 \\ 0 & 4 \end{bmatrix} = \begin{bmatrix} y - 2 & 4 - x \\ 0 & w + 7 \end{bmatrix}$$

Perform each of the following operations whenever possible.

2. $3\begin{bmatrix} 2 & 3 \\ 1 & -4 \\ 5 & 9 \end{bmatrix} - \begin{bmatrix} -2 & 6 \\ 3 & -1 \\ 0 & 8 \end{bmatrix}$

3. $\begin{bmatrix} 1 \\ 2 \end{bmatrix} + \begin{bmatrix} 4 \\ -6 \end{bmatrix} + \begin{bmatrix} 2 & 8 \\ -7 & 5 \end{bmatrix}$

4. At a small college, the freshman class consists of 108 women and 142 men. The sophomore class consists of 112 women and 98 men. The junior class consists of 123 women and 117 men. The senior class consists of 110 women and 130 men. Write this information as a 2 × 4 matrix.

Find each of the following matrix products, whenever possible.

5. $\begin{bmatrix} 2 & 1 & -3 \\ 4 & 0 & 5 \end{bmatrix}\begin{bmatrix} 1 & 3 \\ 2 & 4 \\ 3 & -2 \end{bmatrix}$

6. $\begin{bmatrix} 2 & -4 \\ 3 & 5 \end{bmatrix}\begin{bmatrix} 4 \\ 2 \\ 7 \end{bmatrix}$

7. $\begin{bmatrix} 2 & 8 \\ 1 & -2 \end{bmatrix}\begin{bmatrix} 3 & -5 & 4 \\ -4 & 7 & 8 \end{bmatrix}$

8. Decide whether or not the following pair of matrices are inverses.

$$\begin{bmatrix} 1 & 0 & 1 \\ -1 & 0 & 1 \\ 0 & 1 & 0 \end{bmatrix}\begin{bmatrix} \frac{1}{2} & -\frac{1}{2} & 0 \\ 0 & 0 & 1 \\ \frac{1}{2} & \frac{1}{2} & 0 \end{bmatrix}$$

Find the inverse, if it exists, for each of the following matrices.

9. $\begin{bmatrix} -8 & 5 \\ 3 & -2 \end{bmatrix}$

10. $\begin{bmatrix} 4 & 12 \\ 2 & 6 \end{bmatrix}$

11. $\begin{bmatrix} 1 & 3 & 4 \\ 2 & 7 & 8 \\ -2 & -5 & -7 \end{bmatrix}$

Use matrix inverses to solve each of the following.

12. $2x + y = -6$
$3x - y = -29$

13. $x + y \quad\ = 5$
$\quad\ y - 2z = 23$
$x \quad\quad + 3z = -27$

Evaluate each of the following determinants.

14. $\begin{vmatrix} 6 & 8 \\ 2 & -7 \end{vmatrix}$

15. $\begin{vmatrix} 2 & 0 & 8 \\ -1 & 7 & 9 \\ 12 & 5 & -3 \end{vmatrix}$

16. Solve the following determinant equation for x.

$$\begin{vmatrix} 1 & 3x & 3 \\ -2 & 0 & 7 \\ 4 & x & 5 \end{vmatrix} = -202$$

Explain why each of the following statements is true.

17. $\begin{vmatrix} 6 & 7 \\ -5 & 2 \end{vmatrix} = - \begin{vmatrix} -5 & 2 \\ 6 & 7 \end{vmatrix}$

18. $\begin{vmatrix} 7 & 2 & -1 \\ 5 & -4 & 3 \\ 6 & -2 & 1 \end{vmatrix} = \begin{vmatrix} 7 & 2 & -1 \\ 19 & 0 & 1 \\ 13 & 0 & 0 \end{vmatrix}$

Solve each of the following systems by Cramer's rule. Identify any systems with dependent equations or any inconsistent systems.

19. $2x - 3y = -33$
$4x + 5y = 11$

20. $x + y - z = -4$
$2x - 3y - z = 5$
$x + 2y + 2z = 3$

9

Additional Topics

Sequences (lists of numbers) are discussed in this chapter with special attention to two types of sequences, *arithmetic* and *geometric*. The chapter also presents the binomial theorem, a sum of a sequence, and *mathematical induction*, a method that can be used to prove theorems about sequences.

The last three sections of the chapter deal with ways to answer questions such as "How many different poker hands can be dealt with a 52-card deck?" and "How many license plates can be made using two letters followed by three numbers?" or "What is the probability of these events?" Questions like the first two are answered by *counting theory,* which gives methods of counting the number of ways an event can occur. *Probability theory* uses these methods of counting to give the probability of events.

9.1 Sequences

If the domain of a function is the set of positive integers, the range elements can be *ordered,* as $f(1), f(2), f(3)$, and so on. This ordered list of numbers is called a **sequence.** Since the letter x has been used to suggest real numbers, the variable n is used instead with sequences to suggest the positive integer domain. Although a sequence may be defined by $f(n) = 2n + 3$, for example, it is customary to use a_n instead of $f(n)$ and write $a_n = 2n + 3$.

The elements of a sequence, called the **terms** of the sequence, are written in order as a_1, a_2, a_3, ... The **general term,** or ***n*th term,** of the sequence is a_n. The general term of a sequence is used to find any term of the sequence. For example, if $a_n = n + 1$, then $a_2 = 2 + 1 = 3$.

A sequence is a **finite sequence** if the domain is the set $\{1, 2, 3, 4, . . ., n\}$, where n is a positive integer. An **infinite sequence** has the set of all positive integers as its domain. For example, the sequence of positive even integers,

$$2, 4, 6, 8, 10, 12, 14, . . .,$$

is infinite, but the sequence of days in June,

$$1, 2, 3, 4, . . ., 29, 30,$$

is finite.

Example 1   Write the first five terms for each of the following sequences.

(a) $a_n = \dfrac{n + 1}{n + 2}$

Replacing n, in turn, with 1, 2, 3, 4, and 5 gives

$$\frac{2}{3}, \frac{3}{4}, \frac{4}{5}, \frac{5}{6}, \frac{6}{7}.$$

(b) $a_n = (-1)^n \cdot n$

Replace n, in turn, with 1, 2, 3, 4, and 5, to get

$$a_1 = (-1)^1 \cdot 1 = -1$$
$$a_2 = (-1)^2 \cdot 2 = 2$$
$$a_3 = (-1)^3 \cdot 3 = -3$$
$$a_4 = (-1)^4 \cdot 4 = 4$$
$$a_5 = (-1)^5 \cdot 5 = -5.$$

(c) $b_n = \dfrac{(-1)^n}{2^n}$

Here $b_1 = -1/2$, $b_2 = 1/4$, $b_3 - -1/8$, $b_4 = 1/16$, and $b_5 = -1/32$. ✢

Example 2   Find the first four terms for the sequences defined as follows.

(a) $a_1 = 4$; $a_{n+1} = 2a_n + 1$

We have $a_1 = 4$, so

$$a_2 = 2 \cdot a_1 + 1 = 2 \cdot 4 + 1 = 9,$$
$$a_3 = 2 \cdot a_2 + 1 = 2 \cdot 9 + 1 = 19,$$

and $\qquad\qquad a_4 = 2 \cdot a_3 + 1 = 2 \cdot 19 + 1 = 39.$

(b) $a_1 = 2; a_{n+1} = a_n + n - 1$

$$a_1 = 2$$
$$a_2 = a_1 + 2 - 1 = 2 + 1 = 3$$
$$a_3 = a_2 + 3 - 1 = 3 + 2 = 5$$
$$a_4 = a_3 + 4 - 1 = 5 + 3 = 8 \quad \clubsuit$$

The definitions of the sequences in Example 2 are examples of **recursive definitions** in which each term is defined as an expression involving the previous term.

The sum of the terms of a sequence is called a **series.** A compact shorthand notation can be used to write a series. For example,

$$1 + 5 + 9 + 13 + 17 + 21$$

is the sum of the first six terms in the sequence with general term $a_n = 4n - 3$. This series can be written as

$$\sum_{i=1}^{6} (4i - 3)$$

(read "the sum from $i = 1$ to 6 of $4i - 3$").

The Greek letter sigma, Σ, is used to mean "sum." To evaluate this sum, replace i in $4i - 3$ first by 1, then 2, then 3, until finally i is replaced with 6.

$$\sum_{i=1}^{6} (4i - 3) = (4 \cdot 1 - 3) + (4 \cdot 2 - 3) + (4 \cdot 3 - 3)$$
$$+ (4 \cdot 4 - 3) + (4 \cdot 5 - 3) + (4 \cdot 6 - 3)$$
$$= 1 + 5 + 9 + 13 + 17 + 21 = 66$$

In this notation, i is called the **index of summation.** Do not confuse this use of i with the use of i to represent an imaginary number. Other letters may be used for the index of summation.

Example 3   Evaluate each of the following sums.

(a) $\displaystyle\sum_{i=1}^{4} i^2(i + 1) = 1^2(1 + 1) + 2^2(2 + 1) + 3^2(3 + 1) + 4^2(4 + 1)$

$$= 1 \cdot 2 + 4 \cdot 3 + 9 \cdot 4 + 16 \cdot 5$$
$$= 2 + 12 + 36 + 80$$
$$= 130$$

(b) $\displaystyle\sum_{j=3}^{6} \frac{j + 1}{j - 2} = \frac{3 + 1}{3 - 2} + \frac{4 + 1}{4 - 2} + \frac{5 + 1}{5 - 2} + \frac{6 + 1}{6 - 2}$

$$= \frac{4}{1} + \frac{5}{2} + \frac{6}{3} + \frac{7}{4} = \frac{41}{4} \quad \clubsuit$$

In the next few sections, formulas are developed that can be used to evaluate some of these sums without finding each term individually.

In statistics, the **mean** of a set of numbers, $x_1, x_2, x_3, \ldots, x_n$ is defined as

$$\bar{x} = \frac{\displaystyle\sum_{i=1}^{n} x_i}{n}.$$

That is, to find the mean, add all the members of the set of numbers and divide the sum by the number of members.

Example 4 Find the mean for the set of numbers 10, 12, 18, 24, 20, 22, 14.
Here $n = 7$, since there are seven numbers in the list.

$$\sum_{i=1}^{7} x_i = 10 + 12 + 18 + 24 + 20 + 22 + 14 = 120$$

Thus, the mean is

$$\bar{x} = \frac{\displaystyle\sum_{i=1}^{7} x_i}{7} = \frac{120}{7} = 17.1,$$

to the nearest tenth. ✤

9.1 Exercises

Write the first five terms of each of the following sequences. See Example 1.

1. $a_n = 6n + 4$ **2.** $a_n = 3n - 2$ **3.** $a_n = 2^n$ **4.** $a_n = 3^n$

5. $a_n = (-1)^{n+1}$ **6.** $a_n = (-1)^n(n + 2)$ **7.** $a_n = \dfrac{2n}{n + 3}$ **8.** $a_n = \dfrac{-4}{n + 5}$

9. $a_n = \dfrac{8n - 4}{2n + 1}$ **10.** $a_n = \dfrac{-3n + 6}{n + 1}$ **11.** $a_n = (-2)^n(n)$

12. $a_n = (-1/2)^n(n^{-1})$ **13.** $a_n = \dfrac{n^2 + 1}{n^2 + 2}$ **14.** $a_n = \dfrac{(-1)^{n-1}(n + 1)}{n + 2}$

15. $a_n = x^n$ **16.** $a_n = n \cdot x^{-n}$

Find the first ten terms for the sequences defined as follows. See Example 2.

17. $a_1 = 4, \quad a_n = a_{n-1} + 5$ **18.** $a_1 = -3, \quad a_n = a_{n-1} + 2$ **19.** $a_1 = -9, \quad a_n = a_{n-1} - 4$

20. $a_1 = -8, \quad a_n = a_{n-1} - 7$ **21.** $a_1 = 2, \quad a_n = 2 \cdot a_{n-1}$ **22.** $a_1 = -3, \quad a_n = 2 \cdot a_{n-1}$

23. $a_1 = 1, \quad a_2 = 1, \quad a_n = a_{n-1} + a_{n-2}$ (the Fibonacci sequence)

24. $a_1 = 3, \quad a_2 = 2, \quad a_n = a_{n-1} - a_{n-2}$

Evaluate each of the following sums. See Example 3.

25. $\displaystyle\sum_{i=1}^{5} (2i + 1)$ **26.** $\displaystyle\sum_{i=1}^{6} (3i - 2)$ **27.** $\displaystyle\sum_{j=1}^{9} j$ **28.** $\displaystyle\sum_{j=1}^{4} \dfrac{1}{j}$

29. $\displaystyle\sum_{i=1}^{5} (i + 1)^{-1}$ **30.** $\displaystyle\sum_{i=1}^{4} i^i$ **31.** $\displaystyle\sum_{i=1}^{3} i^{-i}$ **32.** $\displaystyle\sum_{k=1}^{4} (k + 1)^2$

33. $\displaystyle\sum_{k=1}^{6} (-1)^k \cdot k$ **34.** $\displaystyle\sum_{i=1}^{7} (-1)^{i+1} \cdot i^2$ **35.** $\displaystyle\sum_{i=3}^{7} \frac{i}{2}$ **36.** $\displaystyle\sum_{i=5}^{9} \frac{i + 1}{3}$

37. $\displaystyle\sum_{i=5}^{8} \frac{i^2 + 1}{2}$ **38.** $\displaystyle\sum_{i=3}^{5} \frac{i + 3}{i}$ **39.** $\displaystyle\sum_{j=0}^{4} 2^j$ **40.** $\displaystyle\sum_{i=0}^{4} 3^{-i}$

Write out each of the following. Do not try to evaluate.

41. $\displaystyle\sum_{i=1}^{4} x^i$ **42.** $\displaystyle\sum_{i=3}^{7} x^{-i}$ **43.** $\displaystyle\sum_{i=1}^{5} i \cdot x^i$ **44.** $\displaystyle\sum_{i=1}^{5} \frac{i^2}{x^i}$

Find the mean for each of the following sets of numbers. See Example 4.

45. 10, 9, 6, 12, 8, 7, 7, 7, 8, 6

46. 124, 132, 129, 135, 130

47. 28.1, 29.2, 31.5, 26.0, 28.9, 30.4

48. .002, .005, .001, .003, .004, .002

9.2   Arithmetic Sequences

A sequence in which each term after the first is obtained by adding a fixed number to the preceding term is called an **arithmetic sequence** (or **arithmetic progression**). The fixed number that is added is called the **common difference.** The sequence

$$5, 9, 13, 17, 21, \ldots$$

is an arithmetic sequence since each term after the first is obtained by adding 4 to the previous term. That is,

$$9 = 5 + 4$$
$$13 = 9 + 4$$
$$17 = 13 + 4$$
$$21 = 17 + 4,$$

and so on. The common difference is 4.

Example 1   Find the common difference, d, for the arithmetic sequence

$$-9, -7, -5, -3, -1, \ldots.$$

Since this sequence is arithmetic, d can be found by choosing any two adjacent terms and subtracting the first from the second. Choosing -7 and -5 gives

$$d = -5 - (-7) = 2.$$

Choosing -9 and -7 would give $d = -7 - (-9) = 2$, the same result.   ✦

If a_1 and d are known, then all the terms of an arithmetic sequence can be found.

Example 2   Write the first five terms for each of the following arithmetic sequences.

(a) The first term is 7, and each succeeding term is found by adding -3 to the preceding term.

Here $a_1 = 7,$

$$a_2 = 7 + (-3) = 4,$$
$$a_3 = 4 + (-3) = 1,$$
$$a_4 = 1 + (-3) = -2,$$

and $a_5 = -2 + (-3) = -5.$

(b) $a_1 = -12, d = 5$

Starting with a_1, add d to each term to get the next term.

$$a_1 = -12$$
$$a_2 = -12 + d = -12 + 5 = -7$$
$$a_3 = -7 + d = -7 + 5 = -2$$
$$a_4 = -2 + d = -2 + 5 = 3$$
$$a_5 = 3 + d = 3 + 5 = 8$$ ✛

If a_1 is the first term of an arithmetic sequence and d is the common difference, then the terms of the sequence are given by

$$a_1 = a_1$$
$$a_2 = a_1 + d$$
$$a_3 = a_2 + d = a_1 + d + d = a_1 + 2d$$
$$a_4 = a_3 + d = a_1 + 2d + d = a_1 + 3d$$
$$a_5 = a_1 + 4d$$
$$a_6 = a_1 + 5d,$$

and, by this pattern, $a_n = a_1 + (n - 1)d.$

---

*n*th Term of an
Arithmetic
Sequence

> In an arithmetic sequence with first term a_1 and common difference d, the nth term, a_n, is given by
>
> $$a_n = a_1 + (n - 1)d.$$

Example 3   Find a_{13} and a_n for the arithmetic sequence

$$-3, 1, 5, 9, \ldots .$$

Here $a_1 = -3$ and $d = 1 - (-3) = 4$. To find a_{13}, substitute 13 for n in the preceding formula.

$$a_{13} = a_1 + (13 - 1)d$$
$$a_{13} = -3 + (12)4$$
$$a_{13} = -3 + 48$$
$$a_{13} = 45$$

Find a_n by substituting values for a_1 and d in the formula for a_n.

$$a_n = -3 + (n - 1) \cdot 4$$
$$a_n = -3 + 4n - 4$$
$$a_n = 4n - 7 \quad \clubsuit$$

Example 4   Find a_{18} and a_n for the arithmetic sequence having $a_2 = 9$ and $a_3 = 15$. Find d first: $d = a_3 - a_2 = 15 - 9 = 6$.

Since
$$a_2 = a_1 + d,$$
$$9 = a_1 + 6 \quad \text{and} \quad a_1 = 3.$$
Then
$$a_{18} = 3 + (18 - 1) \cdot 6$$
$$a_{18} = 105,$$
and
$$a_n = 3 + (n - 1) \cdot 6$$
$$a_n = 3 + 6n - 6$$
$$a_n = 6n - 3. \quad \clubsuit$$

Example 5   Suppose that an arithmetic sequence has $a_8 = -16$ and $a_{16} = -40$. Find a_1.
Since $a_8 = a_1 + (8 - 1)d$, replacing a_8 with -16 gives $-16 = a_1 + 7d$ or $a_1 = -16 - 7d$. Similarly, $-40 = a_1 + 15d$ or $a_1 = -40 - 15d$. From these two equations, using the substitution method from Chapter 7,

$$-16 - 7d = -40 - 15d,$$

so $d = -3$. To find a_1, substitute -3 for d in $-16 = a_1 + 7d$:

$$-16 = a_1 + 7d$$
$$-16 = a_1 + 7(-3)$$
$$a_1 = 5. \quad \clubsuit$$

Example 6   A child building a tower with blocks uses 15 for the first row. Each row has 2 fewer blocks than the previous row. If there are 8 rows in the tower, how many blocks are used for the top row?
The number of blocks in each row forms an arithmetic sequence with $a_1 = 15$ and $d = -2$. Find a_n for $n = 8$ by using the formula $a_n = a_1 + (n - 1)d$.

$$a_8 = 15 + (8 - 1)(-2) = 1$$

There is just one block in the top row. ✚

Sum of an Arithmetic Sequence Suppose someone borrows $3000 and agrees to pay $100 per month plus interest of 1% per month on the unpaid balance until the loan is paid off. The first month he pays $100 to reduce the loan, plus interest of $(.01)3000 = 30$ dollars. The second month he pays another $100 toward the loan and interest of $(.01)2900 = 29$ dollars. Since the loan is reduced by $100 each month, his interest payments decrease by $(.01)100 = 1$ dollar each month, forming the arithmetic sequence

$$30, 29, 28, \ldots, 3, 2, 1.$$

The total amount of interest paid is given by the sum of the terms of this sequence. A formula can be developed to find this sum without adding all thirty numbers directly.

If an arithmetic sequence has terms $a_1, a_2, a_3, a_4 \ldots, a_n$, and S_n is defined as the sum of the first n terms of the sequence, then

$$S_n = a_1 + a_2 + a_3 + \ldots + a_n.$$

A formula for S_n can be found by writing the sum of the first n terms as follows.

$$S_n = a_1 + [a_1 + d] + [a_1 + 2d] + \ldots + [a_1 + (n - 1)d]$$

Next, write this same sum in reversed order.

$$S_n = [a_1 + (n - 1)d] + [a_1 + (n - 2)d] + \ldots + [a_1 + d] + a_1$$

Now add the corresponding sides of these two equations.

$$S_n + S_n = (a_1 + [a_1 + (n - 1)d]) + ([a_1 + d] + [a_1 + (n - 2)d]) \\ + \ldots + ([a_1 + (n - 1)d] + a_1)$$

From this,

$$2S_n = [2a_1 + (n - 1)d] + [2a_1 + (n - 1)d] + \ldots + [2a_1 + (n - 1)d].$$

Since there are n of the $[2a_1 + (n - 1)d]$ terms on the right,

$$2S_n = n[2a_1 + (n - 1)d]$$

$$S_n = \frac{n}{2}[2a_1 + (n - 1)d].$$

Using the formula $a_n = a_1 + (n - 1)d$, S_n can also be written as

$$S_n = \frac{n}{2}[a_1 + a_1 + (n - 1)d],$$

or

$$S_n = \frac{n}{2}(a_1 + a_n).$$

A summary of this work with sums of arithmetic sequences follows.

Sum of the First
n Terms of an
Arithmetic
Sequence

If an arithmetic sequence has first term a_1 and common difference d, then the sum of the first n terms is given by

$$S_n = \frac{n}{2}(a_1 + a_n)$$

or

$$S_n = \frac{n}{2}[2a_1 + (n-1)d].$$

The first formula is used when the first and last terms are known; otherwise the second formula is used.

Either one of these formulas can be used to find the total interest on the $3000 loan discussed above. In the sequence of interest payments, $a_1 = 30$, $d = -1$, $n = 30$, and $a_n = 1$. Choosing the first formula,

$$S_n = \frac{n}{2}(a_1 + a_n),$$

gives

$$S_{30} = \frac{30}{2}(30 + 1) = 15(31) = 465,$$

so a total of $465 interest will be paid over the 30 months.

Example 7   Find S_{12} for the arithmetic sequence

$$-9, -5, -1, 3, 7, \ldots.$$

Using $a_1 = -9$, $n = 12$, and $d = 4$ in the second formula,

$$S_n = \frac{n}{2}[2a_1 + (n-1)d],$$

gives

$$S_{12} = \frac{12}{2}[2(-9) + 11(4)] = 6(-18 + 44) = 156. \quad \clubsuit$$

Example 8   Find the sum of the first 60 positive integers.
Here $n = 60$, $a_1 = 1$, and $a_{60} = 60$. Use the first formula for S_n.

$$S_n = \frac{n}{2}(a_1 + a_n)$$

$$S_{60} = \frac{60}{2}(1 + 60) = 30 \cdot 61 = 1830 \quad \clubsuit$$

Example 9   The sum of the first 17 terms of an arithmetic sequence is 187. If $a_{17} = -13$, find a_1 and d.

Use the first formula for S_n, with $n = 17$, to find a_1.

$$S_{17} = \frac{17}{2}(a_1 + a_{17}) \qquad \text{Let } n = 17$$

$$187 = \frac{17}{2}(a_1 - 13) \qquad \text{Let } S_{17} = 187, a_{17} = -13$$

$$374 = 17(a_1 - 13)$$

$$22 = a_1 - 13$$

$$a_1 = 35$$

Since $a_{17} = a_1 + (17 - 1)d$,

$$-13 = 35 + 16d \qquad \text{Let } a_{17} = -13, a_1 = 35$$

$$-48 = 16d$$

$$d = -3. \quad \clubsuit$$

Any sum of the form

$$\sum_{i=1}^{n} (mi + p),$$

where m and p are real numbers, represents the sum of the terms of an arithmetic sequence having first term

$$a_1 = m(1) + p = m + p$$

and common difference $d = m$. These sums can be evaluated by the formulas in this section, as shown by the next example.

Example 10 Find each of the following sums.

(a) $\displaystyle\sum_{i=1}^{10} (4i + 8)$

This sum represents the sum of the first ten terms of the arithmetic sequence having

$$a_1 = 4 \cdot 1 + 8 = 12,$$

$$n = 10,$$

and

$$a_n = a_{10} = 4 \cdot 10 + 8 = 48.$$

Thus

$$\sum_{i=1}^{10} (4i + 8) = S_{10} = \frac{10}{2}(12 + 48) = 5(60) = 300.$$

(b) $\displaystyle\sum_{j=1}^{15} (9 - j)$

Since $n = 15$, $a_1 = 9 - 1 = 8$, and $a_{15} = 9 - 15 = -6$,

$$\sum_{j=1}^{15} (9 - j) = S_{15} = \frac{15}{2}[8 + (-6)] = \frac{15}{2}(2) = 15. \quad \clubsuit$$

9.2 Exercises

For each of the following arithmetic sequences, write the indicated number of terms. See Example 2.

1. $a_1 = 4, d = 2, n = 5$ **2.** $a_1 = 6, d = 8, n = 4$

3. $a_2 = 9, d = -2, n = 4$ **4.** $a_3 = 7, d = -4, n = 4$

5. $a_3 = -2, d = -4, n = 4$ **6.** $a_2 = -12, d = -6, n = 5$

7. $a_3 = 6, a_4 = 12, n = 6$ **8.** $a_5 = 8, a_6 = 5, n = 6$

For each of the following sequences that are arithmetic, find d and a_n. See Examples 1 and 3.

9. 12, 17, 22, 27, 32, 37, . . .   **10.** 8, 17, 26, 35, 44, 53, . . .

11. 18, 15, 12, 9, 6, . . .   **12.** 30, 24, 18, 12, . . .

13. $-19, -12, -5, 2, 9, . . .$ **14.** $-30, -20, -12, -6, -2, . . .$

15. $x, x + m, x + 2m, x + 3m, x + 4m, . . .$ **16.** $k + p, k + 2p, k + 3p, k + 4p, . . .$

17. $2z + m, 2z, 2z - m, 2z - 2m, 2z - 3m, . . .$

18. $3r - 4z, 3r - 3z, 3r - 2z, 3r - z, 3r, 3r + z, . . .$

Find a_8 and a_n for each of the following arithmetic sequences. See Example 4.

19. $a_1 = 5, d = 2$ **20.** $a_1 = -3, d = -4$ **21.** $a_3 = 2, d = 1$

22. $a_4 = 5, d = -2$ **23.** $a_1 = 8, a_2 = 6$ **24.** $a_1 = 6, a_2 = 3$

25. $a_{10} = 6, a_{12} = 15$ **26.** $a_{15} = 8, a_{17} = 2$ **27.** $a_1 = x, a_2 = x + 3$

28. $a_2 = y + 1, d = -3$ **29.** $a_6 = 2m, a_7 = 3m$ **30.** $a_5 = 4p + 1, a_7 = 6p + 7$

Find the sum of the first ten terms for each of the following arithmetic sequences. See Example 7.

31. $a_1 = 8, d = 3$ **32.** $a_1 = -9, d = 4$ **33.** $a_3 = 5, a_4 = 8$

34. $a_2 = 9, a_4 = 13$ **35.** 5, 9, 13, . . .   **36.** 8, 6, 4, . . .

37. $3\dfrac{1}{2}, 5, 6\dfrac{1}{2}, . . .$ **38.** $2\dfrac{1}{2}, 3\dfrac{3}{4}, 5, . . .$ **39.** $a_1 = 10, a_{10} = 5\dfrac{1}{2}$

40. $a_1 = -8, a_{10} = -5/4$ **41.** $a_1 = 9.428, d = -1.723$ **42.** $a_1 = -3.119, d = 2.422$

43. $a_4 = 2.556, a_5 = 3.004$ **44.** $a_7 = 11.192, a_9 = 4.812$

Evaluate each of the following sums. See Example 10.

45. $\displaystyle\sum_{i=1}^{3} (i + 4)$ **46.** $\displaystyle\sum_{i=1}^{5} (i - 8)$ **47.** $\displaystyle\sum_{j=1}^{10} (2j + 3)$ **48.** $\displaystyle\sum_{j=1}^{15} (5j - 9)$

49. $\displaystyle\sum_{i=1}^{12} (7i + 2)$ **50.** $\displaystyle\sum_{j=1}^{10} (8j - 3)$ **51.** $\displaystyle\sum_{i=1}^{12} (-5 - 8i)$ **52.** $\displaystyle\sum_{k=1}^{19} (-3 - 4k)$

53. $\displaystyle\sum_{i=1}^{1000} i$ **54.** $\displaystyle\sum_{k=1}^{2000} k$ **55.** $\displaystyle\sum_{i=6}^{15} (4i - 2)$

56. $\displaystyle\sum_{i=9}^{20} (8i + 3)$ **57.** $\displaystyle\sum_{j=7}^{12} (6 - 2j)$ **58.** $\displaystyle\sum_{i=4}^{17} (-3 - 5i)$

Find a_1 for each of the following arithmetic sequences. See Examples 5 and 9.

59. $a_9 = 47, \quad a_{15} = 77$ **60.** $a_{10} = 50, \quad a_{20} = 110$ **61.** $a_{15} = 168, \quad a_{16} = 180$

62. $a_{10} = -54, \quad a_{17} = -89$ **63.** $S_{20} = 1090, \quad a_{20} = 102$ **64.** $S_{31} = 5580, \quad a_{31} = 360$

65. $S_{12} = -108, \quad a_{12} = -19$ **66.** $S_{25} = 650, \quad a_{25} = 62$

67. Find the sum of all the integers from 51 to 71.

68. Find the sum of all the integers from -8 to 30.

69. If a clock strikes the proper number of chimes each hour on the hour, how many times will it chime in a month of 30 days?

70. A stack of telephone poles has 30 in the bottom row, 29 in the next, and so on, with one pole in the top row. How many poles are in the stack?

71. A sky diver falls 10 m during the first second, 20 m during the second, 30 m during the third, and so on. How many meters will the diver fall during the tenth second? During the first ten seconds?

72. Deepwell Drilling Company charges a flat $100 set-up charge, plus $5 for the first foot, $6 for the second, $7 for the third, and so on. Find the total charge for a 70-foot well.

73. An object falling under the force of gravity falls about 16 ft the first second, 48 ft during the next second, 80 ft during the third second, and so on. How far would the object fall during the eighth second? What is the total distance the object would fall in eight seconds?

74. The population of a city was 49,000 five years ago. Each year the zoning commission permits an increase of 580 in the population. What will the population be five years from now?

75. A super slide of uniform slope is to be built on a level piece of land. There are to be twenty equally spaced supports, with the longest support 15 m long and the shortest 2 m long. Find the total length of all the supports.

76. How much material would be needed for the rungs of a ladder of 31 rungs, if the rungs taper uniformly from 18 in to 28 in?

Explain why each of the following sequences is arithmetic.

77. log 2, log 4, log 8, log 16, log 32, . . . **78.** log 12, log 36, log 108, log 324, . . .

Decide which of the following statements are true for every real number x_i, real constant k, and natural number n.

79. $\displaystyle\sum_{i=1}^{n} k \cdot x_i = k^n \cdot \sum_{i=1}^{n} x_i$ **80.** $\displaystyle\sum_{i=1}^{n} k \cdot x_i = k \cdot \sum_{i=1}^{n} x_i$ **81.** $\displaystyle\sum_{i=1}^{n} k = nk$

82. $\displaystyle\sum_{i=1}^{n} k = k^n$ **83.** $\displaystyle\left[\sum_{i=1}^{n} x_i\right]^2 = \sum_{i=1}^{n} x_i^2$ **84.** $\displaystyle\sum_{i=1}^{n} (x_i - k) = \sum_{i=1}^{n} x_i - nk$

9.3   *Geometric Sequences*

A **geometric sequence** (or **geometric progression**) is a sequence in which each term after the first is obtained by multiplying the preceding term by a constant nonzero real number, called the **common ratio.** An example of a geometric sequence is 2, 8, 32, 128, . . . in which the first term is 2 and the common ratio is 4.

If the common ratio of a geometric sequence is r, then by the definition of a geometric sequence,

$$\frac{a_{n+1}}{a_n} = r$$

for every positive integer n. By this definition, the common ratio can be found by choosing any term except the first and dividing it by the preceding term.

Example 1   Find the common ratio, r, for the geometric sequence

$$15, \frac{15}{2}, \frac{15}{4}, \frac{15}{8}, \; . \; . \; . \;$$

Find r by choosing *any* two adjacent terms and dividing the second one by the first. Choosing the second and third terms of the sequence gives

$$r = \frac{15}{4} \div \frac{15}{2} = \frac{1}{2}.$$

Additional terms of the sequence can be found by multiplying each successive term by 1/2.   ❖

As mentioned above, the geometric sequence, 2, 8, 32, 128, . . . has $r = 4$. Notice that

$$8 = 2 \cdot 4$$
$$32 = 8 \cdot 4 = (2 \cdot 4) \cdot 4 = 2 \cdot 4^2$$
$$128 = 32 \cdot 4 = (2 \cdot 4^2) \cdot 4 = 2 \cdot 4^3.$$

To generalize this result, assume that a geometric sequence has first term a_1 and common ratio r. The second term can be written as $a_2 = a_1 r$, the third as $a_3 = a_2 r = (a_1 r) r = a_1 r^2$, and so on. Following this pattern, the nth term is $a_n = a_1 r^{n-1}$.

nth Term of a Geometric Sequence	In the geometric sequence with first term a_1 and common ratio r, the nth term is $$a_n = a_1 r^{n-1}.$$

Example 2   Find a_5 and a_n for each of the following geometric sequences.

(a) 4, 12, 36, 108, . . .

The first term, a_1, is 4. Find r by choosing any term except the first and dividing it by the preceding term. For example,

$$r = 36/12 = 3.$$

Since $a_4 = 108$, $a_5 = 3 \cdot 108 = 324$. The fifth term also could be found using the formula for a_n, $a_n = a_1 r^{n-1}$, and replacing n with 5, r with 3, and a_1 with 4.

$$a_5 = 4 \cdot (3)^{5-1} = 4 \cdot 3^4 = 324$$

Also, $a_n = 4 \cdot 3^{n-1}$.

(b) 64, 32, 16, 8, . . .

Here $r = 8/16 = 1/2$, and $a_4 = 8$, so $a_5 = (1/2)(8) = 4$. Alternatively,

$$a_5 = 64\left(\frac{1}{2}\right)^{5-1} = 64\left(\frac{1}{16}\right) = 4.$$

Also, $a_n = 64\left(\frac{1}{2}\right)^{n-1}$.   ✠

Example 3   Find a_1 and r for each of the following geometric sequences.

(a) The third term is 20 and the sixth term is -160.

Use the formula for the nth term of a geometric sequence.

$$\text{For } n = 3, \quad a_3 = a_1 r^2 = 20;$$
$$\text{for } n = 6, \quad a_6 = a_1 r^5 = -160.$$

We have $a_1 r^2 = 20$, so that $a_1 = 20/r^2$. Substitute this in the second equation and solve for r.

$$a_1 r^5 = -160$$
$$\left(\frac{20}{r^2}\right) r^5 = -160$$
$$20r^3 = -160$$
$$r^3 = -8$$
$$r = -2$$

Since $a_1 r^2 = 20$ and $r = -2$, we have $a_1 = 5$.

(b) $a_5 = 15$ and $a_7 = 375$

First substitute $n = 5$ and then $n = 7$ into $a_n = a_1 r^{n-1}$.

$$a_5 = a_1 r^4 = 15 \qquad \text{and} \qquad a_7 = a_1 r^6 = 375$$

Solve the first equation for a_1 to get $a_1 = 15/r^4$. Then substitute for a_1 in the second equation.

$$a_1 r^6 = 375$$

$$\frac{15}{r^4} \cdot r^6 = 375$$

$$15r^2 = 375$$

$$r^2 = 25$$

$$r = \pm 5$$

Either 5 or -5 can be used for r. To find a_1, use

$$a_1 = \frac{15}{r^4}.$$

Replace r with ± 5.

$$a_1 = \frac{15}{(\pm 5)^4} = \frac{15}{625} = \frac{3}{125}$$

There are two sequences that satisfy the given conditions: one with $a_1 = 3/125$ and $r = 5$ and the other with $a_1 = 3/125$ and $r = -5$. ❖

Example 4   An insect population is growing in such a way that each generation is 1.5 times as large as the previous generation. Suppose there are 100 insects in the first generation. How many will there be in the fourth generation?

The population can be written as a geometric sequence with a_1 as the first-generation population, a_2 the second-generation population, and so on. Then the fourth-generation population will be a_4. Using the formula for a_n, where $n = 4$, $r = 1.5$, and $a_1 = 100$, gives

$$a_4 = a_1 r^3 = 100(1.5)^3 = 100(3.375) = 337.5.$$

In the fourth generation, the population will number about 338 insects. ❖

Sum of a Geometric Sequence   In applications of geometric sequences, it is often necessary to know the sum of the first n terms of a sequence. For example, it might be necessary to know the total number of insects in the four generations of the population just discussed.

A formula for the sum of the first n terms of a geometric sequence, S_n, can be found by first writing the sum as

$$S_n = a_1 + a_2 + a_3 + \ldots + a_n.$$

This can also be written as

$$S_n = a_1 + a_1 r + a_1 r^2 + \cdots + a_1 r^{n-1}. \tag{1}$$

If $r = 1$, then $S_n = na_1$, which is a correct formula for this case. If $r \neq 1$, multiply both sides of (1) by r, obtaining

$$rS_n = a_1r + a_1r^2 + a_1r^3 + \cdots + a_1r^n \qquad (2)$$

If equation (2) is subtracted from equation (1), the result is

$$S_n = a_1 + a_1r + a_1r^2 + \ldots + a_1r^{n-1}$$
$$rS_n = \qquad a_1r + a_1r^2 + \ldots + a_1r^{n-1} + a_1r^n$$
$$\overline{\rule{0pt}{0pt}\hspace{5cm}}$$
$$S_n - rS_n = a_1 \qquad\qquad\qquad\qquad\qquad\quad - a_1r^n$$

or $\qquad S_n(1 - r) = a_1(1 - r^n),$

which finally gives

$$S_n = \frac{a_1(1 - r^n)}{1 - r}, \quad \text{where } r \neq 1.$$

This discussion is summarized below.

Sum of the First n Terms of a Geometric Sequence

If a geometric sequence has first term a_1 and common ratio r, then the sum of the first n terms is given by

$$S_n = \frac{a_1(1 - r^n)}{1 - r}, \quad \text{where } r \neq 1.$$

This formula can be used to find the total insect population in Example 4 over the four-generation period. With $n = 4$, $a_1 = 100$, and $r = 1.5$,

$$S_4 = \frac{100(1 - 1.5^4)}{1 - 1.5} = \frac{100(1 - 5.0625)}{-.5} = 812.5,$$

so the total population for the four generations will amount to about 813 insects.

Example 5 (a) Find S_5 for the geometric sequence

$$3, 6, 12, 24, 48.$$

Here $a_1 = 3$ and $r = 2$. Using the formula above,

$$S_5 = \frac{3(1 - 2^5)}{1 - 2} = \frac{3(1 - 32)}{-1} = \frac{3(-31)}{-1} = 93.$$

(b) Find S_4 for the sequence

$$10, 2, \frac{2}{5}, \ldots .$$

Here $a_1 = 10$, $r = 1/5$, and $n = 4$. Substitute these values into the formula for S_n.

$$S_4 = \frac{10\left[1 - \left(\frac{1}{5}\right)^4\right]}{1 - \frac{1}{5}}$$

$$= \frac{10\left(1 - \frac{1}{625}\right)}{\frac{4}{5}}$$

$$= 10\left(\frac{624}{625}\right) \cdot \frac{5}{4}$$

$$S_4 = \frac{312}{25} \quad \clubsuit$$

A sum of the form

$$\sum_{i=1}^{n} m \cdot p^i$$

represents the sum of the terms of a geometric sequence having first term

$$a_1 = m \cdot p^1 = mp$$

and common ratio $r = p$. These sums can be found by the formula for S_n given above.

Example 6   Find each of the following sums.

(a) $\displaystyle\sum_{i=1}^{7} 2 \cdot 3^i$

In this sum, $a_1 = 2 \cdot 3^1 = 6$ and $r = 3$. Thus,

$$\sum_{i=1}^{7} 2 \cdot 3^i = S_7 = \frac{6(1 - 3^7)}{1 - 3} = \frac{6(1 - 2187)}{-2} = \frac{6(-2186)}{-2} = 6558.$$

(b) $\displaystyle\sum_{i=1}^{5} \left(\frac{3}{4}\right)^i = S_5 = \frac{\frac{3}{4}\left[1 - \left(\frac{3}{4}\right)^5\right]}{1 - \frac{3}{4}}$

$$= \frac{\frac{3}{4}\left(1 - \frac{243}{1024}\right)}{\frac{1}{4}}$$

$$= 3\left(\frac{781}{1024}\right) = \frac{2343}{1024} \quad \clubsuit$$

A sequence of equal payments made at equal periods of time is called an **annuity.** The sum of the payments and interest on the payments is the **future value** of the annuity.

Example 7   To save money for a trip to Europe, Marge deposited $1000 at the end of each year for four years in an account paying 6% interest. What is the future value of this annuity?

The first payment will earn interest for 3 years, the second payment for 2 years, and the third payment for 1 year. The last payment earns no interest. The total amount is

$$1000(1.06)^3 + 1000(1.06)^2 + 1000\,(1.06) + 1000.$$

This is the sum of a geometric sequence with first term (starting at the end of the sum as written above) $a_1 = 1000$ and common ratio $r = 1.06$. Using the formula for S_4, the sum of four terms, gives

$$S_4 = \frac{1000[1 - (1.06)^4]}{1 - 1.06}$$

$$\approx \frac{1000[1 - 1.2625]}{-.06}$$

$$= 4374.62.$$

The future value of the annuity is $4374.62. $\clubsuit$

9.3   Exercises

For each of the following, write the terms of the geometric sequence that satisfies the given conditions.

1. $a_1 = 2$, $r = 3$, $n = 4$

2. $a_1 = 4$, $r = 2$, $n = 5$

3. $a_1 = 1/2$, $r = 4$, $n = 4$

4. $a_1 = 2/3$, $r = 6$, $n = 3$

5. $a_1 = -2$, $r = -3$, $n = 4$

6. $a_1 = -4$, $r = 2$, $n = 5$

7. $a_1 = 3125$, $r = 1/5$, $n = 7$

8. $a_1 = 729$, $r = 2/3$, $n = 5$

9. $a_1 = -1$, $r = -1$, $n = 6$
10. $a_1 = -1$, $r = 1$, $n = 5$
11. $a_3 = 6$, $a_4 = 12$, $n = 5$
12. $a_2 = 9$, $a_3 = 3$, $n = 4$

Find a_5 and a_n for each of the following geometric sequences. See Example 2.

13. $a_1 = 4$, $r = 3$
14. $a_1 = 8$, $r = 4$
15. $a_1 = -2$, $r = 3$
16. $a_1 = -5$, $r = 4$
17. $a_1 = -3$, $r = -5$
18. $a_1 = -4$, $r = -2$
19. $a_2 = 3$, $r = 2$
20. $a_3 = 6$, $r = 3$
21. $a_4 = 64$, $r = -4$
22. $a_4 = 81$, $r = -3$

For each of the following sequences that are geometric, find r and a_n. See Examples 2 and 3.

23. 6, 12, 24, 48, . . .
24. 4, 16, 64, 256, . . .
25. 3/4, 3/2, 3, 6, 12, . . .
26. 5/6, 5/3, 10/3, 20/3, 40/3, . . .
27. -4, 2, -1, 1/2, . . .
28. 49, -7, 1, $-1/7$, . . .
29. 18, 20, 24, 32, 48, . . .
30. -7, -5, -3, -1, 1, 3, . . .
31. $a_3 = 9$, $r = 2$
32. $a_5 = 6$, $r = 1/2$
33. $a_3 = -2$, $r = 3$
34. $a_2 = 5$, $r = 1/2$

Find the sum of the first five terms for each of the following geometric sequences. See Example 5.

35. 3, 6, 12, 24, . . .
36. 5, 20, 80, 320, . . .
37. 12, -6, 3, $-3/2$, . . .
38. 18, -3, 1/2, $-1/12$, . . .
39. 9, -6, 4, . . .
40. 50, -10, 2, . . .
41. $a_1 = 4$, $r = 2$
42. $a_1 = 3$, $r = 3$
43. $a_2 = 1/3$, $r = 3$
44. $a_2 = -1$, $r = 2$
45. $a_1 = 8.423$, $r = 2.859$
46. $a_1 = -3.772$, $r = -1.553$
47. $a_1 = -5$, $r = .2033$
48. $a_1 = 4$, $r = -.8765$

Find a_1 and r for each of the following geometric sequences. See Example 3.

49. $a_2 = 6$, $a_6 = 486$
50. $a_3 = -12$, $a_6 = 96$
51. $a_2 = 64$, $a_8 = 1$
52. $a_2 = 100$, $a_5 = 1/10$

Find each of the following sums. See Example 6.

53. $\sum\limits_{i=1}^{4} 2^i$
54. $\sum\limits_{j=1}^{6} 3^j$
55. $\sum\limits_{i=1}^{4} (-3)^i$
56. $\sum\limits_{i=1}^{4} (-3^i)$

57. $\sum\limits_{i=1}^{8} 64(1/2)^i$
58. $\sum\limits_{i=1}^{6} 81(2/3)^i$
59. $\sum\limits_{j=1}^{4} (2/5)^j$
60. $\sum\limits_{i=1}^{4} (-3/4)^i$

61. $\displaystyle\sum_{i=1}^{4} 6(3/2)^i$ **62.** $\displaystyle\sum_{k=1}^{5} 9(5/3)^k$ **63.** $\displaystyle\sum_{i=3}^{6} 2^i$ **64.** $\displaystyle\sum_{i=4}^{7} 3^i$

65. $\displaystyle\sum_{i=1}^{4} 5(3/5)^{i-1}$ **66.** $\displaystyle\sum_{i=1}^{5} 256(-3/4)^{i-1}$

67. Suppose you could save $1 on January 1, $2 on January 2, $4 on January 3, and so on. What amount would you save on January 31?

68. See Exercise 67. What would be the total amount of your savings during January? (*Hint:* $2^{31} = 2,147,483,648$.)

69. Richland Oil has a well that produced $4,000,000 of income its first year. Each year thereafter, the well produced half as much as it did the previous year. What total amount of income would be produced by the well in 6 years?

70. Fruit and vegetable dealer Olive Greene paid 10¢ per lb for 10,000 lb of onions. Each week the price she charges increases by .1¢ per lb, while her onions lose 5% of their weight. If she sells all the onions after 6 weeks, does she make or lose money? How much?

71. The final step in processing a black and white photographic print is to immerse the print in a chemical called "fixer." The print is then washed in running water. Under certain conditions, 98% of the fixer in a print will be removed with 15 min of washing. How much of the original fixer would then be left after one hour?

72. A scientist has a vat containing 100 liters of a pure chemical. Twenty liters are drained and replaced with water. After complete mixing, 20 liters of the mixture are drained and replaced with water. What will be the strength of the mixture after nine such drainings?

73. The half-life of a radioactive substance is the time it takes for half the substance to decay. Suppose that the half-life of a substance is 3 yr and that 10^{15} molecules of the substance are present initially. How many molecules will be present after 15 yr?

74. Each year a machine loses 20% of the value it had at the beginning of the year. Find the value of a machine at the end of 6 yr if it cost $100,000 new.

Find the future value of each of the following annuities. See Example 7.

75. Payments of $1000 at the end of each year for 9 years at 8% interest compounded annually.

76. Payments of $800 at the end of each year for 12 years at 10% interest compounded annually.

77. Payments of $2430 at the end of each year for 10 years at 11% interest compounded annually.

78. Payments of $1500 at the end of each year for 6 years at 12% interest compounded annually.

Explain why the following sequences are geometric.

79. $\log 6$, $\log 6^2$, $\log 6^4$, $\log 6^8$, . . .   **80.** $\log 2$, $\log 4$, $\log 16$, $\log 256$, . . .

9.4 Sums of Infinite Geometric Sequences

In the preceeding section the sum of the first n terms of a geometric sequence was given as

$$S_n = \sum_{i=1}^{n} a_i = \frac{a_1(1 - r^n)}{1 - r},$$

where a_1 is the first term and $r (r \neq 1)$ is the common ratio. Now consider an infinite geometric sequence such as

$$2, 1, \frac{1}{2}, \frac{1}{4}, \frac{1}{8}, \frac{1}{16}, \ldots$$

with first term 2 and common ratio 1/2. Using the formula for S_n gives the following sequence.

$$S_1 = 2, \quad S_2 = 3, \quad S_3 = \frac{7}{2}, \quad S_4 = \frac{15}{4}, \quad S_5 = \frac{31}{8}, \quad S_6 = \frac{63}{16}$$

These sums seem to be getting closer and closer to the number 4. In fact, by selecting a value of n large enough, it is possible to make S_n as close as desired to 4. This idea is expressed as

$$\lim_{n \to \infty} S_n = 4.$$

(Read: "the limit of S_n as n increases without bound is 4.") For no value of n is $S_n = 4$. However, if n is large enough, then S_n is as close to 4 as desired.*

Since

$$\lim_{n \to \infty} S_n = 4,$$

the number 4 is called the *sum* of the infinite geometric sequence

$$2, 1, \frac{1}{2}, \frac{1}{4}, \ldots$$

and

$$2 + 1 + \frac{1}{2} + \frac{1}{4} + \frac{1}{8} + \ldots = 4.$$

Example 1 Find $1 + \frac{1}{3} + \frac{1}{9} + \frac{1}{27} + \ldots$.

Use the formula for the first n terms of a geometric sequence to get

$$S_1 = 1, \quad S_2 = \frac{4}{3}, \quad S_3 = \frac{13}{9}, \quad S_4 = \frac{40}{27},$$

---

*These phrases "large enough" and "as close as desired" are not nearly precise enough for mathematicians; much of a standard calculus course is devoted to making them more precise.

and in general $$S_n = \frac{1\left[1 - \left(\frac{1}{3}\right)^n\right]}{1 - \frac{1}{3}}.$$

The chart below shows the value of $(1/3)^n$ for larger and larger values of n.

n	1	10	100	200
$\left(\frac{1}{3}\right)^n$	$\frac{1}{3}$	1.69×10^{-5}	1.94×10^{-48}	3.76×10^{-96}

As n gets larger and larger, $(1/3)^n$ gets closer and closer to 0. That is,

$$\lim_{n \to \infty} \left(\frac{1}{3}\right)^n = 0,$$

making it reasonable that

$$\lim_{n \to \infty} S_n = \lim_{n \to \infty} \frac{1\left[1 - \left(\frac{1}{3}\right)^n\right]}{1 - \frac{1}{3}} = \frac{1(1 - 0)}{1 - \frac{1}{3}} = \frac{1}{\frac{2}{3}} = \frac{3}{2}.$$

Hence, $$1 + \frac{1}{3} + \frac{1}{9} + \frac{1}{27} + \ldots = \frac{3}{2}. \quad \clubsuit$$

If a geometric sequence has a first term a_1 and a common ratio r, then

$$S_n = \frac{a_1(1 - r^n)}{1 - r} \quad (r \neq 1)$$

for every positive integer n. If $-1 < r < 1$, then $\lim_{n \to \infty} r^n = 0$, and

$$\lim_{n \to \infty} S_n = \frac{a_1(1 - 0)}{1 - r} = \frac{a_1}{1 - r}.$$

This quotient, $a_1/(1 - r)$, is called the **sum of an infinite geometric sequence.**
The limit $\lim_{n \to \infty} S_n$ is often expressed as S_∞ or $\sum_{i=1}^{\infty} a_i$. These results lead to the following definition.

Sum of an Infinite Geometric Sequence	The sum of an infinite geometric sequence with first term a_1 and common ratio r, where $-1 < r < 1$, is given by $$S_\infty = \frac{a_1}{1 - r}.$$

If $|r| > 1$, the terms get larger and larger, so there is no limit as $n \rightarrow \infty$. Hence the sequence will not have a sum.

Example 2   (a) Find the sum

$$-\frac{3}{4} + \frac{3}{8} - \frac{3}{16} + \frac{3}{32} - \frac{3}{64} + \ldots .$$

The first term is $a_1 = -3/4$. To find r, divide any two adjacent terms. For example,

$$r = \frac{-\dfrac{3}{16}}{\dfrac{3}{8}} = -\frac{1}{2}.$$

Since $-1 < r < 1$, the formula in this section applies, and

$$S_\infty = \frac{a_1}{1 - r} = \frac{-\dfrac{3}{4}}{1 - \left(-\dfrac{1}{2}\right)} = -\frac{1}{2}.$$

(b) $\displaystyle\sum_{i=1}^{\infty} \left(\frac{3}{5}\right)^i = \frac{\dfrac{3}{5}}{1 - \dfrac{3}{5}} = \frac{3}{2}$ ✜

The formula in this section can be used to convert repeating decimals (which represent rational numbers) to fractions of the form p/q, where p and q are integers, with $q \neq 0$.

Example 3   Write each repeating decimal in the form p/q, where p and q are integers.
(a) .090909 . . .
This decimal can be written as

$$.090909 \ldots = .09 + .0009 + .000009 + \ldots ,$$

which is the sum of the terms of an infinite geometric sequence having $a_1 = .09$ and $r = .01$. The sum of this sequence is given by

$$S_\infty = \frac{a_1}{1 - r} = \frac{.09}{1 - .01} = \frac{.09}{.99} = \frac{1}{11}.$$

Thus, .090909 . . . = 1/11.
(b) 2.5121212 . . .
Write the number as

$$2.5121212 \ldots = 2.5 + .012 + .00012 + .0000012 + \ldots$$
$$= 2.5 + (.012 + .00012 + .0000012 + \ldots).$$

Beginning with the second term, this is the sum of the terms of an infinite geometric sequence with $a_1 = .012$ and $r = .01$, so $2.5121212 \ldots$ can be written as

$$2.5 + \frac{.012}{1 - .01} = 2.5 + \frac{.012}{.990} = 2.5 + \frac{2}{165} = \frac{829}{330}. \quad \clubsuit$$

9.4 Exercises

Find r for each of the following infinite geometric sequences. Identify any whose sums would exist. See Example 2.

1. 9, 18, 36, 72, 144, . . .

2. 3, 9, 27, 81, . . .

3. 10, 100, 1000, 10,000, . . .

4. $-8, -4, -2, -1, -1/2, \ldots$

5. 12, 6, 3, 3/2, . . .

6. $-8, -16, -32, -64, \ldots$

7. 1, 1.1, 1.21, 1.331, . . .

8. $-1, 1.2, -1.44, 1.728, \ldots$

9. $1, -.9, .81, -.729, \ldots$

10. 1, .7, .49, .343, . . .

Find each of the following sums by using the formula of this section where it applies. See Example 2.

11. $16 + 4 + 1 + \ldots$

12. $81 + 27 + 9 + 3 + 1 + \ldots$

13. $100 + 10 + 1 + \ldots$

14. $128 + 64 + 32 + \ldots$

15. $90 + 30 + 10 + \ldots$

16. $25 + 5 + 1 + \ldots$

17. $256 - 128 + 64 - 32 + 16 - \ldots$

18. $120 - 60 + 30 - 15 + \ldots$

19. $108 - 36 + 12 - 4 + \ldots$

20. $10,000 - 1000 + 100 - 10 + 1 - \ldots$

21. $\dfrac{3}{4} + \dfrac{3}{8} + \dfrac{3}{16} + \ldots$

22. $\dfrac{4}{5} + \dfrac{2}{5} + \dfrac{1}{5} + \ldots$

23. $3 - \dfrac{3}{2} + \dfrac{3}{4} - \ldots$

24. $9 - 3 + 1 - \ldots$

25. $\dfrac{1}{3} - \dfrac{2}{9} + \dfrac{4}{27} - \dfrac{8}{81} + \ldots$

26. $1 + \dfrac{1}{1.01} + \dfrac{1}{(1.01)^2} + \ldots$

27. $\dfrac{1}{36} + \dfrac{1}{30} + \dfrac{1}{25} + \ldots$

28. $1 + \dfrac{1}{2^2} + \dfrac{1}{2^4} + \ldots$

29. $\displaystyle\sum_{i=1}^{\infty} (1/4)^i$

30. $\displaystyle\sum_{i=1}^{\infty} (9/10)^i$

31. $\displaystyle\sum_{i=1}^{\infty} (1.2)^i$

32. $\displaystyle\sum_{i=1}^{\infty} (1.001)^i$

33. $\displaystyle\sum_{i=1}^{\infty} (-1/4)^i$

34. $\displaystyle\sum_{i=1}^{\infty} (.3)^i$

35. $\displaystyle\sum_{i=1}^{\infty} 1/5^i$

36. $\displaystyle\sum_{i=1}^{\infty} (-1/2)^i$

37. $\displaystyle\sum_{i=1}^{\infty} 10^{-i}$

38. $\displaystyle\sum_{i=1}^{\infty} 4^{-i}$

39. $\displaystyle\sum_{i=1}^{\infty} (1/2)^{-i}$

40. $\displaystyle\sum_{i=1}^{\infty} (3/4)^{-i}$

Express each repeating decimal in the form p/q, where p and q are integers. See Example 3.

41. $0.55555\ldots$ **42.** $0.33333\ldots$ **43.** $0.121212\ldots$ **44.** $0.858585\ldots$

45. $0.313131\ldots$ **46.** $0.909090\ldots$ **47.** $0.508508508\ldots$ **48.** $0.613613613\ldots$

49. $0.3455555\ldots$ **50.** $0.5688888\ldots$

51. Mitzi drops a ball from a height of 10 m and notices that on each bounce the ball returns to about 3/4 of its previous height. About how far will the ball travel before it comes to rest? (*Hint:* Consider the sum of two sequences.)

52. A sugar factory receives an order for 1000 units of sugar. The production manager thus orders production of 1000 units of sugar. He forgets, however, that the production of sugar requires some sugar (to prime the machines, for example), and so he ends up with only 900 units of sugar. He then orders an additional 100 units, and receives only 90 units. A further order for 10 units produces 9 units. Finally seeing he is wrong, the manager decides to try mathematics. He views the production process as an infinite geometric progression with $a_1 = 1000$ and $r = .1$. Using this, find the number of units of sugar that he should have ordered originally.

53. After a person pedaling a bicycle removes his or her feet from the pedals, the wheel rotates 400 times the first minute. As it continues to slow down, each minute it rotates only 3/4 as many times as in the previous minute. How many times will the wheel rotate before coming to a complete stop?

54. A pendulum bob swings through an arc 40 cm long on its first swing. Each swing thereafter, it swings only 80% as far as on the previous swing. How far will it swing altogether before coming to a complete stop?

55. A sequence of equilateral triangles is constructed. The first triangle has sides 2 m in length. To get the second triangle, midpoints of the sides of the original triangle are connected. If this process could be continued indefinitely, what would be the total perimeter of all the triangles?

56. What would be the total area of all the triangles in Exercise 55, disregarding the overlapping?

9.5   The Binomial Theorem

In this section we introduce a method of evaluating expressions of the form $(x + y)^n$, where n is a natural number. Some expansions of $(x + y)^n$, for various positive integer values of n, are given below:

$$(x + y)^1 = x + y$$
$$(x + y)^2 = x^2 + 2xy + y^2$$
$$(x + y)^3 = x^3 + 3x^2y + 3xy^2 + y^3$$
$$(x + y)^4 = x^4 + 4x^3y + 6x^2y^2 + 4xy^3 + y^4$$
$$(x + y)^5 = x^5 + 5x^4y + 10x^3y^2 + 10x^2y^3 + 5xy^4 + y^5$$

and so on. Studying these results reveals a pattern that can be used to write a general expression for $(x + y)^n$.

First, notice that each expression begins with x raised to the same power as the binomial itself. That is, the expansion of $(x + y)^1$ has a first term of x^1, $(x + y)^2$ has a first term of x^2, $(x + y)^3$ has a first term of x^3, and so on. Also, the last term in each expansion is y to the same power as the binomial. Thus the expansion of $(x + y)^n$ should begin with the term x^n and end with the term y^n.

Also, the exponents on x decrease by one in each term after the first, while the exponents on y, beginning with y in the second term, increase by one in each succeeding term. That is, the *variables* in the terms of the expansion of $(x + y)^n$ have the following pattern.

$$x^n, \quad x^{n-1}y, \quad x^{n-2}y^2, \quad x^{n-3}y^3, \ldots, xy^{n-1}, \quad y^n$$

This pattern suggests that the sum of the exponents on x and y in each term is n. For example, in the third term in the list above, the variable is $x^{n-2}y^2$, and the sum of the exponents is $n - 2 + 2 = n$.

Now examine the *coefficients* in the terms of the expansions shown above. Writing the coefficients alone gives the following pattern.

**Pascal's
Triangle**

$$
\begin{array}{ccccccccccc}
 & & & & 1 & & 1 & & & & \\
 & & & 1 & & 2 & & 1 & & & \\
 & & 1 & & 3 & & 3 & & 1 & & \\
 & 1 & & 4 & & 6 & & 4 & & 1 & \\
1 & & 5 & & 10 & & 10 & & 5 & & 1 \\
\end{array}
$$

With the coefficients arranged in this way, it can be seen that each number in the triangle is the sum of the two numbers directly above it (one to the right and one to the left.) For example, in the fourth row from the top, 1 is the sum of 1, the only number above it, 4 is the sum of 1 and 3, 6 is the sum of 3 and 3, and so on. This triangular array of numbers is called **Pascal's triangle,** in honor of the seventeenth-century mathematician Blaise Pascal (1623-1662), one of the first to use it extensively.

To get the coefficients for $(x + y)^6$, include a sixth row in the array of numbers given above. Adding adjacent numbers, we find that the sixth row is

$$1 \quad 6 \quad 15 \quad 20 \quad 15 \quad 6 \quad 1.$$

Using these coefficients, the expansion of $(x + y)^6$ is

$$(x + y)^6 = x^6 + 6x^5y + 15x^4y^2 + 20x^3y^3 + 15x^2y^4 + 6xy^5 + y^6.$$

Although it is possible to use Pascal's triangle to find the coefficients of $(x + y)^n$ for any positive integer value of n, this becomes impractical for large values of n because of the need to write out all the preceding rows. A more efficient way of finding these coefficients uses factorial notation. The number $n!$ (read "n-factorial"), is defined as follows.

n-factorial

> For any positive integer *n*:
>
> $$n! = n(n - 1)(n - 2) \ldots (3)(2)(1),$$
>
> and
>
> $$0! = 1$$

For example, $5! = 5 \cdot 4 \cdot 3 \cdot 2 \cdot 1 = 120$, $7! = 7 \cdot 6 \cdot 5 \cdot 4 \cdot 3 \cdot 2 \cdot 1 = 5040$, $2! = 2 \cdot 1 = 2$, and so on.

Now look at the coefficients of the expression

$$(x + y)^5 = x^5 + 5x^4y + 10x^3y^2 + 10x^2y^3 + 5xy^4 + y^5.$$

The coefficient on the second term, $5x^4y$, is 5, and the exponents on the variables are 4 and 1. Note that

$$5 = \frac{5!}{4!1!}.$$

The coefficient on the third term is 10, with exponents of 3 and 2, and

$$10 = \frac{5!}{3!2!}.$$

The last term (the sixth term) can be written as $y^5 = 1x^0y^5$, with coefficient 1, and exponents of 0 and 5. Since $0! = 1$, check that

$$1 = \frac{5!}{0!5!}.$$

Generalizing from these examples, we find that the coefficient for the term of the expansion of $(x + y)^n$ in which the variable part is $x^r y^{n-r}$ (where $r \le n$) will be

$$\frac{n!}{r!(n - r)!}.$$

This number, called a **binomial coefficient,** is often written as $\binom{n}{r}$ (read "*n* above *r*").

Binomial
Coefficient

> For nonnegative integers *n* and *r*, with $r \le n$, the symbol $\binom{n}{r}$ is defined as
>
> $$\binom{n}{r} = \frac{n!}{r!(n - r)!}.$$

These binomial coefficients are just numbers from Pascal's triangle. For example, $\binom{3}{0}$ is the first number in the third row, and $\binom{7}{4}$ is the fifth number in the seventh row.

Example 1   **(a)** $\binom{6}{2} = \dfrac{6!}{2!(6-2)!} = \dfrac{6!}{2!4!} = \dfrac{6 \cdot 5 \cdot 4 \cdot 3 \cdot 2 \cdot 1}{2 \cdot 1 \cdot 4 \cdot 3 \cdot 2 \cdot 1} = 15$

(b) $\binom{8}{0} = \dfrac{8!}{0!(8-0)!} = \dfrac{8!}{0!8!} = \dfrac{8!}{1 \cdot 8!} = 1$

(c) $\binom{10}{10} = \dfrac{10!}{10!(10-10)!} = \dfrac{10!}{10!0!} = 1$ ✦

Our conjectures about the expansion of $(x + y)^n$ are summarized below.

Expansion of $(x + y)^n$

1. There are $(n + 1)$ terms in the expansion.
2. The first term is x^n, and the last term is y^n.
3. The exponent on x decreases by 1 and the exponent on y increases by 1 in each succeeding term.
4. The sum of the exponents on x and y in any term is n.
5. The coefficient of $x^r y^{n-r}$ is $\binom{n}{r}$.

These observations about the expansion of $(x + y)^n$ for any positive integer value of n suggest the **binomial theorem.**

Binomial Theorem

For any positive integer n:

$$(x + y)^n = x^n + \binom{n}{n-1}x^{n-1}y + \binom{n}{n-2}x^{n-2}y^2 + \binom{n}{n-3}x^{n-3}y^3$$

$$+ \ldots + \binom{n}{n-r}x^{n-r}y^r + \ldots + \binom{n}{1}xy^{n-1} + y^n$$

The binomial theorem can be proved using mathematical induction (discussed in Section 9.6). Note that the expansion of $(x + y)^n$ has $n + 1$ terms.

Example 2   Write out the binomial expansion of $(x + y)^9$.

Using the binomial theorem,

$$(x + y)^9 = x^9 + \binom{9}{8} x^8 y + \binom{9}{7} x^7 y^2 + \binom{9}{6} x^6 y^3 + \binom{9}{5} x^5 y^4$$

$$+ \binom{9}{4} x^4 y^5 + \binom{9}{3} x^3 y^6 + \binom{9}{2} x^2 y^7 + \binom{9}{1} xy^8 + y^9.$$

Now evaluate each of the combinations.

$$(x + y)^9 = x^9 + \frac{9!}{8!1!} x^8 y + \frac{9!}{7!2!} x^7 y^2 + \frac{9!}{6!3!} x^6 y^3 + \frac{9!}{5!4!} x^5 y^4$$

$$+ \frac{9!}{4!5!} x^4 y^5 + \frac{9!}{3!6!} x^3 y^6 + \frac{9!}{2!7!} x^2 y^7 + \frac{9!}{1!8!} xy^8 + y^9$$

$$= x^9 + 9x^8 y + 36x^7 y^2 + 84x^6 y^3 + 126x^5 y^4 + 126x^4 y^5$$

$$+ 84x^3 y^6 + 36x^2 y^7 + 9xy^8 + y^9 \quad \clubsuit$$

Example 3   Expand $\left(a - \dfrac{b}{2} \right)^5$.

Again, use the binomial theorem.

$$\left(a - \frac{b}{2} \right)^5 = a^5 + \binom{5}{4} a^4 \left(-\frac{b}{2} \right) + \binom{5}{3} a^3 \left(-\frac{b}{2} \right)^2 + \binom{5}{2} a^2 \left(-\frac{b}{2} \right)^3$$

$$+ \binom{5}{1} a \left(-\frac{b}{2} \right)^4 + \left(-\frac{b}{2} \right)^5$$

$$= a^5 + 5a^4 \left(-\frac{b}{2} \right) + 10a^3 \left(-\frac{b}{2} \right)^2 + 10a^2 \left(-\frac{b}{2} \right)^3$$

$$+ 5a \left(-\frac{b}{2} \right)^4 + \left(-\frac{b}{2} \right)^5$$

$$= a^5 - \frac{5}{2} a^4 b + \frac{5}{2} a^3 b^2 - \frac{5}{4} a^2 b^3 + \frac{5}{16} ab^4 - \frac{1}{32} b^5 \quad \clubsuit$$

rth Term   Any single term of a binomial expansion can be determined without writing out the whole expansion. For example, the 7th term of $(x + y)^9$ has y raised to the 6th power (since y has the power 1 in the second term, the power 2 in the third term, and so on). The exponents on x and y in each term must have a sum of 9, so the exponent on x in the 7th term is $9 - 6 = 3$. Thus, writing the coefficient as given in the binomial theorem, the 7th term should be

$$\frac{9!}{3!6!} x^3 y^6.$$

This is in fact the 7th term of $(x + y)^9$ found in Example 2 above. This discussion suggests the next theorem.

***r*th Term of the Binomial Expansion**	The rth term of the binomial expansion of $(x + y)^n$, where $n \geq r - 1$, is $$\binom{n}{n - (r - 1)} x^{n-(r-1)}y^{r-1}.$$

Example 4 Find the fourth term of $(a + 2b)^{10}$.

In the fourth term $2b$ has an exponent of 3, while a has an exponent of $10 - 3$, or 7. Thus, the fourth term is

$$\binom{10}{7}a^7(2b)^3 = 120a^7(8b^3) = 960a^7b^3. \quad \clubsuit$$

The binomial theorem can be used to find powers of numbers that are close to some integer.

Example 5 Use the binomial theorem to find $(1.01)^8$ correct to three decimal places.

Write 1.01 as $1 + .01$. Then use the binomial theorem.

$$(1.01)^8 = (1 + .01)^8$$

$$= 1^8 + \binom{8}{7}(1)^7(.01) + \binom{8}{6}(1)^6(.01)^2 + \ldots + (.01)^8$$

$$= 1 + 8(1)(.01) + 28(1)(.0001) + \ldots + (.0000000000000001)$$

$$= 1 + .08 + .0028 + \ldots + .0000000000000001$$

$$(1.01)^8 \approx 1.083$$

Since only three decimal places of accuracy are required, there is no need to evaluate all the terms of the expansion. Additional terms, however, would give the answer to more decimal places of accuracy. $\clubsuit$

9.5 Exercises

Write out the binomial expansion for each of the following. See Examples 2 and 3.

1. $(x + y)^6$ **2.** $(m + n)^4$ **3.** $(p - q)^5$ **4.** $(a - b)^7$

5. $(r^2 + s)^5$ **6.** $(m + n^2)^4$ **7.** $(p + 2q)^4$ **8.** $(3r - s)^6$

9. $(7p + 2q)^4$ **10.** $(4a - 5b)^5$ **11.** $(3x - 2y)^6$ **12.** $(7k - 9j)^4$

13. $\left(\dfrac{m}{2} - 1\right)^6$ **14.** $\left(3 + \dfrac{y}{3}\right)^5$ **15.** $\left(2p + \dfrac{q}{3}\right)^4$ **16.** $\left(\dfrac{r}{6} - \dfrac{m}{2}\right)^3$

For each of the following write the indicated term of the binomial expansion. See Example 4.

17. 5th term of $(m - 2p)^{12}$ **18.** 4th term of $(3x + y)^6$

19. 6th term of $(x + y)^9$ **20.** 12th term of $(a - b)^{15}$

21. 9th term of $(2m + n)^{10}$ **22.** 7th term of $(3r - 5s)^{12}$

23. 17th term of $(p^2 + q)^{20}$

24. 10th term of $(2x^2 + y)^{14}$

25. 8th term of $(x^3 + 2y)^{14}$

26. 13th term of $(a + 2b^3)^{12}$

Use the binomial expansion to evaluate each of the following to three decimal places. See Example 5.

27. $(1.10)^{10}$

28. $(.99)^{15}$ (*Hint:* $.99 = 1 - .01$)

29. $(1.99)^8$

30. $(2.99)^3$

31. $(3.02)^6$

32. $(1.01)^9$

In calculus it is shown that

$$(1 + x)^n = 1 + nx + \frac{n(n - 1)}{2!} x^2 + \frac{n(n - 1)(n - 2)}{3!} x^3 + \dots$$

for any real number n (not just positive integers) and any real number x, where $|x| < 1$. This result, a generalized binomial theorem, may be used to find approximate values of powers or roots. For example,

$$\sqrt[4]{630} = (625 + 5)^{1/4} = \left[625 \left(1 + \frac{5}{625} \right) \right]^{1/4} = 625^{1/4} \left(1 + \frac{5}{625} \right)^{1/4}$$

33. Use the expression given above for $(1 + x)^n$ to approximate $(1 + 5/625)^{1/4}$ to the nearest thousandth. Then approximate $\sqrt[4]{630}$.

34. Approximate $\sqrt[3]{9.42}$, using this method.

35. Approximate $(1.02)^{-3}$.

36. Approximate $1/(1.04)^5$.

9.6   Mathematical Induction

Many results in mathematics are claimed to be true for any positive integer. Any of these results could be checked for $n = 1$, $n = 2$, $n = 3$, and so on, but since the set of positive integers is infinite it would be impossible to check every possible case. For example, let S_n represent the statement that the sum of the first n positive integers is $n(n + 1)/2$,

$$S_n: 1 + 2 + 3 + \dots + n = \frac{n(n + 1)}{2}.$$

The truth of this statement can be checked quickly for the first few values of n.

If $n = 1$, S_1 is

$$1 = \frac{1(1 + 1)}{2}, \text{ a true statement, since } 1 = 1.$$

If $n = 2$, S_2 is

$$1 + 2 = \frac{2(2 + 1)}{2}, \text{ a true statement, since } 3 = 3.$$

If $n = 3$, S_3 is $\qquad 1 + 2 + 3 = \dfrac{3(3 + 1)}{2}$, a true statement, since $6 = 6$.

If $n = 4$, S_4 is $1 + 2 + 3 + 4 = \dfrac{4(4 + 1)}{2}$, a true statement, since $10 = 10$.

Continuing in this way would never prove that S_n is true for *every* positive integer value of n. To prove that such statements are true for every positive integer value of n, the following principle is used.

Principle of Mathematical Induction

Let S_n be a statement concerning the positive integer n. Suppose that

(1) S_1 is true;

(2) for any positive integer k, $k \leq n$, if S_k is true, then S_{k+1} is also true.

Then S_n is true for every positive integer value of n.

A proof by mathematical induction can be explained as follows. By assumption (1) above, the statement is true when $n = 1$. By (2) above, the fact that the statement is true for $n = 1$ implies that it is true for $n = 1 + 1 = 2$. Using (2) again, the statement is thus true for $2 + 1 = 3$, for $3 + 1 = 4$, for $4 + 1 = 5$, and so on. Continuing in this way shows that the statement must be true for *every* positive integer, no matter how large.

The situation is similar to that of a number of dominoes lined up as shown in Figure 9.1. If the first domino is pushed over, it pushes the next, which pushes the next, and so on until all are down.

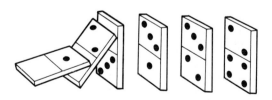

Figure 9.1

Another example of the principle of mathematical induction might be an infinite ladder. Suppose the rungs are spaced so that, whenever you are on a rung, you know you can move to the next rung. Then *if* you can get to the first rung, you can go as high up the ladder as you wish.

Two separate steps are required for a proof by mathematical induction:

Proof by Mathematical Induction	*Step 1* Prove that the statement is true for $n = 1$.
	Step 2 Show that, for any positive integer k, $k \leq n$, if S_k is true, then S_{k+1} is also true.

Mathematical induction is used in the next example to prove the statement discussed above: $S_n = 1 + 2 + 3 + \ldots + n = \dfrac{n(n + 1)}{2}$.

Example 1 Let S_n represent the statement

$$1 + 2 + 3 + \ldots + n = \frac{n(n + 1)}{2}.$$

Prove that S_n is true for every positive integer n.

The proof by mathematical induction is as follows.

Step 1 Show that the statement is true when $n = 1$. If $n = 1$, S_1 becomes

$$1 = \frac{1(1 + 1)}{2},$$

which is true.

Step 2 Show that S_k implies S_{k+1}, where S_k is the statement

$$1 + 2 + 3 + \ldots + k = \frac{k(k + 1)}{2},$$

and S_{k+1} is the statement

$$1 + 2 + 3 + \ldots + k + (k + 1) = \frac{(k + 1)[(k + 1) + 1]}{2}.$$

Start with S_k.

$$1 + 2 + 3 + \ldots + k = \frac{k(k + 1)}{2}$$

Add $k + 1$ to both sides of this equation.

$$1 + 2 + 3 + \ldots + k + (k + 1) = \frac{k(k + 1)}{2} + (k + 1)$$

Now factor out the common factor $k + 1$ on the right to get

$$= (k + 1)\left(\frac{k}{2} + 1\right)$$

$$= (k + 1)\left(\frac{k + 2}{2}\right)$$

$$1 + 2 + 3 + \ldots + k + (k + 1) = \frac{(k + 1)[(k + 1) + 1]}{2}.$$

This final result is the statement for $n = k + 1$; it has been shown that if S_k is true, then S_{k+1} is also true. The two steps required for a proof by mathematical induction have now been completed, so the statement S_n is true for every positive integer value of n. ✤

Example 2   Prove: $4 + 7 + 10 + \ldots + (3n + 1) = \dfrac{n(3n + 5)}{2}$.

Step 1   Show that the statement is true for S_1. S_1 is

$$4 = \frac{1(3 \cdot 1 + 5)}{2}.$$

Since the right side equals 4, S_1 is a true statement.

Step 2   Show that if S_k is true, then $S_{k + 1}$ is true, where S_k is

$$4 + 7 + 10 + \ldots + (3k + 1) = \frac{k(3k + 5)}{2},$$

and S_{k+1} is

$$4 + 7 + 10 + \ldots + (3k + 1) + [3(k + 1) + 1]$$
$$= \frac{(k + 1)[3(k + 1) + 5]}{2}.$$

Start with S_k:

$$4 + 7 + 10 + \ldots + (3k + 1) = \frac{k(3k + 5)}{2}.$$

Adding $[3(k + 1) + 1]$ to both sides of S_k gives

$$4 + 7 + 10 + \ldots + (3k + 1) + [3(k + 1) + 1]$$
$$= \frac{k(3k + 5)}{2} + [3(k + 1) + 1].$$

Clear the parentheses in the new term on the right side of the equals sign and simplify.

$$= \frac{k(3k + 5)}{2} + 3k + 3 + 1$$

$$= \frac{k(3k + 5)}{2} + 3k + 4$$

Now combine the two terms on the right.

$$= \frac{k(3k + 5)}{2} + \frac{2(3k + 4)}{2}$$

$$= \frac{k(3k + 5) + 2(3k + 4)}{2}$$

$$= \frac{3k^2 + 5k + 6k + 8}{2}$$

$$= \frac{3k^2 + 11k + 8}{2}$$

$$= \frac{(k + 1)(3k + 8)}{2}$$

Since $3k + 8$ can be written as $3(k + 1) + 5$,

$$4 + 7 + 10 + \ldots + (3k + 1) + [3(k + 1) + 1]$$
$$= \frac{(k + 1)[3(k + 1) + 5]}{2}.$$

The final result is the statement S_{k+1}. Therefore, if S_k is true, then S_{k+1} is true. The two steps required for a proof by mathematical induction are completed, so the general statement S_n is true for every positive integer value of n. ✦

Example 3   Prove that if x is a real number between 0 and 1, then for every positive integer n,

$$0 < x^n < 1.$$

Here S_1 is the statement

$$\text{if } 0 < x < 1, \text{ then } 0 < x^1 < 1,$$

which is true. S_k is the statement

$$\text{if } 0 < x < 1, \text{ then } 0 < x^k < 1.$$

To show that this implies that S_{k+1} is true, multiply all members of $0 < x^k < 1$ by x to get

$$x \cdot 0 < x \cdot x^k < x \cdot 1.$$

(Here the fact that $0 < x$ is used.) Simplify to get

$$0 < x^{k+1} < x.$$

Since $x < 1$,

$$x^{k+1} < x < 1$$

and

$$0 < x^{k+1} < 1.$$

By this work, if S_k is true, then S_{k+1} is true, so the given statement is true for every positive integer n. ✦

9.6 Exercises

Use the method of mathematical induction to prove the following statements. Assume that
n is a positive integer. See Examples 1–3.

1. $2 + 4 + 6 + \ldots + 2n = n(n + 1)$

2. $1 + 3 + 5 + \ldots + (2n - 1) = n^2$

3. $3 + 6 + 9 + \ldots + 3n = \dfrac{3n(n + 1)}{2}$

4. $5 + 10 + 15 + \ldots + 5n = \dfrac{5n(n + 1)}{2}$

5. $2 + 4 + 8 + \ldots + 2^n = 2^{n+1} - 2$

6. $3 + 3^2 + 3^3 + \ldots + 3^n = \dfrac{3(3^n - 1)}{2}$

7. $1^2 + 2^2 + 3^2 + \ldots + n^2 = \dfrac{n(n + 1)(2n + 1)}{6}$

8. $1^3 + 2^3 + 3^3 + \ldots + n^3 = \dfrac{n^2(n + 1)^2}{4}$

9. $5 \cdot 6 + 5 \cdot 6^2 + 5 \cdot 6^3 + \ldots + 5 \cdot 6^n = 6(6^n - 1)$

10. $7 \cdot 8 + 7 \cdot 8^2 + 7 \cdot 8^3 + \ldots + 7 \cdot 8^n = 8(8^n - 1)$

11. $\dfrac{1}{1 \cdot 2} + \dfrac{1}{2 \cdot 3} + \dfrac{1}{3 \cdot 4} + \ldots + \dfrac{1}{n(n + 1)} = \dfrac{n}{n + 1}$

12. $\dfrac{1}{1 \cdot 4} + \dfrac{1}{4 \cdot 7} + \dfrac{1}{7 \cdot 10} + \ldots + \dfrac{1}{(3n - 2)(3n + 1)} = \dfrac{n}{3n + 1}$

13. $\dfrac{1}{2} + \dfrac{1}{2^2} + \dfrac{1}{2^3} + \ldots + \dfrac{1}{2^n} = 1 - \dfrac{1}{2^n}$

14. $\dfrac{4}{5} + \dfrac{4}{5^2} + \dfrac{4}{5^3} + \ldots + \dfrac{4}{5^n} = 1 - \dfrac{1}{5^n}$

15. $x^{2n} + x^{2n-1}y + \ldots + xy^{2n-1} + y^{2n} = \dfrac{x^{2n+1} - y^{2n+1}}{x - y}$

16. $x^{2n-1} + x^{2n-2}y + \ldots + xy^{2n-2} + y^{2n-1} = \dfrac{x^{2n} - y^{2n}}{x - y}$

17. $(a^m)^n = a^{mn}$ (Assume that a and m are constant.)

18. $(ab)^n = a^n b^n$ (Assume that a and b are constant.)

19. If $a > 1$, then $a^n > 1$.

20. If $a > 1$, then $a^n > a^{n-1}$.

21. If $0 < a < 1$, then $a^n < a^{n-1}$.

22. $2^n > n^2$

23. $4^n > n^4$, for $n \geq 7$

24. What is wrong with the following proof by mathematical induction?

Prove: Any natural number equals the next natural number; that is, $n = n + 1$.

Proof. To begin, we assume the statement true for some natural number $n = k$:

$$k = k + 1.$$

We must now show that the statement is true for $n = k + 1$. If we add 1 to both
sides, we have

$$k + 1 = k + 1 + 1$$
$$k + 1 = k + 2.$$

Hence, if the statement is true for $n = k$, it is also true for $n = k + 1$. Thus, the theorem is proved.

25. Suppose that n straight lines (with $n \geq 2$) are drawn in a plane, where no two lines are parallel and no three lines pass through the same point. Show that the number of points of intersection of the lines is $(n^2 - n)/2$.

26. The series of sketches below starts with an equilateral triangle having sides of length 1. In the following steps, equilateral triangles are constructed on each side of the preceding figure. The lengths of the sides of these new triangles are 1/3 the length of the sides of the preceding triangles. Develop a formula for the number of sides of the nth figure. Use mathematical induction to prove your answer.

27. Find the perimeter of the nth figure in Exercise 26.

28. Show that the area of the nth figure in Exercise 26 is

$$\sqrt{3} \left[\frac{2}{5} - \frac{3}{20} \left(\frac{4}{9} \right)^{n-1} \right].$$

9.7 Permutations

If there are 3 roads from Albany to Baker and 2 roads from Baker to Creswich, in how many ways can one travel from Albany to Creswich by way of Baker? For each of the 3 roads from Albany to Baker, there are 2 different roads from Baker to Creswich, so that there are $3 \cdot 2 = 6$ different ways to make the trip, as shown in the **tree diagram** in Figure 9.2.

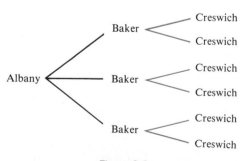

Figure 9.2

This example illustrates the following property.

Fundamental Principle of Counting	If one event can occur in *m* ways and a second event can occur in *n* ways, then both events can occur in *mn* ways, provided the outcome of the first event does not influence the outcome of the second.

The fundamental principle of counting can be extended to any number of events, provided the outcome of no one event influences the outcome of another. Such events are called **independent events.**

Example 1   A restaurant offers a choice of 3 salads, 5 main dishes, and 2 desserts. Use the fundamental principle of counting to find the number of different 3-course meals that can be selected.

Three independent events are involved: selecting a salad, selecting a main dish, and selecting a dessert. The first event can occur in 3 ways, the second event can occur in 5 ways, and the third event can occur in 2 ways; thus there are

$$3 \cdot 5 \cdot 2 = 30 \text{ possible meals.} \quad \clubsuit$$

Example 2   Janet Branson has 5 different books that she wishes to arrange on her desk. How many different arrangements are possible?

Five events are involved: selecting a book for the first spot, selecting a book for the second spot, and so on. Here the outcome of the first event *does* influence the outcome of the other events (since one book has already been chosen). For the first spot Branson has 5 choices, for the second spot 4 choices, for the third spot 3 choices, and so on. Now use the fundamental principle of counting to find that there are

$$5 \cdot 4 \cdot 3 \cdot 2 \cdot 1 \text{ or } 120 \text{ different arrangements.} \quad \clubsuit$$

In using the fundamental principle of counting we often encounter such products as $5 \cdot 4 \cdot 3 \cdot 2 \cdot 1$ from Example 2. For convenience in writing these products, use the symbol *n*! (read "*n* factorial"), which was defined in Section 9.5 and is repeated here.

Definition of *n*!	For any counting number *n*, $$n! = n(n - 1)(n - 2)(n - 3) \ldots (2)(1)$$ and $$0! = 1.$$

By the definition, $5 \cdot 4 \cdot 3 \cdot 2 \cdot 1$ is written as 5!. Also, $3! = 3 \cdot 2 \cdot 1 = 6$. The definition of *n*! means that $n[(n - 1)!] = n!$ for all natural numbers $n \geq 2$. It is useful to have this relation hold also for $n = 1$, so, by definition, $0! = 1$.

Example 3 Suppose Branson wishes to place only 3 of the 5 books on her desk. How many arrangements of 3 books are possible?

She still has 5 ways to fill the first spot, 4 ways to fill the second spot, and 3 ways to fill the third. Since she wants to use 3 books, there are only 3 spots to be filled (3 events) instead of 5, so there are

$$5 \cdot 4 \cdot 3 = 60 \text{ arrangements.} \quad \clubsuit$$

The number 60 in the example above is called the number of *permutations* of 5 things taken 3 at a time, written $P(5, 3) = 60$. Example 2 showed that the number of ways of arranging 5 elements from a set of 5 elements, written $P(5, 5)$, is 120.

A **permutation** of n elements taken r at a time is one of the ways of arranging r elements taken from a set of n elements ($r \le n$). Generalizing from the examples above, the number of permutations of n elements, taken r at a time, denoted by $P(n, r)$, is

$$P(n, r) = n(n - 1)(n - 2) \cdots (n - r + 1)$$

$$= \frac{n(n - 1)(n - 2) \cdots (n - r + 1)(n - r)(n - r - 1) \cdots (2)(1)}{(n - r)(n - r - 1) \cdots (2)(1)}$$

$$= \frac{n!}{(n - r)!}.$$

This proves the following result.

Permutations of *n* **Elements** *r* **at a Time**	If $P(n, r)$ denotes the number of permutations of n elements taken r at a time, $r \le n$, then $$P(n, r) = \frac{n!}{(n - r)!}.$$

Example 4 Suppose 8 people enter an event in a swim meet. Assuming there are no ties, in how many ways could the gold, silver, and bronze prizes be awarded?

Using the fundamental principle of counting, there are 3 choices to be made giving $8 \cdot 7 \cdot 6 = 336$. However, we can also use the formula for $P(n, r)$ to get the same result.

$$P(8, 3) = \frac{8!}{5!} = \frac{8 \cdot 7 \cdot 6 \cdot 5 \cdot 4 \cdot 3 \cdot 2 \cdot 1}{5 \cdot 4 \cdot 3 \cdot 2 \cdot 1} = 8 \cdot 7 \cdot 6 = 336 \quad \clubsuit$$

Example 5 In how many ways can 6 students be seated in a row of 6 desks?

$$P(6, 6) = 6! = 6 \cdot 5 \cdot 4 \cdot 3 \cdot 2 \cdot 1 = 720 \quad \clubsuit$$

9.7 Exercises

Evaluate each of the following. See Examples 4 and 5.

1. $P(7, 7)$ **2.** $P(5, 3)$ **3.** $P(6, 5)$ **4.** $P(4, 2)$

5. $P(10, 2)$ **6.** $P(8, 2)$ **7.** $P(8, 3)$ **8.** $P(11, 4)$

9. $P(7, 1)$ **10.** $P(18, 0)$ **11.** $P(9, 0)$ **12.** $P(14, 1)$

13. In how many ways can 8 people be seated in a row of 8 seats?

14. In how many ways can 7 out of 10 people be assigned to 7 seats?

15. In how many ways can 5 players be assigned to the 5 positions on a basketball team, assuming that any player can play any position? In how many ways can 10 players be assigned to the 5 positions?

16. A couple has narrowed down their choice of names for a new baby to 3 first names and 5 middle names. How many different first and middle name arrangements are possible?

17. How many different homes are available if a builder offers a choice of 5 basic plans, 3 roof styles, and 2 types of siding?

18. An automaker produces 7 models, each available in 6 colors, with 4 upholstery fabrics and 5 interior colors. How many varieties of the auto are available?

19. A concert is to consist of five works: two modern, two romantic, and one classical. In how many ways can the program be arranged?

20. If the program in Exercise 19 must be shortened to 3 works chosen from the 5, how many arrangements are possible?

21. In Exercise 19, how many different programs are possible if the 2 modern works are to be played first, then the 2 romantic, and then the classical?

22. How many 4-letter radio station call letters can be made if the first letter must be K or W and no letter may be repeated? How many if repeats are allowed?

23. How many of the 4-letter call letters in Exercise 22 with no repeats end in K?

24. How many 7-digit telephone numbers are possible if the first digit cannot be zero and
(a) only odd digits can be used;
(b) the number must be a multiple of 10 (that is, must end in zero);
(c) the number must be a multiple of 100;
(d) the first three digits are 481;
(e) no repetitions are allowed?

25. (a) How many different license numbers consisting of 3 letters followed by 3 digits are possible?
(b) How many are possible if either the 3 digits or the 3 letters can come first?

26. A business school gives courses in typing, shorthand, transcription, business English, technical writing, and accounting. How many ways can a student arrange his program if he takes 3 courses?

27. How many different license plate numbers can be formed using 3 letters followed by 3 digits if no repeats are allowed? How many if there are no repeats and either letters or numbers come first?

28. Show that $P(n, n - 1) = P(n, n)$.

9.8 Combinations

In the preceding section we discussed a method for finding the number of ways to arrange r elements taken from a set of n elements. Sometimes however, the arrangement (or order) of the elements is not important.

For example, suppose three people (Ms. Opelka, Mr. Adams, and Ms. Jacobs) apply for 2 identical jobs. Ignoring all other factors, in how many ways can the personnel officer select 2 people from the 3 applicants? Here the arrangement or order of the people is unimportant. Selecting Ms. Opelka and Mr. Adams is the same as selecting Mr. Adams and Ms. Opelka. Therefore, there are only 3 ways to select 2 of the 3 applicants:

Ms. Opelka and Mr. Adams;

Ms. Opelka and Ms. Jacobs;

Mr. Adams and Ms. Jacobs.

These three choices are called the *combinations* of 3 elements taken 2 at a time. A **combination** of n elements taken r at a time is one of the ways in which r elements can be chosen from n elements.

Each combination of r elements forms $r!$ permutations. Therefore, the number of combinations of n elements taken r at a time is found by dividing the number of permutations, $P(n, r)$, by $r!$ to get

$$\frac{P(n, r)}{r!}$$

combinations. This expression can be rewritten as follows

$$\frac{P(n, r)}{r!} = \frac{\dfrac{n!}{(n - r)!}}{r!} = \frac{n!}{(n - r)!r!}$$

The symbol $\binom{n}{r}$ is used to represent the number of combinations of n things taken r at a time. (Recall that the quantity expressed by $\binom{n}{r}$ also gives the coefficients in the binomial theorem.) With this symbol the results above are stated as follows.

Combinations of n Elements r at a Time	If $\binom{n}{r}$ represents the number of combinations of n elements taken r at a time, $r \le n$, then $$\binom{n}{r} = \frac{n!}{(n - r)!r!}.$$

In the discussion above, it was shown that $\binom{3}{2} = 3$. The same result can be found using the formula.

$$\binom{3}{2} = \frac{3!}{(3 - 2)!2!} = \frac{3 \cdot 2 \cdot 1}{1 \cdot 2 \cdot 1} = 3$$

Example 1   How many different committees of 3 people can be chosen from a group of 8 people?

Since the order in which the members of the committee are chosen does not affect the result, use combinations to get

$$\binom{8}{3} = \frac{8!}{5!3!} = \frac{8 \cdot 7 \cdot 6 \cdot 5 \cdot 4 \cdot 3 \cdot 2 \cdot 1}{5 \cdot 4 \cdot 3 \cdot 2 \cdot 1 \cdot 3 \cdot 2 \cdot 1} = 56. \quad \clubsuit$$

Example 2   From a group of 30 employees, 3 are to be selected to work on a special project.

(a) In how many different ways can the employees be selected?

The number of 3-element combinations from a set of 30 elements must be found. (Use combinations, not permutations, because order within the group of 3 does not affect the result.) Using the formula gives

$$\binom{30}{3} = \frac{30!}{27!3!} = 4060.$$

There are 4060 ways to select the project group.

(b) In how many different ways can the group of 3 be selected if it has already been decided that a certain employee must work on the project?

Since one employee has already been selected to work on the project, the problem is reduced to selecting 2 more employees from the 29 employees that are left:

$$\binom{29}{2} = \frac{29!}{27!2!} = 406.$$

In this case, the project group can be selected in 406 different ways. $\clubsuit$

Example 3   Find $\binom{n}{n}$.

Use the formula with $r = n$.

$$\binom{n}{n} = \frac{n!}{(n-n)!n!} = \frac{n!}{0!n!} = \frac{n!}{1 \cdot n!} = 1$$

This is reasonable since there is only one way of selecting a group of n objects from a set of n objects when order is disregarded. $\clubsuit$

Example 4   Find $\binom{n}{1}$.

Again use the formula.

$$\binom{n}{1} = \frac{n!}{(n-1)!1!} = \frac{n \cdot (n-1)!}{(n-1)!1!} = n$$

This result agrees with what is already known: there are n distinct one-element subsets of an n-element set. $\clubsuit$

The results of Examples 3 and 4 can be stated as properties and applied directly to special cases. Thus, from Example 3,

$$\binom{5}{5} = 1, \qquad \binom{8}{8} = 1,$$

and so on. Also, from Example 4,

$$\binom{6}{1} = 6 \quad \text{and} \quad \binom{9}{1} = 9.$$

Example 5   Show that $\binom{5}{3} = \binom{5}{2}$.

From the formula,

$$\binom{5}{3} = \frac{5!}{3!2!} \quad \text{and} \quad \binom{5}{2} = \frac{5!}{2!3!}$$

Since multiplication is commutative,

$$\binom{5}{3} = \binom{5}{2}. \quad \clubsuit$$

Example 5 shows that there are the same number of ways to select three things from a group of five as to select two things from a group of five. This makes sense, because each time you choose a set of two, you are also choosing a set of three, the three not selected.

Example 5 is a special case of the following result.

If $\binom{n}{r}$ represents the number of combinations of n things taken r at a time, then

$$\binom{n}{r} = \binom{n}{n-r}.$$

The proof of this result is left as an exercise.

9.8   Exercises

Evaluate each of the following. See Examples 1 and 2.

1. $\binom{6}{5}$ 2. $\binom{4}{2}$ 3. $\binom{8}{5}$ 4. $\binom{10}{2}$

5. $\binom{15}{4}$ 6. $\binom{9}{3}$ 7. $\binom{10}{7}$ 8. $\binom{10}{3}$

9. $\binom{14}{1}$ 10. $\binom{20}{2}$ 11. $\binom{18}{0}$ 12. $\binom{13}{0}$

13. A club has 30 members. If a committee of 4 is selected at random, how many committees are possible?

14. How many different samples of 3 apples can be drawn from a crate of 25 apples?

15. Three students are to be selected from a group of 12 students to participate in a special class. In how many ways can this be done? In how many ways can the group that will not participate be selected?

16. Hal's Hamburger Heaven sells hamburgers with cheese, relish, lettuce, tomato, mustard, or ketchup. How many different hamburgers can be made using any 3 of the extras?

17. How many different 2-card hands can be dealt from a deck of 52 cards?

18. How many different 13-card bridge hands can be dealt from a deck of 52 cards?

19. Five cards are marked with the numbers 1, 2, 3, 4, and 5, shuffled, and 2 cards are then drawn. How many different 2-card combinations are possible?

20. If a bag contains 15 marbles, how many samples of 2 marbles can be drawn from it? How many samples of 4 marbles?

21. In Exercise 20, if the bag contains 3 yellow, 4 white, and 8 blue marbles, how many samples of 2 can be drawn in which both marbles are blue?

22. In Exercise 14, if it is known that there are 5 rotten apples in the crate:
   (a) How many samples of 3 could be drawn in which all 3 are rotten?
   (b) How many samples of 3 could be drawn in which there are 1 rotten apple and 2 good apples?

23. A city council is composed of 5 liberals and 4 conservatives. Three members are to be selected randomly as delegates to a convention.
   (a) How many delegations are possible?
   (b) How many delegations could have all liberals?
   (c) How many delegations could have 2 liberals and 1 conservative?
   (d) If 1 member of the council serves as mayor, how many delegations are possible which include the mayor?

24. Seven workers decide to send a delegation of 2 to their supervisor to discuss their grievances.
   (a) How many different delegations are possible?
   (b) If it is decided that a certain employee must be in the delegation, how many different delegations are possible?
   (c) If there are 2 women and 5 men in the group, how many delegations would include at least one woman?

A poker hand is made up of 5 cards drawn at random from a deck of 52 cards. Any 5 cards in one suit are called a flush. The 5 highest cards, that is, the A, K, Q, J, and 10, of any one suit are called a royal flush. Use combinations to set up each of the following. Do not evaluate.

25. Find the total number of all possible poker hands.

26. How many royal flushes in hearts are possible?

27. How many royal flushes in any of the four suits are possible?

28. How many flushes in hearts are possible?

29. How many flushes in any of the four suits are possible?

30. How many combinations of 3 aces and 2 eights are possible?

Solve the following problems by using either combinations or permutations.

31. How many ways can the letters of the word TOUGH be arranged?

32. If Matthew has 8 courses to choose from, how many ways can he arrange his schedule if he must pick 4 of them?

33. How many samples of 3 pineapples can be drawn from a crate of 12?

34. Velma specializes in making different vegetable soups with carrots, celery, beans, peas, mushrooms, and potatoes. How many different soups can she make using any 4 ingredients?

35. Show that $\binom{n}{r} = \binom{n}{n-r}$.

36. Find $\binom{n}{n-1}$.

9.9 Probability

The study of probability has become increasingly popular because it has a wide range of practical applications. The basic ideas of probability are introduced in this section.

Consider an experiment that has one or more possible **outcomes,** each of which is equally likely to occur. For example, the experiment of tossing a coin has two equally likely possible outcomes: landing heads up *(H)* or landing tails up *(T)*. Also, the experiment of rolling a die has 6 equally likely outcomes: landing so the face that is up shows 1, 2, 3, 4, 5, or 6 points.

The set *S* of all possible outcomes of a given experiment is called the **sample space** for the experiment. (In this text all sample spaces are finite.) A sample space for the experiment of tossing a coin consists of the outcomes *H* and *T*. This sample space can be written in set notation as

$$S = \{H, T\}.$$

Similarly, a sample space for the experiment of rolling a single die is

$$S = \{1, 2, 3, 4, 5, 6\}.$$

Any subset of the sample space is called an **event.** In the experiment with the die, for example, "the number showing is a three" is an event, say E_1, such that $E_1 = \{3\}$. "The number showing is greater than three" is also an event, say E_2, such that $E_2 = \{4, 5, 6\}$. To represent the number of outcomes that belong to event *E*, the notation *n(E)* is used. Then $n(E_1) = 1$ and $n(E_2) = 3$.

The notation *P(E)* is used to define the *probability* of an event. If the outcomes in the sample space for an experiment are equally likely, then the probability of event *E* occurring is found in the following way.

Definition of Probability of Event E	In a sample space with equally likely outcomes, the **probability** of an event E, written $P(E)$, is the ratio of the number of outcomes in sample space S that belong to event E, $n(E)$, to the total number of outcomes in sample space S, $n(S)$. That is, $$P(E) = \frac{n(E)}{n(S)}.$$

To use this definition to find the probability of the event E_1 given above, start with the sample space for the experiment, $S = \{1, 2, 3, 4, 5, 6\}$, and the desired event, $E_1 = \{3\}$. Since $n(E_1) = 1$ and since there are 6 outcomes in the sample space,

$$P(E_1) = \frac{n(E_1)}{n(S)} = \frac{1}{6}.$$

Example 1   A single die is rolled. Write the following events in set notation and give the probability for each event.

(a) E_3: the number showing is even

Use the definition above. Since $E_3 = \{2, 4, 6\}$, $n(E_3) = 3$. As shown above, $n(S) = 6$, so

$$P(E_3) = \frac{3}{6} = \frac{1}{2}.$$

(b) E_4: the number showing is greater than 4

Again $n(S) = 6$. Event $E_4 = \{5, 6\}$, with $n(E_4) = 2$. By the definition,

$$P(E_4) = \frac{2}{6} = \frac{1}{3}.$$

(c) $E_5 = \{1, 2, 3, 4, 5, 6\}$ and $P(E_5) = \frac{6}{6} = 1$

(d) $E_6 = \emptyset$ and $P(E_6) = \frac{0}{6} = 0$ ✣

In part (c), $E_5 = S$. Therefore the event E_5 is certain to occur every time the experiment is performed. An event that is certain to occur, such as E_5, always has a probability of 1. On the other hand, $E_6 = \emptyset$ and $P(E_6)$ is 0. The probability of an impossible event, such as E_6, is always 0, since none of the outcomes in the sample space satisfy the event. For any event E, $P(E)$ is between 0 and 1 inclusive.

The set of all outcomes in the sample space that do *not* belong to event E is called the **complement** of E, written E'. For example, in the experiment of drawing a single card from a standard deck of 52 cards, let E be the event "the card is an ace." Then E' is the event "the card is not an ace." From the definition of E', for any event E,

$$E \cup E' = S \quad \text{and} \quad E \cap E' = \emptyset.$$

Probability concepts can be illustrated using **Venn diagrams,** as shown in Figure 9.3. The rectangle in Figure 9.3 represents the sample space in an experiment. The area inside the circle represents event E, while the area inside the rectangle, but outside the circle, represents event E'.

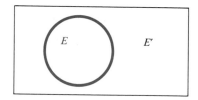

Figure 9.3

Example 2   In the experiment of drawing a card from a well-shuffled deck, find the probability of event E, the card is an ace, and event E'.

Since there are four aces in the deck of 52 cards, $n(E) = 4$ and $n(S) = 52$. Therefore,

$$P(E) = \frac{n(E)}{n(S)} = \frac{4}{52} = \frac{1}{13}.$$

Of the 52 cards, 48 are not aces, so

$$P(E') = \frac{n(E')}{n(S)} = \frac{48}{52} = \frac{12}{13}. \quad \clubsuit$$

In Example 2, $P(E) + P(E') = (1/13) + (12/13) = 1$. This is always true for any event E and its complement E'. That is,

$$P(E) + P(E') = 1.$$

This can be restated as

$$P(E) = 1 - P(E') \quad \text{or} \quad P(E') = 1 - P(E).$$

These two equations suggest an alternate way to compute the probability of an event. For example, if it is known that $P(E) = 1/10$, then

$$P(E') = 1 - \frac{1}{10} = \frac{9}{10}.$$

Sometimes probability statements are expressed in terms of odds, a comparison of $P(E)$ with $P(E')$. The **odds** in favor of an event E are expressed as the ratio of $P(E)$ to $P(E')$ or as the fraction $P(E)/P(E')$. For example, if the probability of rain can be established as 1/3, the odds that it will rain are

$$P(\text{rain}) \text{ to } P(\text{no rain}) = \frac{1}{3} \text{ to } \frac{2}{3} = \frac{1/3}{2/3} = \frac{1}{2} \quad \text{or} \quad 1 \text{ to } 2.$$

On the other hand, the odds that it will not rain are 2 to 1 (or 2/3 to 1/3). If the odds in favor of an event are, say, 3 to 5, then the probability of the event is 3/8, while the probability of the complement of the event is 5/8. If the odds favoring event E are m to n, then

$$P(E) = \frac{m}{m + n} \quad \text{and} \quad P(E') = \frac{n}{m + n}$$

A **compound event** involves an *alternative,* such as E or F, where E and F are events. For example, in the experiment of rolling a die, suppose H is the event "the result is a 3," and K is the event "the result is an even number." What is the probability of "the result is a 3 or an even number"? From the results earlier in this section,

$$H = \{3\} \qquad\qquad P(H) = \frac{1}{6}$$

$$K = \{2, 4, 6\} \qquad\qquad P(K) = \frac{3}{6} = \frac{1}{2}$$

$$H \text{ or } K = \{2, 3, 4, 6\} \qquad P(H \text{ or } K) = \frac{4}{6} = \frac{2}{3}$$

Notice that $P(H) + P(K) = P(H \text{ or } K)$.

Before assuming that this relationship is true in general, consider another event for this experiment, "the result is a 2," event G.

$$G = \{2\} \qquad\qquad P(G) = \frac{1}{6}$$

$$K = \{2, 4, 6\} \qquad\qquad P(K) = \frac{3}{6} = \frac{1}{2}$$

$$K \text{ or } G = \{2, 4, 6\} \qquad P(K \text{ or } G) = \frac{3}{6} = \frac{1}{2}$$

In this case $P(K) + P(G) \neq P(K \text{ or } G)$. See Figure 9.4.

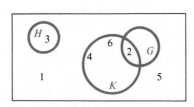

Figure 9.4

As Figure 9.4 suggests, the difference in the two examples above comes from the fact that events H and K cannot occur simultaneously. Such events are called **mutually exclusive events.** In fact, $H \cap K = \emptyset$, which is true for any two mutually exclusive events. Events K and G, however, can occur simultaneously. Both are satisfied if the result of the roll is a 2, the element in their intersection $(K \cap G = \{2\})$, but $P(2)$ should not be counted twice. That is why $P(E \cap F) = P(2)$ must be subtracted.

Probability of Alternate Events	For any events E and F: $$P(E \text{ or } F) = P(E \cup F) = P(E) + P(F) - P(E \cap F)$$

Example 3 One card is drawn from a well-shuffled deck of 52 cards. What is the probability of the following outcomes?

(a) The card is an ace or a spade.

The events "drawing an ace" and "drawing a spade" are not mutually exclusive since it is possible to draw the ace of spades, an outcome satisfying both events. The probability is

$$P(\text{ace or spade}) = P(\text{ace}) + P(\text{spade}) - P(\text{ace and spade})$$

$$= \frac{4}{52} + \frac{13}{52} - \frac{1}{52} = \frac{16}{52} = \frac{4}{13}.$$

(b) The card is a three or a king.

"Drawing a 3" and "drawing a king" are mutually exclusive events because it is impossible to draw one card that is both a 3 and a king. Using the rule given above,

$$P(3 \text{ or } K) = P(3) + P(K) - P(3 \text{ and } K)$$

$$= \frac{4}{52} + \frac{4}{52} - 0 = \frac{8}{52} = \frac{2}{13}. \quad \clubsuit$$

Example 4 For the experiment consisting of one roll of a pair of dice, find the probability that the sum of the points showing is at most 4.

The statement "at most 4" can be rewritten as "2 or 3 or 4." (A sum of 1 is meaningless here.) Then

$$P(\text{at most 4}) = P(2 \text{ or } 3 \text{ or } 4)$$

$$= P(2) + P(3) + P(4), \tag{1}$$

since the events represented by "2," "3," and "4" are mutually exclusive.

The sample space for this experiment includes the 36 possible pairs of numbers from 1 to 6: (1, 1), (1, 2), (1, 3), (1, 4), (1, 5), (1, 6), (2, 1), (2, 2), and so on. The pair $(1, 1)$ is the only one with a sum of 2, so $P(2) = 1/36$. Also $P(3) = 2/36$ since both $(1, 2)$ and $(2, 1)$ give a sum of 3. The pairs, $(1, 3)$, $(2, 2)$, and $(3, 1)$ have a sum of 4, so $P(4) = 3/36$. Substituting into equation (1) above gives

$$P(\text{at most } 4) = \frac{1}{36} + \frac{2}{36} + \frac{3}{36} = \frac{6}{36} = \frac{1}{6}. \quad \clubsuit$$

The properties of probability discussed in this section are summarized as follows.

Properties of Probability

For any events E and F:

1. $0 \leq P(E) \leq 1$
2. $P(\text{a certain event}) = 1$
3. $P(\text{an impossible event}) = 0$
4. $P(E') = 1 - P(E)$
5. $P(E \text{ or } F) = P(E \cup F) = P(E) + P(F) - P(E \cap F)$

9.9 Exercises

Write a sample space with equally likely outcomes for each of the following experiments.

1. A two-headed coin is tossed once.
2. Two ordinary coins are tossed.
3. Three ordinary coins are tossed.
4. Slips of paper marked with the numbers 1, 2, 3, 4, and 5 are placed in a box. After mixing well, two slips are drawn.
5. An unprepared student takes a three-question true/false quiz in which he guesses the answer to all three questions.
6. A die is rolled and then a coin is tossed.

Write the following events in set notation and give the probability of each event. See Examples 1–4.

7. In the experiment described in Exercise 2:
   (a) both coins show the same face;
   (b) at least one coin turns up heads.
8. In Exercise 1:
   (a) the result of the toss is heads;
   (b) the result of the toss is tails.
9. In Exercise 4:
   (a) both slips are marked with even numbers;
   (b) both slips are marked with odd numbers;
   (c) both slips are marked with the same number;
   (d) one slip is marked with an odd number, the other with an even number.

10. In Exercise 5:
    (a) the student gets all three answers correct;
    (b) he gets all three answers wrong;
    (c) he gets exactly two answers correct;
    (d) he gets at least one answer correct.

11. A marble is drawn at random from a box containing 3 yellow, 4 white, and 8 blue marbles. Find the following probabilities.
    (a) A yellow marble is drawn.
    (b) A blue marble is drawn.
    (c) A black marble is drawn.
    (d) What are the odds in favor of drawing a yellow marble?
    (e) What are the odds against drawing a blue marble?

12. A baseball player with a batting average of .300 comes to bat. What are the odds in favor of his getting a hit?

13. In Exercise 4, what are the odds that the sum of the numbers on the two slips of paper is 5?

14. If the odds that it will rain are 4 to 5, what is the probability of rain?

15. If the odds that a candidate will win an election are 3 to 2, what is the probability that the candidate will lose?

16. A card is drawn from a well-shuffled deck of 52 cards. Find the probability that the card is
    (a) a 9;   (b) black;   (c) a black 9;   (d) a heart;
    (e) the 9 of hearts;   (f) a face card (K, Q, J of any suit).

17. Mrs. Elliott invites 10 relatives to a party: her mother, two uncles, three brothers, and four cousins. If the chances of any one guest arriving first are equally likely, find the following probabilities.
    (a) The first guest is an uncle or a brother.
    (b) The first guest is a brother or cousin.
    (c) The first guest is a brother or her mother.

18. One card is drawn from a standard deck of 52 cards. What is the probability that the card is
    (a) a 9 or a 10?   (b) red or a 3?   (c) a heart or black?
    (d) less than a 4? (Consider aces as 1's.)

19. Two dice are rolled. Find the probability of the following events.
    (a) The sum of the points is at least 10.
    (b) The sum of the points is either 7 or at least 10.
    (c) The sum of the points is 2 or the dice both show the same number.

20. A student estimates that his probability of getting an A in a certain course is .4; a B, .3; a C, .2; and a D, .1.
    (a) Assuming that only the grades A, B, C, D, and F are possible, what is the probability that he will fail the course?
    (b) What is the probability that he will receive a grade of C or better?
    (c) What is the probability that he will receive at least a B in the course?
    (d) What is the probability that he will get at most a C in the course?

21. If a marble is drawn from a bag containing 3 yellow, 4 white, and 8 blue marbles, what is the probability of the following?

   (a) The marble is either yellow or white.

   (b) It is either yellow or blue.

   (c) It is either blue or white.

22. The table shows the probability that a customer of a department store will make a purchase in the indicated price range.

Cost	Probability
Below $5	.25
$5–$19.99	.37
$20–$39.99	.11
$40–$69.99	.09
$70–$99.99	.07
$100–$149.99	.08
$150 or more	.03

Find the probability that a customer makes a purchase that is
(a) less than $20; **(b)** $40 or more; **(c)** more than $99.99; **(d)** less than $100.

Chapter 9 *Summary*

Key Words

9.1 sequence
terms
finite sequence
infinite sequence
recursive definition
series
index of summation
9.2 arithmetic sequence
common difference
9.3 geometric sequence
common ratio
9.5 Pascal's triangle
n-factorial
binomial coefficient
binomial theorem

9.7 tree diagram
independent events
permutation
9.8 combination
9.9 outcome
sample space
probability
complement
Venn diagram
odds
compound event
mutually exclusive events

Key Ideas

9.2 *n*th Term of an Arithmetic Sequence

In an arithmetic sequence with first term a_1 and common difference d, the nth term, a_n, is given by

$$a_n = a_1 + (n - 1)d.$$

9.2 Sum of the First *n* Terms of an Arithmetic Sequence

If an arithmetic sequence has first term a_1 and common difference d, then the sum of the first n terms is given by

$$S_n = \frac{n}{2}(a_1 + a_n)$$

or

$$S_n = \frac{n}{2}[2a_1 + (n - 1)d].$$

9.3 *n*th Term of a Geometric Sequence

In the geometric sequence with first term a_1 and common ratio r, the nth term is

$$a_n = a_1 r^{n-1}.$$

9.3 Sum of the First n Terms of a Geometric Sequence	If a geometric sequence has first term a_1 and common ratio r, then the sum of the first n terms is given by
	$$S_n = \frac{a_1(1 - r^n)}{1 - r}, \quad \text{where } r \neq 1.$$

9.4 Sum of an Infinite Geometric Sequence	The sum of an infinite geometric sequence with first term a_1 and common ratio r, where $-1 < r < 1$, is given by
	$$S_\infty = \frac{a_1}{1 - r}.$$

9.5 Binomial Theorem	For any positive integer n:
	$$(x + y)^n = x^n + \binom{n}{n-1}x^{n-1}y + \binom{n}{n-2}x^{n-2}y^2 + \binom{n}{n-3}x^{n-3}y^3$$ $$+ \ldots + \binom{n}{n-r}x^{n-r}y^r + \ldots + \binom{n}{1}xy^{n-1} + y^n$$

9.6 Principle of Mathematical Induction	Let S_n be a statement concerning the positive integer n. Suppose that
	(1) S_1 is true;
	(2) for any positive integer k, $k \leq n$, if S_k is true, then S_{k+1} is also true.
	Then S_n is true for every positive integer value of n.

9.7 Fundamental Principle of Counting	If one event can occur in m ways and a second event can occur in n ways, then both events can occur in mn ways, provided the outcome of the first event does not influence the outcome of the second.

9.7 Permutations of n Elements r at a time	If $P(n, r)$ denotes the number of permutations of n elements taken r at a time, $r \leq n$, then
	$$P(n, r) = \frac{n!}{(n - r)!}.$$

9.8 Combinations of n Elements r at a Time	If $\binom{n}{r}$ represents the number of combinations of n elements taken r at a time, $r \leq n$, then
	$$\binom{n}{r} = \frac{n!}{(n - r)!r!}.$$

9.9 Probability of Event *E*

In a sample space with equally likely outcomes, the probability of an event E, written $P(E)$, is the ratio of the number of outcomes in sample space S that belong to event E, $n(E)$, to the total number of outcomes in sample space S, $n(S)$. That is,

$$P(E) = \frac{n(E)}{n(S)}.$$

9.9 Properties of Probability

For any events E and F:

1. $0 \le P(E) \le 1$

2. $P(\text{a certain event}) = 1$

3. $P(\text{an impossible event}) = 0$

4. $P(E') = 1 - P(E)$

5. $P(E \text{ or } F) = P(E \cup F) = P(E) + P(F) - P(E \cap F)$

Chapter 9   *Review Exercises*

Write the first five terms for each of the following sequences.

1. $a_n = \dfrac{n}{n+1}$

2. $a_n = (-2)^n$

3. $a_n = 2(n+3)$

4. $a_n = n(n+1)$

5. $a_1 = 5, a_n = a_{n-1} - 3$

6. $a_1 = -2, a_n = 3a_{n-1}$

7. $a_1 = 5, a_2 = 3, a_n = a_{n-1} - a_{n-2}$ for $n \geq 2$

8. $b_1 = -2, b_2 = 2, b_3 = -4, b_n = -2 \cdot b_{n-2}$ if n is even, and $b_n = 2 \cdot b_{n-2}$ if n is odd.

9. Arithmetic, $a_1 = 6, d = -4$

10. Arithmetic, $a_3 = 9, a_4 = 7$

11. Arithmetic, $a_1 = 3 - \sqrt{5}, a_2 = 4$

12. Arithmetic, $a_3 = \pi, a_4 = 0$

13. Geometric, $a_1 = 4, r = 2$

14. Geometric, $a_4 = 8, r = 1/2$

15. Geometric, $a_1 = -3, a_2 = 4$

16. Geometric, $a_3 = 8, a_5 = 72$

17. A certain arithmetic sequence has $a_6 = -4$ and $a_{17} = 51$. Find a_1 and a_{20}.

18. For a given geometric sequence, $a_1 = 4$ and $a_5 = 324$. Find a_6.

Find a_8 for each of the following arithmetic sequences.

19. $a_1 = 6, d = 2$

20. $a_1 = -4, d = 3$

21. $a_1 = 6x - 9, a_2 = 5x + 1$

22. $a_3 = 11m, a_5 = 7m - 4$

Find S_{12} for each of the following arithmetic sequences.

23. $a_1 = 2, d = 3$

24. $a_2 = 6, d = 10$

25. $a_1 = -4k, d = 2k$

Find a_5 for each of the following geometric sequences.

26. $a_1 = 3, r = 2$

27. $a_2 = 3125, r = 1/5$

28. $a_1 = 5x, a_2 = x^2$

29. $a_2 = \sqrt{6}, a_4 = 6\sqrt{6}$

Find S_4 for each of the following geometric sequences.

30. $a_1 = 1, r = 2$

31. $a_1 = 3, r = 3$

32. $a_1 = 2k, a_2 = -4k$

Evaluate each of the following sums that exist.

33. $18 + 9 + 9/2 + 9/4 + \ldots$

34. $20 + 15 + 45/4 + 135/16 + \ldots$

35. $-5/6 + 5/9 - 10/27 + \ldots$

36. $.9 + .09 + .009 + .0009 + \ldots$

Convert each of the following repeating decimals to rational numbers.

37. $.6666 \ldots$

38. $.512512512 \ldots$

Evaluate each of the following sums that exist.

39. $\displaystyle\sum_{i=1}^{4} \dfrac{2}{i}$

40. $\displaystyle\sum_{i=1}^{7} (-1)^{i+1} \cdot 6$

41. $\displaystyle\sum_{i=4}^{8} 3i(2i - 5)$

42. $\displaystyle\sum_{i=1}^{6} i(i + 2)$

43. $\displaystyle\sum_{i=1}^{6} 4 \cdot 3^i$

44. $\displaystyle\sum_{i=1}^{4} 8 \cdot 2^i$

45. $\displaystyle\sum_{i=1}^{\infty} \left(\dfrac{5}{8}\right)^i$

46. $\displaystyle\sum_{i=1}^{\infty} -10\left(\dfrac{5}{2}\right)^i$

47. A stack of canned goods in a market display requires 15 cans on the bottom, 13 in the next layer, 11 in the next layer, and so on. How many cans are needed for the display?

48. Gale Stockdale borrows $6000 at simple interest of 12% per year. He will repay the loan and interest in monthly payments of $260, $258, $256, and so on. If he makes 30 payments, what is the total amount required to pay off the loan plus the interest?

49. On a certain production line, new employees during their first week turn out 5/4 as many items each day as on the previous day. If a new employee produces 48 items the first day, how many will she produce on the fifth day of work?

50. The half-life of a radioactive substance is 20 years. If 600 g are present at the start, how much will be left after 100 years?

Use the binomial theorem to expand each of the following.

51. $(x + 2y)^4$ **52.** $(3z - 5w)^3$ **53.** $\left(\dfrac{k}{2} - g\right)^5$

Find the indicated term or terms for each of the following expansions.

54. Fifth term of $(3x - 2y)^6$ **55.** Eighth term of $(2m + n^2)^{12}$

56. First four terms of $(3 + x)^{16}$ **57.** Last three terms of $(2m - 3n)^{15}$

Use the binomial expansion to evaluate the following numbers to three decimal places.

58. $(1.02)^{12}$ **59.** $(.98)^7$

Show that each of the following is true for positive integers n and r with r ≤ n.

60. $P(n, n - 1) = P(n, n)$ **61.** $P(n, 1) = n$ **62.** $P(n, 0) = 1$

63. $\dbinom{n}{n} = 1$ **64.** $\dbinom{n}{0} = 1$ **65.** $\dbinom{n}{n - 1} = n$ **66.** $\dbinom{n}{n - r} = \dbinom{n}{r}$

Use mathematical induction to prove that each of the following is true for every positive integer n.

67. $2 + 6 + 10 + 14 + \ldots + (4n - 2) = 2n^2$

68. $2^2 + 4^2 + 6^2 + \ldots + (2n)^2 = \dfrac{2n(n + 1)(2n + 1)}{3}$

69. $2 + 2^2 + 2^3 + \ldots + 2^n = 2(2^n - 1)$

70. $1^3 + 3^3 + 5^3 + \ldots + (2n - 1)^3 = n^2(2n^2 - 1)$

Find each of the following.

71. $P(5, 5)$ **72.** $P(9, 2)$ **73.** $P(6, 0)$ **74.** $\dbinom{8}{3}$

75. $\dbinom{10}{5}$ **76.** $\dbinom{6}{0}$ **77.** $\dbinom{7}{1}$

78. Two people are planning their wedding. They can select from 2 different chapels, 4 soloists, 3 organists, and 2 ministers. How many different wedding arrangements will be possible?

79. John Jacobs, who is furnishing his apartment, wants to buy a new couch. He can select from 5 different styles, each available in 3 different fabrics, with 6 color choices. How many different couches are available?

80. Four students are to be assigned to 4 different summer jobs. Each student is qualified for all 4 jobs. In how many ways can the jobs be assigned?

81. A student body council consists of a president, vice-president, secretary-treasurer, and 3 representatives at large. Three members are to be selected to attend a conference.
(a) How many different such delegations are possible?
(b) How many if the president must attend?

82. Nine football teams are competing for first-, second-, and third-place titles in a state-wide tournament. In how many ways can the winners be determined?

83. How many different license plates can be formed with a letter followed by 3 digits and then 3 letters? How many such license plates have no repeats?

Solve each of the following equations for n.

84. $P(n, 3) = 8 \cdot P(n - 1, 2)$

85. $30 \cdot P(n, 2) = P(n + 2, 4)$

86. $\binom{n + 2}{4} = 15 \cdot \binom{n}{4}$

87. $\binom{n}{n - 2} = 66$

A card is drawn from a standard deck of 52 cards. Find the probability that each of the following is drawn.

88. A black king

89. A face card or an ace

90. An ace or a diamond

91. A card that is not a diamond

92. A card that is not a diamond or not black

A sample shipment of 5 swimming pool filters is chosen. The probability of exactly 0, 1, 2, 3, 4, or 5 filters being defective is given in the following table.

Number defective	0	1	2	3	4	5
Probability	.31	.25	.18	.12	.08	.06

Find the probability that the following numbers of filters are defective.

93. No more than 3

94. At least 2

Chapter 9   *Test*

Write the first five terms for each of the following sequences.

1. $a_n = (-1)^n(n^2 + 1)$

2. $a_1 = 5, a_n = n + a_{n-1}$ for $n \geq 2$

3. $a_n = n + 1$ if n is odd and $a_n = a_{n-1} + 2$ if n is even.

4. Arithmetic, $a_2 = 1, a_3 = 25$

5. Geometric, $a_1 = 81, r = 2/3$

6. A certain arithmetic sequence has $a_7 = -6$ and $a_{15} = -2$. Find a_{31}.

7. For a given geometric sequence, $a_1 = 12$ and $a_6 = -3/8$. Find a_3.

8. Find S_{10} for the arithmetic sequence with $a_1 = 37$ and $d = 13$.

9. Find a_7 for the geometric sequence with $a_1 = 2x^3$ and $a_3 = 18x^7$.

10. Find S_4 for the geometric sequence with $a_1 = 4$ and $r = 1/2$.

Evaluate each of the following sums that exist.

11. $75 + 30 + 12 + \dfrac{24}{5} + \ldots$

12. $\dfrac{1}{48} + \dfrac{1}{32} + \dfrac{3}{64} + \dfrac{9}{128} + \dfrac{27}{256} + \ldots$

Evaluate each of the following sums that exist.

13. $\displaystyle\sum_{i=1}^{30} (5i + 2)$

14. $\displaystyle\sum_{i=1}^{\infty} 54(2/9)^i$

15. Fred Meyers deposited $50 in a new savings account on February 1. On the first of each month thereafter he deposited $5 more than the previous month's deposit. Find the total amount of money he deposited after 20 months.

16. The number of bacteria in a certain culture doubles every 30 minutes. If 50 are present at noon, how many will be present at 4:30 P.M.?

17. Use mathematical induction to prove that
$$8 + 14 + 20 + 26 + \ldots + (6n + 2) = 3n^2 + 5n$$
for every positive integer n.

Find each of the following.

18. $P(11, 3)$
19. $P(7, 7)$
20. $\dbinom{10}{2}$
21. $\dbinom{12}{0}$

22. A clothing manufacturer makes women's coats in four different styles. Each coat can be made from one of three fabrics. Each fabric comes in five different colors. How many different coats can be made?

23. A club with 30 members is to elect a president, secretary, and treasurer from its membership. If a member can hold at most one position, in how many ways can the offices be filled?

A card is drawn from a standard deck of 52 cards. Find the probability that each of the following is drawn.

24. A red three

25. A card that is not a face card

26. A king or a spade

A sample of 4 transistors is chosen. The probability of exactly 0, 1, 2, 3 or 4 transistors being defective is given in the following table.

Number defective	0	1	2	3	4
Probability	.19	.43	.30	.07	.01

Find the probability that the following numbers of transistors are defective.

27. At most 2
28. More than 1

29. Use the binomial theorem to expand $(2x - 3y)^4$.

30. Find the middle term in the expansion of $(w - 2y)^6$.

Appendices

Before small calculators became readily available, computations involving multiplication, division, powers, and roots were performed by using the properties of logarithms and tables of base 10 logarithms. Now many of those problems can be worked on a calculator. It still is not possible to calculate numbers like $1.4^{17.3}$ and $12.8^{.004}$ with nonscientific calculators, however. This section shows how to do such calculations with logarithms.

As mentioned earlier, logarithms to base 10, called common logarithms, are most convenient for numerical calculations. With $\log_{10} x$ abbreviated as $\log x$,

$$\log 1000 = \log 10^3 = 3$$
$$\log 100 = \log 10^2 = 2$$
$$\log 10 = \log 10^1 = 1$$
$$\log 1 = \log 10^0 = 0$$
$$\log .1 = \log 10^{-1} = -1$$
$$\log .01 = \log 10^{-2} = -2$$
$$\log .001 = \log 10^{-3} = -3,$$

and so on. A calculator with a log key or a table of logarithms is needed to find logarithms of numbers other than powers of 10. For example, to find log 27.31 with a calculator, enter 27.31 and press the log key. (This may be a second function key on some calculators.) The result should be 1.4363217. Logarithms of numbers between 0 and 1 are negative numbers. For example, a calculator will give log 0.483 as $-.31605287$.

In this appendix we give a brief explanation of how to use the table of common logarithms (Table 3 in Appendix B). Since there is no rational number x such that $10^x = 6$, there is no rational number x such that $x = \log 6$. A decimal approximation can be found for log 6, however. The following excerpt from Table 3 shows that a decimal approximation for log 6 is given by .7782:

$$\log 6 \approx .7782$$

(or, equivalently, $10^{.7782} \approx 6$). Since most logarithms are approximations, it is common to replace $\approx$ with $=$ and write $\log 6 = .7782$.

	0	1	2	3	4	5	6	7	8	9
5.7	.7559	.7566	.7574	.7582	.7589	.7597	.7604	.7612	.7619	.7627
5.8	.7634	.7642	.7649	.7657	.7664	.7672	.7679	.7686	.7694	.7701
5.9	.7709	.7716	.7723	.7731	.7738	.7745	.7752	.7760	.7767	.7774
6.0	.7782	.7789	.7796	.7803	.7810	.7818	.7825	.7832	.7839	.7846
6.1	.7853	.7860	.7868	.7875	.7882	.7889	.7896	.7903	.7910	.7917
6.2	.7924	.7931	.7938	.7945	.7952	.7959	.7966	.7973	.7980	.7987
6.3	.7993	.8000	.8007	.8014	.8021	.8028	.8035	.8041	.8048	.8055
6.4	.8062	.8069	.8075	.8082	.8089	.8096	.8102	.8109	.8116	.8122
6.5	.8129	.8136	.8142	.8149	.8156	.8162	.8169	.8176	.8182	.8189
6.6	.8195	.8202	.8209	.8215	.8222	.8228	.8235	.8241	.8248	.8254

To find log 6240 using the table, start with

$$1000 < 6240 < 10,000,$$

so that log 1000 < log 6240 < log 10,000. Since log 1000 = 3 and log 10,000 = 4,

$$3 < \log 6240 < 4.$$

Therefore, log 6240 equals 3 plus some decimal. This decimal part of the logarithm is found in the table. The decimal part of a logarithm is called the **mantissa** and the integer part, the **characteristic.** The entries in the table of logarithms are all positive, so the mantissa must always be positive. The characteristic may be any integer, however.

To determine the characteristic, first write the number in scientific notation. For example, for log 6240, write 6240 as 6.24×10^3. Then, by the properties of logarithms discussed in Chapter 6,

$$\begin{aligned} \log 6240 &= \log (6.24 \times 10^3) \\ &= \log 6.24 + \log 10^3 \\ &= \log 6.24 + 3 \log 10 \\ &= \log 6.24 + 3. \end{aligned}$$

To find the decimal part of log 6240 in the table, locate the first two significant figures, 62, in the left column of the table. Next, find the third digit, 4, across the top. The intersection of this row and column gives the mantissa, .7952. Since the characteristic is 3,

$$\log 6240 = 3.7952.$$

Example 1   Find log .00587.

Use scientific notation and the properties of logarithms to get

$$\begin{aligned} \log .00587 &= \log (5.87 \times 10^{-3}) \\ &= \log 5.87 + \log 10^{-3} \\ &= .7686 + (-3) \\ &= .7686 - 3. \end{aligned}$$

The logarithm is usually left in this form. A calculator would give the answer as -2.2314, the algebraic sum of .7686 and -3. The decimal portion in the calculator answer is a negative number. This is not the best form for the logarithm when using tables, since it is not clear which number is the mantissa. The best choice depends on the anticipated use of the logarithm.   ❖

In Example 1, the number .00587 whose logarithm is .7686 $-$ 3 is called the **antilogarithm** (abbreviated *antilog*) of .7686 $-$ 3.

Example 2   Find the antilogarithm of the following logarithms.

(a) 1.8075

Look in Table 3 for a mantissa of .8075. The number with a mantissa of .8075 is 6.42. Since the characteristic is 1, the antilogarithm of 1.8075 is $6.42 \times 10^1 =$ 64.2.

With a calculator, the antilogarithm N is found using the fact that

$$\log_{10} N = 1.8075,$$

so
$$N = 10^{1.8075}.$$

Enter 1.8075, then press the 10^x key. On some calculators, antilogarithms are found using INV and log keys instead.

(b) -1.2381

To find the mantissa, which *must* be a positive number, add and subtract 2 as follows.

$$-1.2381 + 2 - 2 - .7619 - 2$$

This form shows that the mantissa is .7619 and the characteristic is -2. From Table 3, the number with a mantissa of .7619 is 5.78, so the antilogarithm is

$$5.78 \times 10^{-2} = .0578.$$

Verify that log .0578 $= -1.2381$.   ❖

The following examples show how to use Table 3 and the properties of logarithms to do some numerical calculations.

Example 3   Calculate each of the following using logarithms.

(a) $\dfrac{93.1}{2.69}$

By a property of logarithms,

$$\log \frac{93.1}{2.69} = \log 93.1 - \log 2.69$$
$$= 1.9689 - .4298$$
$$= 1.5391.$$

From Table 3, the number whose logarithm has a mantissa closest to .5391 is 3.46, and

$$\frac{93.1}{2.69} \approx 3.46 \times 10^1 = 34.6.$$

(b) $(2.73)^4$

Use a property of logarithms to get

$$\log(2.73)^4 = 4 \log 2.73$$
$$= 4(.4362)$$
$$= 1.7448.$$

From the table or a calculator, the antilogarithm of 1.7448 is 55.6, so

$$(2.73)^4 \approx 55.6.$$

(c) $\sqrt[3]{.0762}$

$$\log \sqrt[3]{.0762} = \log (.0762)^{1/3}$$
$$= \frac{1}{3} \log .0762$$
$$= \frac{1}{3}(.8820 - 2)$$

To preserve the characteristic as an integer, change the characteristic to a multiple of 3 before multiplying by 1/3. One way to do this is to add and subtract 1 (which adds 0 to the characteristic) as follows.

$$.8820 - 2 = 1 - 1 + .8820 - 2$$
$$= 1.8820 - 3$$

Now complete the work above.

$$\log \sqrt[3]{.0762} = \frac{1}{3}(1.8820 - 3)$$
$$= .6273 - 1$$

The antilogarithm is .424, and

$$\sqrt[3]{.0762} \approx .424 \quad \clubsuit$$

Interpolation   Table 3 gives logarithms to four decimal places of numbers with three significant digits. **Linear interpolation** is used to find logarithms of numbers with four significant digits, as shown in the next example.

Example 4   Find log 75.37.
Since 75.3 < 75.37 < 75.4,

$$\log 75.3 < \log 75.37 < \log 75.4$$

Find log 75.3 and log 75.4 in the table.

$$10 \left\{ 7 \left\{ \begin{array}{l} \log 75.3 \;\; = 1.8768 \\ \log 75.37 = \\ \log 75.4 \;\; = 1.8774 \end{array} \right\} .0006 \right.$$

Since 75.37 is 7/10 of the way between 75.3 and 75.4, take 7/10 of .0006, the difference between the two logarithms.

$$\frac{7}{10}(.0006) \approx .0004$$

Add .0004 to log 75.3 to get

$$\log 75.37 = 1.8768 + .0004 = 1.8772. \quad \clubsuit$$

Interpolation also can be used when finding an antilogarithm, as shown in the next example.

Example 5   Find *x* such that log *x* = .3275.
The mantissas closest to .3275 are .3263 and .3284. Get the antilogarithms of these mantissas.

$$.0021 \left\{ .0012 \left\{ \begin{array}{l} .3263 = \log 2.12 \\ .3275 = \\ .3284 = \log 2.13 \end{array} \right\} .01 \right.$$

$$\frac{.0012}{.0021}(.01) \approx .006$$

Add .006 to 2.12, getting

$$.3275 = \log 2.126,$$

with $x = 2.126$. ✛

Appendix A Exercises

Find the characteristic of the logarithm of each of the following.

1. 875   **2.** 9462   **3.** 2,400,000   **4.** 875,000
5. .00023   **6.** .098   **7.** .000042   **8.** .000000257

Find the logarithms of the following numbers. Use interpolation if necessary. See Examples 1 and 4.

9. 875   **10.** 3,750   **11.** 12,800   **12.** 653,000
13. 7.63   **14.** 9.37   **15.** .000893   **16.** .00376
17. 3.876   **18.** 2.975   **19.** 68,250   **20.** 103,200
21. .0003824   **22.** .00008632

Find the antilogarithms of each of the following logarithms. See Examples 2 and 5.

23. 1.5366   **24.** 2.9253   **25.** -1.1267 **26.** $-.7496$
27. 3.4947   **28.** 4.6863   **29.** $.8039 - 3$ **30.** $.8078 - 2$

Use logarithms (and interpolation) to find approximations to three-digit accuracy (four digits in Exercises 41, 42, 45, and 46) for each of the following. See Example 3.

31. $(79.6)(4.83)$ **32.** $(123)(7.89)$ **33.** $\dfrac{9.03}{43.8}$ **34.** $\dfrac{2.81}{.436}$

35. $(8.16)^{1/3}$ **36.** $\sqrt{276}$ **37.** $(.463)^{.4}$ **38.** π^{π} (Use $\pi = 3.14$)

39. $\dfrac{(21.3)^2}{(.86)^3}$ **40.** $\dfrac{(.893)^2}{(.624)^3}$ **41.** $\dfrac{(38.47)(258.6)}{\sqrt{124.8}}$ **42.** $\dfrac{(968.4)\sqrt[3]{25.68}}{29.83}$

43. $\dfrac{(2.51)^2}{(\sqrt{1.52})(3.94)}$ **44.** $\dfrac{(7.06)^3}{(31.7)(\sqrt{1.09})}$

45. $(49.83)^{1/2} + (2.917)^2$ **46.** $(38.42)^{1/3} + (86.13)^{1/4}$

Use common logarithms and Table 3 to find each of the following logarithms. Round to the nearest hundredth. Use the change of base theorem from Section 6.3.

47. $\ln 125$ **48.** $\ln 63.1$ **49.** $\ln .98$ **50.** $\ln 1.53$
51. $\ln 275$ **52.** $\ln 39.8$ **53.** $\log_2 10$ **54.** $\log_5 8$
55. $\log_4 12$ **56.** $\log_7 4$

Use common logarithms to solve the following equations. Round to the nearest thousandth.

57. $10^x = 2.5$ **58.** $10^{x-1} = 143$ **59.** $100^{2x+1} = 17$ **60.** $1000^{5x} = 2010$
61. $50^{x+2} = 8$ **62.** $200^{x+3} = 12$ **63.** $.01^{3x} = .005$ **64.** $.001^{2x} = .0004$

Appendix B: Tables

Table 1   Powers and Roots

n	n^2	n^3	$\sqrt{n}$	$\sqrt{10n}$	$\sqrt[3]{n}$	n	n^2	n^3	$\sqrt{n}$	$\sqrt{10n}$	$\sqrt[3]{n}$
1	1	1	1.000	3.162	1.000	51	2601	132,651	7.141	22.583	3.708
2	4	8	1.414	4.472	1.260	52	2704	140,608	7.211	22.804	3.733
3	9	27	1.732	5.477	1.442	53	2809	148,877	7.280	23.022	3.756
4	16	64	2.000	6.325	1.587	54	2916	157,464	7.348	23.238	3.780
5	25	125	2.236	7.071	1.710	55	3025	166,375	7.416	23.452	3.803
6	36	216	2.449	7.746	1.817	56	3136	175,616	7.483	23.664	3.826
7	49	343	2.646	8.367	1.913	57	3249	185,193	7.550	23.875	3.849
8	64	512	2.828	8.944	2.000	58	3364	195,112	7.616	24.083	3.871
9	81	729	3.000	9.487	2.080	59	3481	205,379	7.681	24.290	3.893
10	100	1000	3.162	10.000	2.154	60	3600	216,000	7.746	24.495	3.915
11	121	1331	3.317	10.488	2.224	61	3721	226,981	7.810	24.698	3.936
12	144	1728	3.464	10.954	2.289	62	3844	238,328	7.874	24.900	3.958
13	169	2197	3.606	11.402	2.351	63	3969	250,047	7.937	25.100	3.979
14	196	2744	3.742	11.832	2.410	64	4096	262,144	8.000	25.298	4.000
15	225	3375	3.873	12.247	2.466	65	4225	274,625	8.062	25.495	4.021
16	256	4096	4.000	12.649	2.520	66	4356	287,496	8.124	25.690	4.041
17	289	4913	4.123	13.038	2.571	67	4489	300,763	8.185	25.884	4.062
18	324	5832	4.243	13.416	2.621	68	4624	314,432	8.246	26.077	4.082
19	361	6859	4.359	13.784	2.688	69	4761	328,509	8.307	26.268	4.102
20	400	8000	4.472	14.142	2.714	70	4900	343,000	8.367	26.458	4.121
21	441	9261	4.583	14.491	2.759	71	5041	357,911	8.426	26.646	4.141
22	484	10,648	4.690	14.832	2.802	72	5184	373,248	8.485	26.833	4.160
23	529	12,167	4.796	15.166	2.844	73	5329	389,017	8.544	27.019	4.179
24	576	13,824	4.899	15.492	2.884	74	5476	405,224	8.602	27.203	4.198
25	625	15,625	5.000	15.811	2.924	75	5625	421,875	8.660	27.386	4.217
26	676	17,576	5.099	16.125	2.962	76	5776	438,976	8.718	27.568	4.236
27	729	19,683	5.196	16.432	3.000	77	5929	456,533	8.775	27.749	4.254
28	784	21,952	5.292	16.733	3.037	78	6084	474,552	8.832	27.928	4.273
29	841	24,389	5.385	17.029	3.072	79	6241	493,039	8.888	28.107	4.291
30	900	27,000	5.477	17.321	3.107	80	6400	512,000	8.944	28.284	4.309
31	961	29,791	5.568	17.607	3.141	81	6561	531,441	9.000	28.460	4.327
32	1024	32,768	5.657	17.889	3.175	82	6724	551,368	9.055	28.636	4.344
33	1089	35,937	5.745	18.166	3.208	83	6889	571,787	9.110	28.810	4.362
34	1156	39,304	5.831	18.439	3.240	84	7056	592,704	9.165	28.983	4.380
35	1225	42,875	5.916	18.708	3.271	85	7225	614,125	9.220	29.155	4.397
36	1296	46,656	6.000	18.974	3.302	86	7396	636,056	9.274	29.326	4.414
37	1369	50,653	6.083	20.235	3.332	87	7569	658,503	9.327	29.496	4.431
38	1444	54,872	6.164	19.494	3.362	88	7744	681,472	9.381	29.665	4.448
39	1521	59,319	6.245	19.748	3.391	89	7921	704,969	9.434	29.833	4.465
40	1600	64,000	6.325	20.000	3.420	90	8100	729,000	9.487	30.000	4.481
41	1681	68,921	6.403	20.248	3.448	91	8281	753,571	9.539	30.166	4.498
42	1764	74,088	6.481	20.494	3.476	92	8464	778,688	9.592	30.332	4.514
43	1849	79,507	6.557	20.736	3.503	93	8649	804,357	9.644	30.496	4.531
44	1936	85,184	6.633	20.976	3.530	94	8836	830,584	9.695	30.659	4.547
45	2025	91,125	6.708	21.213	3.557	95	9025	857,375	9.747	30.822	4.563
46	2116	97,336	6.782	21.448	3.583	96	9216	884,736	9.798	30.984	4.579
47	2209	103,823	6.856	21.679	3.609	97	9409	912,673	9.849	31.145	4.595
48	2304	110,592	6.928	21.909	3.634	98	9604	941,192	9.899	31.305	4.610
49	2401	117,649	7.000	22.136	3.659	99	9801	970,299	9.950	31.464	4.626
50	2500	125,000	7.071	22.361	3.684	100	10,000	1,000,000	10.000	31.623	4.642

Table 2 Powers of e and Natural Logarithms

x	e	e^{-x}	ln x	x	e	e^{-x}	ln x
.00	1.00000	1.00000		1.0	2.71828	.36787	.0000
.01	1.01005	.99004	− 4.6052	1.1	3.00416	.33287	.0953
.02	1.02020	.98019	− 3.9120	1.2	3.32011	.30119	.1823
.03	1.03045	.97044	− 3.5066	1.3	3.66929	.27253	.2624
.04	1.04081	.96078	− 3.2189	1.4	4.05519	.24659	.3365
.05	1.05127	.95122	− 2.9957	1.5	4.48168	.22313	.4055
.06	1.06183	.94176	− 2.8134	1.6	4.95302	.20189	.4700
.07	1.07250	.93239	− 2.6593	1.7	5.47394	.18268	.5306
.08	1.08328	.92311	− 2.5257	1.8	6.04964	.16529	.5878
.09	1.09417	.91393	− 2.4079	1.9	6.68589	.14956	.6419
.10	1.10517	.90483	− 2.3026	2.0	7.38905	.13533	.6931
.11	1.11628	.89583	− 2.2073	2.1	8.16616	.12245	.7419
.12	1.12750	.88692	− 2.1203	2.2	9.02500	.11080	.7885
.13	1.13883	.87810	− 2.0402	2.3	9.97417	.10025	.8329
.14	1.15027	.86936	− 1.9661	2.4	11.0231	.09071	.8755
.15	1.16183	.86071	− 1.8971	2.5	12.1824	.08208	.9163
.16	1.17351	.85214	− 1.8326	2.6	13.4637	.07427	.9555
.17	1.18530	.84366	− 1.7720	2.7	14.8797	.06720	.9933
.18	1.19722	.83527	− 1.7148	2.8	16.4446	.06081	1.0296
.19	1.20925	.82696	− 1.6607	2.9	18.1741	.05502	1.0647
.2	1.22140	.81873	− 1.6094	3.0	20.0855	.04978	1.0986
.3	1.34985	.74081	− 1.2040	3.1	22.1979	.04505	1.1314
.4	1.49182	.67031	− .9163	3.2	24.5325	.04076	1.1632
.5	1.64872	.60653	− .6931	3.3	27.1126	.03688	1.1939
.6	1.82211	.54881	− .5108	3.4	29.9641	.03337	1.2238
.7	2.01375	.49658	− .3567	3.5	33.1154	.03020	1.2528
.8	2.22554	.44932	− .2231	3.6	36.5982	.02732	1.2809
.9	2.45960	.40656	− .1054	3.7	40.4473	.02472	1.3083
				3.8	44.7012	.02237	1.3350
				3.9	49.4024	.02024	1.3610

Table 2   Powers of e and Natural Logarithms (continued)

x	e	e^{-x}	ln x	x	e	e^{-x}	ln x
4.0	54.5981	.01832	1.3863	7.0	1096.63	.00091	1.9459
4.1	60.3402	.01657	1.4110	7.1	1211.96	.00083	1.9601
4.2	66.6863	.01500	1.4351	7.2	1339.43	.00075	1.9741
4.3	73.6997	.01357	1.4586	7.3	1480.29	.00068	1.9879
4.4	81.4508	.01228	1.4816	7.4	1635.98	.00061	2.0015
4.5	90.0170	.01111	1.5041	7.5	1808.03	.00055	2.0149
4.6	99.4842	.01005	1.5261	7.6	1998.19	.00050	2.0281
4.7	109.947	.00910	1.5476	7.7	2208.34	.00045	2.0412
4.8	121.510	.00823	1.5686	7.8	2440.59	.00041	2.0541
4.9	134.290	.00745	1.5892	7.9	2697.27	.00037	2.0669
5.0	148.413	.00674	1.6094	8.0	2980.94	.00034	2.0794
5.1	164.022	.00610	1.6292	8.1	3294.45	.00030	2.0919
5.2	181.272	.00552	1.6487	8.2	3640.94	.00027	2.1041
5.3	200.336	.00499	1.6677	8.3	4023.86	.00025	2.1163
5.4	221.406	.00452	1.6864	8.4	4447.05	.00022	2.1282
5.5	244.691	.00445	1.7047	8.5	4914.75	.00020	2.1401
5.6	270.426	.00370	1.7228	8.6	5431.65	.00018	2.1518
5.7	298.867	.00335	1.7405	8.7	6002.90	.00017	2.1633
5.8	330.299	.00303	1.7579	8.8	6634.23	.00015	2.1748
5.9	365.036	.00274	1.7750	8.9	7331.96	.00014	2.1861
6.0	403.428	.00248	1.7918	9.0	8103.08	.00012	2.1972
6.1	445.856	.00224	1.8083	9.1	8955.29	.00011	2.2083
6.2	492.748	.00203	1.8245	9.2	9897.13	.00010	2.2192
6.3	544.570	.00184	1.8405	9.3	10938.0	.00009	2.2300
6.4	601.843	.00166	1.8563	9.4	12088.4	.00008	2.2407
6.5	665.139	.00150	1.8718	9.5	13359.7	.000075	2.2513
6.6	735.093	.00136	1.8871	9.6	14764.8	.0000768	2.2618
6.7	812.403	.00123	1.9021	9.7	16317.6	.0000661	2.2721
6.8	897.844	.00111	1.9169	9.8	18033.8	.0000555	2.2824
6.9	992.271	.00101	1.9315	9.9	19930.4	.0000550	2.2925
				10.0	22026.5	.000045	2.3026

Table 3   Common Logarithms

n	0	1	2	3	4	5	6	7	8	9
1.0	.0000	.0043	.0086	.0128	.0170	.0212	.0253	.0294	.0334	.0374
1.1	.0414	.0453	.0492	.0531	.0569	.0607	.0645	.0682	.0719	.0755
1.2	.0792	.0828	.0864	.0899	.0934	.0969	.1004	.1038	.1072	.1106
1.3	.1139	.1173	.1206	.1239	.1271	.1303	.1335	.1367	.1399	.1430
1.4	.1461	.1492	.1523	.1553	.1584	.1614	.1644	.1673	.1703	.1732
1.5	.1761	.1790	.1818	.1847	.1875	.1903	.1931	.1959	.1987	.2014
1.6	.2041	.2068	.2095	.2122	.2148	.2175	.2201	.2227	.2253	.2279
1.7	.2304	.2330	.2355	.2380	.2405	.2430	.2455	.2480	.2504	.2529
1.8	.2553	.2577	.2601	.2625	.2648	.2672	.2695	.2718	.2742	.2765
1.9	.2788	.2810	.2833	.2856	.2878	.2900	.2923	.2945	.2967	.2989
2.0	.3010	.3032	.3054	.3075	.3096	.3118	.3139	.3160	.3181	.3201
2.1	.3222	.3243	.3263	.3284	.3304	.3324	.3345	.3365	.3385	.3404
2.2	.3424	.3444	.3464	.3483	.3502	.3522	.3541	.3560	.3579	.3598
2.3	.3617	.3636	.3655	.3674	.3692	.3711	.3729	.3747	.3766	.3784
2.4	.3802	.3820	.3838	.3856	.3874	.3892	.3909	.3927	.3945	.3962
2.5	.3979	.3997	.4014	.4031	.4048	.4065	.4082	.4099	.4116	.4133
2.6	.4150	.4166	.4183	.4200	.4216	.4232	.4249	.4265	.4281	.4298
2.7	.4314	.4330	.4346	.4362	.4378	.4393	.4409	.4425	.4440	.4456
2.8	.4472	.4487	.4502	.4518	.4533	.4548	.4564	.4579	.4594	.4609
2.9	.4624	.4639	.4654	.4669	.4683	.4698	.4713	.4728	.4742	.4757
3.0	.4771	.4786	.4800	.4814	.4829	.4843	.4857	.4871	.4886	.4900
3.1	.4914	.4928	.4942	.4955	.4969	.4983	.4997	.5011	.5024	.5038
3.2	.5051	.5065	.5079	.5092	.5105	.5119	.5132	.5145	.5159	.5172
3.3	.5185	.5198	.5211	.5224	.5237	.5250	.5263	.5276	.5289	.5302
3.4	.5315	.5328	.5340	.5353	.5366	.5378	.5391	.5403	.5416	.5428
3.5	.5441	.5453	.5465	.5478	.5490	.5502	.5514	.5527	.5539	.5551
3.6	.5563	.5575	.5587	.5599	.5611	.5623	.5635	.5647	.5658	.5670
3.7	.5682	.5694	.5705	.5717	.5729	.5740	.5752	.5763	.5775	.5786
3.8	.5798	.5809	.5821	.5832	.5843	.5855	.5866	.5877	.5888	.5899
3.9	.5911	.5922	.5933	.5944	.5955	.5966	.5977	.5988	.5999	.6010
4.0	.6021	.6031	.6042	.6053	.6064	.6075	.6085	.6096	.6107	.6117
4.1	.6128	.6138	.6149	.6160	.6170	.6180	.6191	.6201	.6212	.6222
4.2	.6232	.6243	.6253	.6263	.6274	.6284	.6294	.6304	.6314	.6325
4.3	.6335	.6345	.6355	.6365	.6375	.6385	.6395	.6405	.6415	.6425
4.4	.6435	.6444	.6454	.6464	.6474	.6484	.6493	.6503	.6513	.6522
4.5	.6532	.6542	.6551	.6561	.6571	.6580	.6590	.6599	.6609	.6618
4.6	.6628	.6637	.6646	.6656	.6665	.6675	.6684	.6693	.6702	.6712
4.7	.6721	.6730	.6739	.6749	.6758	.6767	.6776	.6785	.6794	.6803
4.8	.6812	.6821	.6830	.6839	.6848	.6857	.6866	.6875	.6884	.6893
4.9	.6902	.6911	.6920	.6928	.6937	.6946	.6955	.6964	.6972	.6981
5.0	.6990	.6998	.7007	.7016	.7024	.7033	.7042	.7050	.7059	.7067
5.1	.7076	.7084	.7093	.7101	.7110	.7118	.7126	.7135	.7143	.7152
5.2	.7160	.7168	.7177	.7185	.7193	.7202	.7210	.7218	.7226	.7235
5.3	.7243	.7251	.7259	.7267	.7275	.7284	.7292	.7300	.7308	.7316
5.4	.7324	.7332	.7340	.7348	.7356	.7364	.7372	.7380	.7388	.7396

Table 3   Common Logarithms (continued)

n	0	1	2	3	4	5	6	7	8	9
5.5	.7404	.7412	.7419	.7427	.7435	.7443	.7451	.7459	.7466	.7474
5.6	.7482	.7490	.7497	.7505	.7513	.7520	.7528	.7536	.7543	.7551
5.7	.7559	.7566	.7574	.7582	.7589	.7597	.7604	.7612	.7619	.7627
5.8	.7634	.7642	.7649	.7657	.7664	.7672	.7679	.7686	.7694	.7701
5.9	.7709	.7716	.7723	.7731	.7738	.7745	.7752	.7760	.7767	.7774
6.0	.7782	.7789	.7796	.7803	.7810	.7818	.7825	.7832	.7839	.7846
6.1	.7853	.7860	.7868	.7875	.7882	.7889	.7896	.7903	.7910	.7917
6.2	.7924	.7931	.7938	.7945	.7952	.7959	.7966	.7973	.7980	.7987
6.3	.7993	.8000	.8007	.8014	.8021	.8028	.8035	.8041	.8048	.8055
6.4	.8062	.8069	.8075	.8082	.8089	.8096	.8102	.8109	.8116	.8122
6.5	.8129	.8136	.8142	.8149	.8156	.8162	.8169	.8176	.8182	.8189
6.6	.8195	.8202	.8209	.8215	.8222	.8228	.8235	.8241	.8248	.8254
6.7	.8261	8267	.8274	.8280	.8287	.8293	.8299	.8306	.8312	.8319
6.8	.8325	.8331	.8338	.8344	.8351	.8357	.8363	.8370	.8376	.8382
6.9	.8388	.8395	.8401	.8407	.8414	.8420	.8426	.8432	.8439	.8445
7.0	.8451	.8457	.8463	.8470	.8476	.8482	.8488	.8494	.8500	.8506
7.1	.8513	.8519	.8525	.8531	.8537	.8543	.8549	.8555	.8561	.8567
7.2	.8573	.8579	.8585	.8591	.8597	.8603	.8609	.8615	.8621	.8627
7.3	.8633	.8639	.8645	.8651	.8657	.8663	.8669	.8675	.8681	.8686
7.4	.8692	.8698	.8704	.8710	.8716	.8722	.8727	.8733	.8739	.8745
7.5	.8751	.8756	.8762	.8768	.8774	.8779	.8785	.8791	.8797	.8802
7.6	.8808	.8814	.8820	.8825	.8831	.8837	.8842	.8848	.8854	.8859
7.7	.8865	.8871	.8876	.8882	.8887	.8893	.8899	.8904	.8910	.8915
7.8	.8921	.8927	.8932	.8938	.8943	.8949	.8954	.8960	.8965	.8971
7.9	.8976	.8982	.8987	.8993	.8998	.9004	.9009	.9015	.9020	.9025
8.0	.9031	.9036	.9042	.9047	.9053	.9058	.9063	.9069	.9074	.9079
8.1	.9085	.9090	.9096	.9101	.9106	.9112	.9117	.9122	.9128	.9133
8.2	.9138	.9143	.9149	.9154	.9159	.9165	.9170	.9175	.9180	.9186
8.3	.9191	.9196	.9201	.9206	.9212	.9217	.9222	.9227	.9232	.9238
8.4	.9243	.9248	.9253	.9258	.9263	.9269	.9274	.9279	.9284	.9289
8.5	.9294	.9299	.9304	.9309	.9315	.9320	.9325	.9330	.9335	.9340
8.6	.9345	.9350	.9355	.9360	.9365	.9370	.9375	.9380	.9385	.9390
8.7	.9395	.9400	.9405	.9410	.9415	.9420	.9425	.9430	.9435	.9440
8.8	.9445	.9450	.9455	.9460	.9465	.9469	.9474	.9479	.9484	.9489
8.9	.9494	.9499	.9504	.9509	.9513	.9518	.9523	.9528	.9533	.9538
9.0	.9542	.9547	.9552	.9557	.9562	.9566	.9571	.9576	.9581	.9586
9.1	.9590	.9595	.9600	.9605	.9609	.9614	.9619	.9624	.9628	.9633
9.2	.9638	.9643	.9647	.9652	.9657	.9661	.9666	.9671	.9675	.9680
9.3	.9685	.9689	.9694	.9699	.9703	.9708	.9713	.9717	.9722	.9727
9.4	.9731	.9736	.9741	.9745	.9750	.9754	.9759	.9763	.9768	.9773
9.5	.9777	.9782	.9786	.9791	.9795	.9800	.9805	.9809	.9814	.9818
9.6	.9823	.9827	.9832	.9836	.9841	.9845	.9850	.9854	.9859	.9863
9.7	.9868	.9872	.9877	.9881	.9886	.9890	.9894	.9899	.9903	.9908
9.8	.9912	.9917	.9921	.9926	.9930	.9934	.9939	.9943	.9948	.9952
9.9	.9956	.9961	.9965	.9969	.9974	.9978	.9983	.9987	.9991	.9996

Appendix C: Symbols Used in This Book

$\{a, b\}$	set containing the elements a and b
$\emptyset$	empty set
$\{x \mid x \text{ has property } P\}$	set of all elements x such that x has property P
$=$	equals
$\neq$	is not equal to
$\lvert x \rvert$	absolute value of x
$\cap$	set intersection
$\cup$	set union
$<$	is less than
$\leq$	is less than or equal to
$>$	is greater than
$\geq$	is greater than or equal to
$\nless$	is not less than
$\ngtr$	is not greater than
a^m	a to the power m
a^{-n}	$1/a^n$
$\sqrt[n]{a}$	nth root of a
$a^{1/n}$	$\sqrt[n]{a}$
$\sqrt{a}$	square root of a
$\approx$	is approximately equal to
$f(x)$	f of x
i	$\sqrt{-1}$
$a + bi$	complex number
$\pm$	plus or minus
(a, b)	ordered pair of numbers
m	slope

f^{-1}	inverse of function f
$f \circ g$	composite function
$\begin{vmatrix} a & b \\ c & d \end{vmatrix}$	two-by-two determinant
$\begin{bmatrix} a & b & c \\ d & e & f \end{bmatrix}$	matrix of two rows, three columns
$\log_a x$	logarithm of x to the base a
$\log x$	common, or base 10, logarithm of x
$\ln x$	natural (base e) logarithm of x
e	a number, approximately 2.7182818
a_n	nth term of a sequence
$\displaystyle\sum_{i=1}^{n} x_i$	summation
$n \rightarrow +\infty$	n increases without bound
$\displaystyle\lim_{x \to a} f(x)$	limit of $f(x)$ as x approaches a
S_n	sum of the first n terms of a sequence
$n!$	n factorial
$P(n, r)$	number of permutations of n elements r at a time
$\dbinom{n}{r}$	number of combinations of n elements r at a time
$n(E)$	number of elements in set E
$P(E)$	probability of event E
E'	complement of set E

Answers to Selected Exercises

CHAPTER 1

Section 1.1 (page 12)

1. 1, 3 **3.** -6, $-12/4$ (or -3), 0, 1, 3 **5.** $-\sqrt{3}$, 2π, $\sqrt{12}$ **7.** Natural, whole, integer, rational, real **9.** Rational, real **11.** Irrational, real **13.** 81 **15.** -243 **17.** $(-2)^7$ **19.** $8m^7$ **21.** $\dfrac{1}{a^4}$ **23.** $9k^3$ **25.** 5 **27.** -148 **29.** -6 **31.** -60 **33.** -12 **35.** $-25/36$ **37.** $-6/7$ **39.** 2.22 **41.** -54 **43.** $-1/2$ **45.** 2 **47.** -2.463 **49.** $-.979$ **51.** Commutative **53.** Associative **55.** Distributive **57.** Inverse **59.** Identity **61.** Distributive **63.** $6x + 6y$ **65.** $-3z + 3y$ **67.** $-8r + k$ **69.** $ar + as - at$ **71.** $36m$ **73.** $20y$ **75.** $-9r$ **77.** $8y - 4z + 12q$ **79.** $2y/3 + 4z/9 - 5/3$ **81.** For example, $11 - 2 \neq 2 - 11$. **83.** For example, $(12 - 3) - 1 \neq 12 - (3 - 1)$. **85.** No; for example, $6 + (5 \cdot 4) \neq (6 + 5) \cdot (6 + 4)$. **87.** (a) Multiply both sides of $a^2/b^2 = 2$ by b^2. (b) $2b^2$ is a multiple of 2. (c) a^2 is even because a^2 equals the even number $2b^2$; since the square of an odd number is odd, a must be even. (d) Replace a with $2c$. (e) $(2c)^2 = 4c^2$; divide both sides by 2. (f) $2c^2$ is a multiple of 2. (g) If $2c^2$ is even, b^2 is even, since $b^2 = 2c^2$. (h) We have both a and b even, but we claimed that the fraction a/b was in lowest terms.

Section 1.2 (page 20)

1. -9, -4, -2, 3, 8 **3.** $-|9|$, $-|-6|$, $|-8|$ **5.** -5, -4, -2, $-\sqrt{3}$, $\sqrt{6}$, $\sqrt{8}$, 3 **7.** $3/4$, $7/5$, $\sqrt{2}$, $22/15$, $8/5$ **9.** $-|-8| - |-6|$, $-|-2| + (-3)$, $-|3|$, $-|-2|$, $|-8 + 2|$ **11.** $-6 < 15$ **13.** $30 > 20$ **15.** $1 \leq 2$ **17.** $x < 3$ **19.** $k < -7$ **21.** $k > 25$ **23.** 8 **25.** 6 **27.** 4 **29.** 16 **31.** 8 **33.** -16 **35.** 6 **37.** -6 **39.** $8 - \sqrt{50}$ **41.** $5 - \sqrt{7}$ **43.** $\pi - 3$ **45.** $x - 4$ **47.** $8 - 2k$ **49.** $56 - 7m$ **51.** $8 + 4m$ or $4m + 8$ **53.** $y - x$ **55.** $3 + x^2$ **57.** $1 + p^2$ **59.** First part of multiplication property **61.** Second part of multiplication property **63.** First part of multiplication property **65.** Triangle inequality, $|a + b| \leq |a| + |b|$ **67.** Property of absolute value, $|a| \cdot |b| = |ab|$ **69.** Property of absolute value, $|a/b| = |a|/|b|$ **71.** If $x = y$ or $x = -y$ **73.** If $y = 0$ or if $|x| \geq |y|$ and x and y have opposite signs **75.** If $|x| \geq |y|$ and either both positive or both negative **77.** -1 if $x < 0$ and 1 if $x > 0$

Section 1.3 (page 27)

1. $-1/64$ **3.** 8 **5.** $x^4/625$ **7.** 1 **9.** 1 **11.** 2^{10} or 1024 **13.** $x^{15}y^{12}$ **15.** p^8/q^2 **17.** z^9/y^9 **19.** $2^3a^9/b^9$ or $8a^9/b^9$ **21.** r^6/s^{15} **23.** $1/(ab^3)$ **25.** $x^6/(2^2y^4)$ or $x^6/(4y^4)$ **27.** $1/6^4$ or $1/1296$ **29.** m **31.** $(3m)/n$ **33.** $216m^9$ **35.** $x^2/(2y)$ **37.** $(4 + s)^6$ **39.** $x/2$ **41.** $1(r^{10}s^8t^{12})$ **43.** $k^2/(p + q)^7$ **45.** 0 **47.** $73/8$ **49.** 1 **51.** $1/6$ **53.** $-13/66$ **55.** $35/18$ **57.** 6.93×10^4

59. 1.27×10^7 **61.** 1×10^{-4} **63.** 5.4×10^{-2} **65.** 820,000   **67.** 298   **69.** .0361
71. .00615   **73.** 4.232   **75.** 1.8×10^{-9} **77.** 7.025×10^4 or 70,250   **79.** 4.86×10^7
81. 1.29×10^{-4} **83.** -6.04×10^{11} **85.** 3.2×10^4 hr, or about 3 yr and 8 mo
87. About 8.33 min   **89.** About 1.25×10^{30} bacteria   **91.** $2k$ **93.** $30p^{3-r}$ **95.** k^{-7} or $1/k^7$
97. $1/(8b^y)$ **99.** $-2m^{2-n}$

Section 1.4 (page 35)
1. Polynomial; degree 11; monomial   **3.** Polynomial; degree 6; binomial   **5.** Polynomial; degree 6;
binomial   **7.** Polynomial; degree 6; trinomial   **9.** Not a polynomial   **11.** Not a polynomial
13. $x^2 - x + 3$ **15.** $9y^2 - 4y + 4$ **17.** $6a^2 + a - 1$ **19.** $3b^2 - 8b - 1$
21. $-14q^2 + 11q - 14$ **23.** $4p^2 - 16$ **25.** $11y^3 - 18y^2 + 4y$ **27.** $28r^2 + r - 2$
29. $15x^2 - (7/3)x - (2/9)$ **31.** $(6/25)y^2 + (11/40)yz + (1/16)z^2$ **33.** $25r^2 - 4$
35. $16m^2 + 16mn + 4n^2$ **37.** $4p^2 - 12p + 9 + 4pq - 6q + q^2$ **39.** $9q^2 + 30q + 25 - p^2$
41. $8z^3 - 12z^2 + 6z - 1$ **43.** $125r^3 + 225r^2t^2 + 135rt^4 + 27t^6$ **45.** $12x^5 + 8x^4 - 20x^3 + 4x^2$
47. $15m^4 - 10m^3 + 5m^2 - 5m$ **49.** $-2z^3 + 7z^2 - 11z + 4$ **51.** $8m^3 + 1$
53. $m^2 + mn - 2n^2 - 2km + 5kn - 3k^2$ **55.** $a^2 - 2ab + b^2 + 4ac - 4bc + 4c^2$
57. $2m^2 - 4m + 8$ **59.** $5y^2 - 3xy + 8x^2$ **61.** $3y + 2$ **63.** $5p - 1$
65. $4z - 3 + [-5/(2z + 5)]$ **67.** $x^2 - 3x + 2$ **69.** $5x^2 - 3x + 4 + 4/(3x + 4)$
71. $3m^3 - 5m^2 + 7m - 1 + 3/(m - 3)$ **73.** $x^2 + 2x + 3$ **75.** $2z^3 - z^2 - (1/2)z - 5/4 +$
$[(-13/4)z + 17/4]/(2z^2 + z + 1)$ **77.** $3x^2 - 4xy + y^2$ **79.** $y^2 - 2yz + z^2$
81. $18p^2 - 27pq - 35q^2$ **83.** $27x^3 - 108x^2y + 144xy^2 - 64y^3$ **85.** $36k^2 - 36k + 9$
87. $3x + 2 + [-2/(4x - 5)]$ **89.** $16x^2 - 9y^2$ **91.** $27p^3 - 1$ **93.** $p^3 - 7p^2 - p - 7$
95. $5x^3 + 10x^2 + 4x$ **97.** $2p^3 - 5p^2 - p + 1/3 + [(-7/3)p^2 - 7p + (7/3)]/(3p^3 + p^2 - 1)$
99. $3x^3 - 3x^2 + 6x + 2$ **101.** -8 **103.** 1   **105.** -24 **107.** $k^{2m} - 4$
109. $b^{2r} + b^r - 6$ **111.** $3p^{2x} - 5p^x - 2$ **113.** $m^{2x} - 4m^x + 4$ **115.** $q^{2p} - 10q^pp^q + 25p^{2q}$
117. $27k^{3a} - 54k^{2a} + 36k^a - 8$ **119.** m **121.** $m + n$ **123.** $x^2 + y^2 + z^2 + 2xy + 2xz + 2yz$

Section 1.5 (page 44)
1. $4m(3n - 2)$ **3.** $3r(4r^2 + 2r - 1)$ **5.** $4k^2m^3(1 + 2k^2 - 3m)$ **7.** $2(a + b)(1 + 2m)$
9. $(y + 2)(3y + 2)$ **11.** $(r + 3)(3r - 5)$ **13.** $(m - 1)(2m^2 - 7m + 7)$
15. $(q + 2)(r + 2s)$ **17.** $(2s + 3)(3t - 5)$ **19.** $(t^3 + s^2)(r - p)$ **21.** $(3p - 7)(2p + 5)$
23. $(5z - 2x)(4z - 9x)$ **25.** $(3 - m^2)(5 - r^2)$ **27.** $(y + 9)(y - 8)$ **29.** $(s + 7t)(s - 5t)$
31. $6(a - 10)(a + 2)$ **33.** $3m(m + 1)(m + 3)$ **35.** $3(2r - 1)(2r + 5)$
37. $(3k - 2p)(2k + 3p)$ **39.** $(5a + 3b)(a - 2b)$ **41.** $x^2(3 - x)^2$ **43.** $2a^2(4a - b)(3a + 2b)$
45. $(p + 10)(p - 10)$ **47.** $(3a + 4)(3a - 4)$ **49.** $(5s^2 + 3t)(5s^2 - 3t)$
51. $(xy + 5a)(xy - 5a)$ **53.** $(a + b + 4)(a + b - 4)$ **55.** $(p^2 + 25)(p + 5)(p - 5)$
57. $(3m - 2)^2$ **59.** $2(4a - 3b)^2$ **61.** $(2xy + 7)^2$ **63.** $(a - 3b - 3)^2$
65. $(2 - a)(4 + 2a + a^2)$ **67.** $(5x - 3)(25x^2 + 15x + 9)$ **69.** $(3y^3 + 5z^2)(9y^6 - 15y^3z^2 + 25z^4)$
71. $r(r^2 + 18r + 108)$ **73.** $(3 - m - 2n)(9 + 3m + 6n + m^2 + 4mn + 4n^2)$
75. $(m^2 - 5)(m^2 + 2)$ **77.** $9(7k - 3)(k + 1)$ **79.** $(3a - 7)^2$ **81.** $(7m + 2n)(3m + n)$
83. $(r + s)(a^3 + b^2)$ **85.** $(2y - 3)^2$ **87.** Prime   **89.** $(6a + 5)^2$
91. $(2 + x)(2 - x)(4 + 2x + x^2)(4 - 2x + x^2)$ **93.** $(3p^2 - 1)(2p^2 + 3)$
95. $(b + 4 + a)(b + 4 - a)$ **97.** $(q + 3 + p)(q + 3 - p)$
99. $(6p + 5q)(36p^2 - 30pq + 25q^2)$ **101.** $(4p - 1)(p + 1)$ **103.** Prime   **105.** $4xy$
107. $p^{-4}(1 - p^2)$ **109.** $4k^{-3}(3 + k - 2k^2) = 4k^{-3}(1 + k)(3 - 2k)$ **111.** $25p^{-6}(4 - 2p^4 + 3p^8)$
113. $(m + n^p)(m - 2n^p)$ **115.** $(3z^a + x^b)(2z^a - x^b)$ **117.** $(3a^{2k} + b^{4k})(3a^{2k} - b^{4k})$
119. $(2y^a - 3)^2$ **121.** $[3(m + p)^k + 5][2(m + p)^k - 3]$ **123.** ± 36 **125.** 9

Section 1.6 (page 52)
1. $\{x \mid x \neq -6\}$ **3.** $\{x \mid x \neq 3/5\}$ **5.** No restrictions **7.** $5p/2$ **9.** $8/9$ **11.** $3/(t-3)$
13. $(2x+4)/x$ **15.** $(m-2)/(m+3)$ **17.** $(2m+3)/(4m+3)$ **19.** $(y+2)/2$
21. $25p^2/9$ **23.** $2/9$ **25.** $3/10$ **27.** $5x/y$ **29.** $2(a+4)/(a-3)$ **31.** 1
33. $(k+2)/(k+3)$ **35.** $(m+6)/(m+3)$ **37.** $(m-3)/(2m-3)$ **39.** $(x^2-1)/x^2$
41. $(x+y)/(x-y)$ **43.** $(x^2-xy+y^2)/(x^2+xy+y^2)$ **45.** $14/r$ **47.** $19/(6k)$
49. $5/(12y)$ **51.** 1 **53.** $(6+p)/(2p)$ **55.** $(8-y)/(4y)$ **57.** $137/(30m)$
59. $(2y+1)/[y(y+1)]$ **61.** $3/[2(a+b)]$ **63.** $-2/[(a+1)(a-1)]$
65. $(2m^2+2)/[m-1)(m+1)]$ **67.** $4/(a-2)$ **69.** $(3x+y)/(2x-y)$
71. $5/[(a-2)(a-3)(a+2)]$ **73.** $(x-11)/[(x+4)(x-4)(x-3)]$
75. $a(a+5)/[(a+6)(a+1)(a-1)]$ **77.** $(6x^2-6x-5)/[(x+5)(x-2)(x-1)]$ **79.** $a+b$
81. -1 **83.** $1/(ab)$ **85.** $y(xy-9)/(x^2y^2-9)$ **87.** $(x+1)/(x-1)$ **89.** $-1/(x+1)$
91. $(2-b)(1+b)/[b(1-b)]$ **93.** $(m^3-4m-1)/(m-2)$ **95.** $(3z^2-75z+8)(z+1)/[5(z-25)]$
97. $(p+5)/[p(p+1)]$ **99.** $-1/[x(x+h)]$

Section 1.7 (page 57)
1. 2 **3.** 4 **5.** $1/9$ **7.** $27/8$ **9.** $1/25$ **11.** $16/9$ **13.** $256/81$ **15.** 1.552
17. $.321$ **19.** $4p^2$ **21.** $9x^4$ **23.** $5y^3z^4$ **25.** 2^2 or 4 **27.** $27^{1/3}$ or 3 **29.** 4^2 or 16
31. $m^{7/3}$ **33.** $(1+n)^{5/4}$ **35.** $\dfrac{6z^{2/3}}{y^{5/4}}$ **37.** $64a^{1/4}b^{37/2}$ **39.** r **41.** $1/(12k^{5/2})$
43. $a^{2/3}b^2$ **45.** $h^{1/3}t^{1/5}/k^{2/5}$ **47.** 1 **49.** $x^{13/12}$ **51.** $16/y^{11/12}$ **53.** $1/(k+5)^{1/2}$
55. $4z^2/(x^5y)$ **57.** r^{6+p} **59.** $m^{3/2}$ **61.** $x^{(2n^2-1)/n}$ **63.** $p^{(m+n+m^2)/(mn)}$ **65.** $y-10y^2$
67. $-4k^{10/3}+24k^{4/3}$ **69.** x^2-x **71.** $r-2+r^{-1}$ or $r-2+1/r$
73. $x^{1/2}+x^{-1/2}$ or $x^{1/2}+1/x^{1/2}$ **75.** $k^{3/4}(4k+1)$ **77.** $z^{-1/2}(9+2z)$ **79.** $p^{-7/4}(p-2)$
81. $(p+4)^{-3/2}(p^2+9p+21)$ **83.** 64 **85.** $\$64{,}000{,}000$ **87.** About $\$10{,}000{,}000$
89. 29 **91.** 177

Section 1.8 (page 55)
1. $\sqrt[3]{m^2}$ or $(\sqrt[3]{m})^2$ **3.** $5\sqrt[5]{m^4}$ or $5(\sqrt[5]{m})^4$ **5.** $\dfrac{-4}{\sqrt[3]{z}}$ **7.** $\sqrt[3]{(2m+p)^2}$ or $(\sqrt[3]{2m+p})^2$
9. $k^{2/5}$ **11.** $-a^{2/3}$ **13.** $-3(5p^3)^{1/2}=-3\cdot5^{1/2}p^{3/2}$ **15.** $18mn^{3/2}p^{1/2}$ **17.** 5 **19.** 6
21. -5 **23.** $5\sqrt{2}$ **25.** $20\sqrt{5}$ **27.** $50\sqrt{2}$ **29.** $3\sqrt[3]{3}$ **31.** $5\sqrt[3]{2}$ **33.** $-2\sqrt[4]{2}$
35. $-3\sqrt{5}/5$ **37.** $-\sqrt[3]{12}/2$ **39.** $\sqrt[4]{24}/2$ **41.** $-24\sqrt{6}$ **43.** $32\sqrt[3]{2}$ **45.** $2x^2z^4\sqrt{2x}$
47. $2zx^2y\sqrt[3]{2z^2x^2y}$ **49.** $np^2\sqrt[4]{m^2n^3}$ **51.** $5rz\sqrt[6]{5^3r^2z^4}$ or $5rz\sqrt[6]{125r^2z^4}$ **53.** $\sqrt{6x}/(3x)$
55. $x^2y\sqrt{xy}/z$ **57.** $2\sqrt[3]{x}/x$ **59.** $-km\sqrt[3]{k^2}/r^2$ **61.** $h\sqrt[4]{9g^3hr^2}/(3r^2)$ **63.** $m\sqrt[3]{n^2}/n$
65. $2\sqrt[4]{x^3y^3}$ **67.** $\sqrt[6]{4}$ or $\sqrt[3]{2}$ **69.** $\sqrt[18]{x}$ **71.** $9\sqrt{3}$ **73.** $\sqrt{2}$ **75.** $-2\sqrt{7p}$
77. $2\sqrt[3]{2}$ **79.** $7\sqrt[3]{3}$ **81.** $2\sqrt{3}$ **83.** $13\sqrt[3]{4}/6$ **85.** $ab\sqrt{ab}(b-2a^2+b^3)$ **87.** -7
89. 10 **91.** $11+4\sqrt{6}$ **93.** $5\sqrt{6}$ **95.** $2\sqrt[3]{9}-7\sqrt[3]{3}-4$ **97.** $-3(1+\sqrt{2})$
99. $(\sqrt{15}-3)/2$ **101.** $(10\sqrt{3}+15\sqrt{2})/6$ **103.** $(3\sqrt{5}+3\sqrt{15}-2\sqrt{3}-6)/33$
105. $p(\sqrt{p}-2)/(p-4)$ **107.** $a(\sqrt{a+b}+1)/(a+b-1)$ **109.** $-2x-2\sqrt{x(x+1)}-1$
111. $-1/[2(1-\sqrt{2})]$ **113.** $x/(\sqrt{x}+x)$ **115.** $-1/(2x-2\sqrt{x(x+1)}+1)$
117. 5.36×10^3 **119.** 1.40×10^{-1} **121.** 2.64×10^{-1} **123.** 9.93 **125.** $|m+n|$
127. $|z-3x|$ **129.** 3.14626437

Chapter 1   Review Exercises (page 70)

These review exercises cover the material in the chapter as follows: 1–42, Section 1.1;
43–56, Section 1.2; 57–78, Section 1.4; 79–98, Section 1.5; 99–112, Section 1.6;
113–132, Section 1.3; 133–144, Section 1.7, 145–162, Section 1.8.

1. 6   **3.** $-12, -6, -\sqrt{4}$ (or -2), 0, 6   **5.** $-\sqrt{7}, \pi/4, \sqrt{11}$ **7.** Natural, whole, integer,
rational, real   **9.** Integers, rational numbers, real numbers   **11.** Undefined   **13.** -15
15. 6   **17.** 32   **19.** $-37/18$ **21.** $-3/20$ **23.** 17   **25.** -32 **27.** -13
29. Commutative   **31.** Inverse   **33.** Associative   **35.** Identity   **37.** $9p + 9q$
39. $-15 + 3r$ **41.** $a(b + c)$ **43.** $-7, -3, -2, 0, \pi, 8$
45. $-|3 - (-2)|, -|-2|, |6 - 4|, |8 + 1|$ **47.** -3 **49.** -1 **51.** $3 - \sqrt{8}$ **53.** $m - 3$
55. $4 - \pi$ **57.** $-7m^2 - m - 3$ **59.** $-8z^3 + 4z^2 - 11z$ **61.** $16y^2 + 42y - 49$
63. $9k^2 - 30km + 25m^2$ **65.** $15w^3 - 22w^2 + 11w - 2$ **67.** $a^2 - b^2 - ac + 3bc - 2c^2$
69. $p^{2q} - 2p^q - 3$ **71.** $3m^3n/2 - m^2n^3 + 2m^6/n$ **73.** $3p + 8$ **75.** $5m - 7$
77. $k^3 - 2k^2 + 3$ **79.** $z(7z - 9z^2 + 1)$ **81.** $4p^3(3p^2 - 2p + 5)$ **83.** $(r + 7p)(r - 6p)$
85. $(3m + 1)(2m - 5)$ **87.** $5m^3(3m - 5n)(2m + n)$ **89.** $(13y^2 + 1)(13y^2 - 1)$
91. $8(y - 5z^2)(y^2 + 5yz^2 + 25z^4)$ **93.** $3(9r - 10)(2r + 1)$ **95.** $(r - 3s)(a + 5b)$
97. $(4m - 7 + 5a)(4m - 7 - 5a)$ **99.** $x/(2y)$ **101.** $1/[2k^2(k - 1)]$ **103.** $(x + 1)/(x + 4)$
105. $(3m + n)(3m - n)$ **107.** $37/(20y)$ **109.** $(x + 9)/[(x - 3)(x - 1)(x + 1)]$
111. $(q + p)/(pq - 1)$ **113.** $1/64$ **115.** $16/25$ **117.** $-10z^8$ **119.** $(-16p^7)/q$
121. $y^6/(36x^4z^4)$ **123.** $1/a^{23}$ **125.** $(36s)/r$ **127.** $1/[p^2(m + n)]$ **129.** $r^2/(s^4t^4)$
131. $-20/(m^5n^6)$ **133.** $p^{1/3}$ **135.** $-14r^{17/12}$ **137.** $y^{1/2}$ **139.** k^{2-9p}
141. $10z^{7/3} - 4z^{1/3}$ **143.** $3p^2 + 3p^{3/2} - 5p - 5p^{1/2}$ **145.** $10\sqrt{2}$ **147.** $5\sqrt[4]{2}$
149. $\sqrt{21r}/(3r)$ **151.** $2^3y^4\sqrt{2m}/m^2$ or $8y^4\sqrt{2m}/m^2$ **153.** $\sqrt[12]{m}$ **155.** 66
157. $m(-9\sqrt{2m} + 5\sqrt{m})$ **159.** $6(3 + \sqrt{2})/7$ **161.** $k(\sqrt{k} + 3)/(k - 9)$

Chapter 1   Test (page 73)

1. $\dfrac{4}{\pi}$ **2.** $-10, -\dfrac{9}{3}, 0, \dfrac{2}{3}, \dfrac{4}{\pi}, \sqrt{25}$ **3.** $-10, -\dfrac{9}{3}, 0, \sqrt{25}$ **4.** $\sqrt{25}$ **5.** 2   **6.** 9

7. Commutative   **8.** Distributive   **9.** Identity   **10.** Inverse

11. $-|-2|, -\sqrt{2}, 0, \sqrt{9}, \pi, |4 - 9|, \dfrac{21}{4}$ **12.** $2a^2 - a - 3$ **13.** $4x^{2n} - 12x^n + 9$

14. $3y^3 + 5y^2 + 2y + 8$ **15.** $m^3n - 6m$ **16.** $x^2 - x + \dfrac{-1}{x - 3}$ **17.** $(x^2 + 4)(x + 2)(x - 2)$

18. $2m(4m + 3)(3m - 4)$ **19.** Prime   **20.** $(x - 2)(x^2 + 2x + 4)(y + 3)(y - 3)$
21. $[x^4(x + 1)]/[3(x^2 + 1)]$ **22.** $(4x^2 + x)/[(x + 2)(x + 1)(2x - 3)]$ **23.** $2a/(2a - 3)$
24. $y/(y + 2)$ **25.** $-1/16$ **26.** $x^{2n}y^{2n-2}$ **27.** $4x^2/y^{16}$ **28.** yz/x **29.** $3x^2y^4\sqrt{2x}$

30. $\sqrt[5]{8xy^3}/(2y)$ **31.** $2\sqrt{2x}$ **32.** $3\sqrt{2} + 3$ **33.** $x - y$

CHAPTER 2

Section 2.1 (page 80)

1. Identity   **3.** Conditional   **5.** Identity   **7.** Contradiction   **9.** Not equivalent
11. Equivalent   **13.** Not equivalent   **15.** Not equivalent   **17.** $\{4\}$ **19.** $\{12\}$
21. $\{-2/7\}$ **23.** $\{-7/8\}$ **25.** $\{-1\}$ **27.** $\{3\}$ **29.** $\{4\}$ **31.** $\{3/4\}$ **33.** $\{-12/5\}$
35. $\emptyset$ **37.** $\{27/7\}$ **39.** $\{-59/6\}$ **41.** $\emptyset$ **43.** $\{0\}$ **45.** $\{-32\}$ **47.** $\{-4\}$
49. $\{-19/75\}$ **51.** $x = -3a + b$ **53.** $x = (3a + b)/(3 - a)$ **55.** $x = (3 - 3a)/(a^2 - a - 1)$

57. $x = 2a^2/(a^2 + 3)$ **59.** $x = (m + 4)/(5 + 2m)$ **61.** 68°F   **63.** 15°C   **65.** 37.8°C
67. 104°F   **69.** 13%   **71.** $432   **73.** $1500   **75.** $205.41   **77.** $66.50   **81.** 6
83. -2 **85.** {.72}   **87.** {6.53}   **89.** $\{-13.26\}$

Section 2.2 (page 89)

1. $V = k/P$ **3.** $l = V/(wh)$ **5.** $g = (V - V_0)/t$ **7.** $g = 2s/t^2$ **9.** $B = 2A/h - b$
11. $r_1 = S/(2\pi h) - r_2$ **13.** $l = gt^2/(4\pi^2)$ **15.** $h = (S - 2\pi r^2)/(2\pi r)$ **17.** $f = AB(p + 1)/24$
19. $R = r_1 r_2/(r_2 + r_1)$ **21.** 6 cm   **23.** 7 quarters   **25.** 2 liters   **27.** 38/3 kg
29. 400/3 liters   **31.** About 840 mi   **33.** 15 min   **35.** 35 kph   **37.** 2.7 mi   **39.** 3.6 hr
41. 39 hr   **43.** 40/3 hr   **45.** $57.65   **47.** $11,000 at 6% and $9000 at 10%   **49.** $54,000 at
8.5% and $6000 at 16%   **51.** $70,000 for land that makes a profit and $50,000 for land that produces a
loss   **53.** $402.70   **55.** $3000 in passbook accounts and $17,000 in long-term deposits
57. $20,000 at 9% and $40,000 at 10%

Section 2.3 (page 97)

1. Imaginary   **3.** Real   **5.** Imaginary   **7.** Imaginary   **9.** $10i$ **11.** $-20i$
13. $-i\sqrt{39}$ **15.** $5 + 2i$ **17.** $-6 - 14i$ **19.** $9 - 5i\sqrt{2}$ **21.** -5
23. -4 **25.** 2   **27.** -2 **29.** $7 - i$ **31.** $2 + 0i$ **33.** $1 - 10i$ **35.** $-10 + 5i$
37. $8 - i$ **39.** $-14 + 2i$ **41.** $5 - 12i$ **43.** $-8 - 6i$ **45.** $13 + 0i$ **47.** $7 + 0i$
49. $0 + 25i$ **51.** $12 + 9i$ **53.** $0 + i$ **55.** $7/25 - (24/25)i$ **57.** $26/29 + (7/29)i$
59. $-2 + i$ **61.** $0 - 2i$ **63.** $\dfrac{3 - 2\sqrt{5}}{13} + \dfrac{-2 - 3\sqrt{5}}{13}i$ **65.** $5/2 + i$
67. $-16/65 - (37/65)i$ **69.** $27/10 + (11/10)i$ **71.** i **73.** i **75.** 1   **77.** $-i$
79. 1   **81.** i **83.** $a = 0$ or $b = 0$

Section 2.4 (page 106)

1. $\{\pm 4\}$ **3.** $\{\pm 3\sqrt{3}\}$ **5.** $\{(1 \pm 2\sqrt{3})/3\}$ **7.** $\{2, 3\}$ **9.** $\{-5/2, 10/3\}$
11. $\{-3/2, -1/4\}$ **13.** $\{3, 5\}$ **15.** $\{1 \pm \sqrt{5}\}$ **17.** $\left\{-\dfrac{1}{2} \pm \dfrac{1}{2}i\right\}$ **19.** $\{(1 \pm \sqrt{5})/2\}$

21. $\{3 \pm \sqrt{2}\}$ **23.** $\left\{\dfrac{3}{2} \pm \dfrac{\sqrt{2}}{2}i\right\}$ **25.** $\left\{\dfrac{1}{6} \pm \dfrac{\sqrt{23}}{6}i\right\}$ **27.** $\{(2 \pm \sqrt{10})/3\}$

29. $\{4.024, -.690\}$ **31.** $\{(\sqrt{2} \pm \sqrt{6})/2\}$ **33.** $\{\sqrt{2}, \sqrt{2}/2\}$ **35.** $\{(-\sqrt{5} \pm 1)/2\}$

37. $\{(1 \pm \sqrt{13 + 8\sqrt{2}})/2(1 + \sqrt{2})\}$ **39.** $\left\{1, -\dfrac{1}{2} \pm \dfrac{\sqrt{3}}{2}i\right\}$ **41.** $\left\{-3, \dfrac{3}{2} \pm \dfrac{3\sqrt{3}}{2}i\right\}$

43. $\{4, -2 \pm 2i\sqrt{3}\}$ **45.** $\left\{-\dfrac{5}{2}, \dfrac{5}{4} \pm \dfrac{5\sqrt{3}}{4}i\right\}$ **47.** $\{3 \pm \sqrt{5}\}$ **49.** $\left\{\dfrac{\sqrt{3}}{3} \pm \dfrac{\sqrt{6}}{3}i\right\}$
51. $\{(-1 \pm \sqrt{7})/2\}$ **53.** $\{1, 3/2\}$ **55.** $x \neq 7, x \neq -2/3$ **57.** $y \neq -1/4$
59. $k \neq (-1 + \sqrt{17})/8, k \neq (-1 - \sqrt{17})/8$ **61.** None   **63.** 0; one rational solution
65. 1; two rational solutions   **67.** 84; two irrational solutions   **69.** -23; two imaginary solutions
71. ± 12 **73.** 121/4   **75.** 1, 9   **77.** $t = \pm\sqrt{2sg}/g$ **79.** $h = \pm\sqrt{d^4 kL}/L$ or $\pm d^2\sqrt{kL}/L$
81. $t = \pm\sqrt{(s - s_0 - k)g}/g$ **83.** $R = (E^2 - 2Pr \pm E\sqrt{E^2 - 4Pr})/(2P)$
85. (a) $x = (y \pm \sqrt{8 - 11y^2})/4$ (b) $y = (x \pm \sqrt{6 - 11x^2})/3$

Section 2.5 (page 111)
1. 16, 18 or $-18, -16$ **3.** 12, -2 **5.** 100 yd by 400 yd   **7.** 1 in   **9.** 10 in by 13 in
11. 2 cm   **13.** Bob: 36 hr, Paul: 45 hr   **15.** 4 hr   **17.** 50 mph   **19.** 2.5 hr
21. Don't wait; pick now.   **23.** 80   **25.** 40 ft   **27.** 16.95 cm, 20.95 cm, 26.95 cm

Section 2.6 (page 118)
1. $\{\pm\sqrt{3}, \pm i\sqrt{5}\}$ **3.** $\{\pm 1, \pm\sqrt{10}/2\}$ **5.** $\{4, 6\}$ **7.** $\{(-5 \pm \sqrt{21})/2\}$
9. $\{(-6 \pm 2\sqrt{3})/3, (-4 \pm \sqrt{2})/2\}$ **11.** $\{7/2, -1/3\}$ **13.** $\{-63, 28\}$ **15.** $\{\pm\sqrt{5}/5\}$
17. $\{\pm i\sqrt{3}, \pm i\sqrt{6}/3\}$ **19.** $\{-1\}$ **21.** $\{5\}$ **23.** $\{9\}$ **25.** $\{9\}$ **27.** Ø
29. $\{2, -2\}$ **31.** $\{8\}$ **33.** $\{-2\}$ **35.** $\{2, -2/9\}$ **37.** $\{4\}$ **39.** $\{-2\}$ **41.** $\{2\}$
43. $\{3/2\}$ **45.** $\{31\}$ **47.** $\{-3, 1\}$ **49.** $\{-27, 3\}$ **51.** $\{1/4, 1\}$ **53.** $\{0, 8\}$
55. $\{\pm 2.121, \pm 1.528\}$ **57.** $\{\pm .917, \pm .630i\}$ **59.** $h = (d/k)^2$ **61.** $L = P^2g/4$
63. $y = (a^{2/3} - x^{2/3})^{3/2}$

Section 2.7 (page 126)
1. $(-1, 4)$

3. $(-\infty, 0)$

5. $[1, 2)$

7. $(-\infty, -9)$

9. $-4 < x < 3$

11. $x \le -1$

13. $-2 \le x < 6$

15. $x \le -4$

17. $(-\infty, 4]$

19. $[-1, +\infty)$

21. $(-\infty, 6]$

23. $(-\infty, 4)$

25. $[-11/5, +\infty)$

27. $[1, 4]$

29. $(-6, -4)$

31. $(-16, 19]$

33. $[-3, 3]$

35. $(-\infty, -3] \cup [-1, +\infty)$

37. $[-2, 3]$

39. $(-\infty, 1/2) \cup (4, +\infty)$

41. $(-\infty, 0) \cup (0, +\infty)$

43. $((-5 - \sqrt{33})/2, (-5 + \sqrt{33})/2)$ **45.** $[1 - \sqrt{2}, 1 + \sqrt{2}]$

47. $[-2, 3/2] \cup [3, +\infty)$ **49.** $(-\infty, -2] \cup [0, 2]$ **51.** $(-2, 0) \cup (1/4, +\infty)$
53. 140 points   **55.** $[500, +\infty)$ **57.** $[45, +\infty)$ **59.** $(0, 5/4) \cup (6, +\infty)$
61. For any x in $[0, 5/3) \cup (10, +\infty)$ **63.** 32°F to 86°F   **65.** $(-5, 3]$ **67.** $(-\infty, -2)$
69. $(-\infty, 6) \cup [15/2, +\infty)$ **71.** $(-\infty, 1) \cup (9/5, +\infty)$ **73.** $(-\infty, -3/2) \cup [-1/2, +\infty)$
75. $(-2, +\infty)$ **77.** $(0, 4/11) \cup (1/2, +\infty)$ **79.** $(-\infty, -2] \cup (1, 2)$ **81.** $(-\infty, 5)$
83. $[3/2, +\infty)$ **85.** $(5/2, +\infty)$ **87.** $[-8/3, 3/2] \cup (6, +\infty)$
89. If k is in $(-\infty, -4\sqrt{2}] \cup [4\sqrt{2} +\infty)$ **91.** If k is in $(-\infty, 0] \cup [8, +\infty)$

Section 2.8 (page 132)

1. $\{1, 3\}$ **3.** $\{-1/3, 1\}$ **5.** $\{2/3, 8/3\}$ **7.** $\{-6, 14\}$ **9.** $\{5/2, 7/2\}$ **11.** $\{-4/3, 2/9\}$
13. $\{-7/3, -1/7\}$ **15.** $\{-3/5, 11\}$ **17.** $\{-1/2\}$ **19.** $[-3, 3]$ **21.** $(-\infty, -1) \cup (1, +\infty)$
23. $\varnothing$ **25.** $[-10, 10]$ **27.** $(-4, -1)$ **29.** $(-\infty, -2/3) \cup (2, +\infty)$
31. $(-\infty, -8/3] \cup [2, +\infty)$ **33.** $(-3/2, 13/10)$ **35.** $|x - 2| \le 4$ **37.** $|z - 12| \ge 2$
39. $|k - 1| = 6$ **41.** If $|x - 2| \le .0004$, then $|y - 7| \le .00001$.   **43.** $m = 2; n = 20$

Chapter 2   Review Exercises (page 135)

*These review exercises cover the material in the chapter as follows: 1–14, Section 2.1;
15–32, Section 2.2; 33–54, Section 2.3; 55–70, Section 2.4; 71–76, Section 2.5; 77–92,
Section 2.6; 93–112, Section 2.7; 113–128, Section 2.8.*

1. $\{6\}$ **3.** $\{-11/3\}$ **5.** $\varnothing$ **7.** $\left\{\dfrac{1}{60}\right\}$ **9.** $\{13\}$ **11.** $x = -6b - a - 6$

13. $x = (6 - 3m)/(1 + 2k - km)$ **15.** $y = 6a/(4 - a)$ **17.** $C = (5/9)(F - 32)$
19. $j = n - nA/I$ **21.** $r_1 = kr_2/(r_2 - k)$ **23.** $L = V/(\pi r^2)$ **25.** $x = -4/(y^2 - 5y - 6p)$
27. \$500   **29.** 8 mi   **31.** \$55,000 at 11.5% and \$35,000 at 12%   **33.** $10 - 3i$
35. $-8 + 13i$ **37.** $19 + 17i$ **39.** $146 + 0i$ **41.** $7 - 24i$ **43.** $-30 - 40i$
45. $5/2 + (7/2)i$ **47.** $11/5 + (2/5)i$ **49.** $-i$ **51.** $-i$ **53.** -1 **55.** $\{-7 \pm \sqrt{5}\}$
57. $\{5/2, -3\}$ **59.** $\{7, -3/2\}$ **61.** $\{3, -1/2\}$ **63.** $\{\sqrt{2} \pm 1\}$ **65.** -188; two imaginary
solutions   **67.** 484; two rational solutions   **69.** 0; one rational solution   **71.** 1/2 ft
73. 6.5 hr   **75.** 10 mph   **77.** $\{\pm i, \pm 1/2\}$ **79.** $\{-2, 3\}$ **81.** $\{5/2, -15\}$ **83.** $\{3\}$
85. $\{-2, -1\}$ **87.** $\varnothing$ **89.** $\{1, -4\}$ **91.** $\{-1\}$ **93.** $(-7/13, +\infty)$ **95.** $(-\infty, 1]$
97. $(1, +\infty)$ **99.** $[4, 5]$ **101.** $[-4, 1]$ **103.** $(-2/3, 5/2)$ **105.** $(-\infty, -4] \cup [0, 4]$
107. $(-2, 0)$ **109.** $(-3, 1) \cup [7, +\infty)$ **111.** 0 to 30 (Remember: x must start at 0.)
113. $\{3, -11\}$ **115.** $\{11/27, 25/27\}$ **117.** $\{4/3, -2/7\}$ **119.** $[-7, 7]$
121. $(-\infty, -3) \cup (3, +\infty)$ **123.** $[-6, -3]$ **125.** $(-2/7, 8/7)$
127. $(-\infty, -4) \cup (-2/3, +\infty)$

Chapter 2   Test (page 138)

1. $\{5/2\}$ **2.** $\varnothing$ **3.** $x = 2b/(k^2 - 2)$ **4.** $c = ab/(a + b)$ **5.** $13\dfrac{1}{3}$ qt   **6.** 225 mi

7. $1 + 9i$ **8.** $-26 - 7i$ **9.** $-3 + i$ **10.** i **11.** $\left\{1 \pm \dfrac{\sqrt{6}}{6}i\right\}$ **12.** $\{(6 \pm \sqrt{42})/3\}$

13. 1; 2 rational   **14.** 6 ft by 8 ft   **15.** {2}   **16.** {0, 4}   **17.** $\{\pm\sqrt{5}, \pm i\sqrt{2}\}$
18. $\{-4, 2\}$ **19.** $[-3, +\infty)$ **20.** $[-10, 2]$ **21.** $(-\infty, -1] \cup [3/2, +\infty)$
22. $\left(-\infty, \dfrac{5}{2}\right) \cup [4, +\infty)$ **23.** $\{-1/3, -3\}$ **24.** $(-2, 7)$ **25.** $(-\infty, -6] \cup [5, +\infty)$

CHAPTER 3

Section 3.1 (page 153)
1. $8\sqrt{2}$; (9, 3)   **3.** $\sqrt{34}$; $(-11/2, -7/2)$ **5.** $\sqrt{133}$; $(2\sqrt{2}, 3\sqrt{5}/2)$ **7.** 27.203
9. (13, 10)   **11.** No   **13.** No   **15.** Yes   **17.** No
19.   **21.**   **23.**

$x^2 + y^2 = 36$ $(x-2)^2 + y^2 = 36$ $(x+2)^2 + (y-5)^2 = 12$

25. $(x-1)^2 + (y-4)^2 = 9$ **27.** $(x+1)^2 + (y-2)^2 = 25$ **29.** $(x+3)^2 + (y+2)^2 = 4$
31. $x^2 + y^2 = 25$

33.   **35.**   **37.**

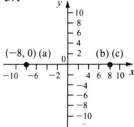

39. x-axis, y-axis, origin   **41.** y-axis   **43.** Origin   **45.** y-axis   **47.** Origin
49. Quadrants I and IV   **51.** Quadrants II and IV   **53.** Quadrants III and IV   **57.**

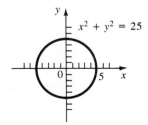

59. $(2 + \sqrt{7}, 2 + \sqrt{7}), (2 - \sqrt{7}, 2 - \sqrt{7})$ **61.** It is reflected about the x-axis.

63.

65. $(x + 2)^2 + (y + 3/2)^2 = 149/4$ **67.** Yes

Section 3.2 (page 166)

1.

3.

5.

7.

9.

11.

13. 1/5   **15.** 7/9   **17.** 0   **19.** 3   **21.** −3/4   **23.** 0   **25.** .5785

27.

29.

31.

33. $2x + y = 5$ **35.** $3x + 2y = -7$ **37.** $x = -8$ **39.** $-x + 4y = 13$
41. $3x + 4y = 12$ **43.** $2x - 3y = 6$ **45.** $x = -6$ **47.** $5.081x + y = -4.69256$
49. Slope 3; y-intercept -1 **51.** Slope 4; y-intercept -7 **53.** Slope $-3/4$; y-intercept 0
55. $x + 3y = 11$ **57.** $5x - 3y = -13$ **59.** $x = -5$ **61.** Yes   **63.** No   **65.** No
67. (a) $-1/2$ **(b)** $-7/2$ **75.** $y = 640x + 1100$ **77.** $y = -1000x + 40,000$
79. $y = 2.5x - 70$

Section 3.3 (page 180)

1.

3.

5. Vertex: (2, 0);
   axis: $x = 2$

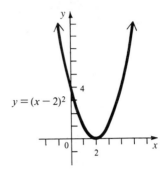

$y = (x - 2)^2$

7. Vertex: $(-3, -4)$;
   axis: $x = -3$

$y = (x + 3)^2 - 4$

9. Vertex: $(-3, 2)$;
   axis: $x = -3$

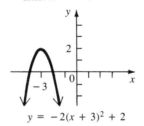

$y = -2(x + 3)^2 + 2$

11. Vertex: $(-1, -3)$;
   axis: $x = -1$

$y = -\frac{1}{2}(x + 1)^2 - 3$

13. Vertex: (1, 2);
   axis: $x = 1$

$y = x^2 - 2x + 3$

15. Vertex: $(-2, 6)$;
   axis: $x = -2$

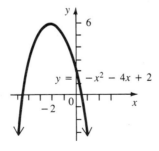

$y = -x^2 - 4x + 2$

17. Vertex: (1, 3);
   axis: $x = 1$

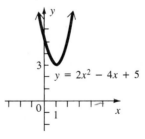

$y = 2x^2 - 4x + 5$

19.

$y = .14x^2 + .56x - .3$

21.

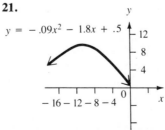

$y = -.09x^2 - 1.8x + .5$

23.

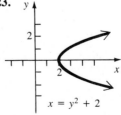

$x = y^2 + 2$

25.

$x = (y + 1)^2$

27.

$x = (y + 2)^2 - 1$

29.

$x = -2(y + 3)^2$

31.

$x = \dfrac{1}{2}(y + 1)^2 + 3$

33.

$x = y^2 + 2y - 8$

35.

$x = -2y^2 + 2y - 3$

37. 80 by 160 ft   **39.** 300 sandwiches; 100   **41.** 5 in   **43.** 10, 10

45. (a) $x(500 - x) = 500x - x^2$

(b)

(c) 250
(d) \$62,500

47.

49. $6\sqrt{3}$ cm   **51.** 25   **53.** $12y = x^2$ **55.** $-8x = y^2$ **57.** $8(y - 4) = (x - 3)^2$ or $y = (1/8)(x - 3)^2 + 4$ **59.** $6(x + 33/6) = (y - 1)^2$ or $x = (1/6)(y - 1)^2 - (11/2)$

Section 3.4 (page 192)

1.

$\dfrac{x^2}{9} + \dfrac{y^2}{4} = 1$

3.

$\dfrac{x^2}{6} + \dfrac{y^2}{9} = 1$

5.

$x^2 + 4y^2 = 16$

7.

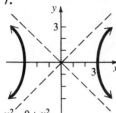

$x^2 = 9 + y^2$

9.

$25x^2 - 4y^2 = -100$

11.

$\dfrac{x^2}{(1/9)} + \dfrac{y^2}{(1/16)} = 1$

13.

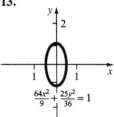

$\dfrac{64x^2}{9} + \dfrac{25y^2}{36} = 1$

15.

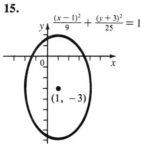

$\dfrac{(x-1)^2}{9} + \dfrac{(y+3)^2}{25} = 1$

$(1, -3)$

17.

$\dfrac{(x-3)^2}{16} - \dfrac{(y+2)^2}{49} = 1$

19.

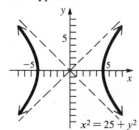

$\dfrac{(y+1)^2}{25} - \dfrac{(x-3)^2}{36} = 1$

21. $x^2/36 + y^2/16 = 1$

23. $y^2/4 - x^2/16 = 1$

25. $(x - 2)^2/16 + (y + 2)^2/9 = 1$ **27.** $(x - 4)^2/9 - (y - 3)^2/4 = 1$

29. Hyperbola

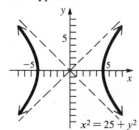

$x^2 = 25 + y^2$

31. Ellipse

$9x^2 + 36y^2 = 36$

33. Circle

$(-3, 2)$

$\dfrac{(x+3)^2}{16} + \dfrac{(y-2)^2}{16} = 1$

35. Parabola

$y^2 - 4y = x + 4$

37. Empty set (no graph)

39. Circle

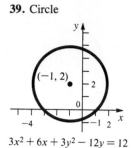

$3x^2 + 6x + 3y^2 - 12y = 12$

41. Parabola

$x^2 - 6x + y = 0$

43. Hyperbola

$4x^2 - 8x - y^2 - 6y = 6$

45. Ellipse

$4x^2 - 8x + 9y^2 + 54y = -84$

47. Empty set (no graph)

49. Circle

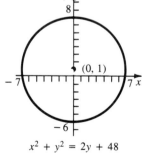

$x^2 + y^2 = 2y + 48$

51. Circle

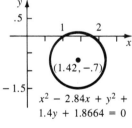

$x^2 - 2.84x + y^2 + 1.4y + 1.8664 = 0$

Section 3.5 (page 200)

1.

$x \leq 3$

3.

$x + 2y \leq 6$

5.

$2x + 3y \geq 4$

7.

$3x - 5y > 6$

9.

$5x \le 4y - 2$

11.

$y < 3x^2 + 2$

13.

$y > (x - 1)^2 + 2$

15.

$x^2 + (y + 3)^2 \le 16$

17.

$4x^2 \le 4 - y^2$

19.

$9x^2 - 16y^2 > 144$

21.

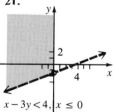

$x - 3y < 4,\ x \le 0$

23.

$3x + 2y \ge 6,\ y \le 2$

25.

$x + 2y < 4,\ 3x - y > 5$

27.

$2x + 3y < 6,\ x - 5y \ge 10$

29.

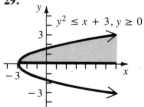

$y^2 \le x + 3,\ y \ge 0$

31.

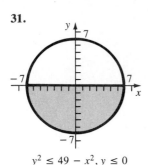

$y^2 \le 49 - x^2, y \le 0$

33.

$\dfrac{x^2}{36} < 1 - \dfrac{y^2}{121}, \quad x \ge 0$

Chapter 3 Review Exercises (page 203)

These review exercises cover the material in the chapter as follows: 1–23, Section 3.1;
25–57, Section 3.2; 59–77, Section 3.3; 79–87, Section 3.4; 89–97, Section 3.5.

1. $\sqrt{85}$; $(-1/2, 2)$ **3.** 5; $(-6, 11/2)$ **5.** -7; -1; 8; 23   **7.** No such points exist.
9. $(x + 2)^2 + (y - 3)^2 = 25$ **11.** $(x + 8)^2 + (y - 1)^2 = 289$ **13.** $(2, -3)$; $r = 1$
15. $(-7/2, -3/2)$; $r = 3\sqrt{6}/2$ **17.** Yes; yes; yes   **19.** No; no; no   **21.** Yes; no; no
23. Yes; yes; yes   **25.** 6/5   **27.** Undefined slope   **29.** 9/4   **31.** 1/5   **33.** 0

35.

37.

39.

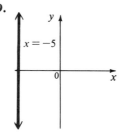

41. $x + 3y = 10$ **43.** $2x + y = 1$ **45.** $15x + 30y = 13$ **47.** $y = 3/4$
49. $5x - 8y = -40$ **51.** $y = -5$

53.

55.

57.

59.

$y = 3(x + 1)^2 - 5$

61.

63. $(-3, -9)$; $x = -3$

65. $(7/2, -41/4)$; $x = 7/2$ **67.** $(-1, 4)$; $x = -1$ **69.** 5.5 and 5.5   **71.** A square, 45 m on a side

73. $(-2, 0)$; $y = 0$ **75.** $(2, -1)$; $y = -1$ **77.** $(29/4, 5/2)$; $y = 5/2$

$x = y^2 - 2$

$x = -(y + 1)^2 + 2$

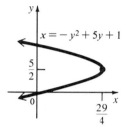

$x = -y^2 + 5y + 1$

79.

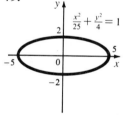

$\dfrac{x^2}{25} + \dfrac{y^2}{4} = 1$

81.

$\dfrac{x^2}{4} - \dfrac{y^2}{9} = 1$

83.

$x^2 = 16 + y^2$

85.

$\dfrac{25x^2}{9} + \dfrac{4y^2}{25} = 1$

87.

$\dfrac{(x + 1)^2}{16} - \dfrac{(y - 2)^2}{4} = 1$

89.

$5x - y \leq 20$

91.

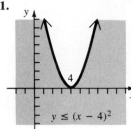

$y \le (x - 4)^2$

93.

$25y^2 - 36x^2 > 900$

95.

$2x - y \ge 4,$
$y \ge -2$

97.

$y \le x^2,$
$x \ge 0$

Chapter 3 Test (page 205)
1. $\sqrt{34}$ **2.** ± 3 **3.** $(x - 5)^2 + (y + 1)^2 = 25$ **4.** Center: $(-2, 3)$; radius: 1
5. Origin only   **6.** $-7/9$ **7.** 5/4   **8.**   **9.** $x + 2y = 4$ **10.** $y = 7$

$4x - 5y = 10$

11. $2x - y = -14$ **12.**   **13.**   **14.** Vertex: $(5/2, -9/4)$;
axis: $x = 5/2$

$(-1, -1)$ $(2, -3)$

$y = 3x^2 - 12x + 9$

15. 800 sq ft   **16.**

$x = -(y + 2)^2 - 4$

17.

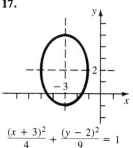

$$\frac{(x + 3)^2}{4} + \frac{(y - 2)^2}{9} = 1$$

18.

$$\frac{y^2}{25} - \frac{x^2}{9} = 1$$

19.

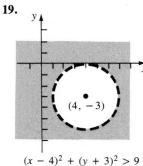

$(x - 4)^2 + (y + 3)^2 > 9$

20.

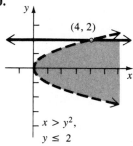

$x > y^2,$
$y \leq 2$

CHAPTER 4

Section 4.1 (page 212)
1. (a) 0 **(b)** 4 **(c)** 2 **(d)** 4   **3. (a)** -3 **(b)** -2 **(c)** 0 **(d)** 2   **5.** -1 **7.** -10
9. 4   **11.** $3a - 1$ **13.** 3   **15.** 72   **17.** $-6m - 1$ **19.** $15a - 7$
21. $25p^2 - 20p + 4$ **23.** -66.8 **25.** -42.464 **27.** $(-\infty, +\infty); (-\infty, +\infty)$
29. $(-\infty, +\infty); [0, +\infty)$ **31.** $[-8, +\infty); [0, +\infty)$ **33.** $[-4, 4]; [0, 4]$ **35.** $[-5, 4];$
$[-2, 6]$ **37.** $(-\infty, +\infty); (-\infty, 12]$ **39.** $[-3, 4]; [-6, 8]$ **41.** 2   **43.** 16
45. 1/4   **47.** 2^m **49.** $\sqrt{2}$ **51. (a)** $x^2 + 2xh + h^2 - 4$ **(b)** $2xh + h^2$ **(c)** $2x + h$
53. (a) $6x + 6h + 2$ **(b)** $6h$ **(c)** 6   **55. (a)** $2x^3 + 6x^2h + 6xh^2 + 2h^3 + x^2 + 2xh + h^2$
(b) $6x^2h + 6xh^2 + 2h^3 + 2xh + h^2$ **(c)** $6x^2 + 6xh + 2h^2 + 2x + h$ **57. (a)** $C(x) = 10x + 500$
(b) $R(x) = 35x$ **(c)** $P(x) = 25x - 500$ **59. (a)** $C(x) = 150x + 2700$ **(b)** $R(x) = 280x$
(c) $P(x) = 130x - 2700$ **61.** 25   **63.** Break-even point is 45 units; do not produce, since 38 units
would result in a loss.   **65.** Break-even point is -50; impossible to make a profit here.

Section 4.2 (page 220)

1. Function;
domain $(-\infty, +\infty)$;
range $(-\infty, +\infty)$

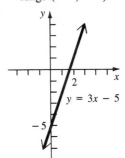

$y = 3x - 5$

3. Function;
domain $(-\infty, +\infty)$;
range $[-1, +\infty)$

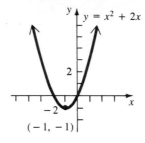

$y = x^2 + 2x$

$(-1, -1)$

5. Function;
domain $(-\infty, +\infty)$;
range $[0, +\infty)$

$y = |x + 1|$

7. Function;
domain $(-\infty, +\infty)$;
range $[0, +\infty)$

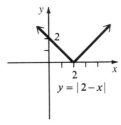

$y = |2 - x|$

9. Function;
domain $(-\infty, +\infty)$;
range $[0, +\infty)$

$y = |5x + 4|$

11. Function;
domain $(-\infty, +\infty)$;
range $[-1, +\infty)$

$y + 1 = |x - 2|$

13. Domain $[0, +\infty)$;
range $(-\infty, +\infty)$

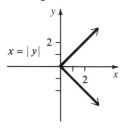

$x = |y|$

15. Function;
domain $(-\infty, +\infty)$;
range $[4, +\infty)$

$y = |x| + 4$

17. Domain $[-1/2, +\infty)$;
range $(-\infty, +\infty)$

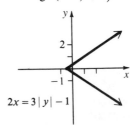

$2x = 3|y| - 1$

19. Function;
domain $(-\infty, +\infty)$;
range $[0, +\infty)$

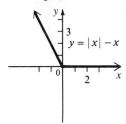

$y = |x| - x$

21. (a) -10 **(b)** -2 **(c)** -1 **(d)** 2 **(e)** 4 **23. (a)** -10 **(b)** 2 **(c)** 5 **(d)** 3 **(e)** 5

25.

$$f(x) = \begin{cases} x - 1 \text{ if } x \le 3 \\ 2 \text{ if } x > 3 \end{cases}$$

27.

$$f(x) = \begin{cases} 4 - x \text{ if } x < 2 \\ 1 + 2x \text{ if } x \ge 2 \end{cases}$$

29.

$$f(x) = \begin{cases} 2x + 1 \text{ if } x \ge 0 \\ x \text{ if } x < 0 \end{cases}$$

31.

$$f(x) = \begin{cases} 2 + x \text{ if } x < -4 \\ -x \text{ if } -4 \le x \le 5 \\ 3x \text{ if } x > 5 \end{cases}$$

33.

$$f(x) = \begin{cases} |x| \text{ if } x > -2 \\ x \text{ if } x \le -2 \end{cases}$$

35.

$f(x) = [-x]$

37.

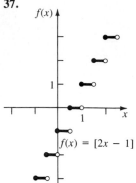

$$f(x) = [2x - 1]$$

39.

$$f(x) = [3x]$$

41.

$$f(x) = [3x] - 1$$

43. Domain $(-\infty, 4]$;
range $[0, +\infty)$

$$y = \sqrt{4 - x}$$

45. Domain $(-\infty, -3] \cup$
$[3, +\infty)$; range $[0, +\infty)$

$$f(x) = \sqrt{x^2 - 9}$$

47. Domain $(-\infty, +\infty)$;
range $(-\infty, -3]$

$$\frac{y}{3} = -\sqrt{1 + \frac{x^2}{16}}$$

49. Domain $[-8, +8]$;
range $[0, 1]$

$$y = \sqrt{1 - \frac{x^2}{64}}$$

51. **(a)** \$11 **(b)** \$18 **(c)** \$32 **(d)**
(e) Domain: $(0, +\infty)$ (at least in theory);
range: $\{11, 18, 25, 32, 39, \ldots\}$

53. $-[-x]$ oz

55.

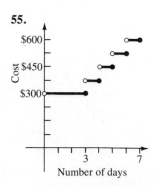

57. (a) 140 **(b)** 220 **(c)** 220 **(d)** 220 **(e)** 220 **(f)** 60 **(g)** 60 **(h)**

Section 4.3 (page 226)
1. (a) $10x + 2$ **(b)** $-2x - 4$ **(c)** $24x^2 + 6x - 3$ **(d)** $(4x - 1)/(6x + 3)$ **(e)** All domains are $(-\infty, +\infty)$, except for f/g, which is $(-\infty, -1/2) \cup (-1/2, +\infty)$. **3. (a)** $4x^2 - 4x + 1$ **(b)** $2x^2 - 1$ **(c)** $(3x^2 - 2x)(x^2 - 2x + 1)$ **(d)** $(3x^2 - 2x)/(x^2 - 2x + 1)$ **(e)** All domains are $(-\infty, +\infty)$, except for f/g, which is $(-\infty, 1) \cup (1, +\infty)$. **5. (a)** $\sqrt{2x + 5} + \sqrt{4x - 9}$ **(b)** $\sqrt{2x + 5} - \sqrt{4x - 9}$ **(c)** $\sqrt{(2x + 5)(4x - 9)}$ **(d)** $\sqrt{(2x + 5)/(4x - 9)}$ **(e)** All domains are $[9/4, +\infty)$, except f/g, which is $(9/4, +\infty)$. **7. (a)** $5x^2 - 11x + 7$ **(b)** $3x^2 - 11x - 3$ **(c)** $(4x^2 - 11x + 2)(x^2 + 5)$ **(d)** $(4x^2 - 11x + 2)/(x^2 + 5)$ **(e)** All domains (including f/g) are $(-\infty, +\infty)$. **9.** 55
11. 1848 **13.** $-6/7$ **15.** $4m^2 + 6m + 1$ **17.** 1122 **19.** 97 **21.** $256k^2 + 48k + 2$
23. $24x + 4$; $24x + 35$; both domains are $(-\infty, +\infty)$. **25.** $-5x^2 + 20x + 18$; $-25x^2 - 10x + 6$; both domains are $(-\infty, +\infty)$. **27.** $-64x^3 + 2$; $-4x^3 + 8$; both domains are $(-\infty, +\infty)$.
29. $1/x^2$; $1/x^2$; domain of $f \circ g$: $(-\infty, 0) \cup (0, +\infty)$; domain of $g \circ f$: $(-\infty, 0) \cup (0, +\infty)$
31. $2\sqrt{2x - 1}$; $8\sqrt{x + 2} - 6$; domain of $f \circ g$: $[1/2, +\infty)$; domain of $g \circ f$: $[-2, +\infty)$ **33.** $x/(2 - 5x)$; $2(x - 5)$; domain of $f \circ g$: $(-\infty, 0) \cup (0, 2/5) \cup (2/5, +\infty)$; domain of $g \circ f$: $(-\infty, 5) \cup (5, +\infty)$
35. $\sqrt{(x - 1)/x}$; $-1/\sqrt{x + 1}$; domain of $f \circ g$: $(-\infty, 0) \cup [1, +\infty)$; domain of $g \circ f$: $(-1, +\infty)$

Other correct answers are possible for Exercises 43–47.

43. $f(x) = x^2$; $g(x) = 6x - 2$ **45.** $f(x) = \sqrt{x}$; $g(x) = x^2 - 1$ **47.** $f(x) = \dfrac{x^2 + 1}{5 - x^2}$; $g(x) = x - 2$ **49.** $18a^2 + 24a + 9$ **51.** $4\pi t^2$

Section 4.4 (page 234)
1. One-to-one **3.** One-to-one **5.** Not one-to-one **7.** One-to-one **9.** One-to-one
11. Not one-to-one **13.** Not one-to-one **15.** Not one-to-one **17.** One-to-one **19.** One-to-one **21.** One-to-one **23.** Not one-to-one **25.** Inverses **27.** Not inverses **29.** Not inverses **31.** Inverses **33.** Inverses **35.** Not inverses **37.** Not inverses **39.** Not inverses **41.** Inverses **43.** Not inverses

45.

47.

49.

51. $f^{-1}(x) = (x + 5)/4$

53. $f^{-1}(x) = (-5/2)x$

55. $f^{-1}(x) = \sqrt[3]{-x - 2}$

57. $f^{-1}(x) = (9 - x)/3$

59. Not one-to-one

61. $f^{-1}(x) = 4/x$

63. $f^{-1}(x) = 2/x$

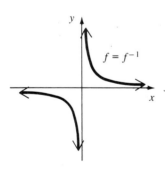

65. $f^{-1}(x) = x^2 - 6$, domain: $[0, +\infty)$

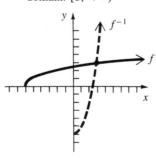

67. $f^{-1}(x) = -\sqrt{x^2 + 9}$, domain: $[0, +\infty)$

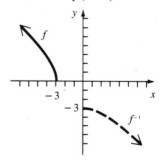

69. $f^{-1}(x) = 1 - \sqrt{x}$, domain: $[0, +\infty)$

Section 4.5 (page 240)

1. $a = kb$ **3.** $x = k/y$ **5.** $r = kst$ **7.** $w = kx^2/y$ **9.** 220/7   **11.** 32/15
13. 18/125   **15.** 1075   **17.** 2304 ft   **19.** .0444 ohms   **21.** 140 kg per cm^2 **23.** $1375
25. 8/9 metric tons   **27.** 7500 lb   **29.** 150 kg per m^2 **31.** 4,000,000   **33.** 1600 calls
35. About 4.94   **37.** About 7.4 km   **39.** 263 lb

Chapter 4   Review Exercises (page 246)

These review exercises cover the material in the chapter as follows: 1–23, Section 4.1;
25–39, Section 4.2; 41–55, Section 4.3; 57–69, Section 4.4; 71–79, Section 4.5.

1. $(-\infty, +\infty)$ **3.** $(-\infty, +\infty)$ **5.** $(-\infty, +\infty)$ **7.** $[-7, 7]$ **9.** $(-\infty, +\infty)$
11. $(-\infty, 7) \cup (7, +\infty)$ **13.** -19 **15.** 2   **17.** 6   **19.** $-z^2 - 2z + 5$
21. $-4x - 2h + 4$ **23.** $-3x^2 - 3xh - h^2 + 4x + 2h$
25.

27.

29.

31.

33.

35.

37.

39.

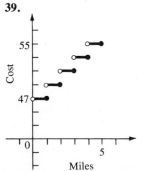

41. $4x^2 - 3x - 8$ **43.** 44   **45.** $16k^2 - 6k - 8$ **47.** $-23/4$ **49.** $(-\infty, +\infty)$
51. $\sqrt{x^2 - 2}$ **53.** $\sqrt{34}$ **55.** 1   **57.** Not one-to-one   **59.** One-to-one
61. Not one-to-one   **63.** Not one-to-one
65. $f^{-1}(x) = (x - 3)/12$ **67.** $f^{-1}(x) = \sqrt[3]{x + 3}$ **69.** $f^{-1}(x) = \sqrt{25 - x^2}$; domain: [0, 5]

71. $m = kz^2$ **73.** $Y = (kMN^2)/X^3$ **75.** 27/2   **77.** 1372/729   **79.** 36 inches

Chapter 4   Test (page 248)

1. $(-\infty, +\infty)$ **2.** $(-\infty, -4] \cup [-3, +\infty)$ **3.** $(-\infty, -4) \cup (-4, +\infty)$ **4.** $(-3, 3)$
5. 8   **6.** -5 **7.** $y - 5$ **8.**   **9.**

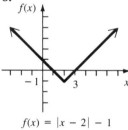

$f(x) = |x - 2| - 1$

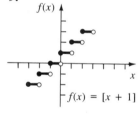

$f(x) = [x + 1]$

10.   **11.**   **12.**

$f(x) = \begin{cases} 3 \text{ if } x < -2 \\ 2 - \dfrac{1}{2}x \text{ if } x \geq -2 \end{cases}$

$f(x) = \sqrt{x^2 - 4}$

$f(x) = \sqrt{3 - x}$

13. $12a^2 - 4a + 8$ **14.** 2   **15.** 23/2   **16.** 6
17. $6p^2 - 8p + 15$ **18.** Not one-to-one   **19.** One-to-one   **20.** One-to-one
21. $f^{-1}(x) = (1 - x)/x$ **22.** $f^{-1}(x) = -\sqrt{x - 2}$ **23.** $f^{-1}(x) = (x - 8)/2$ **24.** Not inverses
25. 36 lb

CHAPTER 5

Section 5.1 (page 258)

1. (a)   (b) y-axis

$P(x) = \frac{1}{4}x^6$

3. (a)   (b) Origin

$P(x) = \frac{-5}{4}x^5$

5. (a)   (b) $(0, 1)$

$P(x) = \frac{1}{2}x^3 + 1$

7. (a)   (b) $(-1, 0)$

$P(x) = -(x + 1)^3$

9. (a)   (b) $x = 1$

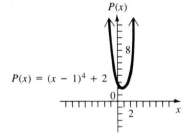

$P(x) = (x - 1)^4 + 2$

11.

$P(x) = 2x(x - 3)(x + 2)$

13.

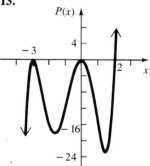

$P(x) = x^2(x - 2)(x + 3)^2$

15.

$P(x) = x^3 - x^2 - 2x$

17.

$P(x) = 3x^4 + 5x^3 - 2x^2$

19.

$P(x) = 2x^3(x^2 - 4)(x - 1)$

21.

$P(x) = x^4 - 4x^2$

23.

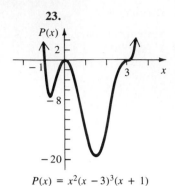

$P(x) = x^2(x - 3)^3(x + 1)$

25. (a)   (b) Between 4 and 5 hr, closer to
5 hr   **(c)** From about 1 hr to about 8 hr

27. (a)   (b) Increasing from $t = 0$
to $t = 6\ 2/3$ and from $t = 10$
on; decreasing from $t = 6\ 2/3$ to
$t = 10$

29. Odd   **31.** Even   **33.** Odd   **35.** Neither   **37.** Even   **39.** Neither
41. Maximum is 26.136 when $x = -3.4$; minimum is 25 when $x = -3$.   **43.** Maximum is 1.048 when
$x = -.1$; minimum is -5 when $x = -1$.   **45.** Maximum is 84 when $x = -2$; minimum is -13 when
$x = -1$.

47.

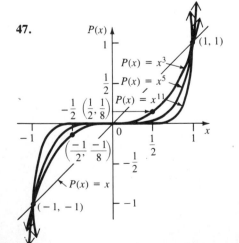

Section 5.2 (page 264)

1. $x^2 - 3x - 2$ **3.** $m^3 - m^2 - 6m$ **5.** $3x^2 + 4x + 3/(x - 5)$
7. $4m^2 - 4m + 1 + [-3/(m + 1)]$ **9.** $x^4 + x^3 + 2x - 1 + 3/(x + 2)$
11. $(1/3)x^2 - (1/9)x + 1/(x - 1/3)$ **13.** $9z^2 + 10z + 27 + 52/(z - 2)$ **15.** $y^2 + y + 1$
17. $x^3 + x^2 + x + 1$ **19.** $P(x) = (x - 1)(x^2 + 2x + 3) - 5$
21. $P(x) = (x + 2)(-x^2 + 4x - 8) + 20$ **23.** $P(x) = (x - 3)(x^3 + 2x + 5) + 20$
25. $P(x) = (x + 1)(3x^3 + x^2 - 11x + 11) + 4$ **27.** -8 **29.** 2 **31.** -30
33. $-6 - i$ **35.** $-2 - 3i$ **37.** Yes **39.** Yes **41.** Yes **43.** Yes **45.** No
47. Yes **49.** No

Section 5.3 (page 269)

1. No **3.** Yes **5.** Yes **7.** No **9.** $P(x) = -(1/6)x^3 + (13/6)x + 2$
11. $P(x) = (-1/2)x^3 - (1/2)x^2 + x$ **13.** $P(x) = -10x^3 + 30x^2 - 10x + 30$
15. $P(x) = x^2 - 10x + 26$ **17.** $P(x) = x^3 - 4x^2 + 6x - 4$ **19.** $P(x) = x^3 - 3x^2 + x + 1$
21. $P(x) = x^4 - 6x^3 + 10x^2 + 2x - 15$ **23.** $P(x) = x^3 - 8x^2 + 22x - 20$
25. $P(x) = x^4 - 4x^3 + 5x^2 - 2x - 2$ **27.** $P(x) = x^4 - 10x^3 + 42x^2 - 82x + 65$
29. $P(x) = x^5 - 12x^4 + 74x^3 - 248x^2 + 445x - 500$ **31.** $P(x) = x^4 - 6x^3 + 17x^2 - 28x + 20$

33. $-1 + i, -1 - i$ **35.** $3, -2 - 2i$ **37.** $-\dfrac{1}{2} + \dfrac{\sqrt{5}}{2}i, -\dfrac{1}{2} - \dfrac{\sqrt{5}}{2}i$ **39.** $i, 2i, -2i$

41. $3, -2, 1 - 3i$ **43.** $2 - i, 1 + i, 1 - i$ **45.** $P(x) = (x - 2)(2x - 5)(x + 3)$
47. $P(x) = (x + 4)(3x - 1)(2x + 1)$ **49.** $P(x) = (x - 3i)(x + 4)(x + 3)$
51. $P(x) = (x - 1 - i)(2x - 1)(x + 3)$ **53.** Zeros are -2 (multiplicity 2), $-1, 3$; $P(x) =$
$(x + 2)^2(x + 1)(x - 3)$ **55.** At $t = 3$ and $t = -2$

Section 5.4 (page 274)

1. $\pm 1, \pm 1/2, \pm 1/3, \pm 1/6$ **3.** $\pm 1, \pm 1/2, \pm 1/3, \pm 1/4, \pm 1/6, \pm 1/12, \pm 2, \pm 2/3$ **5.** $\pm 1, \pm 1/2,$
$\pm 2, \pm 4, \pm 8$ **7.** $\pm 1, \pm 2, \pm 5, \pm 10, \pm 25, \pm 50$ **9.** $-1, -2, 5$ **11.** $2, -3, -5$
13. No rational zeros **15.** $1, -2, -3, -5$ **17.** $-4, 3/2, -1/3$; $P(x) = (x + 4)(2x - 3)(3x + 1)$
19. $-3/2, -2/3, 1/2$; $P(x) = (2x - 1)(3x + 2)(2x + 3)$ **21.** $1/2$; $P(x) = (2x - 1)(x^2 + 4x + 8)$
23. $-2, -1, 1/2, 1$; $P(x) = (x + 2)(x + 1)(2x - 1)(x - 1)$ **25.** $-1, -2, -2/3, 2$;
$P(x) = (x + 1)(x + 2)(x - 2)(3x + 2)$ **27.** $-2, 2/3$; $P(x) = (x + 2)(3x - 2)(x^2 + 1)$
29. $-2, -1, 1$ (multiplicity 2), $5/2$; $P(x) = (x + 2)(x + 1)(x - 1)^2(2x - 5)$ **31.** $-2/3, -1, 3$
33. $1, -5/4$ **35.** 1

Section 5.5 (page 282)

1. 2 or 0 positive real zeros; 1 negative real zero **3.** 1 positive real zero; 1 negative real zero
5. 2 or 0 positive real zeros; 3 or 1 negative real zeros **23. (a)** Positive: 1; negative: 2 or 0 **(b)** -3,
$-1.4, 1.4$ **25. (a)** Positive: 2 or 0; negative: 1 **(b)** $-2, 1.6, 4.4$ **27. (a)** Positive: 1; negative: 0
(b) 1.5 **29. (a)** Positive: 3 or 1; negative: 1 **(b)** $1.1, -.7$ **31. (a)** Positive: 1; negative: 1
(b) $-1.5, 3.1$

33.

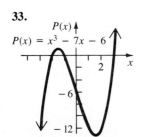

$P(x) = x^3 - 7x - 6$

35.

$P(x) = x^4 - 5x^2 + 6$

37.

$P(x) = 6x^3 + 11x^2 - x - 6$

39.

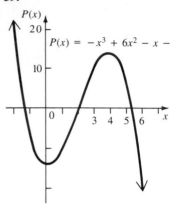

$P(x) = -x^3 + 6x^2 - x - 14$

41. 3.24, −1.24

43. −3.65, −.32, 1.65, 6.32

Section 5.6 (page 292)

1. Vertical asymptote: $x = 5$; horizontal asymptote: $y = 0$ **3.** Vertical asymptote: $x = 7/3$; horizontal asymptote: $y = 0$ **5.** Vertical asymptote: $x = -2$; horizontal asymptote: $y = -1$
7. Vertical asymptote: $x = -9/2$; horizontal asymptote: $y = 3/2$ **9.** Vertical asymptotes: $x = 3, x = 1$; horizontal asymptote: $y = 0$ **11.** Vertical asymptote: $x = -3$; oblique asymptote: $y = x - 3$
13. Vertical asymptotes: $x = -2, x = 5/2$; horizontal asymptote: $y = 1/2$ **15.** No vertical asymptotes; horizontal asymptote: $y = 2$

17.

19.

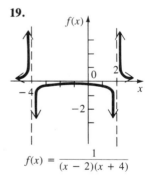

$f(x) = \dfrac{1}{(x - 2)(x + 4)}$

21.

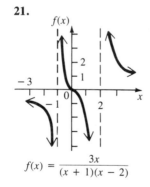

$f(x) = \dfrac{3x}{(x + 1)(x - 2)}$

23.

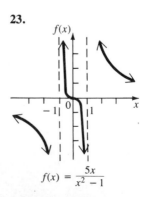

$f(x) = \dfrac{5x}{x^2 - 1}$

25.

$f(x) = \dfrac{3x}{x - 1}$

27.

$f(x) = \dfrac{x + 1}{x - 4}$

29.

$$f(x) = \frac{x}{x^2 - 9}$$

31.

$$f(x) = \frac{x^2 - 3x + 2}{x - 3}$$

33.

$$f(x) = \frac{2x^2 + 3}{x - 4}$$

35.

$$f(x) = \frac{x^2 + 2x}{2x - 1}$$

37.

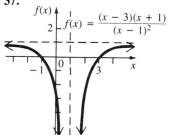

$$f(x) = \frac{(x - 3)(x + 1)}{(x - 1)^2}$$

39.

$$f(x) = \frac{(x - 5)(x - 2)}{x^2 + 9}$$

41.

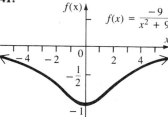

$$f(x) = \frac{-9}{x^2 + 9}$$

43.

$$f(x) = \frac{x^2 + 2x - 3}{x - 1}$$

45.

$$C(x) = \frac{10x}{49(101 - x)}$$

47. (a) 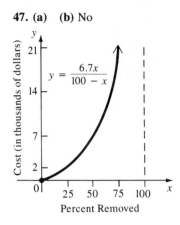 **(b)** No

$$y = \frac{6.7x}{100 - x}$$

Cost (in thousands of dollars)

Percent Removed

49. (a) $65.5 tens of millions, or $655,000,000 **(b)** $64 tens of millions, or $640,000,000 **(c)** $60 tens of millions or $600,000,000 **(d)** $40 tens of millions or $400,000,000 **(e)** $0

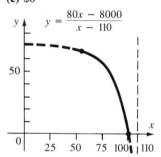

$$y = \frac{80x - 8000}{x - 110}$$

Chapter 5 Review Exercises (page 297)

These review exercises cover the material in the chapter as follows: 1–9, Section 5.1; 11–21, Section 5.2; 23–37, Section 5.3; 39–45, Section 5.4; 47–57, Section 5.5; 59–65, Section 5.6.

1.

$P(x) = x^3 + 5$

3.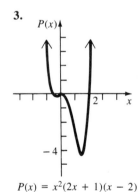

$P(x) = x^2(2x + 1)(x - 2)$

5.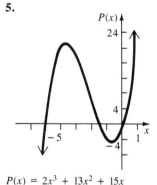

$P(x) = 2x^3 + 13x^2 + 15x$

7. Odd   **9.** Odd   **11.** $2x^2 - 3x + 2 + (-8)/(x + 1)$ **13.** $4m^2 + 9m + 18 + (24)/(m - 2)$
15. $P(z) = (z + 2)(2z^2 + 11z + 8) + 4$ **17.** $P(m) = (m - 4)(2m^2 - 5m - 3) + 3$
19. 11   **21.** 28   **23.** $P(x) = x^3 - 10x^2 + 17x + 28$ **25.** $P(x) = x^4 - x^3 - 9x^2 + 7x + 14$
27. No   **29.** No   **31.** $P(x) = -2x^3 + 6x^2 + 12x - 16$ **33.** $P(x) = x^4 - 3x^2 - 4$
35. $P(x) = x^3 + x^2 - 4x + 6$
37. $3 + i, 3 - i, 2i, -2i; P(x) = (x - 3 - i)(x - 3 + i)(x - 2i)(x + 2i)$ **39.** $1/2, -1, 5$
41. $4, -1/2, -2/3$ **43.** None   **53.** $-2.3, 4.6$

55.

$P(x) = 2x^3 - 11x^2 - 2x + 2$

57.

$P(x) = x^3 + 3x^2 - 4x - 2$

59.

$f(x) = \dfrac{8}{x}$

61.

$f(x) = \dfrac{4x - 2}{3x + 1}$

63.

$f(x) = \dfrac{2x}{x^2 - 1}$

65.

$f(x) = \dfrac{x^2 - 1}{x}$

67.

$f(x) = \dfrac{4x^2 - 9}{2x + 3}$

Chapter 5   Test (page 299)

1.

$P(x) = x^3 - 3$

2.

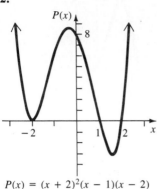

$P(x) = (x + 2)^2(x - 1)(x - 2)$

3. Odd   **4.** $3x^2 - 2x - 5 + 16/(x + 2)$ **5.** $2x^2 - x - 5 + 3/(x - 5)$ **6.** -7

7. $P(x) = x^3 + 3x^2 - 28x$ **8.** Yes, because when $P(x)$ is divided by $x - 3$, the remainder is 0.

9. Yes, because $P(-2) = 0$.   **10.** $P(x) = 5x^3 + 5x^2 + 45x + 45$ **11.** $P(x) = x^3 - 3x^2 + 4x - 12$

12. $3i, -3i, -\dfrac{1}{2} \pm \dfrac{\sqrt{3}}{2}i$ **13.** $5/2, -1$ (multiplicity 2)   **14.** None

15. $1.5, -.6$ **16.**

$P(x) = x^3 - 3x + 4$

17. Alternating signs in the last row of the synthetic division show that -1 is a lower bound.

18. Since $P(-1) = -1$ and $P(0) = 5$, the intermediate value theorem ensures a zero in $[-1, 0]$.

19.

$f(x) = \dfrac{2x^2}{(x - 1)^2}$

20.

$f(x) = \dfrac{1}{x^2 - 3x}$

CHAPTER 6

Section 6.1 (page 310)

1. 8   **3.** 5   **5.** 4/9   **7.** $\sqrt{6}/3$

9. (a)

$f(x) = 2^x + 1$

(b)

$f(x) = 2^x - 4$

(c)

$f(x) = 2^{x+1}$

(d)

$f(x) = 2^{x-4}$

11.

$f(x) = 3^x$

13.

$f(x) = \left(\dfrac{3}{2}\right)^x$

15.

$f(x) = e^{-x}$

17.

$f(x) = e^{x+1}$

19.

$f(x) = 2^{|x|}$

21.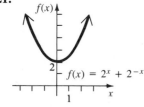

$f(x) = 2^x + 2^{-x}$

23. {1/2} **25.** {−2} **27.** {0} **29.** {3} **31.** {−8, 8} **33.** {1/5}
35. {0} **37.** {3/5} **39.** {−2/3}

43.

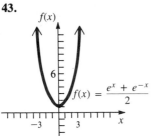

$$f(x) = \frac{e^x + e^{-x}}{2}$$

45.

$$f(x) = x^2 \cdot 2^{-x}$$

47.

$$f(x) = x \cdot e^x$$

49. (a) 500 g **(b)** 409 g **(c)** 335 g **(d)** 184 g **(e)**

$$Q(t) = 500e^{-.05t}$$

51. (a) \$4049.28 **(b)** \$4257.61 **(c)** \$4332.45 **53.** \$3516.56 **55.** \$13,891.16
57. \$76,855.95 **59.** \$5583.95 **61.** \$41,845.63 **63.** 12.0% **65.** 8.0%
67. \$37,764.16 **69.** \$75,780.31 **71. (a)** \$9404.63 **(b)** \$9405.31 **(c)** \$9405.34
73. \$10,917.55 **75.** \$1020.99 **77.** \$6549.85 **79. (a)** About 2,700,000 **(b)** About 3,000,000
(c) About 9,600,000 **(d)** About 38,000,000 **81.** 2.717

Section 6.2 (page 319)
1. $\log_3 81 = 4$ **3.** $\log_{10} 10,000 = 4$ **5.** $\log_{1/2} 16 = -4$ **7.** $\log_{10} .0001 = -4$
9. $6^2 = 36$ **11.** $(\sqrt{3})^8 = 81$ **13.** $10^{-4} = .0001$ **15.** $m^n = k$ **17.** 2 **19.** 1
21. −3 **23.** 1/2 **25.** −1/6 **27.** 9 **29.** 8 **31.** 9 **33.** {5} **35.** {1/5}
37. $\log_3 2 - \log_3 5$ **39.** $\log_2 6 + \log_2 x - \log_2 y$ **41.** $1 + (1/2)\log_5 7 - \log_5 3$ **43.** Cannot
be simplified using the properties of logarithms **45.** $\log_k p + 2\log_k q - \log_k m$
47. $(1/2)(\log_m 5 + 3\log_m r - 5 \cdot \log_m z)$ **49.** $\log_a (xy)/m$ **51.** $\log_m (a^2/b^6)$ **53.** $\log_x (a^{3/2}b^{-4})$
or $\log_x (a^{3/2}/b^4)$ **55.** $\log_b [7x(x + 2)/8]$ **57.** $\log_a [(z - 1)^2(3z + 2)]$
59. $\log_5 (5^{1/3}m^{-1/3})$ or $\log_5 (5^{1/3}/m^{1/3})$
61. (a)

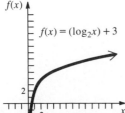

$$f(x) = (\log_2 x) + 3$$

(b)

$$f(x) = \log_2(x + 3)$$

(c)

$$f(x) = |\log_2(x + 3)|$$

63.

$f(x) = \log_3 x$

65.

$f(x) = \log_{1/2}(1-x)$

67.

$f(x) = \log_2 x^2$

69.

$f(x) = x \cdot \log_{10} x$

71. .7781   **73.** .9542   **75.** 1.4771

77. (a) About 239
(b) About 477
(c) About 759   **(d)**

$F(t) = 500 \log_{10}(2t + 3)$

Section 6.3 (page 325)
1. 1.6335   **3.** 3.0959   **5.** −1.8539   **7.** 4.7782   **9.** 6.3630   **11.** −2.5257
13. 11.3022   **15.** 5.8579   **17.** 2636.33   **19.** .0039   **21.** 1.43   **23.** .59   **25.** .96
27. 1.89

29.

$y = \ln x$

31.

$y = x \ln x$

33. 31.2

35. (a) About 350 yr **(b)** About 4000 yr **(c)** About 2300 yr **37. (a)** 21 **(b)** 70 **(c)** 91 **(d)** 120
(e) 140 **39.** $398,000,000I_0$ **41.** 3.2 **43.** 1.8 **45.** 2×10^{-3} **47.** 1.6×10^{-5}
49. 1 **51. (a)** 2.03 **(b)** 2.28 **(c)** 2.17 **(d)** 1.21 **53.** 378 ft **55. (a)** 7,300,000; 77,300,000;
168,000,000; 175,000,000 **(b)** According to the given equation, the population can never increase to
197,273,000.
57.

$p = 86.3 \ln h - 680$
$(h \geq 3000)$

59. .63 **61. (a)** 3 **(b)** $5^2 = 25$ **(c)** $1/e$ or e^{-1} **63. (a)** 5 **(b)** ln 3 **(c)** 2 ln 3 or ln 9

Section 6.4 (page 334)
1. {1.631} **3.** {−.535} **5.** {−.080} **7.** {2.386} **9.** {−.123} **11.** Ø **13.** {2}
15. {.500} **17.** {4} **19.** {17.475} **21.** {11} **23.** {5} **25.** {10} **27.** {4}
29. Ø **31.** {11} **33.** {3} **35.** Ø **37.** {8} **39.** {−2, 2} **41.** {1, 10}
43. {2.423} **45.** {−.204} **47.** About 34.7 sec **49. (a)** About 495 rabbits
(b) About 2.75 mo **51.** About 5600 yr **53.** About 15,000 yr **55.** About 3280 yr
57. 23 days **59.** $t = (1000/k)\ln (P/P_0)$ **61.** $t = (-1/k)\log_{10} [(T - T_0)/(T_1 - T_0)]$
63. $I = I_0 \cdot 10^{d/10}$ **65.** 4.9 **67.** 5.30 **69.** 3.8 **71.** 9.3% **73. (a)** 611 million
(b) 1007 million **(c)** 2028 million

Chapter 6 Review Exercises (page 339)

These review exercises cover the material in the chapter as follows: 1–13, Section 6.1;
15–27, Section 6.2; 29–47, Section 6.3; 49–83, Section 6.4.

1. **3.**

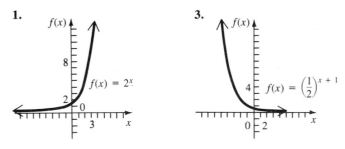

5. {5/3} **7.** {3/2} **9.** 800 g **11.** $1790.19 **13.** About 12 yr **15.** $\log_2 32 = 5$
17. $\log_{1/16} 1/2 = 1/4$ **19.** $\log_{10} 3 = .4771$ **21.** $10^{-3} = .001$ **23.** $10^{.537819} = 3.45$
25. $\log_3 m + \log_3 n - \log_3 5 - \log_3 r$ **27.** $2 \log_5 x + 4 \log_5 y + (1/5)(3 \log_5 m + \log_5 p)$
29. 1.659 **31.** −3.252 **33.** 3.555 **35.** 11.878 **37.** 2.255 **39.** 6.049
41. 3147.748 **43.** .015 **45.** .8m **47.** 1.5m **49.** $1999.00 **51.** $3875.79
53. $3576.92 **55.** $5101.93 **57.** $2663.54 **59.** $1293.68 **61.** $3006.16
63. About 6.9 yr **65.** About 5.3 yr **67.** 15.8% **69.** 2 yr **71.** $12,590.76
73. $580,792.63 **75.** {1.490} **77.** {1.303} **79.** {4} **81.** {3} **83.** {−3}

Chapter 6   Test (page 341)

1.

$f(x) = (1.5)^{x+2}$

2.

$f(x) = \log_{1/2} x$

3. $\{5\}$ **4.** $\{1/32\}$ **5.** 329.3 g   **6.** $6529.24

7. 17.3 yr   **8.** $\log_a b = 2$ **9.** $\ln 4.82 = c$ **10.** $3^{3/2} = \sqrt{27}$ **11.** $e^a = 5$
12. $2 \log_7 x + (1/4)\log_7 y - 3 \log_7 z$ **13.** 7.741   **14.** 4.581   **15.** 238.013   **16.** $5838.08
17. 10.7   **18.** $\{3.107\}$ **19.** $\{7\}$ **20.** $\{20.125\}$

CHAPTER 7

Section 7.1 (page 350)
1. $\{(6, 15)\}$ **3.** $\{(2, -3)\}$ **5.** $\{(4, 3)\}$ **7.** $\{(1, 2)\}$ **9.** $\{(4.35, -.986)\}$ (rounded)
11. $\{(3, 6)\}$ **13.** $\{(-1, 4)\}$ **15.** $\emptyset$ **17.** $\{(1, 3)\}$ **19.** $\{(4, -2)\}$
21. Dependent equations   **23.** $\{(12, 6)\}$ **25.** $\{(4, 3)\}$ **27.** $\{(2, 2)\}$ **29.** $\{(1/5, 1)\}$
31. $\{(-4, -1)\}$ **33.** $\{(.38, -.47)\}$ **35.** $a = -3; b = -1$ **37.** $m = (1/3)b = 13/3$
39. $16,000 at 12%; $8000 at 11%   **41.** A: 400 kg; B: 300 kg   **43.** 40/3 gal of 98 octane; 80/3 gal
of 92 octane   **45.** Goats cost $30; sheep cost $25.   **47.** 32 days at $28   **49.** $18,000 at 9%;
$12,000 at 10%   **51.** $x = 63; R = C = 346.5$ **53.** $x = 2100; R = C = 52,000$
55. $x = 370; R = C = 6900$ **57.** 32; 80   **59.** 28/3; 28   **61.** 5; 50/3
63. (a) 16   **(b)** 11   **(c)** 6   **(d)** 8   **(e)** 4   **(f)** 0   **65. (a)** See graph below.   **(b)** 125   **(c)** 50
(g) and **(k)** are graphed below.   **(h)** 0   **(i)** 40/3
(j) 80/3   **(l)** 8   **(m)** 6

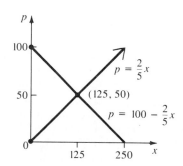

Section 7.2 (page 357)

1. $\{(1, 2, -1)\}$ **3.** $\{(2, 0, 3\}$ **5.** $\emptyset$ **7.** $\{(1, 2, 3)\}$ **9.** $\{(-1, 2, 1)\}$ **11.** $\{(4, 1, 2)\}$
13. $\{(1/2, 2/3, -1)\}$ **15.** $\{(2, 4, 2)\}$ **17.** $\{(-1, 1, 1/3)\}$ **19.** $\{(.4, -.8, .7)\}$
21. $\emptyset$ **23.** $\{(x, -4x + 3, -3x + 4)\}$ **25.** $\{(x, -x + 15, -9x + 69)\}$
27. $\{(x, (-15 + x)/3, (24 + x)/3\}$ **29.** $a = 3/4; b = 1/4; c = -1/2$ **31.** $x^2 + y^2 + x - 7y = 0$
33. 6 ones; 9 fives; 15 twenties   **35.** 30 barrels each of \$150 and \$190 glue; 210 barrels of \$120
glue   **37.** 15 cm; 12 cm; 6 cm   **39.** \$50,000 at 10%; \$10,000 at 9%; and \$40,000 at 7.5%

Section 7.3 (page 366)

1. $\begin{bmatrix} 2 & 4 \\ 0 & -1 \end{bmatrix}$ **3.** $\begin{bmatrix} 1 & \frac{1}{5} \\ 0 & 1 \end{bmatrix}$ **5.** $\begin{bmatrix} 1 & 5 & 6 \\ 0 & 13 & 11 \\ 4 & 7 & 0 \end{bmatrix}$ **7.** $\begin{bmatrix} -3 & 1 & -4 \\ 2 & 1 & 3 \\ -17 & 0 & -13 \end{bmatrix}$

9. $\left[\begin{array}{cc|c} 2 & 3 & 11 \\ 1 & 2 & 8 \end{array}\right]$ **11.** $\left[\begin{array}{cc|c} 1 & 5 & 6 \\ 1 & 2 & 8 \end{array}\right]$ **13.** $\left[\begin{array}{ccc|c} 2 & 1 & 1 & 3 \\ 3 & -4 & 2 & -7 \\ 1 & 1 & 1 & 2 \end{array}\right]$ **15.** $\left[\begin{array}{ccc|c} 1 & 1 & 0 & 2 \\ 0 & 2 & 1 & -4 \\ 0 & 0 & 1 & 2 \end{array}\right]$

17. $\left[\begin{array}{ccc|c} 1 & 0 & 0 & 5 \\ 0 & 1 & 0 & -2 \\ 0 & 0 & 1 & 3 \end{array}\right]$ **19.** $x = 2, y = 3$ **21.** $2x + y = 1, 3x - 2y = -9$

23. $x = 2, y = 3, z = -2$ **25.** $3x + 2y + z = 1, 2y + 4z = 22, -x - 2y + 3z = 15$
27. $\{(2, 3)\}$ **29.** $\{(-3, 0)\}$ **31.** $\{(7/2, -1)\}$ **33.** $\{(5/2 - 1)\}$ **35.** $\emptyset$
37. $\{(3y + 1)/6, y)\}$ **39.** $\{(-2, 1, 3)\}$ **41.** $\{(-1, 23, 16)\}$ **43.** $\{(3, 2, -4)\}$
45. $\{(2, 1, -1)\}$ **47.** $\emptyset$ **49.** $\{(-1, 2, 5, 1)\}$ **51.** $\{(1.7, 2.4)\}$ **53.** $\{(.5, -.7, 6)\}$
55. \$3000 at 8%; \$1000 at 5%; \$6000 at 9%   **57.** \$1.25 for unleaded; \$1.22 for regular; \$1.23 for premium
59. $\{(z + 14/5, z - 12/5, z)\}$ **61.** $\{((12 - z)/7, (4z - 6)/7, z)\}$ **63.** $\emptyset$ **65. (a)** $x_3 + x_4 = 600$
(b) The system has no unique solution; equations $x_2 + x_3 = 700$ and $x_3 + x_4 = 600$ are dependent.
(c) $x_2 - x_4 = 100$ **(d)** $x_4 = 1000 - x_1$; 1000   **(e)** $x_4 = x_2 - 100$; 100   **(f)** $x_4 = 600 - x_3$; 600, 600
(g) 600; 600; 700; 1000

Section 7.4 (page 376)

1. $\{(1, 1), (-2, 4)\}$ **3.** $\{(2, 1), (1/3, 4/9)\}$ **5.** $\{(2, 12), (-4, 0)\}$ **7.** $\{(2, -3), (-6/5, 17/5)\}$
9. $\{(-3/5, 7/5), (-1, 1)\}$ **11.** $\{(2, 2), (2, -2), (-2, 2), (-2, -2)\}$ **13.** $\{(0, 0)\}$
15. $\{(i, \sqrt{6}), (-i, \sqrt{6}), (i, -\sqrt{6}), (-i, -\sqrt{6})\}$ **17.** $\{(1, -1), (-1, 1), (1, 1), (-1, -1)\}$
19. $\{(3, 6), (-3, 6), (3, -6), (-3, -6)\}$ **21.** Same circles or $\{(x, y)|x^2 + y^2 = 10\}$
23. $\{(i\sqrt{15}/5, 2\sqrt{10}/5), (i\sqrt{15}/5, -2\sqrt{10}/5), (-i\sqrt{15}/5, 2\sqrt{10}/5), (-i\sqrt{15}/5, -2\sqrt{10}/5)\}$
25. $\{(2, 3), (3, 2)\}$ **27.** $\{(-3, 5), (15/4, -4)\}$ **29.** $\{(4, -1/8), (-2, 1/4)\}$
31. $\{(3, 2), (-3, -2), (4, 3/2), (-4, -3/2)\}$ **33.** $\{(3, 5), (-3, -5), (5i, -3i), (-5i, 3i)\}$
35. $\{(\sqrt{5}, 0), (-\sqrt{5}, 0), (\sqrt{5}, \sqrt{5}), (-\sqrt{5}, -\sqrt{5})\}$ **37.** $\{(3, -3), (3, 3)\}$
39. $\{((1 + \sqrt{13})/2, (-1 + \sqrt{13})/2), ((-1 - \sqrt{21})/2, (3 + \sqrt{21})/2)\}$ **41.** -2 and 8   **43.** yes
45. 24 cm and 18 cm   **47.** $4 - \sqrt{3}$ or about 2.3 sec ($4 + \sqrt{3}$ makes the height of the disc negative)

Section 7.5 (page 381)

1.

$x + y \le 4$
$x - 2y \ge 6$

3.

$4x + 3y < 12$
$y + 4x > -4$

5.

$x + y \le 6$
$2x + 2y \ge 12$

Only the points of the line are included.

7.

$x + 2y \le 4$
$y \ge x^2 - 1$

9.

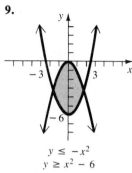

$y \le -x^2$
$y \ge x^2 - 6$

11.

$x^2 - y^2 < 1$
$-1 < y < 1$

13.

$2x^2 - y^2 > 4$
$2y^2 - x^2 > 4$

15.

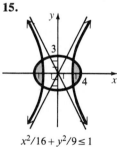

$x^2/16 + y^2/9 \le 1$
$x^2/4 - y^2/16 \ge 1$

17.

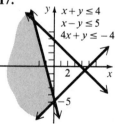

$x + y \le 4$
$x - y \le 5$
$4x + y \le -4$

19. No solution

21.

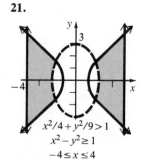

$x^2/4 + y^2/9 > 1$
$x^2 - y^2 \ge 1$
$-4 \le x \le 4$

23.

$y \ge 3^x$
$y \ge 2$

25.

$|x| \geq 2$
$|y| \geq 4$
$y < x^2$

27.

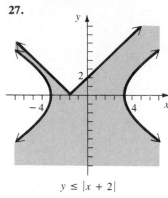

$y \leq |x + 2|$
$x^2/16 - y^2/9 \leq 1$

29.

$y \geq |4 - x|$
$y \geq |x|$

31. (6/5, 6/5); 42/5   **33.** (17/3, 5); 49/3   **35.** $112 with 4 pigs, 12 geese   **37.** 8 of #1; 3 of #2   **39.** 6.4 million gallons of gasoline and 3.2 million gallons of fuel oil, for maximum revenue of $16,960,000

Chapter 7   Review Exercises (page 385)

These review exercises cover the material in the chapter as follows: 1–17, Section 7.1; 19–27, Section 7.2; 29–31, Section 7.3; 33–43, Section 7.4; 45–49, Section 7.5.

1. {(5, 7)}   **3.** {(−5/4, −79/40)}   **5.** Dependent equations; {((2y − 11)/5, y)}   **7.** {(−1, 3)}
9. {(6, −12)}   **11.** {(2, 5)}   **13.** 10 at 25¢, 12 at 50¢   **15.** $10,000 at 6%, $20,000 at 7%, $20,000 at 9%   **17.** 17.5 g of 12-carat, 7.5 g of 22-carat   **19.** {(−1, 2, 3)}   **21.** {(1, 3, −1)}
23. 10 lb of $4.60 tea; 8 lb of $5.75 tea; 2 lb of $6.50 tea   **25.** 164 fives; 86 tens; 40 twenties
27. {((15z + 6)/11, (14z − 1)/11, z)}   **29.** {(−4, 6)}   **31.** {(−3, 2)}   **33.** {(0, 1, 0)}
35. {(−2, 3), (1, 0)}   **37.** {(2, 3), (−2, 3), (2, −3), (−2, −3)}   **39.** {(6, 2/3), (−4, −1)}
41. Yes; {((8 − 8√41)/5, (16 + 4√41)/5), ((8 + 8√41)/5, (16 − 4√41)/5)}   **43.** 10 and 6 or −10 and −6

45.

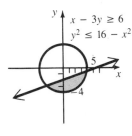

$x - 3y \geq 6$
$y^2 \leq 16 - x^2$

47. $2x + 3y/2 \leq 16$;
$3x + 2y/3 \leq 12$;
$x \geq 0; y \geq 0$

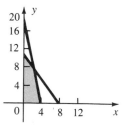

49. Maximum of 24 at (0, 6)

Chapter 7   Test (page 387)

1. {(1, 2)}   **2.** {(4, 3)}   **3.** {((−3y − 7)/2, y}; dependent   **4.** $\left\{ \left(4, \dfrac{5}{2} \right) \right\}$ **5.** {(1, 2)}
6. Ø; inconsistent   **7.** 10 nickels, 13 dimes   **8.** $7,200 at 4%, $8,200 at 5%   **9.** {(5, 3, 6)}

10. $\{(2, 0, -1)\}$ **11.** First number, 1; second number, 1; third number, 0   **12.** $\{(5, 1)\}$
13. $\{(2, 2z - 1, z)\}$ **14.** $\emptyset$ **15.** $\{(2, 3), (-2, 1)\}$ **16.** $\{(3, 4), (4, 3)\}$ **17.** No real number
solutions   **18.**

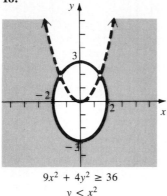

$$9x^2 + 4y^2 \geq 36$$
$$y < x^2$$

19. A minimum of 8 at $(1, 6)$.
20. $\{((4 - z)/9, (-7 + 13z)/9, z)\}$

CHAPTER 8

Section 8.1 (page 394)
1. 2×2; square   **3.** 3×4 **5.** 2×1; column   **7.** 4×1; column   **9.** 3×3; square
11. $x = 2; y = 4; z = 8$ **13.** $m = 8; n = -2; z = 2; y = 5; w = 6$ **15.** $x = -15; y = 5$;
$k = 3$ **17.** $m = 6; n = -9$ **19.** $z = 18; r = 3; s = 3; p = 3; a = 3/4$

21. $\begin{bmatrix} 14 & -11 & -3 \\ -2 & 4 & -1 \end{bmatrix}$ **23.** $\begin{bmatrix} -12 & 16 \\ 4 & 2 \end{bmatrix}$ **25.** $\begin{bmatrix} 22 & -43 \\ 14 & -27 \\ -11 & 52 \end{bmatrix}$ **27.** Not possible

29. $\begin{bmatrix} 13 & 3 & 0 & -2 \\ 9 & -12 & 4 & 8 \\ 12 & -11 & -1 & 9 \end{bmatrix}$ **31.** $\begin{bmatrix} -12x + 8y & -x + y \\ x & 8x - y \end{bmatrix}$ **33.** $\begin{bmatrix} 10.874 & 17.011 \\ 12.609 & 15.573 \\ 8.892 & 14.517 \end{bmatrix}$

35. $\begin{bmatrix} -4 & 8 \\ 0 & 6 \end{bmatrix}$ **37.** $\begin{bmatrix} 8 & -16 \\ 0 & -12 \end{bmatrix}$ **39.** $\begin{bmatrix} 2 & 6 \\ -4 & 6 \end{bmatrix}$ **41.** $\begin{bmatrix} 60 & -10 \\ -44 & 9 \end{bmatrix}$ **43.** $\begin{bmatrix} 6 & 8 \\ -8 & 9 \end{bmatrix}$

45. $[18.2 \quad 19.0 \quad 17.6 \quad 18.5]$; $\begin{bmatrix} 18.2 \\ 19.0 \\ 17.6 \\ 18.5 \end{bmatrix}$ **47.** $\begin{bmatrix} 7 & 2 \\ 9 & 0 \\ 8 & 6 \end{bmatrix}$; $\begin{bmatrix} 7 & 9 & 8 \\ 2 & 0 & 6 \end{bmatrix}$ **49.** Yes, always true

51. Yes, always true   **53.** Yes, always true

Section 8.2 (page 403)

1. 2×2; 2×2 **3.** 4×4; 2×2 **5.** 3×2; BA cannot be found   **7.** AB cannot be found;
3×2 **9.** AB cannot be found; BA cannot be found.   **11.** $\begin{bmatrix} 13 \\ 25 \end{bmatrix}$ **13.** $\begin{bmatrix} -17 \\ -1 \end{bmatrix}$

15. $\begin{bmatrix} -17 & -10 \\ 1 & 2 \end{bmatrix}$ **17.** $\begin{bmatrix} -2 & 10 \\ 0 & 8 \end{bmatrix}$ **19.** $\begin{bmatrix} -2 & 5 & 0 \\ 6 & 6 & 1 \\ 12 & 2 & -3 \end{bmatrix}$ **21.** $[2 \quad 7 \quad -4]$

23. $\begin{bmatrix} 20 & 0 & 8 \\ 8 & 0 & 4 \end{bmatrix}$ **25.** Cannot be multiplied   **27.** $[-8]$ **29.** $\begin{bmatrix} -6 & 12 & 18 \\ 4 & -8 & -12 \\ -2 & 4 & 6 \end{bmatrix}$

31. $\begin{bmatrix} 11.7144 & 22.199 \\ 25.8752 & 53.5427 \end{bmatrix}$ **33.** $\begin{bmatrix} 16 & 22 \\ 7 & 19 \end{bmatrix}$ **35.** $\begin{bmatrix} -10 & 16 & 26 \\ 5 & 7 & 22 \end{bmatrix}$ **37.** No; no   **39.** Always
true   **41.** Always true   **43.** Always true   **45.** Not true. $(A + B)^2 = (A + B)(A + B) =$
$A^2 + AB + BA + B^2$. Since multiplication is not commutative, $AB + BA$ is not equal to $2AB$.
47. (a) $[3 \quad 1 \quad 2 \quad 4]$ **(b)** $[10.44 \quad 10.51 \quad 10.17]$

Section 8.3 (page 415)

1. Yes   **3.** No   **5.** Yes   **7.** No   **9.** Yes   **11.** $\begin{bmatrix} 0 & \frac{1}{2} \\ -1 & \frac{1}{2} \end{bmatrix}$ **13.** Does not exist

15. $\begin{bmatrix} 2 & 1 \\ -\frac{3}{2} & -\frac{1}{2} \end{bmatrix}$ **17.** $\begin{bmatrix} -2.5 & 5 \\ 12.5 & -15 \end{bmatrix}$ **19.** $\begin{bmatrix} -5 & 15 \\ 10 & -20 \end{bmatrix}$ **21.** $\begin{bmatrix} 1 & 0 & 0 \\ 0 & -1 & 0 \\ -1 & 0 & 1 \end{bmatrix}$

23. $\begin{bmatrix} -\frac{1}{24} & \frac{3}{16} & -\frac{1}{2} \\ \frac{1}{8} & -\frac{1}{16} & \frac{1}{2} \\ \frac{1}{8} & -\frac{1}{16} & -\frac{1}{2} \end{bmatrix}$ **25.** $\begin{bmatrix} 15 & 4 & -5 \\ -12 & -3 & 4 \\ -4 & -1 & 1 \end{bmatrix}$ **27.** $\begin{bmatrix} -\frac{10}{3} & \frac{5}{9} & -\frac{10}{9} \\ \frac{20}{3} & \frac{5}{9} & \frac{80}{9} \\ -5 & \frac{5}{6} & -\frac{20}{3} \end{bmatrix}$

29. $A = \begin{bmatrix} 1 & 1 \\ 2 & -1 \end{bmatrix}$; $X = \begin{bmatrix} x \\ y \end{bmatrix}$; $B = \begin{bmatrix} 8 \\ 4 \end{bmatrix}$ **31.** $A = \begin{bmatrix} 4 & 5 \\ 2 & 3 \end{bmatrix}$; $X = \begin{bmatrix} x \\ y \end{bmatrix}$; $B = \begin{bmatrix} 7 \\ 5 \end{bmatrix}$

33. $A = \begin{bmatrix} 1 & 1 & 1 \\ 2 & 1 & 3 \\ 5 & -2 & 2 \end{bmatrix}$; $X = \begin{bmatrix} x \\ y \\ z \end{bmatrix}$; $B = \begin{bmatrix} 9 \\ 17 \\ 16 \end{bmatrix}$ **35.** $\{(2, 3)\}$ **37.** $\{(-2, 4)\}$ **39.** $\{(4, -6)\}$

41. $\{(5/2, -1)\}$ **43.** $\{(1, 1, 2)\}$ **45.** $\{(1/4, 1/4, 13/4)\}$ **47.** $\{(12, -15/11, -65/11)\}$
49. 6 plates; 3 cups

Section 8.4 (page 422)

1. -36 **3.** 7 **5.** 0 **7.** -26 **9.** -16 **11.** 0 **13.** $2x - 32$ **15.** $y^2 - 16$
17. $x^2 - y^2$ **19.** $-.57$ **21.** -1 **23.** 8 **25.** 17 **27.** 166 **29.** 0 **31.** 0
33. $4x$ **35.** $-.043804$ **37.** $\{-4\}$ **39.** $\{13\}$ **41.** $0, -5, 0$ **43.** $-6, 0, -6$
45. -88 **47.** 1 **49.** $5/2$ **51.** 7 **53.** 8 **55.** $x + 3y - 11 = 0$

Section 8.5 (page 431)

1. Property 5   **3.** Property 1   **5.** By Property 4, multiply each element in the second row by $-1/4$.
The determinant of the resulting matrix is 0, by Property 5.   **7.** Use Property 5 after multiplying each
element in the third column by 1/2.   **9.** Use Property 5 after multiplying each element in the third column
by 1/2.   **11.** Property 2   **13.** Property 3   **15.** Property 4   **17.** Property 6 (first row added to
second row)   **19.** Property 6 (-2 times the elements of the first column added to the second column)

21. Property 6 (3 times the elements of the second row added to the third row)   **23.** Property 6
25. 0   **27.** 0   **29.** -19 **31.** 32   **33.** -32

Section 8.6 (page 437)

1. $\{(2, 2)\}$ **3.** $\{(2, -5)\}$ **5.** $\{(2, 0)\}$ **7.** $\{(2, 3)\}$ **9.** Can't use Cramer's rule, $D = 0$;
$\{((4 - 2y)/3, y)\}$ **11.** Can't use Cramer's rule, $D = 0$; $\emptyset$ **13.** $\{(-1, 2, 1)\}$
15. $\{(-3, 4, 2)\}$ **17.** $\{(0, 0, -1)\}$ **19.** Can't use Cramer's rule, $D = 0$; $\emptyset$ **21.** Can't use
Cramer's rule, $D = 0$; $\{((-32 + 19z)/4, (24 - 13z)/4, z)\}$ **23.** $\{(2, 3, 1)\}$ **25.** $\{(0, 4, 2)\}$
27. $\{(31/5, 19/10, -29/10)\}$ **29.** $\{(1, 0, -1, 2)\}$ **31.** $\{(-.2, -.8)\}$ **33.** $\{(.2, .5, -.2)\}$
35. $\{(-a - b, a^2 + ab + b^2)\}$ **37.** $\{(1, 0)\}$

Chapter 8   Review Exercises (page 442)

*These review exercises cover the material in the chapter as follows: 1–11, Section 8.1;
13–21, Section 8.2; 23–35, Section 8.3; 37–43, Section 8.4; 45–49, Section 8.5; 51–53,
Section 8.6.*

1. $m = -8; p = -7; x = 2; y = 9; z = 5$ **3.** $a = 5; x = 3/2; y = 0; z = 9$

5. $\begin{bmatrix} 0 & -2 & 7 \\ 6 & -1 & 10 \end{bmatrix}$ **7.** Cannot be done   **9.** Cannot be done   **11.** $\begin{bmatrix} 87 & 6.4 \\ 70 & 4.0 \\ 55 & 3.3 \end{bmatrix}$; $\begin{bmatrix} 87 & 70 & 55 \\ 6.4 & 4.0 & 3.3 \end{bmatrix}$

13. $\begin{bmatrix} 11 & 20 \\ 14 & 40 \end{bmatrix}$ **15.** $\begin{bmatrix} -3 \\ 10 \end{bmatrix}$ **17.** $\begin{bmatrix} -2 & 22 & 31 \\ 25 & 12 & 9 \end{bmatrix}$ **19.** Yes   **21.** No

23. $\begin{bmatrix} 3 & -1 \\ -5 & 2 \end{bmatrix}$ **25.** $\begin{bmatrix} \frac{1}{2} & 0 \\ \frac{1}{10} & \frac{1}{5} \end{bmatrix}$ **27.** $\begin{bmatrix} \frac{2}{3} & 0 & -\frac{1}{3} \\ \frac{1}{3} & 0 & -\frac{2}{3} \\ -\frac{2}{3} & 1 & \frac{1}{3} \end{bmatrix}$ **29.** $\{(2, 2)\}$ **31.** $\{(2, 1)\}$

33. $\{(-1, 0, 2)\}$ **35.** Not possible; dependent equations   **37.** -25 **39.** -44
41. $\{-7/3\}$ **43.** Any number   **45.** One row is all zeros.   **47.** Multiply the elements of row 1 of
the second matrix by 2.   **49.** In the first matrix, add to the elements of row 2 the results of multiplying
each element in row 1 by 1.   **51.** $\{(-2, 5)\}$ **53.** $\{(-4, 6, 2)\}$

Chapter 8   Test (page 444)

1. $w = -3, x = -1, y = 7$ **2.** $\begin{bmatrix} 8 & 3 \\ 0 & -11 \\ 15 & 19 \end{bmatrix}$ **3.** Sum does not exist.

4. $\begin{bmatrix} 108 & 112 & 123 & 110 \\ 142 & 98 & 117 & 130 \end{bmatrix}$ **5.** $\begin{bmatrix} -5 & 16 \\ 19 & 2 \end{bmatrix}$ **6.** Product does not exist.

7. $\begin{bmatrix} -26 & -46 & 72 \\ 11 & -19 & -12 \end{bmatrix}$ **8.** Inverses   **9.** $\begin{bmatrix} -2 & -5 \\ -3 & -8 \end{bmatrix}$ **10.** inverse does not exist.

11. $\begin{bmatrix} -9 & 1 & -4 \\ -2 & 1 & 0 \\ 4 & -1 & 1 \end{bmatrix}$ **12.** $\{(-7, 8)\}$ **13.** $\{(0, 5, -9)\}$ **14.** -58 **15.** -844

16. $\{-2\}$ **17.** Exchanging two rows of a matrix reverses the sign of the determinant.   **18.** The value
of a determinant is unchanged if a multiple of a row of the matrix is added to the corresponding elements of
another row.   **19.** $\{(-6, 7)\}$ **20.** $\{(1, -2, 3)\}$

CHAPTER 9

Section 9.1 (page 449)
1. 10, 16, 22, 28, 34   **3.** 2, 4, 8, 16, 32   **5.** 1, -1, 1, -1, 1   **7.** 1/2, 4/5, 1, 8/7, 5/4
9. 4/3, 12/5, 20/7, 28/9, 36/11   **11.** -2, 8, -24, 64, -160 **13.** 2/3, 5/6, 10/11, 17/18, 26/27
15. x, x^2, x^3, x^4, x^5 **17.** 4, 9, 14, 19, 24, 29, 34, 39, 44, 49   **19.** -9, -13, -17, -21, -25,
-29, -33, -37, -41, -45 **21.** 2, 4, 8, 16, 32, 64, 128, 256, 512, 1024   **23.** 1, 1, 2, 3, 5, 8,
13, 21, 34, 55   **25.** 35   **27.** 45   **29.** 29/20   **31.** 139/108   **33.** 3   **35.** 25/2
37. 89   **39.** 31   **41.** $x + x^2 + x^3 + x^4$ **43.** $x + 2x^2 + 3x^3 + 4x^4 + 5x^5$ **45.** 8
47. 29.0

Section 9.2 (page 456)
1. 4, 6, 8, 10, 12   **3.** 11, 9, 7, 5   **5.** 6, 2, -2, -6 **7.** -6, 0, 6, 12, 18, 24
9. Arithmetic; $d = 5$, $a_n = 7 + 5n$ **11.** Arithmetic; $d = -3$, $a_n = 21 - 3n$ **13.** Arithmetic;
$d = 7$, $a_n = -26 + 7n$ **15.** Arithmetic; $d = m$, $a_n = x + nm - m$ **17.** Arithmetic;
$d = -m$, $a_n = 2z + 2m - mn$ **19.** $a_8 = 19$; $a_n = 3 + 2n$ **21.** $a_8 = 7$; $a_n = n - 1$
23. $a_8 = -6$; $a_n = 10 - 2n$ **25.** $a_8 = -3$; $a_n = -39 + 9n/2$
27. $a_8 = x + 21$; $a_n = x + 3n - 3$ **29.** $a_8 = 4m$; $a_n = mn - 4m$ **31.** 215
33. 125   **35.** 230   **37.** 205/2   **39.** 155/2   **41.** 16.745   **43.** 32.28   **45.** 18
47. 140   **49.** 570   **51.** -684 **53.** 500,500   **55.** 400   **57.** -78 **59.** $a_1 = 7$
61. $a_1 = 0$ **63.** $a_1 = 7$ **65.** $a_1 = 1$ **67.** 1281   **69.** 4680 chimes   **71.** 100 m; 550 m

73. 240 ft; 1024 ft   **75.** 170 m   **79.** Not true; $\sum_{i=1}^{n} kx_i = k\sum_{i=1}^{n} x_i$ **81.** True

83. Not true; $\left(\sum_{i=1}^{n} x_i\right)^2 = (x_1 + x_2 + x_3 + \ldots + x_n)^2$ but $\sum_{i=1}^{n} x_i^2 = (x_1^2 + x_2^2 + x_3^2 + \ldots + x_n^2)$

Section 9.3 (page 463)
1. 2, 6, 18, 54   **3.** 1/2, 2, 8, 32   **5.** -2, 6, -18, 54   **7.** 3125, 625, 125, 25, 5, 1, 1/5
9. -1, 1, -1, 1, -1, 1   **11.** 3/2, 3, 6, 12, 24   **13.** $a_5 = 324$; $a_n = 4 \cdot 3^{n-1}$
15. $a_5 = -162$; $a_n = -2 \cdot 3^{n-1}$ **17.** $a_5 = -1875$; $a_n = -3(-5)^{n-1}$
19. $a_5 = 24$; $a_n = (3/2)2^{n-1}$ **21.** $a_5 = -256$; $a_n = -1(-4)^{n-1}$
23. Geometric; $r = 2$; $a_n = 6 \cdot 2^{n-1}$ **25.** Geometric; $r = 2$; $a_n = (3/4)2^{n-1}$
27. Geometric; $r = -1/2$; $a_n = -4(-1/2)^{n-1}$ **29.** Not geometric   **31.** Geometric; $a_n = (9/4)2^{n-1}$
33. Geometric; $a_n = (-2/9)3^{n-1}$ **35.** 93   **37.** 33/4   **39.** 55/9   **41.** 124   **43.** 121/9
45. 860.95   **47.** -6.2737 **49.** $a_1 = 2$, $r = 3$ or $a_1 = -2$, $r = -3$
51. $a_1 = 128$, $r = 1/2$ or $a_1 = -128$, $r = -1/2$ **53.** 30   **55.** 60   **57.** 255/4
59. 406/625   **61.** 585/8   **63.** 120   **65.** 272/25   **67.** 2^{30} or $1,073,741,824$
69. $7,875,000   **71.** .00002%   **73.** $(1/32) \times 10^{15}$ molecules   **75.** $12,487.56
77. $40,634.48

Section 9.4 (page 469)
1. 2   **3.** 10   **5.** 1/2; sum would exist   **7.** 1.1   **9.** $-.9$; sum would exist   **11.** 64/3
13. 1000/9   **15.** 135   **17.** 512/3   **19.** 81   **21.** 3/2   **23.** 2   **25.** 1/5
27. Cannot be found   **29.** 1/3   **31.** Cannot be found   **33.** $-1/5$ **35.** 1/4   **37.** 1/9
39. Cannot be found   **41.** 5/9   **43.** 4/33   **45.** 31/99   **47.** 508/999   **49.** 311/900
51. 70 m   **53.** 1600   **55.** 12 m

Section 9.5 (page 475)

1. $x^6 + 6x^5y + 15x^4y^2 + 20x^3y^3 + 15x^2y^4 + 6xy^5 + y^6$ **3.** $p^5 - 5p^4q + 10p^3q^2 - 10p^2q^3 + 5pq^4 - q^5$
5. $r^{10} + 5r^8s + 10r^6s^2 + 10r^4s^3 + 5r^2s^4 + s^5$ **7.** $p^4 + 8p^3q + 24p^2q^2 + 32pq^3 + 16q^4$
9. $2401p^4 + 2744p^3q + 1176p^2q^2 + 224pq^3 + 16q^4$ **11.** $729x^6 - 2916x^5y + 4860x^4y^2 - 4320x^3y^3 +$
$2160x^2y^4 - 576xy^5 + 64y^6$ **13.** $m^6/64 - 3m^5/16 + 15m^4/16 - 5m^3/2 + 15m^2/4 - 3m + 1$
15. $16p^4 + 32p^3q/3 + 8p^2q^2/3 + 8pq^3/27 + q^4/81$ **17.** $7920m^8p^4$ **19.** $126x^4y^5$ **21.** $180m^2n^8$
23. $4845p^8q^{16}$ **25.** $439,296x^{21}y^7$ **27.** 2.594   **29.** 245.937   **31.** 758.650
33. 5.010   **35.** .942

Section 9.6 (page 481)

Exercises 1–25 are proofs, so answers are not supplied.   **27.** $P_n = \dfrac{4^{n-1}}{3^{n-2}}$ or $3\left(\dfrac{4}{3}\right)^{n-1}$

Section 9.7 (page 485)

1. 5040   **3.** 720   **5.** 90   **7.** 336   **9.** 7   **11.** 1   **13.** 40,320   **15.** 120; 30,240
17. 30   **19.** 120   **21.** 4   **23.** 552   **25. (a)** $26^3 \cdot 10^3 = 17,576,000$ **(b)** 35,152,000
27. $26 \cdot 25 \cdot 24 \cdot 10 \cdot 9 \cdot 8 = 11,232,000$; $2(26 \cdot 25 \cdot 24 \cdot 10 \cdot 9 \cdot 8) = 22,464,000$

Section 9.8 (page 488)

1. 6   **3.** 56   **5.** 1365   **7.** 120   **9.** 14   **11.** 1   **13.** 27,405   **15.** 220; 220
17. 1326   **19.** 10   **21.** 28   **23. (a)** 84   **(b)** 10   **(c)** 40   **(d)** 28   **25.** $\dbinom{52}{5} = 2,598,960$
27. 4   **29.** $4 \cdot \dbinom{13}{5} = 5148$ **31.** $P(5, 5) = 120$ **33.** $\dbinom{12}{3} = 220$

Section 9.9 (page 495)

1. Let h = heads, t = tails. S = {h}   **3.** S = {hhh, hht, hth, thh, htt, tht, tth, ttt}   **5.** Let c =
correct, w = wrong. S = {ccc, ccw, cwc, wcc, wwc, wcw, cww, www}   **7. (a)** {hh, tt}, 1/2
(b) {hh, ht, th}, 3/4   **9. (a)** {2 and 4}, 1/10   **(b)** {1 and 3, 1 and 5, 3 and 5}, 3/10   **(c)** Ø, 0
(d) {1 and 2, 1 and 4, 2 and 3, 2 and 5, 3 and 4, 4 and 5}, 3/5   **11. (a)** 1/5   **(b)** 8/15   **(c)** 0   **(d)** 1 to 4
(e) 7 to 8   **13.** 1 to 4   **15.** 2/5   **17. (a)** 1/2   **(b)** 7/10   **(c)** 2/5   **19. (a)** 1/6   **(b)** 1/3
(c) 1/6   **21. (a)** 7/15   **(b)** 11/15   **(c)** 4/5

Chapter 9   Review Exercises (page 501)

These review exercises cover the material in the chapter as follows: 1–7, Section 9.1;
9–49, Sections 9.2–9.4; 51–59, Section 9.5; 67–69, Section 9.6; 61–65 and 71–87,
Sections 9.7 and 9.8; 89–93, Section 9.9.

1. 1/2, 2/3, 3/4, 4/5, 5/6   **3.** 8, 10, 12, 14, 16   **5.** 5, 2, -1, -4, -7 **7.** 5, 3, -2, -5, -3
9. 6, 2, -2, -6, -10 **11.** $3 - \sqrt{5}$, 4, $5 + \sqrt{5}$, $6 + 2\sqrt{5}$, $7 + 3\sqrt{5}$ **13.** 4, 8, 16, 32, 64
15. -3, 4, $-16/3$, 64/9, $-256/27$ **17.** $a_1 = -29$; $a_{20} = 66$ **19.** 20   **21.** $-x + 61$
23. 222   **25.** $84k$ **27.** 25   **29.** ± 36 **31.** 120   **33.** 36   **35.** $-1/2$ **37.** 2/3
39. 25/6   **41.** 690   **43.** 4368   **45.** 5/3   **47.** 64 cans   **49.** About 117 items
51. $x^4 + 8x^3y + 24x^2y^2 + 32xy^3 + 16y^4$ **53.** $k^5/32 - 5k^4g/16 + 5k^3g^2/4 - 5k^2g^3/2 + 5kg^4/2 - g^5$
55. $25,344m^5n^{14}$ **57.** $-420 \cdot 3^{13}m^2n^{13} + 30 \cdot 3^{14}mn^{14} - 3^{15}n^{15}$ **59.** .868   **71.** 120   **73.** 1
75. 252   **77.** 7   **79.** 90   **81. (a)** 20   **(b)** 10   **83.** $26 \cdot 10 \cdot 10 \cdot 10 \cdot 26 \cdot 26 \cdot 26 =$
$456,976,000$; $26 \cdot 10 \cdot 9 \cdot 8 \cdot 25 \cdot 24 \cdot 23 = 258,336,000$ **85.** {4}   **87.** {12}   **89.** 4/13
91. 3/4   **93.** .86

Chapter 9   Test (page 503)
1. $-2, 5, -10, 17, -26$ **2.** 5, 7, 10, 14, 19   **3.** 2, 4, 4, 6, 6   **4.** $-11, 1, 13, 25, 37$
5. 81, 54, 36, 24, 16   **6.** 6   **7.** 3   **8.** 955   **9.** $1458x^{15}$ **10.** 15/2   **11.** 125
12. The sum does not exist.   **13.** 2385   **14.** 108/7   **15.** $1950   **16.** 25,600   **18.** 990
19. 5040   **20.** 45   **21.** 1   **22.** 60   **23.** 24,360   **24.** 1/26   **25.** 10/13
26. 4/13   **27.** .92   **28.** .38   **29.** $16x^4 - 96x^3y + 216x^2y^2 - 216xy^3 + 81y^4$
30. $-160w^3y^3$

Appendix A (page 510)
1. 2   **3.** 6   **5.** -4 **7.** -5 **9.** 2.9420   **11.** 4.1072   **13.** .8825
15. $.9509 - 4$ **17.** .5884   **19.** 4.8341   **21.** $.5825 - 4$ **23.** 34.4   **25.** .0747
27. 3124   **29.** .006367   **31.** 384   **33.** .206   **35.** 2.01   **37.** .735   **39.** 713
41. 890.4   **43.** 1.30   **45.** 15.57   **47.** 4.83   **49.** $-.02$ **51.** 5.62   **53.** 3.32
55. 1.79   **57.** {.398}   **59.** {$-.192$}   **61.** {-1.468}   **63.** {.384}

Answers to Appendix on Sets

1. 12, 13, 14, 15, 16, 17, 18, 19, 20   **3.** 1, 1/2, 1/4, 1/8, 1/16, 1/32   **5.** 17, 22, 27, 32, 37, 42, 47
7. 8, 9, 10, 11, 12, 13, 14   **9.** Finite   **11.** Infinite   **13.** Infinite   **15.** Infinite   **17.** $\in$
19. $\notin$ **21.** $\in$ **23.** $\notin$ **25.** $\notin$ **27.** $\notin$ **29.** False   **31.** True   **33.** True
35. True   **37.** False   **39.** True   **41.** True   **43.** False   **45.** True   **47.** False
49. True   **51.** True   **53.** True   **55.** True   **57.** False   **59.** True   **61.** True
63. False   **65.** $\subseteq$ **67.** $\not\subseteq$ **69.** $\subseteq$ **71.** {0, 2, 4}
73. {0, 1, 2, 3, 4, 5, 6, 7, 8, 9, 11, 13}   **75.** M or {0, 2, 4, 6, 8}
77. {0, 1, 2, 3, 4, 5, 7, 9, 11, 13}   **79.** {0, 2, 4, 6, 8, 10, 12}   **81.** {10, 12}
83. $\emptyset$; $\emptyset$ and R are disjoint   **85.** N or {1, 3, 5, 7, 9, 11, 13}   **87.** R or {0, 1, 2, 3, 4}
89. {0, 1, 2, 3, 4, 6, 8}   **91.** R or {0, 1, 2, 3, 4}   **93.** $\emptyset$; Q' and $(N' \cap U)$ are disjoint
95. All students in this school who are not taking this course   **97.** All students in this school who are
taking calculus and history   **99.** All students in this school who are taking this course or history
101. {1, 2, 3, 4, 5, 6}   **103.** {*hhh, hht, hth, thh, htt, tht, tth, ttt*}   **105.** X and Y are disjoint.
107. $X \subseteq Y$ **109.** $X = \emptyset$

Index

Appendix on Sets

Think of a **set** as a collection of objects. The objects that belong to a set are called the **elements** or **members** of the set. In algebra, the elements of a set are usually numbers. Sets are commonly written using **set braces,** { }. For example, the set containing the elements 1, 2, 3, and 4 is written

$$\{1, 2, 3, 4\}.$$

Since the order in which the elements are listed is not important, this same set can also be written as $\{4, 3, 2, 1\}$ or with any other arrangement of the four numbers.

To show that 4 is an element of the set $\{1, 2, 3, 4\}$, use the symbol $\in$ and write

$$4 \in \{1, 2, 3, 4\}.$$

Also, $2 \in \{1, 2, 3, 4\}$. To show that 5 is *not* an element of this set, place a slash through the symbol:

$$5 \notin \{1, 2, 3, 4\}.$$

It is customary to name sets with capital letters. If S is used to name the set above, then

$$S = \{1, 2, 3, 4\}.$$

Set S was written by listing its elements. It is sometimes easier to describe a set in words. For example, set S might be described as "the set containing the first four counting numbers." In this example, the notation $\{1, 2, 3, 4\}$, with the elements listed between set braces, is briefer than the verbal description. However, the set F, consisting of all fractions between 0 and 1, could not be described by listing its elements. (Try it.)

Set F is an example of an *infinite set,* one that has an unending list of distinct elements. A *finite set* is one that has a limited number of elements. Some infinite sets, unlike F, can be described by a listing process. For example, the set of numbers used for counting, called the **natural numbers,** or the **counting numbers,** can be written as

$$N = \{1, 2, 3, 4, \ldots\},$$

where the three dots show that the list of the elements of the set continues in the same way.

Sets are often written using a **variable,** a letter that is used to represent one element from a set of numbers. The set of numbers is called the **domain** of the variable. For example,

$$\{x | x \text{ is a natural number between 2 and 7}\}$$

(read "the set of all elements x such that x is a natural number between 2 and 7") represents the set $\{3, 4, 5, 6\}$. The numbers 2 and 7 are *not* between 2 and 7. The notation used here, $\{x | x \text{ is a natural number between 2 and 7}\}$, is called **set-builder notation.**

EXAMPLE 1 Write the elements belonging to each of the following sets.

(a) $\{x | x \text{ is a counting number less than 5}\}$
The counting numbers less than 5 make up the set $\{1, 2, 3, 4\}$.

(b) $\{x | x \text{ is a state that touches Florida}\}$
The states touching Florida make up the set $\{\text{Alabama, Georgia}\}$. ● *

When discussing a particular situation or problem, we can usually identify a **universal set** (whether expressed or implied) that contains all the elements ap-

---

*The symbol ● represents the end of an example.

pearing in any set used in the given problem. The letter U is used to represent the universal set.

At the other extreme from the universal set is the **null set,** or **empty set,** the set containing no elements. The set of all people twelve feet tall is an example of the null set. Write the null set in either of two ways: use the special symbol $\emptyset$ or else write set braces enclosing no elements, $\{\ \ \}$. (Be careful not to combine these symbols; $\{\emptyset\}$ is *not* the null set.)

Every element of the set $S = \{1, 2, 3, 4\}$ is a natural number. Because of this, set S is a *subset* of the set N of natural numbers, written $S \subseteq N$. By definition, set A is a **subset** of set B if every element of set A is also an element of set B. For example, if $A = \{2, 5, 9\}$ and $B = \{2, 3, 5, 6, 9, 10\}$, then $A \subseteq B$. However, there are some elements of B that are not in A, so B is not a subset of A, written $B \not\subseteq A$. By the definition, every set is a subset of itself.

Figure 1.1 shows a set A that is a subset of set B. The rectangle of the drawing represents the universal set U. Such diagrams are called **Venn diagrams.** These diagrams are used as an aid in clarifying and discussing the relationships among sets.

By the definition of subset, the null set is a subset of every set. That is,

if A is any set, then $\emptyset \subseteq A$.

Two sets A and B are **equal** whenever $A \subseteq B$ and $B \subseteq A$. In other words, $A = B$ if the two sets contain exactly the same elements. For example,

$$\{1, 2, 3\} = \{3, 1, 2\},$$

since both sets contain exactly the same elements. However,

$$\{1, 2, 3\} \neq \{0, 1, 2, 3\},$$

since the set $\{0, 1, 2, 3\}$ contains the element 0, which is not an element of $\{1, 2, 3\}$.

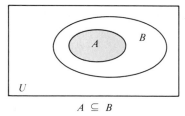

$A \subseteq B$

Figure 1.1

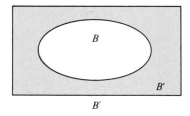

B'

Figure 1.2

Operations on Sets   Given a set A and a universal set U, the set of all elements of U that do not belong to set A is called the **complement** of set A. For example, if set A is the set of all the female students in your class, and U is the set of all students in the class, then the complement of A would be the set of all the male students in the class. The complement of set A is written A' (read "A-prime"). The Venn diagram in Figure 1.2 shows a set B. Its complement, B', is colored.

EXAMPLE 2   Let $U = \{1, 2, 3, 4, 5, 6, 7\}$, $A = \{1, 3, 5, 7\}$, and $B = \{3, 4, 6\}$. Find each of the following sets.

(a) A'

Set A' contains the elements of U that are not in A.

$$A' = \{2, 4, 6\}$$

(b) $B' = \{1, 2, 5, 7\}$

(c) $\emptyset' = U$ and $U' = \emptyset$ ●

Given two sets A and B, the set of all elements belonging both to set A and to set B is called the **intersection** of the two sets, written $A \cap B$. For example, the elements that belong to $A = \{1, 2, 4, 5, 7\}$ and $B = \{2, 4, 5, 7, 9, 11\}$ are 2, 4, 5, and 7, so

$$A \cap B = \{1, 2, 4, 5, 7\} \cap \{2, 4, 5, 7, 9, 11\} = \{2, 4, 5, 7\}.$$

The Venn diagram in Figure 1.3 shows two sets A and B; their intersection, $A \cap B$, is shown in color.

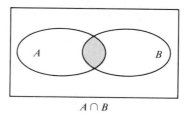

$A \cap B$

Figure 1.3

EXAMPLE 3   (a) $\{9, 15, 25, 36\} \cap \{15, 20, 25, 30, 35\} = \{15, 25\}$

The elements 15 and 25 are the only ones belonging to both sets.

(b) $\{2, 3, 4, 5, 6\} \cap \{1, 2, 3, 4\} = \{2, 3, 4\}$ ●

Two sets that have no elements in common are called **disjoint sets.** For example, there are no elements common to both $\{50, 51, 54\}$ and $\{52, 53, 55, 56\}$, so these two sets are disjoint, and

$$\{50, 51, 54\} \cap \{52, 53, 55, 56\} = \emptyset.$$

This result can be generalized:

if A and B are any two disjoint sets, then $A \cap B = \emptyset$.

The set of all elements belonging to set A or to set B is called the **union** of the two sets, written $A \cup B$. For example,

$$\{1, 3, 5\} \cup \{3, 5, 7, 9\} = \{1, 3, 5, 7, 9\}.$$

The Venn diagram in Figure 1.4 shows two sets A and B; the union of the sets, $A \cup B$, is shown in color.

$A \cup B$

Figure 1.4

EXAMPLE 4   **(a)** Find the union of $\{1, 2, 5, 9, 14\}$ and $\{1, 3, 4, 8\}$.

Begin by listing the elements of the first set, $\{1, 2, 5, 9, 14\}$. Then include any elements from the second set that are not already listed. Doing this gives

$$\{1, 2, 5, 9, 14\} \cup \{1, 3, 4, 8\} = \{1, 2, 3, 4, 5, 8, 9, 14\}.$$

(b) $\{1, 3, 5, 7\} \cup \{2, 4, 6\} = \{1, 2, 3, 4, 5, 6, 7\}$ ●

The processes of finding the complement of a set, the intersection of two sets, and the union of two sets are called **set operations.** These operations are similar to operations on numbers, such as addition, subtraction, multiplication, or division.

The various operations on sets are summarized below.

Set Operations

For all sets A and B, and universal sets U:

The **complement** of set A is the set A' of all elements in the universal set that do not belong to set A.

$$A' = \{x | x \in U, \quad x \notin A\}$$

The **intersection** of sets A and B, written $A \cap B$, is made up of all the elements belonging to set A and set B at the same time.

$$A \cap B = \{x | x \in A \text{ and } x \in B\}$$

The **union** of sets A and B, written $A \cup B$, is made up of all the elements belonging to set A or to set B.

$$A \cup B = \{x | x \in A \text{ or } x \in B\}$$

1.1 EXERCISES

List all the elements of each of the following sets.

1. $\{12, 13, 14, \ldots, 20\}$ **2.** $\{8, 9, 10, \ldots, 17\}$ **3.** $\{1, 1/2, 1/4, \ldots, 1/32\}$

4. $\{3, 9, 27, \ldots, 729\}$ **5.** $\{17, 22, 27, \ldots, 47\}$ **6.** $\{74, 68, 62, \ldots, 38\}$

7. {all natural numbers greater than 7 and less than 15}

8. {all natural numbers not greater than 4}

Identify the sets in Exercises 9–16 as finite *or* infinite.

9. $\{4, 5, 6, \ldots, 15\}$

10. $\{4, 5, 6, \ldots\}$

11. $\{1, 1/2, 1/4, 1/8, \ldots\}$

12. $\{0, 1, 2, 3, 4, 5, \ldots, 75\}$

13. $\{x | x$ is a natural number larger than 5$\}$

14. $\{x | x$ is a person alive on the earth now$\}$

15. $\{x | x$ is a fraction between 0 and 1$\}$

16. $\{x | x$ is a positive even natural number$\}$

Complete the blanks with either $\in$ or $\notin$ so that the resulting statement is true.

17. 6 _____ $\{3, 4, 5, 6\}$

18. 9 _____ $\{3, 2, 5, 9, 8\}$

19. -4 _____ $\{4, 6, 8, 10\}$

20. -12 _____ $\{3, 5, 12, 14\}$

21. 0 _____ $\{2, 0, 3, 4\}$

22. 0 _____ $\{5, 6, 7, 8, 10\}$

23. $\{3\}$ _____ $\{2, 3, 4, 5\}$

24. $\{5\}$ _____ $\{3, 4, 5, 6, 7\}$

25. $\{0\}$ _____ $\{0, 1, 2, 5\}$

26. $\{2\}$ _____ $\{2, 4, 6, 8\}$

27. 0 _____ $\emptyset$

28. $\emptyset$ _____ $\emptyset$

Tell whether each statement is true *or* false.

29. $3 \in \{2, 5, 6, 8\}$

30. $6 \in \{-2, 5, 8, 9\}$

31. $1 \in \{3, 4, 5, 11, 1\}$

32. $12 \in \{18, 17, 15, 13, 12\}$

33. $9 \notin \{2, 1, 5, 8\}$

34. $3 \notin \{7, 6, 5, 4\}$

35. $\{2, 5, 8, 9\} = \{2, 5, 9, 8\}$

36. $\{3, 0, 9, 6, 2\} = \{2, 9, 0, 3, 6\}$

37. $\{5, 8, 9\} = \{5, 8, 9, 0\}$

38. $\{3, 7, 12, 14\} = \{3, 7, 12, 14, 0\}$

39. $\{x | x$ is a natural number less than 3$\} = \{1, 2\}$

40. $\{x | x$ is a natural number greater than 10$\} = \{11, 12, 13, \ldots\}$

41. $\{5, 7, 9, 19\} \cap \{7, 9, 11, 15\} = \{7, 9\}$

42. $\{8, 11, 15\} \cap \{8, 11, 19, 20\} = \{8, 11\}$

43. $\{2, 1, 7\} \cup \{1, 5, 9\} = \{1\}$

44. $\{6, 12, 14, 16\} \cup \{6, 14, 19\} = \{6, 14\}$

45. $\{3, 2, 5, 9\} \cap \{2, 7, 8, 10\} = \{2\}$

46. $\{8, 9, 6\} \cup \{9, 8, 6\} = \{8, 9\}$

47. $\{3, 5, 9, 10\} \cap \emptyset = \{3, 5, 9, 10\}$

48. $\{3, 5, 9, 10\} \cup \emptyset = \{3, 5, 9, 10\}$

49. $\{1, 2, 4\} \cup \{1, 2, 4\} = \{1, 2, 4\}$

50. $\{1, 2, 4\} \cap \{1, 2, 4\} = \emptyset$

51. $\emptyset \cup \emptyset = \emptyset$

52. $\emptyset \cap \emptyset = \emptyset$

Let $A = \{2, 4, 6, 8, 10, 12\}$
 $B = \{2, 4, 8, 10\}$
 $C = \{4, 10, 12\}$
 $D = \{2, 10\}$
 $U = \{2, 4, 6, 8, 10, 12, 14\}$.

Tell whether each statement is true *or* false.

53. $A \subseteq U$

54. $C \subseteq U$

55. $D \subseteq B$

56. $D \subseteq A$

57. $A \subseteq B$

58. $B \subseteq C$

59. $\emptyset \subseteq A$

60. $\emptyset \subseteq \emptyset$

61. $\{4, 8, 10\} \subseteq B$

62. $\{0, 2\} \subseteq D$

63. $B \subseteq D$

64. $A \not\subseteq C$

Insert ⊆ or ⊄ in each blank to make the resulting statement true.

65. {2, 4, 6} _____ {3, 2, 5, 4, 6}

66. {1, 5} _____ {0, −1, 2, 3, 1, 5}

67. {0, 1, 2} _____ {1, 2, 3, 4, 5}

68. {5, 6, 7, 8} _____ {1, 2, 3, 4, 5, 6, 7}

69. ∅ _____ {1, 4, 6, 8}

70. ∅ _____ ∅

Let $U = \{0, 1, 2, 3, 4, 5, 6, 7, 8, 9, 10, 11, 12, 13\}$
$\quad M = \{0, 2, 4, 6, 8\}$
$\quad N = \{1, 3, 5, 7, 9, 11, 13\}$
$\quad Q = \{0, 2, 4, 6, 8, 10, 12\}$
$\quad R = \{0, 1, 2, 3, 4\}.$

Use these sets to find each of the following. Identify any disjoint sets. See Examples 2–4.

71. $M \cap R$	**72.** $M \cup R$	**73.** $M \cup N$
74. $M \cap N$	**75.** $M \cap U$	**76.** $M \cup Q$
77. $N \cup R$	**78.** $U \cap N$	**79.** N'
80. Q'	**81.** $M' \cap Q$	**82.** $Q \cap R'$
83. $\emptyset \cap R$	**84.** $\emptyset \cap Q$	**85.** $N \cup \emptyset$
86. $R \cup \emptyset$	**87.** $(M \cap N) \cup R$	**88.** $(N \cup R) \cap M$
89. $(Q \cap M) \cup R$	**90.** $(N \cup R) \cap M$	**91.** $(M' \cup Q) \cap R$
92. $Q \cap (M \cup N)$	**93.** $Q' \cap (N' \cap U)$	**94.** $(U \cap \emptyset') \cup R$

Let $U = \{$all students in this school$\}$
$\quad M = \{$all students taking this course$\}$
$\quad N = \{$all students taking calculus$\}$
$\quad P = \{$all students taking history$\}.$

Describe each of the following sets in words.

95. M'	**96.** $M \cup N$	**97.** $N \cap P$
98. $N' \cap P'$	**99.** $M \cup P$	**100.** $P' \cup M'$

In the study of probability, the set of all possible outcomes for an experiment is called the sample space for the experiment. For example, the sample space for tossing an honest coin is the set {h, t}. *Find the sample space for each of these exercises.*

101. Rolling an honest die

102. Tossing a coin twice

103. Tossing a coin three times

104. Having three children

State the conditions on sets X and Y for which the following are true.

105. $X \cap Y = \emptyset$	**106.** $X \cup Y = \emptyset$	**107.** $X \cap Y = X$
108. $X \cup Y = X$	**109.** $X \cap \emptyset = X$	**110.** $X \cup \emptyset = X$